OECD Environmental Outlook to 2030

ORGANISATION FOR ECONOMIC CO-OPERATION AND DEVELOPMENT

The OECD is a unique forum where the governments of 30 democracies work together to address the economic, social and environmental challenges of globalisation. The OECD is also at the forefront of efforts to understand and to help governments respond to new developments and concerns, such as corporate governance, the information economy and the challenges of an ageing population. The Organisation provides a setting where governments can compare policy experiences, seek answers to common problems, identify good practice and work to co-ordinate domestic and international policies.

The OECD member countries are: Australia, Austria, Belgium, Canada, the Czech Republic, Denmark, Finland, France, Germany, Greece, Hungary, Iceland, Ireland, Italy, Japan, Korea, Luxembourg, Mexico, the Netherlands, New Zealand, Norway, Poland, Portugal, the Slovak Republic, Spain, Sweden, Switzerland, Turkey, the United Kingdom and the United States. The Commission of the European Communities takes part in the work of the OECD.

OECD Publishing disseminates widely the results of the Organisation's statistics gathering and research on economic, social and environmental issues, as well as the conventions, guidelines and standards agreed by its members.

Also available in French under the title:
Perspectives de l'environnement de l'OCDE à l'horizon 2030

Corrigenda to OECD publications may be found online at: *www.oecd.org/publishing/corrigenda.*

Preface

The environmental challenges we face over the coming few decades are daunting. They will require concerted policy action, and co-operation among countries, different ministries within countries, and with stakeholder partners. Climate change is particularly high on the political agenda now, but we also face the challenges of halting biodiversity loss, ensuring clean water and adequate sanitation for all, and reducing the health impacts of environmental degradation.

The analysis presented in this OECD Environmental Outlook *shows that the necessary policies and solutions are available, that they are achievable and that they are affordable. But we need to act now, while it is still relatively inexpensive, particularly in the rapidly emerging economies. One scenario in this* Outlook *found that if we are willing to accept a 98% increase in global GDP from now to 2030 – rather than the 99% in our Baseline – we could achieve significant improvements in air and water quality, and progress towards climate targets. This is not a lot to pay (you can call it the cost of insurance). The consequences and costs of inaction, on the other hand, would be much higher.*

This Outlook *provides policy-makers with guidance on how to address the more complex and long-term global environmental challenges, in a way that is cost-effective and can also deal with the shorter-term concerns of their local constituencies. The OECD is well positioned to provide this guidance. The analysis in this* Outlook *is based on an economic and environmental modelling framework, drawing on expert inputs from across the Organisation – macroeconomic assumptions from our Economics Department, energy projections from our sister organisation, the International Energy Agency, agricultural assumptions from our Trade and Agriculture Directorate – and environmental modelling expertise from the Dutch Environmental Assessment Agency.*

Environment Ministries cannot address these challenges alone. They need the support of other areas, in particular of the Ministries of Finance to provide environmental policy reforms with a strong financial backing. And they need the support of ministries of energy, agriculture, transport and industry to implement the sectoral policies required to reduce the environmental impacts of our production and consumption patterns.

Countries will need to restructure their economies in order to move towards a low carbon, greener and more sustainable future. The costs of this restructuring are affordable, but the transition needs to be managed carefully to address social and competitiveness impacts, and to take advantage of new opportunities, like eco-innovation. Removal of environmentally harmful subsidies, particularly for fossil fuels and agricultural production, is a necessary first step: it would shift the economy away from activities that pollute and over-use natural resources while saving money for tax payers. The focus should be on taxing the "bad", rather than subsidising the "good". The reason is simple: the "bad" is known (e.g. CO_2 emissions), while the "good" of today can becom obsolete or be proven to be inefficient tomorrow. Policy simulations carried out for this OECD Environmental Outlook *demonstrate that widespread use of market-based instruments can considerably lower the cost of action to achieve ambitious environmental goals.*

The most pressing environmental challenges cannot be solved by OECD countries alone. The OECD Environmental Outlook to 2030 *shows that the global cost of action will be much lower if all countries work together to achieve common environmental goals. To implement cost-effective solutions, developed countries will need to work closely with emerging economies – especially Brazil, Russia, India, Indonesia, China and South Africa – as well as with other developing countries.*

Governments, businesses, trade unions, NGOs and all citizens need to join forces to ensure that the ecosystem services that support economic growth and human well-being are not lost. With the size of the world economy expected to double by 2030, while population is expected to increase by one-third, continuing or expanding our current patterns of consumption and production is simply unsustainable. The OECD Environmental Outlook *shows that the policies and solutions to address these challenges over the coming decades are available and affordable. But if we want to avoid irreversible damage to our environment and the very high costs of policy inaction, we'd better start working right away.*

Angel Gurría
Secretary-General

OECD ENVIRONMENTAL OUTLOOK TO 2030 – ISBN 978-92-64-04048-9 – © OECD 2008

Acknowledgements

The *OECD Environmental Outlook to 2030* was developed by a team within the OECD Environment Directorate, under the management of Lorents Lorentsen (Director), Rob Visser (Deputy Director), Helen Mountford (Head of Division), and Jan Bakkes (Netherlands Environmental Assessment Agency).

The OECD Environment Policy Committee (EPOC) was responsible for the oversight of the development of the report. The Working Party on Global and Structural Policies (WPGSP) was responsible for overseeing the modelling framework and application. Other OECD bodies also provided expert input to selected chapters, including: the Working Party on National Environmental Policies (WPNEP); Joint Meeting of the Chemicals Committee and the Working Party on Chemicals, Pesticides and Biotechnology; Working Group on Environmental Information and Outlooks (WGEIO); Working Group on Economic Aspects of Biodiversity (WGEAB); Working Group on Transport; Working Group on Waste Prevention and Recycling (WGWPR); Joint Working Party on Agriculture and Environment (JWPAE); Joint Working Party on Trade and Environment (JWPTE); Committee on Fisheries (COFI); and the Tourism Committee.

Representatives from non-OECD countries – in particular Brazil, China, India, and the Russian Federation – provided input to the report through a Global Forum on Sustainable Development event in May 2007. Stakeholder representatives provided input to draft chapters, in particular environmental citizens' organisations (co-ordinated through the European Environmental Bureau), industry (co-ordinated through the Business and Industry Advisory Committee to the OECD), and trade unions (co-ordinated through the Trade Union Advisory Committee to the OECD).

The main crafters of the chapters of the *OECD Environmental Outlook* were:

Executive Summary	Kumi Kitamori
Introduction	Helen Mountford
The World to 2030 – The Consequences of Policy Inaction	
I. Drivers of Environmental Change	
1. Consumption, Production and Technology	Ysé Serret, Nick Johnstone, Ivan Hascic, Takako Haruyama
2. Population Dynamics and Demographics	Xavier Leflaive
3. Economic Development	Philip Bagnoli, Jean Chateau, Yong Gun Kim
4. Globalisation	Cristina Tébar Less, Philip Bagnoli
5. Urbanisation	Kyung Yong Lee, Carine Barbier (IDDRI)
6. Key Variations to the Standard Expectation to 2030	Philip Bagnoli, Jean Chateau, Yong Gun Kim

II. Environmental Challenges	
7. Climate Change	Jan Corfee-Morlot, Dennis Tirpak, Jane Ellis, Philip Bagnoli, Yong-Gun Kim, Jean Chateau, Detlef van Vuuren (MNP)
8. Air Pollution	Frank de Leeuw, Jan Bakkes, Hans Eerens, Robert Koelemeijer (all MNP)
9. Biodiversity	Philip Bagnoli, Takako Haruyama
10. Freshwater	Gerard Bonnis
11. Waste and Material Flows	Henrik Harjula, Myriam Linster, Soizick de Tilly
12. Health and Environment	Pascale Scapecchi, Nicolas Gagnon, Dian Turnheim, Frank de Leeuw (MNP)
13. Cost of Policy Inaction	Nick Johnstone, Jan Corfee-Morlot, Ivan Hascic
Policy Responses	
III. Sectoral Developments and Policies	
14. Agriculture	Philip Bagnoli, Jean Chateau, Yong Gun Kim, Wilfrid Legg, Olivier Belaud, Elke Stehfest (MNP)
15. Fisheries and Aquaculture	Martha Heitzmann, Helen Mountford, Philip Bagnoli
16. Transport	Tom Jones, Michael Donohue, Nadia Caid
17. Energy	Jan Corfee-Morlot, Jane Ellis, Trevor Morgan
18. Chemicals	Richard Sigman
19. Selected Industries	
Steel and Cement	Nils Axel Braathen
Pulp and Paper	Xavier Leflaive
Tourism	Xavier Leflaive
Mining	Peter Borkey
IV. Putting the Policies Together	
20. Environmental Policy Packages	Helen Mountford, Tom Kram (MNP)
21. Institutions and Approaches for Policy Implementation	Kumi Kitamori, Krzysztof Michalak
22. Global Environmental Co-operation	Roberto Martin-Hurtado
Annexes	
A. Regional environmental implications	Xavier Leflaive
B. Modelling framework and assumptions	Jan Bakkes and Detlef van Vuuren (MNP), Philip Bagnoli

The economic part of the modelling work for the *OECD Environmental Outlook* was undertaken by the OECD team working on ENV-Linkages, and the environmental modelling work was undertaken by the Netherlands Environmental Assessment Agency (MNP). The MNP used IMAGE and related environmental models, including collaboration with LEI at Wageningen University and Research Centre on agriculture-economy modelling, and the Center for Environmental Systems Research (CESR) at Kassel University for modelling of water quantity issues. In addition, environmental modelling drew on results from The World Bank and the European Commission's Joint Research Centres (on air pollution) and the Sustainable Europe Research Institute (on use of material resources).

The modelling teams were:

ENV-Linkages (OECD)	IMAGE and related environmental models	
	Core team:	*Specific contributions:*
Philip Bagnoli	Tom Kram	Annelies Balkema
Jean Chateau	Jan Bakkes	Johannes Bollen
Yong-Gun Kim	Lex Bouwman	Hans Eerens
Sebnem Sahin	Gerard van Drecht	Michel den Elzen
	Bas Eickhout	Henk Hilderink
	Michel Jeuken	Morna Isaac
	Frank de Leeuw	Paul Lucas
	Mark van Oorschot	Jos Olivier
	Elke Stehfest	Ellen Teichert (CESR)
	Detlef van Vuuren	Kerstin Verzano (CESR)
		Jasper van Vliet
		Martina Weiss (CESR)
	From collaborating institutes:	
	Andrzej Tabeau (LEI)	
	Hans van Meijl (LEI)	
	Frank Voss (CESR)	
	Geert Woltjer (LEI)	

The MNP team consisted of the above modelling team as well as Johan Meijer, Kees Klein Goldewijk, Peter Janssen, Rineke Oostenrijk, Ton Manders, Robert Koelemeijer, Peter Bosch, Dick Nagelhout and Jan Bakkes (project management).

Statistical and research assistance was provided by Cuauhtémoc Rebolledo-Gómez, Takako Haruyama, Carla Bertuzzi, Simon Faucher, and Niels Schenk. Jane Kynaston co-ordinated the reviewing and administrative processes of the report. Kathleen Mechali and Stéphanie Simonin-Edwards provided administrative assistance. Fiona Hall edited the report. Support for the publication process was provided by Katherine Kraig-Ernandes, Catherine Candea and the OECD Publishing Division.

A number of OECD countries provided financial or in-kind contributions to support the modelling and *Outlook* work, including: Canada, the Czech Republic, Japan, Korea, the Netherlands, Norway, the United Kingdom, and the United States.

This publication is dedicated to the memory of Takako Haruyama (1976-2007).

Table of Contents

Acronyms and Abbreviations 21
Executive Summary 23
Introduction: Context and Methodology 35

THE WORLD TO 2030 – THE CONSEQUENCES OF POLICY INACTION

I. Drivers of Environmental Change

Chapter 1. **Consumption, Production and Technology** 47
Introduction 49
Key trends and projections: consumption and the environment 49
Key trends and projections: production and the environment 54
Key trends and projections: technology and the environment 57
Notes 62
References 62

Chapter 2. **Population Dynamics and Demographics** 65
Introduction 67
Keys trends and projections 68
Notes 73
References 73

Chapter 3. **Economic Development** 75
Introduction 77
Key trends and projections 79
Policy implications 86
Notes 87
References 88

Chapter 4. **Globalisation** 89
Introduction 91
Key trends and projections 95
Policy implications 103
Notes 104
References 104

Chapter 5. **Urbanisation** 107
Introduction 109

Key trends and projections 110
Policy implications 117
Notes 119
References 119

Chapter 6. **Key Variations to the Standard Expectation to 2030** 121
Introduction 123
Key variations in the drivers 126
Policy implications 133
Notes 134
References 134

II. Impacts of Environmental Change

Chapter 7. **Climate Change** 139
Introduction 141
Key trends and projections 143
Policy implications 147
Policy simulations 154
Summary 170
Notes 171
References 173

Chapter 8. **Air Pollution** 177
Introduction 179
Key trends and projections 182
Policy implications 186
Policy simulations: urban air quality 189
Notes 193
References 194

Chapter 9. **Biodiversity** 197
Introduction 199
Key trends and projections 200
Policy implications 211
Costs of inaction 215
Notes 215
References 215

Chapter 10. **Freshwater** 219
Introduction 221
Key trends and projections 221
Policy implications 226
Notes 231
References 233
Annex 10.A1. Key Assumptions and Uncertainties in the Water Projections 235

Chapter 11. **Waste and Material Flows** 237
Introduction 239
Key trends and projections 239
Policy implications 248
Notes 250
References 250

Chapter 12. **Health and Environment** 253
Introduction 255
Key trends and projections: outdoor air pollution 256
Key trends and projections: water supply, sanitation and hygiene 262
Policy implications 266
Notes 267
References 267

Chapter 13. **Cost of Policy Inaction** 269
Introduction 271
Issues in valuation (key assumptions and uncertainties) 273
Selected examples of the costs of inaction 274
Other issues 284
Concluding remarks 285
Notes 286
References 287

POLICY RESPONSES

III. Sectoral Developments and Policies

Chapter 14. **Agriculture** 295
Introduction 297
Key trends and projections 298
Policy implications 308
Costs of inaction 315
Notes 316
References 316
Annex 14.A1. Biofuels Simulation Results 319

Chapter 15. **Fisheries and Aquaculture** 323
Introduction 325
Key trends and projections 329
Policy implications 333
Notes 338
References 339

Chapter 16. **Transport** 341
Introduction 343
Trends and projections 344

Policy implications 350
References 354

Chapter 17. **Energy** 357
Introduction 359
Key trends and projections 361
Policy implications 368
Climate change policy simulations 371
Notes 374
References 374

Chapter 18. **Chemicals** 377
Introduction 379
Key trends and projections 380
Policy implications 382
Notes 387
References 387

Chapter 19. **Selected Industries** 389
STEEL AND CEMENT 390
Introduction 391
Key trends and projections 392
Policy simulations 393
PULP AND PAPER 401
Introduction 402
Key trends and projections 404
Policy implications 406
TOURISM 409
Introduction 410
Key trends and projections 411
Policy implications 413
MINING 418
Introduction 419
Key trends and projections 420
Policy implications 424
Notes 425
References 426

IV. Putting the Policies Together

Chapter 20. **Environmental Policy Packages** 431
Introduction 433
Designing and implementing effective mixes of policy instruments 433
Policy packages to address the key environmental issues of the *OECD Outlook* 437
Notes 443
References 443

Chapter 21. **Institutions and Approaches for Policy Implementation** 445

Introduction 447
Institutions for policy development and implementation 447
Political economy of environmental policies 453
Notes 459
References 460

Chapter 22. **Global Environmental Co-operation** 461

Introduction 463
Delivering better international environmental governance 465
Aid for environment in a changing development co-operation context 469
The emergence of alternative forms of co-operation 473
Notes 475
References 475

Annex A. **Regional Environmental Implications** 477

Annex B. **Modelling Framework** 496

List of boxes

1.1. Sustainability in the food and beverage industry 52
2.1. Assumptions and key uncertainties 68
3.1. Sources of assumptions for the modelling framework 77
3.2. Interactions between economy and environment 79
4.1. Discussion of globalisation and environment in UNEP 92
4.2. Environmental impacts of China's accession to the World Trade Organization 93
4.3. Regional trade agreements and the environment 97
4.4. Environmental innovation and global markets 102
4.5. Ensuring developing countries benefit from trade liberalisation 103
5.1. Environmental effects of the residential sector in China 117
5.2. Congestion charging 118
7.1. The European Union Emission Trading System (EU ETS) 152
7.2. Examples of voluntary agreements in OECD countries 154
7.3. Description of Baseline and policy simulations 155
7.4. Key uncertainties and assumptions 156
7.5. Co-benefits and the cost-effectiveness of climate and air pollution policy 163
8.1. Indoor air pollution 179
8.2. Travel distances and residence times of various air pollutants 181
8.3. Key uncertainties and assumptions 183
8.4. Urban air quality 190
9.1. Modelling the impact of agricultural tariff reductions 203
9.2. Environmental impacts of forestry 205
9.3. The need to value biodiversity 211
10.1. The advent of water as an international priority 222
10.2. Policies for water management by agriculture 228
10.3. Policy package simulations: impacts on water projections 229

11.1. A common knowledge base on material flows and resource productivity..... 242
11.2. Dealing with ship waste..... 243
11.3. Key uncertainties and assumptions..... 245
11.4. Environmental and economic benefits of recycling..... 247
11.5. Technology development and transfer..... 250
12.1. Children's health and the environment..... 255
12.2. Key uncertainties..... 259
12.3. Effectiveness of interventions to reduce the incidence of diarrhoea..... 264
14.1. Key drivers and some uncertainties..... 300
14.2. Biofuels: the economic and environmental implications..... 301
14.3. Agricultural technologies and the environment..... 308
14.4. Progress toward de-coupling farm payments in the OECD..... 310
14.5. Intensive *versus* extensive agriculture..... 312
15.1. *El Niño* Southern Oscillation..... 327
15.2. China: the world's largest producer and consumer of fish products..... 331
15.3. The evolving nature of fisheries management objectives..... 334
15.4. Policy simulation: economic effects of limiting global fisheries catch..... 335
16.1. Key uncertainties, choices and assumptions..... 344
16.2. Efficient prices for transport..... 350
16.3. Prospects for liquid biofuels for transport..... 352
17.1. Key uncertainties and assumptions..... 362
17.2. Power generation in China..... 363
17.3. Biofuels in the energy mix..... 364
17.4. The outlook for energy technology..... 369
17.5. IEA technology scenarios..... 370
18.1. Key uncertainties, choices and assumptions..... 380
18.2. The OECD and chemicals..... 383
18.3. Nanotechnologies..... 386
19.1. Model specifications and limitations..... 398
19.2. The cement sector..... 400
19.3. The procurement issue in perspective..... 403
19.4. Key uncertainties, choices and assumptions..... 405
19.5. Tourism, transport and the environment..... 410
19.6. Tourism in China..... 411
19.7. Key uncertainties and assumptions..... 413
19.8. The social agenda of sustainable tourism..... 414
19.9. The potential of ecotourism..... 417
19.10. Potential environmental impacts of mining..... 419
19.11. Key uncertainties and assumptions..... 421
19.12. Corporate governance in the mining sector..... 425
20.1. Policy instruments for environmental management..... 434
20.2. More compact agriculture..... 441
21.1. The changing skills base of environment authorities..... 449
21.2. Compliance assurance..... 451
21.3. Good governance for national sustainable development..... 452
22.1. Reaping mutual benefits from co-operation: The OECD MAD system..... 464
22.2. China and international co-operation..... 465

22.3. Towards a world environment organisation? . . . 467
22.4. The Global Environment Facility (GEF) . . . 468
22.5. The environment and the Millennium Development Goals . . . 472
22.6. Who is benefiting from the Clean Development Mechanism? . . . 472
22.7. Business and the environment: trends in MEA implementation . . . 473
22.8. Effectiveness and efficiency of partnerships involving OECD governments . . . 474
A.1. Assumptions and key uncertainties . . . 478

List of tables

0.1. The *OECD Environmental Outlook to 2030* . . . 24
I.1. Mapping of the *OECD Environmental Outlook* policy simulations by chapter . . . 39
1.1. Designation of responsibility for environmental matters in manufacturing facilities . . . 55
3.1. Productivity in historical perspective for the UK and USA: average annual % rate of change . . . 81
3.2. Global annual average GDP growth (%, 2005-2030): Baseline . . . 82
3.3. Shares of sectors in 2001 and 2030 (in gross economic output) . . . 85
5.1. Land, population and GDP of selected cities as a share of the country total . . . 109
5.2. World and urban populations, 1950-2030 . . . 111
5.3. Average density and built-up area per person, 1990-2000 . . . 114
6.1. Main axes of variation of narratives . . . 124
6.2. Variation 1: percentage change from Baseline for GDP using recent (5-year) productivity trends . . . 128
6.3. Change from Baseline in GDP (%) from long-term change in productivity growth . . . 130
6.4. Percentage change from Baseline of implementing a globalisation variation in 2030 . . . 132
6.5. Worldwide growth estimates, 2005-2050 (annual rates) . . . 133
7.1. *Outlook* Baseline global emissions by region and GHG intensity indicators: 2005, 2030 and 2050 . . . 145
7.2. Related aims and co-benefits of sector policies to reduce GHGs . . . 149
7.3. Coverage of impacts and adaptation in National Communications under the UNFCCC (including NC2, NC3, and NC4) . . . 150
7.4. Policy scenarios compared to Baseline: GHG emissions, CO_2 emissions and global temperature change, 2000-2050 . . . 157
7.5. Characteristics of post TAR stabilisation scenarios and resulting long-term equilibrium global average temperature and the sea level rise component from thermal expansion only . . . 159
7.6. Change (%) in GDP relative to Baseline of different scenarios, 2030 and 2050 . . . 166
9.1. Impact on land types in 2030 of agricultural tariff reform (compared to Baseline) . . . 203
9.2. Environmental impact of invasive alien species . . . 207
9.3. Sample economic impact of invasive species . . . 208
10.1. Population and water stress, 2005 and 2030 . . . 223
10.2. Source of river nitrogen exports to coastal waters, 2000 and 2030 . . . 225

11.1. Municipal waste generation within the OECD area and its regions, 1980-2030 244
11.2. Current municipal waste generation in OECD, BRIICS and the rest of the world (ROW) 246
13.1. Selected types of costs related to air and water pollution 275
13.2. Health effects associated with selected water pollutants.... 276
13.3. Health effects associated with selected air pollutants 277
13.4. Types and incidence of health costs from air and water pollution 278
14.1. Change in total land used for agriculture in 2030 (2005 = 100).... 303
14.2. Percentage differences in GHG emissions from land use changes, 2005 to 2030 306
14.3. Sources of greenhouse gas emission/mitigation potential in agriculture 307
14.4. Agricultural input/output-linked payments in selected countries (2001, millions USD) 313
14.5. Impact of policy simulation on agriculture and land use types, relative to Baseline in 2030.... 314
14.6. Impact of a 1 to 2 degree Celsius temperature change 316
14.A1.1. International price of crude oil (2001 USD) 319
14.A1.2. Share of biofuel as a percentage of all transport fuel (volume in gasoline energy equivalent).... 320
14.A1.3. World prices of agricultural products (% differences from the Baseline) 321
17.1. Environmental impact of the energy sector, 1980 to 2030.... 360
17.2. World primary energy consumption in the Baseline (EJ), 1980-2050 361
19.1. Characteristics of different steel production technologies globally (2000) 391
19.2. Estimated impacts on SO_2 emissions.... 399
19.3. Integrated kraft mill wastewater, TSS waste load and BOD5 waste load 403
19.4. International tourist arrivals by tourist receiving region (millions), 1995-2020 412
19.5. Trends for inbound tourism, 1995-2004.... 412
19.6. Production and prices of some major mineral commodities, 2000-2005 421
19.7. Trends in the production of metals, 1995 to 2005.... 423
20.1. Change in selected environmental variables under the Baseline and EO policy package scenario 439
22.1. Environmental aid to developing regions, 1990-2005.... 471
A.1. The 13 regional clusters used in the *Outlook*.... 478
A.2. North America: key figures, 1980-2030.... 479
A.3. OECD Europe: key figures, 1980-2030 480
A.4. OECD Asia: key figures, 1980-2030 481
A.5. OECD Pacific: key figures, 1980-2030 482
A.6. Russia and the Caucasus: key figures, 1980-2030.... 483
A.7. South Asia (including India): key figures, 1980-2030 484
A.8. China region: key figures, 1980-2030 486
A.9. The Middle East: key figures, 1980-2030.... 487
A.10. Brazil: key figures, 1980-2030.... 488
A.11. Other Latin America and the Caribbean: key figures, 1980-2030 489
A.12. Africa: key figures, 1980-2030 490
A.13. Eastern Europe and Central Asia: key figures, 1980-2030 491

A.14. Other Asian countries: key figures, 1980-2030 492
A.15. The world: key figures, 1980-2030.... 493
B.1. Summary of key physical output, by model 509
B.2. Clustering of model results for presentation in the *OECD Environmental Outlook* 511

List of figures

0.1. Average annual GDP growth, 2005-2030.... 24
0.2. Total greenhouse gas emissions (by region), 1970-2050 25
0.3. People living in areas of water stress, by level of stress, 2005 and 2030 26
1.1. Change in household expenditure, 2005-2030 49
1.2. Projected personal transport activity by region to 2050 50
1.3. Baseline forecasts of energy-related industrial nitrogen emissions, 1970-2030 (Mt) 54
1.4. Baseline forecasts of energy-related industrial sulphur emissions, 1970-2030 (Mt) 55
1.5. Estimated private sector pollution abatement and control expenditures (% of GDP) 56
1.6. Annual average % change in renewable energy production, 1990-2004 58
1.7. Share of environmental R&D in total government R&D, 1981-2005.... 59
1.8. Number of TPF patents in the environmental area, 1978-2002 60
1.9. Growth rates in patents in selected environmental areas, 1995-2004.... 61
2.1. Population growth by region, 1970-2030 69
2.2. Fertility rates by region, 1970-2040.... 70
2.3. The grey dependency ratio.... 71
3.1. Domestic material consumption and GDP, 1980-2005.... 78
3.2. Economy and environment, 1961-2003 79
3.3. Growth trends (average % per year), 1980-2001 80
3.4. Labour force projections, 2005-2030.... 81
3.5. Baseline import growth to 2030.... 86
3.6. Baseline gross output growth of natural resource-using sectors (2005 to 2030) 87
4.1. Exports of merchandise and services by selected countries and regions, annual average growth rates, 2000-2006 95
4.2. Total merchandise exports, % of world total per region, 1996 and 2006.... 96
4.3. Share of imports in GDP: Baseline and globalisation variation 98
4.4. Environmental implications: Baseline and globalisation variation in 2030 98
4.5. Projected commercial balance by sectors (in million USD), 2005 and 2030 99
4.6. Foreign Direct Investment flows by selected regions and countries, 1985 2006 (in billion USD).... 100
5.1. World population – total, urban and rural, 1950-2030.... 111
5.2. Trends in urban area expansion, 1950-2000 113
5.3. Incremental increases to population and urban areas, 1950-2000 113
5.4. Energy use per capita in private passenger travel *versus* urban density, selected world cities 116
6.1. CO_2 energy emissions: OECD and SRES results.... 123
6.2. World GDP growth (annual), 1980-2008 127

6.3. Environmental impacts of the globalisation variation to the Baseline 133
7.1. Global temperature, sea level and Northern hemisphere snow cover trends, 1850-2000 142
7.2. Baseline GHG emissions by regions, 1990 to 2050 145
7.3. Total greenhouse gas emissions by gas and CO_2 emissions by source category, 1980-2050 146
7.4. CO_2eq tax by policy case, 2010 to 2050: USD per tonne CO_2 (2001 USD, constant) 155
7.5. Global GHG emission pathways: Baseline and mitigation cases to 2050 compared to 2100 stabilisation pathways 158
7.6. Change in global emissions, GHG atmospheric concentrations, global mean temperature: Baseline and mitigation cases 160
7.7. Change in mean annual temperature levels in 2050 relative to 1990 (degrees C) 161
7.8. Air pollution co-benefits of GHG mitigation: reduction in NOx and SOx emissions – 450 ppm case and Baseline, 2030 164
7.9. Biodiversity effects of the 450 ppm case by 2050 165
7.10. Economic cost of mitigation policy cases by major country group 166
7.11. Change in value added: 450 ppm CO_2eq stabilisation case relative to Baseline, 2030 168
7.12*a*. Greenhouse gas emissions by regions in 2050: Baseline and 450 ppm cap and trade regime 169
7.12*b*. Regional direct cost of greenhouse gas abatement under different mitigation regimes, 2050 170
8.1. Cities included in the assessments, in 2000 and 2030 182
8.2. Annual mean PM_{10} concentrations, Baseline 184
8.3. Distribution of the urban population according to estimated annual mean PM_{10} concentrations in the modelled cities by regional cluster, 2000 and 2030 184
8.4. Current (2000) and future (2030) ozone concentrations at ground level 186
8.5. Potential exposure of urban population to ozone, 2000 and 2030 187
8.6. Emissions of sulphur dioxides and nitrogen oxides: Baseline and policy cases 191
8.7. Sulphur dioxide emissions, 1970-2050 192
8.8. Annual mean PM_{10} concentrations ($\mu g/m^3$) for the 13 regional clusters, 2030, Baseline and three policy cases 192
8.9. Distribution of the urban population according to estimated annual mean PM_{10} concentrations in the modelled cities, 2030, Baseline (left) compared to policy case ppglobal (right) 193
9.1. Historical and projected future changes indicated by mean species abundance, 2000-2050 200
9.2. Sources of losses in mean species abundance to 2030 201
9.3. Change in food crop area, 1980-2030 202
9.4. Change in agricultural activity in arid areas, 2005-2030 210
9.5. Cumulative change in protected areas worldwide, 1872-2003 212
10.1. People not connected to public sewerage systems, 2000 and 2030 224

10.2. Land area under high soil erosion risk by surface water runoff, 2000-2030 225
11.1. Global resource extraction, by major groups of resources and regions, 1980, 2002 and 2020 240
12.1. Premature deaths from PM_{10} air pollution for 2000 and 2030 257
12.2. Premature deaths from ozone exposure for 2000 and 2030 259
12.3. Estimated deaths from PM_{10} exposure for the Baseline and the three policy cases, 2030 262
12.4. Percentage of total mortality and burden of disease due to unsafe water, sanitation and hygiene, 2002 263
13.1. Defining the "costs of inaction" of environmental policy 272
13.2. Status of world fish stocks (2005) 279
13.3. Global mean temperature change under the Baseline, an aggressive mitigation scenario, and delayed action, 1970-2050 281
13.4. Temperature increases and likely impacts on marine and terrestrial ecosystems 283
14.1. Expected growth of world population, GDP per capita, agricultural production and agricultural land use percentage change from 2005-2030 297
14.2. Production of food crops, 2005-2030 298
14.3. Production of animal products, 2005-2030 299
14.4. Surface agricultural nitrogen losses (2000 and change to 2030) 303
14.5. Water stress, 2005 and 2030 305
14.6. Water withdrawals and irrigation 306
15.1. Global trends in the state of world marine stocks, 1974-2006 326
15.2. World fisheries production, 1970-2004 330
15.3. Projected composition of world fisheries to 2030: capture and aquaculture 333
15.4. Alternative fisheries management profiles 334
16.1. Transport externalities in Europe in 2004 (by impacts) 344
16.2. Global air transportation volumes and GDP (1990 = 100) 345
16.3. Annual new vehicle sales by region to 2030 346
16.4. Transport fuel consumption in the US and Canada by mode, 1971-2030 348
16.5. Energy consumption in the transport sector to 2030 349
16.6. Tax rates on petrol and diesel in OECD countries, 2002 and 2007 351
17.1. World primary energy consumption in the Baseline, to 2050 362
17.2. Primary energy consumption and intensity by region in the Baseline, to 2050 365
17.3. Increase in primary energy use in power generation by fuel and region in the Baseline, 2005-2030 366
17.4. Final energy consumption in the Baseline, 1970-2050 368
17.5. Public energy research and development funding in IEA countries 371
17.6. IEA and OECD selected policy scenarios: CO_2 from energy in 2005 and 2050 372
17.7. Change in primary energy use in power generation by fuel and region: policy scenarios compared with Baseline, 2005-2030 373
17.8. 450 ppm CO_2eq emission pathway compared to the Baseline: technology "wedges" of emission reduction 373
18.1. Projected chemicals production by region, 2005-2030 382
19.1. World crude steel production by process, 1970-2006 392
19.2. Real value added in the iron and steel sector, 2006 and 2030 393

19.3. Domestic demand for iron and steel, 2006 and 2030 393
19.4. Balance of trade in iron and steel products, 2006 and 2030 394
19.5. Estimated changes in steel production in response to OECD-wide and unilateral taxes 395
19.6. Effect of carbon tax on CO_2 emissions in the steel sector, 2010 and 2030 396
19.7. Effect of carbon tax on production in the steel sector, 2010 and 2030 397
19.8. Input intensities in the steel and electricity sectors 398
20.1. Change in emissions of sulphur and nitrogen oxides under the Baseline and EO policy package scenarios, 1980-2030 439
20.2. Global agricultural land area changes under Baseline and compact agriculture scenarios, 2000-2030 441
20.3. Average annual GDP growth by region for the Baseline and the EO policy package, 2005-2030 442
22.1. Multilateral environmental agreements, 1960-2004 466
22.2. Aid for environment, 1990-2005 470
B.1. Structure of production in ENV-Linkages 497
B.2. Structure of IMAGE 2.4 501
B.3. Main links between models deployed for the *OECD Environmental Outlook* 502
B.4. Map of regions used in environmental modelling for the *OECD Environmental Outlook* 512

Acronyms and Abbreviations

BRIC	Brazil, Russia, India and China
BRIICS	Brazil, Russia, India, Indonesia, China and South Africa
CBD	Convention on Biological Diversity
CCS	Carbon capture and storage
CDM	Clean Development Mechanism
CFC	Chlorofluorocarbon
CH_4	Methane
CO	Carbon monoxide
CO_2	Carbon dioxide
CO_2eq	Carbon dioxide equivalents
CSD	Commission on Sustainable Development
DAC	OECD Development Assistance Committee
EJ	Exajoules
EU15	Austria, Belgium, Denmark, Finland, France, Germany, Greece, Ireland, Italy, Luxembourg, Netherlands, Portugal, Spain, Sweden, United Kingdom
EU25	Austria, Belgium, Cyprus, Czech Republic, Denmark, Estonia, Finland, France, Germany, Greece, Hungary, Ireland, Italy, Latvia, Lithuania, Luxembourg, Malta, Netherlands, Poland, Portugal, Slovakia, Slovenia, Spain, Sweden, United Kingdom
EUR	Euro (currency of European Union)
FAO	Food and Agriculture Organization of the United Nations
GBP	Pound sterling
GDP	Gross domestic product
GHG	Greenhouse gas
GJ	Gigajoules
GNI	Gross national income
Gt	Giga tonnes
GW	Gigawatt
HFC	Hydrofluorocarbon
IEA	International Energy Agency
IMAGE	Integrated Model to Assess the Global Environment
IPCC	Intergovernmental Panel on Climate Change
LULUCF	Land use, land use change and forestry
MAD	Mutual Acceptance of Data
MDGs	Millennium Development Goals
MEA	Multilateral environmental agreement
MNP	Netherlands Environmental Assessment Agency
MSA	Mean species abundance

Mt	Million tonnes
MWh	Megawatt-hour
NO_2	Nitrogen dioxide
N_2O	Nitrous oxide
NO_x	Nitrogen oxides
ODA	Official development assistance
ppb	Parts per billion
ppm	Parts per million
PFC	Perfluorocarbon
PM	Particulate matter
$PM_{2.5}$	Particulate matter, particles of 2.5 micrometres (μm) or less
PM_{10}	Particulate matter, particles of 10 micrometres (μm) or less
ppmv	Parts per million by volume
ROW	Rest of world
RTA	Regional trade agreement
SO_2	Sulphur dioxide
SO_x	Sulphur oxides
SF_6	Sulphur hexafluoride
TWh	Terawatt hour
UNFCCC	United Nations Framework Convention on Climate Change
USD	United States dollar
VOC	Volatile organic compound
WHO	World Health Organization
WSSD	World Summit on Sustainable Development
WTO	World Trade Organization

ISBN 978-92-64-04048-9
OECD Environmental Outlook to 2030

Executive Summary

KEY MESSAGES

The *OECD Environmental Outlook to 2030* is based on projections of economic and environmental trends to 2030. The key environmental challenges for the future are presented according to a "traffic light" system (see Table 0.1). The *Outlook* also presents simulations of policy actions to address the key challenges, including their potential environmental, economic and social impacts.

Table 0.1. **The OECD Environmental Outlook to 2030**

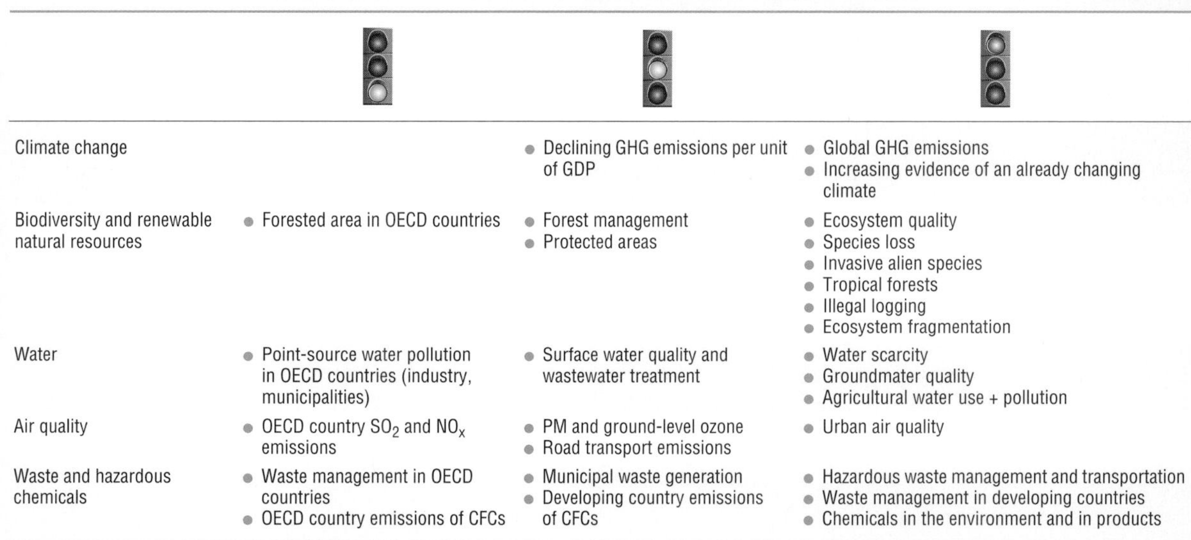

	Green light	Yellow light	Red light
Climate change		• Declining GHG emissions per unit of GDP	• Global GHG emissions • Increasing evidence of an already changing climate
Biodiversity and renewable natural resources	• Forested area in OECD countries	• Forest management • Protected areas	• Ecosystem quality • Species loss • Invasive alien species • Tropical forests • Illegal logging • Ecosystem fragmentation
Water	• Point-source water pollution in OECD countries (industry, municipalities)	• Surface water quality and wastewater treatment	• Water scarcity • Groundmater quality • Agricultural water use + pollution
Air quality	• OECD country SO_2 and NO_x emissions	• PM and ground-level ozone • Road transport emissions	• Urban air quality
Waste and hazardous chemicals	• Waste management in OECD countries • OECD country emissions of CFCs	• Municipal waste generation • Developing country emissions of CFCs	• Hazardous waste management and transportation • Waste management in developing countries • Chemicals in the environment and in products

KEY: ***Green light*** = environmental issues which are being well managed, or for which there have been significant improvements in management in recent years but for which countries should remain vigilant. ***Yellow light*** = environmental issues which remain a challenge but for which management is improving, or for which current state is uncertain, or which have been well managed in the past but are less so now. ***Red light*** = environmental issues which are not well managed, are in a bad or worsening state, and which require urgent attention. All trends are global, unless otherwise specified.

Action is affordable: policy scenarios and costs

The *Outlook* highlights some of the "red light" issues that need to be addressed urgently. The policy scenarios in this *Outlook* indicate that the policies and technologies needed to address the challenges are available and affordable. Ambitious policy actions to protect the environment can increase the efficiency of the economy and reduce health costs. In the long-term, the benefits of early action on many environmental challenges are likely to outweigh the costs.

As an example, a hypothetical global "OECD Environmental Outlook (EO) policy package" (EO policy package, see Chapter 20) was applied. It shows that, by combining specific policy actions, some of the key environmental challenges can be addressed at a cost of just over 1% of world GDP in 2030, or about 0.03 percentage points lower average annual GDP growth to 2030 (Figure 0.1). Thus world GDP would be about 97% higher in 2030 than today, rather than nearly 99% higher. Under such a scenario, emissions of nitrogen oxides and sulphur oxides would be about one-third less in 2030 while little change is projected under a no-new-policy baseline scenario, and by 2030 growth in greenhouse gas emissions would be contained to 13% rather than 37%.

Figure 0.1. **Average annual GDP growth, 2005-2030**

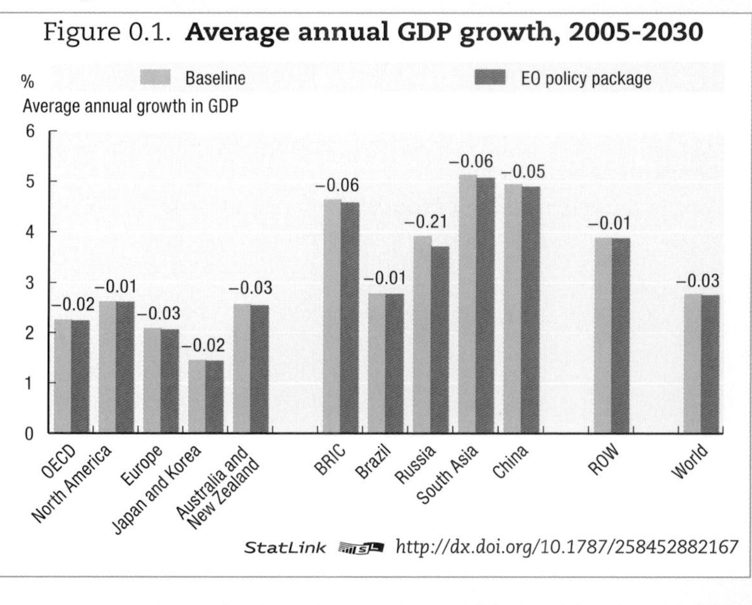

StatLink http://dx.doi.org/10.1787/258452882167

KEY MESSAGES *(cont.)*

More ambitious policy action than the EO policy package would be needed to stabilise greenhouse gas concentrations at the levels being considered in international discussions. Another simulation was run of policies needed to stabilise atmospheric concentration at 450ppm CO_2eq, one of the most ambitious targets being discussed. The simulation shows that to reach this target, actions by all countries are needed to achieve a 39% reduction in global greenhouse gas emissions by 2050 relative to 2000 levels (Figure 0.2). Such action would reduce GDP by 0.5% and 2.5% below Baseline estimates in 2030 and 2050 respectively, equivalent to a reduction in annual GDP growth of about 0.1 percentage points per annum on average. The more countries and sectors that participate in climate change mitigation action, the cheaper and more effective it will be to curb global greenhouse gas emissions. However, these costs are not distributed evenly across regions as seen in Figure 0.1. This suggests the need for burden-sharing mechanisms within an international collaborative framework to protect the global climate. While OECD countries should take the lead, further co-operation with a wider group of emerging economies, the "BRIICS" countries (Brazil, Russia, India, Indonesia, China and South Africa) in particular, can achieve common environmental goals at lower costs.

Figure 0.2. **Total greenhouse gas emissions (by region), 1970-2050**

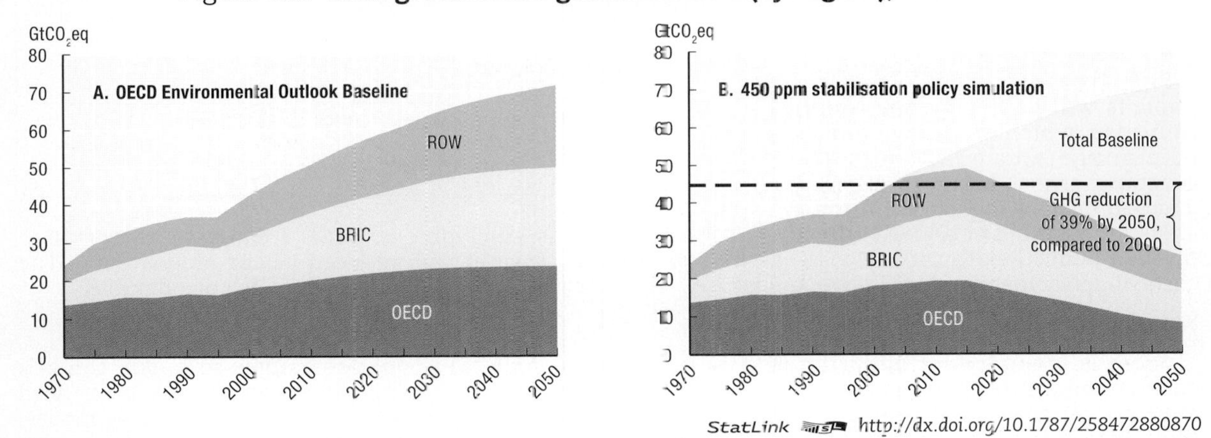

StatLink http://dx.doi.org/10.1787/258472880870

Note: BRIC = Brazil, Russia, India, China. ROW = Rest of world.

The consequences of environmental policy inaction

If no new policy actions are taken, within the next few decades we risk irreversibly altering the environmental basis for sustained economic prosperity. To avoid that, urgent actions are needed to address in particular the "red light" issues of climate change, biodiversity loss, water scarcity and health impacts of pollution and hazardous chemicals (Table 0.1).

Without further policies, by 2030, for example:

- Global emissions of greenhouse gases are projected to grow by a further 37%, and 52% to 2050 (Figure 0.2a). This could result in an increase in global temperature over pre-industrial levels in the range of 1.7-2.4° Celsius by 2050, leading to increased heat waves, droughts, storms and floods, resulting in severe damage to key infrastructure and crops.
- A considerable number of today's known animal and plant species are likely to be extinct, largely due to expanding infrastructure and agriculture, as well as climate change. Food and biofuel production together will require a 10% increase in farmland worldwide with a further loss of wildlife habitat. Continued loss of biodiversity is likely to limit the Earth's capacity to provide the valuable ecosystem services that support economic growth and human well-being.

KEY MESSAGES *(cont.)*

- Water scarcity will worsen due to unsustainable use and management of the resource as well as climate change; the number of people living in areas affected by severe water stress is expected to increase by another 1 billion to over 3.9 billion (Figure 0.3).
- Health impacts of air pollution will increase worldwide, with the number of premature deaths linked to ground-level ozone quadrupling and those linked to particulate matter more than doubling. Chemical production volumes in non-OECD countries are rapidly increasing, and there is insufficient information to fully assess the risks of chemicals in the environment and in products.

Figure 0.3. **People living in areas of water stress, by level of stress, 2005 and 2030**

Millions of people

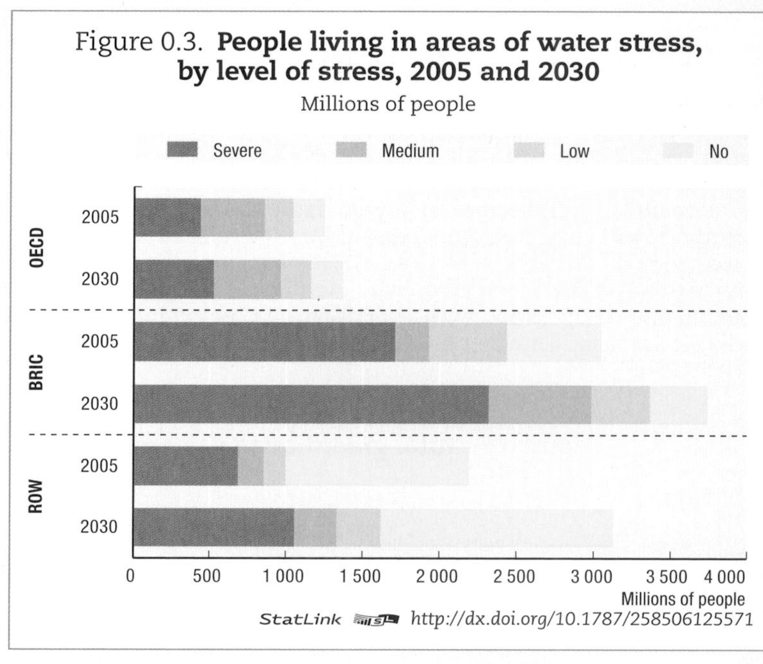

StatLink http://dx.doi.org/10.1787/258506125571

The greatest environmental impacts will be felt by developing countries, which are less equipped to manage and adapt. But the economic and social costs of policy inaction or delaying action in these areas are significant and are already affecting economies – including in OECD countries – directly (*e.g.* through public health service costs) as well as indirectly (*e.g.* through reduced labour productivity). The costs of policy inaction for biodiversity loss (*e.g.* fisheries) and climate change could be considerable.

Key policy options

There is a window of opportunity now to introduce ambitious policy changes to tackle the key environmental problems and promote sustainable development. Investment choices being made today need to be steered towards a better environmental future, particularly choices that will "lock-in" energy modes, transport infrastructure and building stocks for decades to come. The following actions are essential:

- Use a mix of complementary policies to tackle the most challenging and complex environmental problems, with a strong emphasis on market-based instruments, such as taxes and tradable permits, in order to reduce the costs of action.
- Prioritise action in the key sectors driving environmental degradation: energy, transport, agriculture and fisheries. Environmental ministers cannot do this alone. Environmental concerns need to be integrated into all policy-making by relevant ministries including finance, economy and trade, and reflected in all production and consumption decisions.
- Ensure that globalisation can lead to more efficient use of resources and the development and dissemination of eco-innovation. Business and industries need to play a lead role, but governments must provide clear and consistent long-term policy frameworks to encourage eco-innovation and to safeguard environmental and social goals.
- Improve partnerships between OECD and non-OECD countries to address global environmental challenges. Brazil, Russia, India, Indonesia, China and South Africa (BRIICS) in particular are key partners given their growing influence in the world economy and increasing share of global environmental pressures. Further environmental co-operation between OECD and non-OECD countries can help spread knowledge and technological best practices.
- Strengthen international environmental governance to better tackle trans-boundary and global environmental challenges.
- Strengthen attention to the environment in development co-operation programmes, and promote more coherent policies.

What will the environment be like in 2030 if no further action is taken?

Without more ambitious policies, increasing pressures on the environment could cause irreversible damage within the next few decades.

OECD countries have made significant progress in addressing many environmental challenges over the past few decades. Pollution from industrial sources has been reduced, forest coverage and the number and size of natural protected areas have increased (although the quality of protected areas is not always high, and there are still too few marine protected areas), ozone depleting substances have largely been phased-out and the use of natural resources, water and energy has to some extent been decoupled from continuing economic growth (*i.e.* become more efficient per unit of GDP). Policies that successfully led to these achievements should be maintained and scaled-up. However, in most cases, the increasing pressures on the environment from population and economic growth have out-paced the benefits of any efficiency gains.

The remaining environmental challenges (see Table 0.1) are of an increasingly complex or global nature, and their impacts may only become apparent over long timeframes. Among the most urgent of these challenges for both OECD and non-OECD countries are climate change, biodiversity loss, the unsustainable management of water resources and the health impacts of pollution and hazardous chemicals. We are not managing our environment in a sustainable manner.

The picture of economic and environmental trends in the coming decades will differ from region to region. By 2030, the world economy is expected to nearly double and world population to grow from 6.5 billion today to over 8.2 billion people. Most of the growth in both income and population will be in the emerging economies of Brazil, Russia, India, Indonesia, China and South Africa (the BRIICS) and in other developing countries. Rising income and aspirations for better living standards will increase the pressure on the planet's natural resources. The economic prospects of many of the poorest countries are threatened by unsustainable use of natural resources, uncontrolled pollution in rapidly-growing cities and the impacts of climate change. Developing countries are the most vulnerable to climate change as they lack the necessary financial and institutional capacity to adapt.

The global importance of rapidly emerging economies is growing as they become major economic and trade partners, competitors resource users and polluters on a level that compares to the largest of OECD countries. The primary energy consumption of Brazil, Russia, India and China together is expected to grow by 72% between 2005 and 2030, compared with 29% in the 30 OECD countries. Unless ambitious policy action is taken, greenhouse gas emissions from just these four countries will grow by 46% to 2030, surpassing those of the 30 OECD countries combined. Already, 63% of the population in Brazil, Russia, India and China together are living under medium to severe water stress; this share will increase to 80% by 2030 unless new measures to better manage water resources are introduced.

Policy action is affordable, and the cost of inaction is high

A policy package to address some of these key environmental challenges could cost as little as a loss of 0.03 percentage points in annual average GDP growth globally to 2030.

Protecting the environment can go hand-in-hand with continued economic growth. The *Outlook* estimates that world GDP will grow by nearly 99% between 2005 and 2030 under a Baseline projection reflecting no new policies. Without policy changes, the environmental consequences of this growth will be significant. But good environmental policies can lead to "win-win" opportunities for the environment, human health and the economy. To demonstrate this, a hypothetical global "*OECD Environmental Outlook* policy package" (EO policy package) of a number of specific policy actions to address several key environmental challenges simultaneously was put together. The EO policy package would imply a reduction of just over 1% in world GDP in 2030, such that world GDP would be about 97% higher in 2030 than today, instead of nearly 99% higher. On average, this would mean a loss of 0.03 percentage points in annual GDP growth globally to 2030.

Tackling a specific environmental problem can in some cases offer co-benefits in terms of reduction in other environmental pressures, and solutions to global problems can also help to address local environmental problems and *vice versa*. For example, measures to reduce vehicle emissions can both reduce greenhouse gas emissions and improve local air quality, while better insulation for homes and offices can cut energy bills for households and reduce pollution from energy production. For example, the climate policy simulation of a 450ppm CO_2eq stabilisation pathway also found that, in addition to reducing greenhouse gas emissions, the ambitious climate change policies would also lead to reductions in sulphur oxides of 20-30% and in nitrogen oxides of 30-40% by 2030. Similarly, regulations to limit agricultural water pollution from nitrogen fertilisers can also reduce atmospheric emission of nitrous oxide, a potent greenhouse gas.

The cost of inaction is high, while ambitious actions to protect the environment are affordable and can go hand-in-hand with economic growth.

Governments have the responsibility to create appropriate incentives for businesses and consumers to make choices that can help prevent future environmental problems. The investment choices being made today will determine future environmental outcomes. For example, the types of energy infrastructure put in place today will lock-in for decades to come emissions of greenhouse gases. Investments in transport infrastructure today will also affect future mobility options and their environmental impacts. The energy efficiency of our building stock for the coming decades or even centuries is determined by the construction and building efficiency regulations in place today. Fast growing economies offer enormous opportunities for investments in new energy efficiency technologies. For example, China is building new coal-power plants at a rapid pace, and its urban residential building stock is expected to more than double in the next 20 years.

For many of these actions, there will be long delays before their benefits are realised; and in turn, many short-sighted policy decisions taken today may lead to long-term environmental challenges. This makes timing an important issue for the design and implementation of environmental policy over the coming decades. The costs of delaying action, however, could be critical, especially where policy decisions have long-term or irreversible environmental implications or where it is impossible to predict with precision the full extent and character of damage. Biodiversity loss and species extinction are one such example. For climate change, deciding when to act involves balancing the economic costs of more rapid emission reductions now against the future climate risks of delay.

A window of opportunity to act is now open where investments in building, energy and transport infrastructure will be made in the coming decades, especially in fast growing economies.

A window of opportunity to act is now open, but it will not be open for long. We need forward-looking policies today to avoid the high costs of inaction or delayed action over the longer-term.

What action should be taken?

Ensure efficient resource use and eco-innovation

Trade and investment liberalisation can encourage more efficient allocation of resources globally, if sound environmental policy and institutional frameworks are in place. In their absence, globalisation can amplify market and policy failures and intensify environmental pressures. Effective policies are required at local, national, regional and global levels.

Globalisation expands markets and promotes competition, and can motivate businesses to adapt and innovate. Some private sector leaders are already moving ahead, encouraged by stakeholders and consumer demands for "green" innovation and products. Eco-innovation and the wider use of eco-efficient techniques not only improve environmental performance, but can also raise economic productivity, making businesses and leading countries more competitive. The environmental goods and services sector is likely to expand significantly in the future. Businesses can reap the benefits of globalisation if they seize the "first mover" advantage of eco-innovation. Technological solutions have already addressed many environmental problems, and new ones are developing, such as carbon capture and storage and hybrid vehicles, which are likely to become increasingly cost-competitive within the next few decades. For example, if "second generation" biofuel technology (based on biomass waste) becomes widely available by 2030, the projected expansion of agricultural lands to supply biofuels production, the increased use of pesticides, fertilisers and water, and the impacts on biodiversity and ecosystems associated with this land use, could be avoided.

Globalisation provides opportunities to promote efficient use of resources and to spur the development and spread of eco-innovation.

Business has a central role in driving eco-innovation, but governments have an important responsibility to set the appropriate policy frameworks according to national circumstances:

- Long-term policy frameworks that allow environmental costs to be priced into economic activities (*e.g.* through green taxes and tradable permits or regulation) to make green technologies cost-competitive and provide business with the incentives to innovate.
- Well-targeted government support for basic R&D for eco-innovation where justified, including enhanced government-business partnerships.
- Strong policy and institutional frameworks to promote environmental and social objectives alongside efforts to liberalise trade and investment and to level the playing field to make environmental protection and globalisation mutually supportive.

Liberalisation of trade in environmental goods and services could help realise this objective. The number of regional trade agreements is still low but is increasing rapidly, and many now include commitments for environmental co-operation. Multilateral instruments such as the *OECD Recommendation on Environment and Export Credit* and the *OECD Guidelines for Multinational Enterprises* encourage environmentally and socially responsible corporate behaviour and accountability.

While globalisation has a range of potential impacts – both good and bad – on the environment, the state of the environment and natural resources also affects economic development and globalisation. Competition for scarce natural resources, harvesting of some renewable resources such as fish stocks and tropical timber, the impacts of changing climate on agricultural production, energy prices, the search for alternative energy sources, and others, may heavily influence trade and investment patterns in the coming years.

Enhance international environmental co-operation

OECD and non-OECD countries need to work together to achieve common environmental goals.

Economic globalisation, as well as the global nature of many environmental problems, require OECD and non-OECD countries to work together to address the most pressing global environmental challenges and promote sustainable development.

- Developing countries have opportunities to learn from the experience of other countries and "leapfrog" to more energy-efficient, resource-efficient and greener development paths, taking advantage of new know-how and technologies. OECD and non-OECD countries need to work together to spread knowledge, best practices and technologies to mutually benefit from more sustainable production and consumption patterns worldwide.
- Some of the poorest countries in the world have been left behind by globalisation by failing to integrate into the world economy due to their lack of capacity to capture the benefits of globalisation and also due to trade barriers in OECD countries. Further efforts are needed to integrate environmental concerns into development co-operation programmes.
- The BRIICS, in particular, need to be part of international solutions to global environmental challenges, given their increasing role in the world economy and rapidly growing environmental impacts. Also, further environmental co-operation between OECD countries and BRIICS can achieve global environmental goals at lower costs for all.

- For climate change, the more countries that participate in mitigation action, and the more sectors and greenhouse gases that are covered, the cheaper it will be to curb global emissions. The *Outlook* indicates that if OECD countries alone implement a carbon tax starting at USD 25/tonne of CO_2 in 2008, this would lead to a 43% reduction in OECD greenhouse gas emissions. However, global emissions would still be 38% higher in 2050 compared to the 2000 levels. If Brazil, China, India and Russia follow suit with the same policy in 2020, and the rest of the world in 2030, global greenhouse gas emissions in 2050 could be brought down to the 2000 levels (0% increase).
- Stronger international environmental governance is needed to ensure implementation of international agreements to tackle trans-boundary and global environmental challenges.

Prioritise actions in the key sectors affecting the environment: energy, transport, agriculture and fisheries

Most environmental problems can only be solved by coherent government-wide policy actions and co-operation with businesses and civil society. Relevant ministries need to work together to develop better co-ordinated policies so that environmental concerns are integrated into actions by key ministries such as finance, trade, industry, energy, transport, agriculture and health. For example, adaptation to climate change that is already locked-in by past emissions will increasingly need to be integrated into policies governing energy, transport and water infrastructure, land use planning, and development co-operation. Also, the development of biofuels needs to take account of their overall life-cycle impacts on the environment and on food prices. Coherent policy impact assessments need to cover all relevant policy areas, including energy, agriculture, environment, as well as research and technology development, in order to avoid a situation where governments subsidise energy production that can result in dubious environmental benefits and lead to higher agricultural commodity prices. Government authorities increasingly need to work together, including across different levels of government (central, regional, state, local), to successfully ensure the development and implementation of coherent environmental policies.

The OECD *Environmental Outlook* highlights the priority actions needed in key sectors to prevent the environmental damage projected to 2030:

Many environmental challenges cannot be solved by environment ministries alone.

- **Energy**. Fossil fuel use is the main source of carbon dioxide emissions, the principal greenhouse gas that causes climate change. The *Outlook* projects world energy-related carbon dioxide emissions to increase by 52% to 2030 under the no-new-policy Baseline scenario. Meanwhile, world energy sulphur and nitrogen emissions are projected to remain stable around or below recent levels. As investments in energy infrastructure lock-in technologies, fuel needs and related emissions for years to come, an appropriate policy framework is needed now to promote renewable energy and low-carbon alternative processes and fuels, including technologies for carbon capture and storage. Energy pricing that reflects the full cost of carbon is essential, but regulations and support for research and development of new technologies are also needed. Governments should avoid policies that lock-in specific technologies or fuel choices, in particular avoiding technology-specific targets (*e.g.* for biofuels), in order to leave all technology options open and to provide incentives for further innovation. Policies to promote cost-effective energy efficiency measures for buildings, transport and electricity production are needed urgently, particularly in fast growing economies, where infrastructure is being put in place today which will last for many decades.

- **Transport**. Air pollution and greenhouse gas emissions from transport are growing rapidly, from passenger vehicles, aviation and marine transport, contributing to climate change globally and causing health problems in many urban areas. The *Outlook* projects transport-related carbon dioxide emissions to increase by 58% to 2030, while sulphur and nitrogen emissions will fall by a quarter to a third from today's levels. Transport prices rarely reflect their full social and environmental costs, resulting in over-use and sub-optimal choices about the type of transport to use. Transport pricing should fully reflect the costs of environmental damage and health impacts, *e.g.* through taxes on fuels (including the removal of tax exemptions) and road pricing. Research and development of new transport technologies, including vehicles with better fuel economy, hybrid vehicles, etc., should be promoted, especially to help offset projected rapid increases in motorisation in non-OECD countries. The availability, frequency and safety of public transport should be strengthened to provide a viable alternative to private cars. It is mobility and access that need to be ensured, not "transport" *per se*.
- **Agriculture** is by far the largest user of water and is responsible for much of its pollution. The *Outlook* Baseline projects world primary food crop production to grow by 48% and animal products by 46% to 2030. OECD countries will account for large shares, particularly for animal products (37% in 2030 to feed 17% of the world's population). If no new policies are introduced, the conversion of natural land to agricultural use will continue to be a key driver of biodiversity loss. Under current policies, areas for biofuel crops are projected to increase by 242% between 2005 and 2030. Land-related greenhouse gas emissions are smaller than from energy sources, but still important. Production-linked subsidies have in many cases resulted in pollution of water resources and soil, and damaged ecosystems and landscape. Increasingly, production-linked payments are conditional on farmers adopting certain practices to reduce environmental harm. While such "cross-compliance" can help to reduce some of the negative environmental impacts of agricultural production, a more effective approach would be to remove environmentally harmful subsidies in the first place. Taxes on farm chemicals also help limit their use, while appropriate pricing of irrigation water would encourage more rational use of water and cost-recovery for irrigation infrastructure provision.
- **Capture fisheries** exert pressures on ecosystems and biodiversity through depletion of fish stocks, destruction of habitats and pollution. Those environmental pressures can undermine the productivity of affected fisheries and the livelihoods of fishing communities. Fisheries depend on a healthy marine environment. Fishing opportunities are influenced by climate change, natural fluctuations and environmental pressures from other human activities. While progress is already being made in some fisheries towards an ecosystem-based approach, the worrying outlook for capture fisheries highlighted in this report could be reversed by further measures to limit total catch levels, designate fishing seasons and zones, regulate fishing methods and eliminate subsidies for fishing capacity. Stronger international co-operation is needed in this area.

What are the obstacles to change?

While policy reforms are achievable and affordable, some obstacles are preventing the ambitious policy changes needed, including:

- *Fears of impacts on industrial competitiveness.* Possible negative impacts on industrial competitiveness of environmental policies are a key obstacle to decisive policy actions.

Resistance by affected sectors often challenges the political feasibility of introducing environmental measures such as emission standards, targets and green taxes. But concerns about the competitiveness impacts of environmental policies are often overstated. Better information is needed on the actual impacts on affected firms and sectors and this should be compared with the wider and longer term benefits of environmental improvements and potential economy-wide efficiency gains. Nevertheless, some sectors can be adversely affected by environmental measures, especially when such measures are implemented in a non-global manner.

- *Uncertainty about who should take action and who should bear the costs of action.* This is especially so for global environmental challenges like climate change and biodiversity loss, for which the costs and benefits of policy action are unevenly distributed amongst countries and generations. Historically, the majority of greenhouse gas emissions have come from developed countries, but climate change is expected to have the largest impacts on developing countries. Looking forward, CO_2 emissions from non-OECD countries are projected to double to 2030, accounting for almost 73% of the total increase to 2030. However, on a per-capita basis, OECD country emissions will still be three to four times higher than non-OECD countries in 2030. Burden-sharing will be a key issue in the post-2012 climate architecture.
- *Underpricing of natural resource use and pollution.* "Getting the prices right" is often a very efficient way of keeping the costs of environmental policies low and greening the economy. But in practice it is difficult to accurately estimate the full costs of environmental, health and productivity damages caused by economic activities. If the full costs are reflected in their prices, polluting activities will be costlier and there will be clear price incentives for increased resource and energy efficiency. However, in most countries the use of scarce natural resources remains under-priced or even subsidised, and the polluter pays principle is rarely implemented fully. Unsustainable subsidies are pervasive in the industry, agriculture, transport and energy sectors in most OECD countries. They are expensive for governments and tax payers to maintain, and can have harmful environmental and social effects.

Removing the key obstacles to change

The OECD work shows that clean and clever growth need not be expensive. Also, the right policies to protect the environment can lead to long-term net benefits for the economy. To realise this, the following approaches to policy development and implementation could be considered:

- *Phase in the policy to allow for options* such as transitional adjustments, recycling of tax revenues back to affected sectors, border tax adjustments in compliance with World Trade Organization regulations, and international co-operation to harmonise regulations and taxes. Improving public awareness of the overall costs and benefits of the proposed measures will also be important. Transitional measures can be part of the reform package to smooth the transition and soften any unwanted effects from structural changes on particular groups in society, such as increased energy bills for low-income families.
- *Work in partnership with stakeholders,* including business, academia, trade unions and civil society organisations, to find creative and low-cost solutions to many of the environmental challenges. Public support and buy-in, particularly by consumers and affected industries, are often needed to ensure successful implementation of ambitious policies.

- *Bring OECD and non-OECD countries together to identify environmentally effective and economically efficient solutions to common environmental challenges.* OECD countries need to take the lead to mitigate and help developing countries adapt to climate change and realise their mitigation potentials. To stop and reverse biodiversity loss, the need for action is primarily in developing countries where the richest natural resources are located, while the benefits of resource conservation extend globally. The long-term costs to society and the environment of not acting, or of further delaying ambitious action, are likely to outweigh the costs of early action.
- *Make widespread use of market-based approaches to enable efficiency gains and market advantage through innovation.* Market-based instruments – such as taxes, tradable permits and the reform or removal of environmentally harmful subsidies – are a powerful tool for sending price signals to businesses and households to make their production and consumption more sustainable.
- *Develop policy mixes, or combinations of instruments, tailored to specific national circumstances* to tackle many of the urgent remaining environmental problems. Mixes of policy instruments are needed because of the complex and often cross-sectoral nature of environmental issues. This typically means combining a robust regulatory framework with a variety of other instruments, such as strong pricing mechanisms, emissions trading or tradable permits, information-based incentives such as labelling, and infrastructure provision and building codes. In a well-designed mix, instruments can mutually support each other. For example, a labelling scheme can enhance the responsiveness of firms and households to an environmentally related tax, while the existence of the tax helps draw attention to the labelling scheme.

The *OECD Environmental Outlook* demonstrates that meeting the environmental challenges is both economically rational and technologically feasible. Seen from a long-term perspective, the costs of early action are far less than the costs of delaying; the earlier we act, the easier and less expensive the task will be. Policy-makers, businesses and consumers all need to play their part to implement the ambitious policy reforms which will deliver the most cost-effective environmental improvements. In that way, options are left open for future generations to make their own choices about how to enhance their well-being.

ISBN 978-92-64-04048-9
OECD Environmental Outlook to 2030

Introduction: Context and Methodology

Purpose of the report

The purpose of the *OECD Environmental Outlook* is to help government policy-makers to identify the key environmental challenges they face, and to understand the economic and environmental implications of the policies that could be used to address those challenges.

The *Outlook* provides a baseline projection of environmental change to 2030 (referred to as "the Baseline"), based on projected developments in the underlying economic and social factors that drive these changes. The projections are based on a robust general equilibrium economic modelling framework, linked to a comprehensive environmental modelling framework (see below, and Annex B, for more details). Simulations were also run of specific policies and policy packages that could be used to address the main environmental challenges identified, and their economic costs and environmental benefits compared with the Baseline.

This is the second *Environmental Outlook* produced by the OECD. The first *OECD Environmental Outlook* was released in 2001, and provided the analytical basis on which ministers adopted an *OECD Environmental Strategy for the First Decade of the 21st Century*. This second *Outlook*:

- extends the projected baseline used in the first *Outlook* from 2020 to 2030, and even 2050 for some important areas;
- is based on a stronger and more robust modelling framework;
- focuses on the policies that can be used to tackle the main challenges;
- expands the country focus to reflect developments in both OECD and non-OECD regions and their interactions.

Many of the priority issues and sectors identified in this *Outlook* are the same as those highlighted as needing most urgent policy action in the first *OECD Environmental Outlook* (2001) and in the *OECD Environmental Strategy for the First Decade of the 21st Century*. These include the priority issues of climate change, biodiversity loss and water scarcity, and the key sectors exerting pressure on the environment (agriculture, energy and transport). Added to these is a new priority issue: the need to address the health impacts of the build-up of chemicals in the environment. The 2001 *Outlook* indicated the environmental challenges expected in the next couple of decades; this *Outlook* not only deepens and extends this analysis, it also focuses on the policy responses for addressing these challenges. It finds that the solutions are affordable and available if ambitious policy action is implemented today, and if countries work together in partnership to ensure comprehensive action, avoid competitiveness concerns and share the responsibility and costs of action fairly and equitably. This latest *Outlook* analyses the policies that can be used to achieve the *OECD Environmental Strategy*. It will provide the main analytical material to support discussions on further implementation of the *OECD Environmental Strategy* at the OECD Meeting of Environment Ministers planned for early 2008.

Policy context

Why develop an environmental outlook? Many of the economic or social choices that are being made today – for example, investments in transport infrastructure and building construction, fishing fleets, purchase of solar heating panels – will have a direct and lasting affect on the environment in the future. For many of these, the full environmental impacts will not be felt until long after the decisions have been taken. These factors make policy decisions difficult: the costs of policy action to prevent these impacts will hit societies today, but the benefits in terms of improved environmental quality or damage avoided may only be realised in the future. For example, the greenhouse gases released today continue to build up in the atmosphere and will change the future climate, with serious impacts for the environment, the economy and social welfare.

But politicians tend to reflect the short-term interests of the voting public, not the long-term needs of future generations. They also tend to focus on the immediate costs and benefits to their own populations of a given policy approach, rather than on the global impacts. But many of the main environmental challenges countries face in the early 21st century are global or transboundary in nature, including global climate change, biodiversity loss, management of shared water resources and seas, transboundary air pollution, trade in endangered species, desertification, deforestation, etc. Building public understanding and acceptance of the policies that are needed to address these challenges is essential for policy reform.

These political challenges are exacerbated by uncertainty about the future. Often the exact environmental impacts of social and economic developments are poorly understood or disputed. In some cases, scientific uncertainty about environmental or health impacts is a main cause of policy inaction, while in others it is used as a justification for precautionary action. Scientific understanding and consensus about environmental change has been developing rapidly in a number of areas in recent years, for example through the 2005 Millennium Ecosystem Assessment and the 2007 IPCC Fourth Assessment Report on the Science of Climate Change. Despite the improvements in the scientific understanding of such issues, a gap remains in the development and implementation of effective environmental policies based on this scientific understanding.

This *Environmental Outlook* examines the medium to long-term environmental impacts of current economic and social trends, and compares these against the costs of specific policies that could be implemented today to tackle some of the main environmental challenges. The purpose is to provide more rigorous analysis of the costs and benefits of environmental policies to help policy-makers take better, more informed policy decisions now.

Many environmental problems are complex and inter-connected. For example, species loss is often the result of multiple pressures – including hunting, fishing or plant harvesting, loss of habitat through land use change or habitat fragmentation, impacts of pollutants – and thus a mix of policy instruments is needed to tackle the various causes of this loss. These policy packages need to be carefully designed in order to achieve the desired environmental benefits at the lowest economic cost. This *Outlook* examines the policy packages that could be used to tackle some of the key environmental challenges, and the framework conditions needed to ensure their success.

The transboundary or global nature of many of the most pressing environmental challenges identified in this *Outlook* require countries to increasingly work together in partnership to address them. The ways in which OECD environment ministries can work together in partnership with other ministries, stakeholder partners and other countries are explored in this *Outlook*.

A special focus on the emerging economies in the *Outlook*

This *Outlook* identifies the main emerging economies as the most significant partners for OECD countries to work with in the coming decades to tackle global or shared environmental problems. This is because these countries are responsible for an increasingly large share of the global economy and trade, and thus have an increasing capacity to address these challenges, in part because their economies are so dynamic. Moreover, the pressures that they exert on the environment are also growing rapidly.

In some chapters, where data are available and relevant, the BRIICS countries (Brazil, Russia, India, Indonesia, China and South Africa) are highlighted for attention as a country grouping. In other chapters, the smaller country grouping of BRIC (Brazil, Russia, India and China) is examined, or even further disaggregated to each of these four countries individually. The BRIC grouping is used for most of the modelling projections and simulations in the *Outlook*.

Modelling methodology and sources of information

The analysis presented in this *Environmental Outlook* was supported by model-based quantification. On the economic side, the modelling tool used is a new version of the OECD/ World Bank JOBS/Linkages model, operated by a team in the OECD Environment Directorate and called ENV-Linkages. It is a global general equilibrium model containing 26 sectors and 34 world regions and provides economic projections for multiple time periods. It was used to project changes in sector outputs and inputs of each country or region examined to develop the economic baseline to 2030. This was extended to 2050 to examine the impacts of policy simulations in specific areas, such as biodiversity loss and climate change impacts. The economic baseline was developed with expert inputs from, and in co-operation with, other relevant parts of the OECD, such as the Economics Department, the International Energy Agency and the Directorate for Food, Agriculture and Fisheries.

The Integrated Model to Assess the Global Environment (IMAGE) of the Netherlands Environmental Assessment Agency (MNP) was further developed and adjusted to link it to the ENV-Linkages baseline in order to provide the detailed environmental baseline. IMAGE is a dynamic integrated assessment framework to model global change, with the objective of supporting decision-making by quantifying the relative importance of major processes and interactions in the society-biosphere-climate system. The IMAGE suite of models used for the *Outlook* comprises models that also appear in the literature as models in their own right, such as FAIR (specialised to examine burden sharing issues), TIMER (to examine energy), and GLOBIO3 (to examine biodiversity). Moreover, for the *Outlook* the IMAGE suite included the LEITAP model of LEI at Wageningen and the WaterGap model of the Center for Environmental Systems Research at Kassel University. IMAGE and associated models provided the projections of impacts on important environmental endpoints to 2030, such as climate, biodiversity, water stress, nutrient loading of surface water, and air quality. Annex B provides a more detailed description of the modelling framework and main assumptions used for the *Outlook* report.

The Baseline Reference Scenario presents a projection of historical and current trends into the future. This Baseline indicates what the world would be like to 2030 if currently existing policies were maintained, but *no new policies* were introduced to protect the environment. It is an extension of current trends and developments into the future, and as such it does not reflect major new or different developments in either the drivers of environmental change or environmental pressures. A number of major changes are possible in the future, however, that would significantly alter these projections. A few of these were examined as "variations" to the Baseline, and their impacts are described in Chapter 6 to show how these changes might affect the projections presented here.

Because the Baseline reflects no new policies, or in other words it is "policy neutral", it is a reference scenario against which simulations of new policies can be introduced and compared. Simulations of specific policy actions to address key environmental challenges were run in the modelling framework. The differences between the Baseline projections and these policy simulations were analysed to shed light on their economic and environmental impacts.

The simulations undertaken for the *Environmental Outlook* exercise are illustrative rather than prescriptive. They indicate the type and magnitude of the responses that might be expected from the policies examined, rather than representing recommendations to undertake the simulated policy actions. As relevant, some of the policy simulation results are reflected in more than one chapter. The table below summarises the policy simulation analyses and lists the different chapters containing the results.

Sensitivity analysis was undertaken to test the robustness of key assumptions in ENV-Linkages, and some of the results of this analysis are presented in Annex B. This, in conjunction with the Baseline variations described in Chapter 6, provides a clearer picture for the reader of the robustness of the assumptions in the Baseline.

Throughout the *Outlook*, the analysis from the modelling exercise is complemented by extensive data and environmental policy analysis developed at the OECD. Where evidence is available, specific country examples are used to illustrate the potential effects of the policies discussed. Many of the chapters in this *Outlook* have been reviewed by the relevant Committees and Expert Groups of the OECD, and their input has strengthened the analysis.

The *Outlook* is released at about the same time as a number of other forward-looking environmental analyses, such as UNEP's Fourth Global Environment Outlook (GEO-4); the IPCC Fourth Assessment Report (AR-4); the International Assessment of Agricultural Science and Technology for Development supported by the World Bank, FAO and UNEP; and the CGIAR Comprehensive Assessment of Water Use in Agriculture. Through regular meetings and contacts, efforts have been made by the organisations working on these reports to ensure co-ordination and complementarity in the studies, and to avoid overlap. The *OECD Environmental Outlook* differs from most of the others in its emphasis on a single baseline reference scenario against which specific policy simulations are compared for the purpose of policy analysis. Most of the others explore a range of possible "scenarios", which provide a useful communication tool to illustrate the range of possible futures available, but are less amenable to the analysis of specific policy options. The *OECD Environmental Outlook* also looks at developments across the full range of environmental challenges, based strongly on projected developments in the economic and social drivers of environmental change, while many of the other forward-looking analyses focus on a single environmental challenge.

Table I.1. **Mapping of the OECD *Environmental Outlook* policy simulations by chapter**

Simulation title	Simulation description	Chapters in which the results are reflected	Models used
Baseline	The "no new policies" Baseline used throughout the *OECD Environmental Outlook.*	All chapters	ENV-Linkages; IMAGE suite
Globalisation variation	Assumes that past trends towards increasing globalisation continue, including increasing trade margins (increasing demand by lowering prices in importing countries) and reductions in invisible costs (*i.e.* the difference between the price at which an exporter sells a good and the price that an importer pays).	4. Globalisation 6. Key variations to the standard expectation	ENV-Linkages; IMAGE suite
High and low growth scenarios	Variation 1: High economic growth – examines impacts if recent high growth in some countries (*e.g.* China) continues, by extrapolating from trends from the last 5 years of growth rather than the last 20 years. Variation 2: Low productivity growth – assumes productivity growth rates in countries converge towards an annual rate of 1.25% over the long-term, rather than 1.75% as in the Baseline. Variation 3: High productivity growth – assumes productivity growth rates in countries converge towards an annual rate of 2.25% over the long-term.	6. Key variations to the standard expectation	ENV-Linkages
Greenhouse gas taxes	Implementation in participating countries of a tax of USD 25 on CO_2eq, increasing by 2.4% per annum. OECD 2008: only OECD countries impose the tax, starting in 2008. Delayed 2020: all countries apply the tax, but starting only in 2020. Phased 2030: OECD countries implement the tax from 2008; BRIC countries from 2020, and then the rest of the world (ROW) from 2030 onwards. All 2008: in a more aggressive effort to mitigate global GHG emissions, all countries implement the USD 25 tax from 2008.	7. Climate change 13. Cost of policy inaction (Delayed 2020) 17. Energy 20. Environmental policy packages	ENV-Linkages; IMAGE suite
Climate change stabilisation simulation (450 ppm)	Optimised scenario to reach a pathway to stabilise atmospheric concentrations of GHG at 450 ppm CO_2eq over the longer term and limit global mean temperature change to roughly 2 °C. A variation on this case was developed to explore burden-sharing using a cap and trade approach to implementation.	7. Climate change 13. Cost of policy inaction 17. Energy 20. Environmental policy packages	ENV-Linkages; IMAGE suite
Agriculture support and tariff reform	Gradual reduction in agricultural tariffs in all countries to 50% of current levels by 2030. Gradual reduction in production-linked support to agricultural production in OECD countries to 50% of current levels by 2030.	9. Biodiversity 14. Agriculture	ENV-Linkages
Policies to support biofuels production and use	Demand for biofuels growing in line with the IEA *World Energy Outlook* (2006) scenario. DS: a scenario whereby growth in biofuel demand for transport is driven by exogenous changes, keeping total fuel for transport close to the Baseline. OilS: a high crude oil price scenario to determine the profitability of biofuel in the face of increasing costs of producing traditional fossil-based fuels. SubS: a subsidy scenario in which producer prices of biofuels are subsidised by 50%.	14. Agriculture	ENV-Linkages
Fisheries	Global fisheries cap and trade system, representing a 25% reduction in open fisheries catch, with trading allowed within six geographical regions.	15. Fisheries and aquaculture	ENV-Linkages
Steel industry CO_2 tax	Implementation of a carbon tax of 25 USD per tonne CO_2, applied respectively to OECD steel industry only, all OECD sectors, and all sectors worldwide.	19. Selected industries – steel and cement	ENV-Linkages
Policy mix	Three variations of policy packages were modelled, depending on the participating regions: OECD countries only OECD + BRIC Global The policy packages included: • reduction of production-linked support and tariffs in agriculture to 50% of current levels by 2030. • tax on GHG emissions of USD 25 tax CO_2eq, increasing by 2.4% per annum (phased with OECD starting in 2012, BRIC in 2020, ROW in 2030). • moving towards, although not reaching, Maximum Feasible Reduction in air pollution emissions, phased over a long time period depending on GDP/capita. • assuming that the gap to connecting all urban dwellers with sewerage will be closed by 50% by 2030, and installing, or upgrading to the next level, sewage treatment in all participating regions by 2030.	8. Air pollution 10. Freshwater 12. Health and environment 20. Environmental policy packages	ENV-Linkages; IMAGE suite

Structure of the report

The *OECD Environmental Outlook* is divided into two main parts:

i) *The World to 2030 – the Consequences of Policy Inaction:* describes the Baseline, *i.e.* the projected state of the world to 2030 in terms of the key drivers of environmental change and the developing environmental challenges, as well as analysing some possible variations to the Baseline.

ii) *Policy Responses:* focuses on the policy responses at both the sectoral level and in terms of implementing a more comprehensive and coherent policy package.

The first part describes the key elements of the Baseline to 2030, including the main drivers of environmental change (consumption and production patterns, technological innovation, population dynamics and demographic change, economic development, globalisation, and urbanisation) and the key environmental challenges (climate change, air pollution, biodiversity, freshwater, waste and material flows, health and environment). For each of these, the key recent trends and projections to 2030 are presented, as well as some of the policy approaches that are being used to address the environmental challenges. Chapter 6 describes some key variations to the Baseline – for example, how the Baseline would differ if key economic drivers (such as economic growth or global trade) were changing faster than projected in the Baseline. The chapter also explores other sources of uncertainty in the *Outlook* projections. Finally, this first part of the report examines the consequences and costs of policy inaction – essentially the environmental, health and economic impacts embodied in the "no new policies" Baseline scenario.

The second part of the *Outlook* report examines the possible policy responses to address the key environmental challenges, and assesses the economic and environmental impact of these responses. The key sectors whose activities affect the environment are examined, with a brief summary of the trends and outlook for their impacts, followed by an assessment of the policy options that could be applied in that sector to reduce negative environmental impacts. This section assesses the environmental benefits of specific policy options and their potential costs to the sector involved and/or economy-wide (and disaggregated by region where appropriate). This analysis can be used by environment ministries in discussing specific policy options for tackling environmental challenges with their colleagues in other ministries, such as finance, agriculture, energy or transport. The sectors examined include those that were prioritised in the *OECD Environmental Strategy* – agriculture, energy and transport – and also other sectors which strongly affect natural resource use or pollution, such as fisheries, chemicals and selected industries (steel, cement, pulp and paper, tourism and mining).

In addition to analysing sector-specific policies, this part of the *Outlook* also examines the effects of a package of policies (the EO policy package) to tackle the main environmental challenges. The analysis of this EO policy package highlights the potential synergies between policies (*i.e.* where the benefits of combining two or more policies may be greater than the simple sum of their benefits as separate policies), or potential conflicts where policies may undermine each other. Chapter 21 outlines the key framework conditions needed to ensure the successful identification and implementation of appropriate environmental policies at the national level, in particular institutional capacity and policy implementation concerns. Chapter 22, on global environmental co-operation, highlights the issues for which OECD countries will need to work together in partnership with other countries in order to reduce overall costs of policy implementation and maximise benefits. It also assesses the costs of inaction.

Traffic lights in the OECD Environmental Outlook

As with the 2001 *Outlook*, this report uses traffic light symbols to indicate the magnitude and direction of pressures on the environment and environmental conditions. Traffic lights are used to highlight the key trends and projections in the summary table in the Executive Summary, in the Key Messages boxes at the start of each chapter and throughout the chapters. The traffic lights were determined by the experts drafting the chapters, and then refined or confirmed by the expert groups reviewing the report. They represent the following ratings:

Red lights are used to indicate environmental issues or pressures on the environment that require urgent attention, either because recent trends have been negative and are expected to continue to be so in the future without new policies, or because the trends have been stable recently but are expected to worsen.

Yellow lights are given to those pressures or environmental conditions whose impact is uncertain, changing (*e.g.* from a positive or stable trend toward a potentially negative projection), or for which there is a particular opportunity for a more positive outlook with the right policies.

Green lights signal pressures that are stable at an acceptable level or decreasing, or environmental conditions for which the outlook to 2030 is positive.

While the traffic light scheme is simple, thus supporting clear communication, it comes at the cost of sensitivity to the often complex pressures affecting the environmental issues examined in this *Outlook*.

While each of the individual chapters discusses the regional developments for the drivers or environmental impacts analysed, Annex A also provides an easily accessible "summary" of the economic, social and environmental developments in the Baseline for each region. Annex B provides a more detailed analysis of the modelling framework used in the development of the *OECD Environmental Outlook*. A number of background working papers, which provide further information on specific issues addressed in the *Outlook*, were developed to complement the report (see: *www.oecd.org/environment/outlookto2030*).

THE WORLD TO 2030 – THE CONSEQUENCES OF POLICY INACTION

I. DRIVERS OF ENVIRONMENTAL CHANGE

1. Consumption, Production and Technology
2. Population Dynamics and Demographics
3. Economic Development
4. Globalisation
5. Urbanisation
6. Key Variations to the Standard Expectation to 2030

ISBN 978-92-64-04048-9
OECD Environmental Outlook to 2030

Chapter 1

Consumption, Production and Technology

This chapter explores patterns in consumption and production to 2030, as well as developments in technological innovation which can either ameliorate or exacerbate some of the environmental impact of this growth. Environmental pressure from households is projected to significantly increase over the next few decades, in particular in the main emerging economies, as populations and incomes increase and consumption patterns change. Firms are increasingly factoring environmental concerns into their business strategies, but the scale of increasing production outweighs most efficiency gains. The chapter provides a series of policy responses that could help tackle the growing pressures of consumption and production on the environment, including setting clear environmental targets for firms, promoting environmental research and development, and using policy mixes (e.g. energy tax along with an energy-efficiency label).

KEY MESSAGES

- Environmental pressure from households is projected to significantly increase to 2030. Residential energy use in OECD countries is expected to increase on average by 1.4% per year to 2030, while passenger kilometres travelled will increase by about 1% per year.
- Household consumption levels are projected to grow even more rapidly in non-OECD countries, particularly for electricity, personal transport, residential water use and demand for waste management services.
- One of the key determinants of consumption and production patterns is economic growth, with the relative economic importance of countries such as China and India increasing. Population dynamics will also be an important driver of consumption and production to 2030 in non-OECD countries. The trend towards ageing of the population, urbanisation and changing lifestyles will also influence the structure of consumption.

Environmental implications

The bulk of the increase in energy use is expected to come from fossil fuels, which are the main contributors to air pollution and CO_2 emissions. Fossil fuels are expected to represent 90% of total energy supply in 2030.

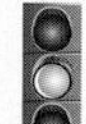

Some promising technologies are emerging which may help to reduce environmental pressure by reducing pollution or encouraging more efficient use of resources. These include hybrid vehicles and solar cells. But some technological developments can also increase pressures on the environment, and it is projected that improvements in energy efficiency of transport vehicles will be more than offset by increases in the number of vehicles owned and in average vehicle utilisation.

While public expenditures on environmental R&D are increasing, their share in total R&D remains small. Environment-related patent activity is also increasing, but no faster than the general rate of patenting.

While far from being universal, a large percentage of companies and firms are increasingly factoring environmental concerns into their business decisions, either in response to government policies, or to improve profits (through increased efficiency, reduced waste, or through a "green" image).

Policy implications

- Use policies that set clear environmental targets, without prescribing specific technologies, to provide the right framework conditions to encourage firms to move towards more efficient pollution abatement and resource use.
- Promote environment related innovations (*e.g.* research capacity, intellectual property rights) by providing the right incentives and using complementarities between instruments. Market-based instruments and well-designed performance standards broaden the potential space for innovation. Firms' investments in environmental R&D increase with the flexibility of the environmental policy instrument.
- Use policy packages, such as economic instruments (*e.g.* an energy tax) along with information-based instruments, to tackle households' growing environmental impact.
- Address equity concerns through general policy reforms, rather than through changes in the design of environmental policy.

Introduction

Production and consumption can have major environmental impacts, such as loss of natural resources, climate change, and other environmental damage caused by emissions and waste. This aspect of sustainable development has been addressed several times at the global level by the United Nations. The 2002 Johannesburg World Summit on Sustainable Development[1] called for the development of a 10-year framework of programmes to promote sustainable consumption and production patterns. This challenging task is co-ordinated under the UN-led Marrakech process. This chapter explores patterns in consumption and production to 2030, as well as patterns in technological innovation which may ameliorate some of the environmental impact of this growth. Each section concludes with some policy implications.

Key trends and projections: consumption and the environment

Environmental pressures from households are significant and their impacts are likely to intensify to 2030 in areas such as residential energy consumption, personal travel, food consumption, waste generation and water use[2] (see Figure 1.1).

Total residential energy use[3] in OECD countries is expected to increase by an average of 1.4% per year from 2003 to 2030 (IEA, 2006a; see also Chapter 17 on energy). This increase will be more rapid in non-OECD countries than in OECD countries. Forecasts indicate that

Figure 1.1. **Change in household expenditure, 2005-2030**

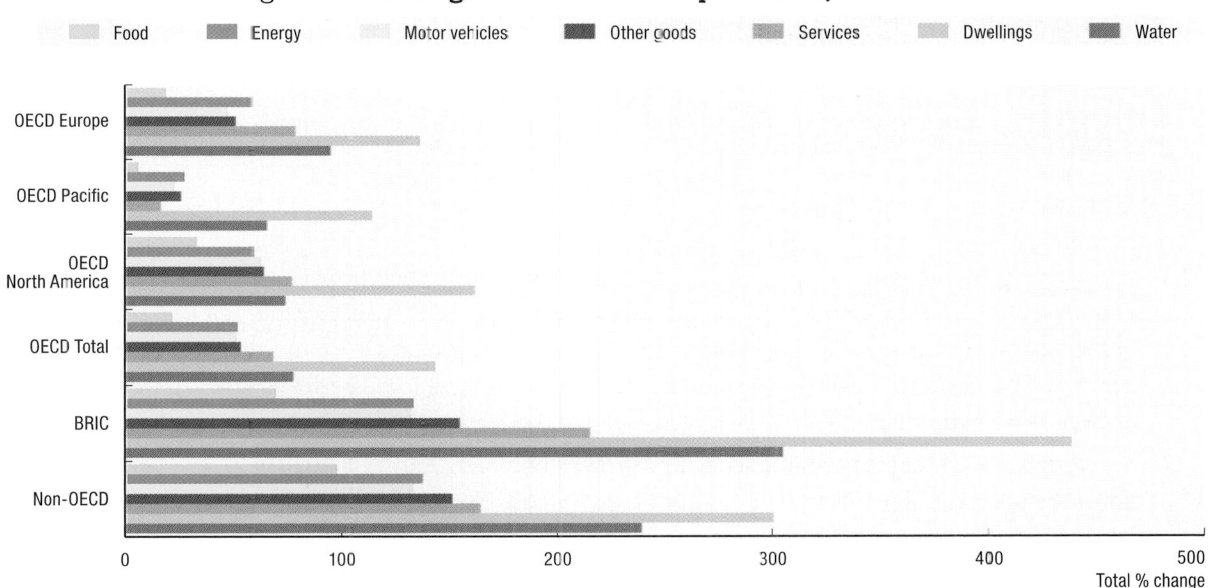

StatLink http://dx.doi.org/10.1787/258506536357

Source: OECD Environmental Outlook Baseline.

non-OECD residential energy use is projected to surpass the OECD total in 2010, and to be nearly 30% higher than the OECD total in 2030. China and India are projected to account for one-half of the total increase in residential energy use in non-OECD countries to 2030. The bulk of the increase in energy use is expected to come from fossil fuels, which are the main contributors to air pollution and CO_2 emissions. Fossil fuels are expected to represent 90% of total energy supply in 2030 (IEA, 2006a).

Passenger kilometres travelled (on rail, air, buses and light duty vehicles) are projected to grow 1.6% per year worldwide to 2030 (see Chapter 16 on transport). Growth rates in passenger transport differ widely by region (Figure 1.2) and are expected to average about 3% in China, 2% in India and about 1% in the three OECD regions (OECD Europe, North America and Pacific) (WBCSD, 2004). Transport-related greenhouse gas (GHG) emissions are also expected to grow significantly, especially in developing countries. Improvements in the energy efficiency of transport vehicles will be more than offset by increases in the number of vehicles owned and in average vehicle use. However, other transport-related air pollution emissions (*e.g.* nitrous oxides [NO_x], volatile organic compounds, carbon monoxide [CO] and particulates) are forecast to decline sharply in developed countries over the next two decades (WBCSD, 2004 and Chapter 8 on air pollution).

Figure 1.2. **Projected personal transport activity by region to 2050**

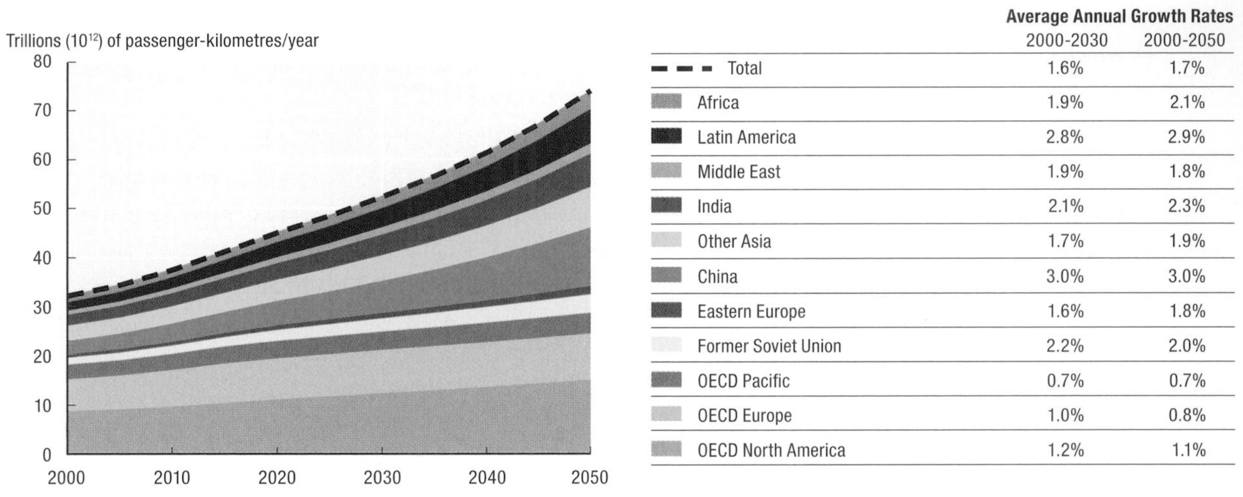

	Average Annual Growth Rates 2000-2030	Average Annual Growth Rates 2000-2050
Total	1.6%	1.7%
Africa	1.9%	2.1%
Latin America	2.8%	2.9%
Middle East	1.9%	1.8%
India	2.1%	2.3%
Other Asia	1.7%	1.9%
China	3.0%	3.0%
Eastern Europe	1.6%	1.8%
Former Soviet Union	2.2%	2.0%
OECD Pacific	0.7%	0.7%
OECD Europe	1.0%	0.8%
OECD North America	1.2%	1.1%

Source: WBCSD, 2004.

Beyond public policies, a number of other factors influence household consumption and its environmental impact. These include economic growth and income, relative prices of goods and services (*e.g.* energy and water pricing), demographics (*e.g.* population growth, ageing of population, household size), and lifestyle changes (*e.g.* the trend towards single occupant households).

Projected per capita annual economic growth between 2001 and 2030 is 2.37% globally, and above 4% for the BRIC group of countries (Brazil, Russia, India and China). A steady increase in per capita disposable income tends to be closely associated with an increase in the consumption of products and services, and consequently more energy consumption, water use and waste generation. Some of these effects may be partly counteracted by technological innovations which improve resource efficiency (see below). For instance, in some OECD countries, there has been a partial decoupling of environmental pressure from economic growth (*e.g.* water consumption).[4]

Consumption patterns are also affected by population dynamics and demographics. The *Outlook* Baseline projects a sharp increase in world population (from approximately 6 billion in 2000 to over 8.2 billion in 2030), which will have a direct impact on consumption levels. Other demographic changes, such as population ageing in OECD countries, will also influence consumption. This trend is driving the demand for tourism travel, as retirees generally have high levels of disposable income and large amounts of free time (see Chapter 2 on population dynamics and demographics). In addition, growing urbanisation in developing regions and changes in household size and composition will affect consumption. The trend in OECD countries towards smaller households and more people living alone intensifies environmental pressures as smaller households tend to use more water and energy per person than larger households.

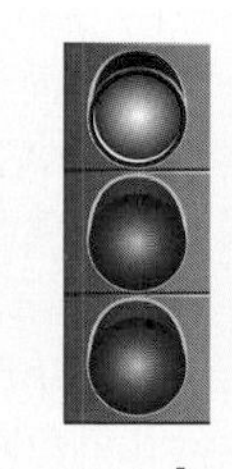

Environmental pressures in the food industry are increasing as a result of the globalisation of food supply chains.

The relative prices of environment-related goods and services are also an important driving force behind household consumption. The effects of price changes on demand will however vary according to the nature of the good. For example, "necessity" goods (such as energy) are less responsive to price changes than "luxury goods".

Some of the specific effects of income on consumption patterns include:

- *Food consumption:* Food consumption is driven by rising per capita incomes, prices and availability. The OECD Baseline projects that world economic development and population growth will cause agricultural production to grow by an average of 1.8% per year between 2001 and 2030 to meet increasing food demand. Global per capita food consumption (kcal/person/day) is projected to rise to 3 050 kcal in 2030, compared to 2 800 for 1997-99. It is likely to reach 3 500 kcal in industrial countries (FAO, 2003). The growth in caloric intake implies a larger share of animal products (meat and dairy products) in the diet, requiring more land per kilo of product. Productivity gains in agriculture and increasingly intensive production partly counteract this trend, but will also have environmental impacts. Demographic changes, such as higher levels of urbanisation, are also associated with higher caloric intakes of animal products, oils and fats, as well as greater demand for processed food (OECD-FAO, 2006; FAO, 2003) (See also Box 1.1).
- *Residential energy use:* As people's incomes grow, so does household energy demand as households increase their stock of electrical appliances. This results in a rise in energy consumption overall, despite energy efficiency gains. Strong economic growth in non-OECD countries is also increasing demand for household appliances, heating, cooling equipment and other energy-consuming devices. Together with household income and population growth, energy prices are considered to be the most important determinants of household energy consumption.
- *Waste generation:* Households with higher incomes tend to dispose of more waste, but do not necessarily invest more (or less) time in recycling activities than poor households.

- *Personal transport:* The number of cars owned by a household tends to increase with incomes, as do car use and total travel. Increased incomes are also linked to longer and more frequent trips. In addition, increased income results *in a higher value of* time, which encourages people to choose faster transport (WBCSD, 2004).

Box 1.1. **Sustainability in the food and beverage industry**

The food and beverage industry in OECD countries has significant environmental impacts. The industry is undergoing structural changes to meet consumer demands for year-round availability of fresh products, greater choice, convenient pre-packed food and as a response to health concerns. It is also increasing its use of energy and other inputs through intensification of the food production and manufacturing system. Food is also travelling further than ever before (both for processing and to find markets), and food-related waste is increasing.

To work towards environmental sustainability, we need to look at the entire life-cycle of the industry (from production to consumption) to identify the environmental impacts and responsibilities of each sector. The nature of distribution and retail systems can play an important role in determining environmental impacts. Agriculture and fisheries production are "upstream" in the food life-cycle. Food crop area is projected under the *Outlook* Baseline to increase by 25% by 2030 in OECD countries. Rapid growth of intensive farming and increasing use of greenhouses will lead to increases in energy, chemical and water use (see Chapter 14 on agriculture). The rapid expansion of aquaculture requires increased energy, feed, and chemical inputs, while by-catch from capture fisheries remains a problem (see Chapter 15 on fisheries and aquaculture). Downstream in the food life-cycle, the final disposal of food wastes requires energy for incineration and causes methane emissions from landfill (see Chapter 11 on waste and material flows). In between, food processing, packaging, transport and consumption each generate other environmental impacts. The globalisation of the food supply system has significantly increased the distance that food travels, known as "food miles". For example, some cod caught in the North Sea are shipped to China, processed there, and shipped back to Europe for consumption, travelling a total of 44 000 km (WWF, 2006). However, the environmental impacts of products must be assessed from a life-cycle perspective,* and on a case-by-case basis.

The food and tobacco industry accounted for more than 8% of the total final energy consumption of the industry sector in 2005 in OECD countries (International Energy Agency, Energy Balances of OECD Countries). Energy and chemical inputs in food manufacturing are increasing as consumers demand more prepared and packaged food. For example, pre-packed salad requires chemical inputs such as chlorine. Other major challenges are the treatment of biodegradable wastes and the generation of associated by-products. A Swiss study estimated that at least one-quarter of total municipal wastes was caused by food consumption (OECD, 2002). The food and beverage industry has a distinct role in influencing household behaviour through product and packaging design, pricing and waste recycling collection.

A number of countries have been looking into the sustainability of the food and beverage industry. For example, Japan has implemented a food recycling law that aims at 20% reduction of food waste. The UK has launched a multi-sectoral strategy aiming to reduce by 10-20% the energy and water use in the food and beverage industry, as well as waste emissions, and environmental costs of food transport (DEFRA, 2006).

* See, for instance, the Environmental Impacts of Products (EIPRO) project to support the development of an EU Integrated Product Policy (*http://susproc.jrc.es/pages/r4.htm*).

Policy implications

Environmental pressures from households are increasing, so better understanding of households' environment-related behaviour (including future trends in household consumption) is essential for successful environmental policies (see OECD, 2008 forthcoming).

Regulatory approaches are predominant in OECD member countries. In the area of personal transport, measures include emission standards (fuel, vehicle) and parking restrictions (*e.g.* access restrictions). To reduce residential energy use, performance and technical standards for appliances are applied, as well as standards associated with the thermal quality of new or existing dwellings.

However, there is ample evidence that households respond to the use of economic instruments, which are increasingly being used by member countries in environmental policy. In the energy sector, residential electricity taxes have been introduced (*e.g.* Germany). Households facing higher energy prices can respond in a number of ways, such as adjusting indoor temperatures, changing heating/cooling systems, making energy conservation investments (*e.g.* insulation), or moving to more energy-efficient housing.

Household personal transport choices are strongly influenced by prices, which are significantly affected by tax policy (*e.g.* petrol taxes, differentiated vehicle taxes). Demand for transport is generally found to be fairly price inelastic in the short-run, but more elastic in the long-run. Indeed, households generally have a much wider range of options available for responding to price increases in the long-run. For example they can buy smaller or more efficient cars, change their place of residence or work, etc. However, the choices made are likely to differ according to household characteristics (*e.g.* income, age). Congestion charges are increasingly being used to influence personal transport choice and manage traffic in urban areas (*e.g.* London, Seoul, Stockholm) (see also Chapter 5 on urbanisation).

User fees that vary according to the amount of waste generated are better at reducing waste and/or at increasing recycling than flat fees, and these are being implemented more widely for household waste (*e.g.* Korea) and for household water use.

Environmental policy, like all public policies, is likely to affect some members of society more than others. Low-income households may therefore bear a disproportionate share of the cost of some environmental policies, whether economic instruments or regulatory approaches. In general, however, these concerns are better addressed in overall economic policy – for example through adjustments to tax and social policies – than in environmental policy measures themselves (Serret and Johnstone, 2006; OECD, 2006).

Important complementarities exist between the different types of policy instruments that can be applied (*e.g.* economic instruments, information-based instruments, direct regulation, integrated product policy and extended producer responsibility). Some mixes are likely to be more effective than others in addressing household consumption patterns. To influence residential energy use, for example, it may be preferable to use an economic instrument such as an energy tax along with an information-based instrument (such as an energy-efficiency label) or more general environmental information awareness programmes, rather than applying either instrument on its own. Similarly, unit-based pricing may be more effective in reducing waste if it is combined with a recycling programme and/or a deposit-refund system (see OECD, 2007c).

Key trends and projections: production and the environment

In OECD countries industrial emissions of air pollutants have been greatly reduced over recent decades. Relevant measures have included the use of low sulphur fuels (*e.g.*, switching from coal to oil to gas) and end-of-pipe measures like flue gas desulphurisation techniques, the use of low nitrogen combustors and particulate capture devices. Energy efficiency improvements have also lowered air pollutant emissions. Less polluting renewable energy sources with low air pollution effects have shown high growth rates, particularly solar and wind, but their share in global electricity production is still only around 2% (IEA, 2007).[5]

Figures 1.3 and 1.4 show energy-related nitrogen and sulphur emissions from industrial sectors in four OECD regions. While the trend in recent decades has generally been downward, the projections to 2030 from the *Outlook* Baseline, which reflects no new policies, indicate that they will not continue at the same rate.

Figure 1.3. **Baseline forecasts of energy-related industrial nitrogen emissions, 1970-2030 (Mt)**

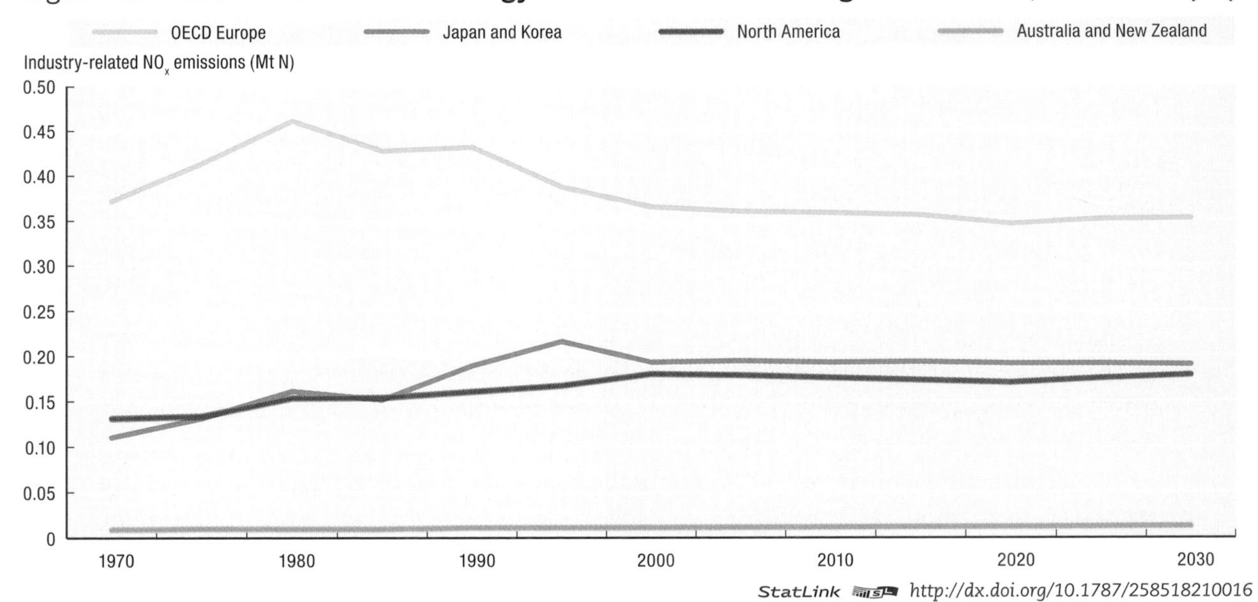

StatLink http://dx.doi.org/10.1787/258518210016

Source: OECD Environmental Outlook Baseline.

Abatement policies are also starting to emerge in several non-OECD countries, particularly for flue-gas desulphurisation, but thus far have been insufficient to decouple emissions from economic growth. Although several industries in China are equipped with flue-gas desulphurisation techniques, until recently they have not always been used (OECD, 2007e). This situation has now improved, as fines have been increased to a level that it is no longer profitable to disable their use. However, effective use will require a system to adequately monitor emissions, and to apply sufficiently high penalties for non-compliance.

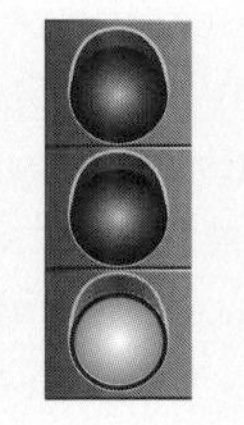

Firms are increasingly factoring environmental concerns into their business strategies.

The extent to which firms address environmental concerns and how they do so differs markedly. A good understanding of

Figure 1.4. **Baseline forecasts of energy-related industrial sulphur emissions, 1970-2030 (Mt)**

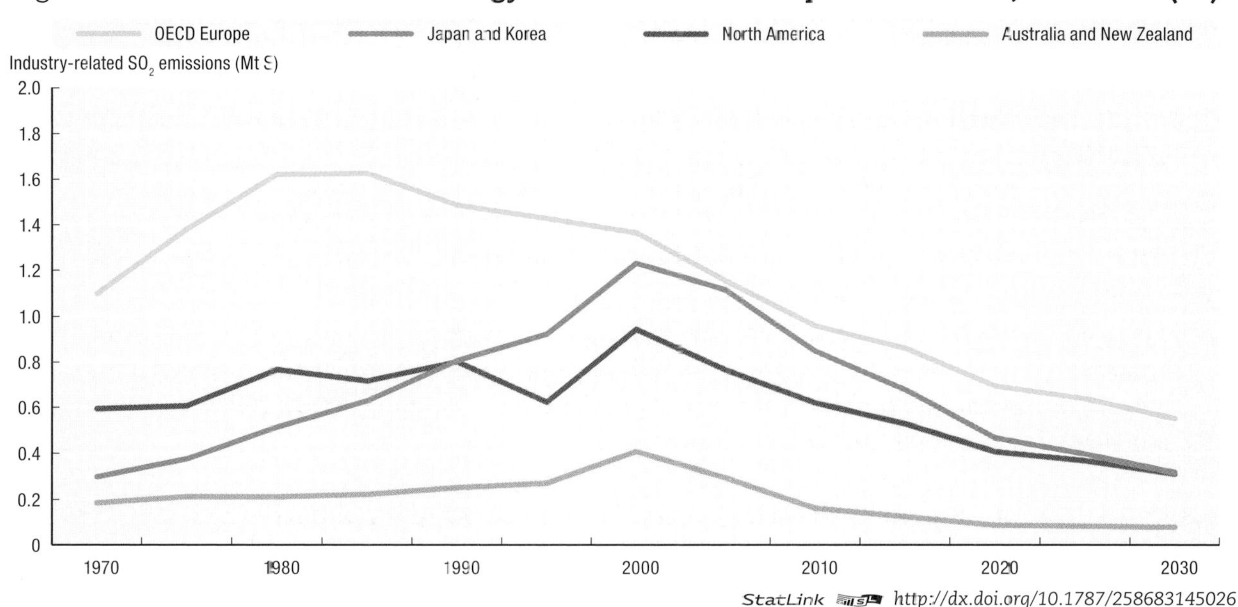

StatLink http://dx.doi.org/10.1787/258683145026

Source: *OECD Environmental Outlook* Baseline.

manufacturing firms' commercial motivations, decision-making procedures and organisational structure is essential for improving the design and implementation of public environmental policies. An OECD survey of 4 000 facilities in seven countries offers lessons about firms' environmental management practices.[6] The percentage of firms reporting that they have introduced an environmental management system (EMS) ranged from 30% in Germany and Hungary to almost 57% in the United States. The total number of ISO 14001 certifications has increased dramatically in recent years – from 14 106 applications at the end of 1999, to 90 569 at the end of 2004, with registered certifications in 127 different countries.[7]

Firms also vary in their institutional set-up for environmental responsibility, depending on their size. Table 1.1 shows the percentages of facilities with a designated employee responsible for environmental matters. It is clear that big firms are much more likely to have designated somebody for this purpose than small firms. However, this person is most frequently located in an environmental health and safety department (and less frequently in senior management, finance or production and operations).

Table 1.1. **Designation of responsibility for environmental matters in manufacturing facilities**

	Number of employees				
	< 100	100-249	250-499	> 500	Total
% of firms with a staff member responsible for env. matters	54.6%	68.0%	87.1%	93.4%	70.3%

StatLink http://dx.doi.org/10.1787/256624034320

Source: Johnstone, 2007.

While interesting, data on environmental management practices reflect intentions, rather than concrete actions, to improve environmental performance. According to the OECD (2007d), most countries have a similar share of their GDP allocated to private sector pollution abatement (approximately 0.5% on average). These percentages have remained relatively constant over time. Figure 1.5 shows the evolution in the share in GDP of total private sector pollution abatement control expenditures for a selection of OECD countries for which some time series data are available.[8] The spikes in the Polish and Czech data are particularly striking.

Figure 1.5. **Estimated private sector pollution abatement and control expenditures (% of GDP)**

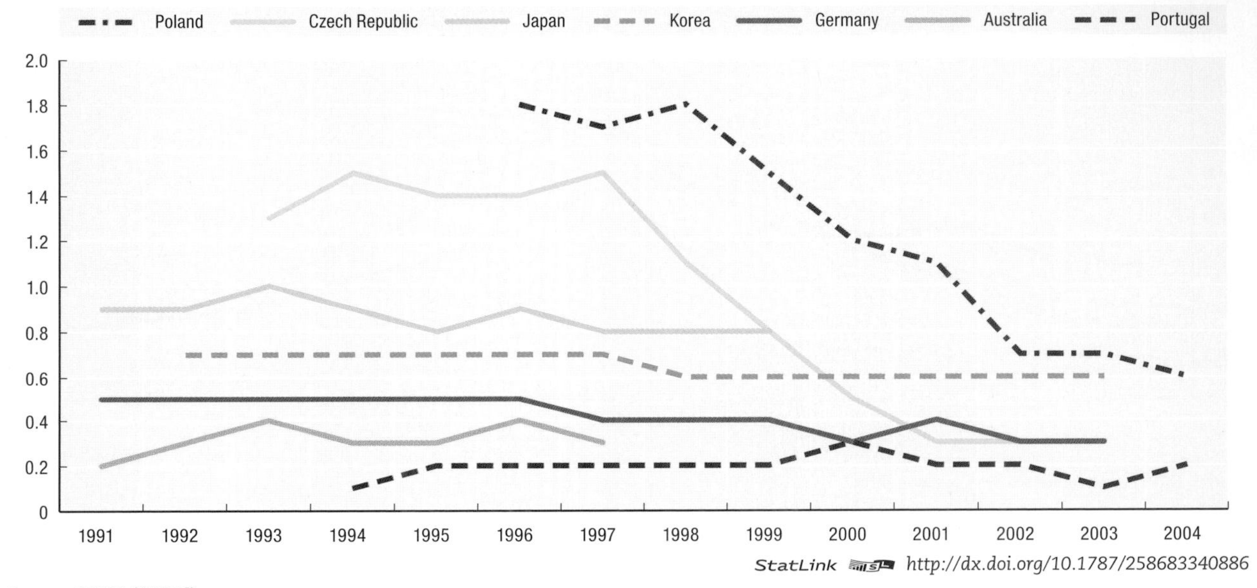

StatLink http://dx.doi.org/10.1787/258683340886

Source: OECD (2007d).

To improve its environmental performance, a facility can decide either to change its production process (CPP) or to treat pollution with an end-of-pipe (EOP) technology. In the early years following the widespread introduction of environmental regulations, firms tended to invest in end-of-pipe technologies, such as flue-gas desulphurisation or membrane technologies, which reduce air pollution emissions or wastewater effluent following production. However, it is often more cost-effective and environmentally effective to change the production process in order to generate fewer unwanted by-products, allowing facilities to adopt such strategies through the use of less prescriptive policies.

The reported share of CPP investment to total investment in the manufacturing sector is relatively high in the UK and Finland: 52% and 49% respectively in 1999 (OECD, 2003). On the other hand, in 1999 Polish and Spanish firms reported that 77% and 73% of their environmental investment expenditures were for EOP investments. There is, however, some evidence that the share of CPP investment has increased over the last two decades. Indeed, the vast majority of firms responding to the OECD survey reported that their main approach for addressing environmental concerns was best described as CPP, rather than EOP. The highest percentages were for the machinery, instruments, motor vehicles and transport equipment sectors, with over 80% reporting that integrated changes in production processes were their primary means for addressing production-related environmental concerns.

Policy implications

A variety of environmental policy measures can be used to reduce the environmental impacts associated with production processes. The two most common types of policies are direct forms of regulation (*i.e.* technology-based standards and performance-based standards) and, increasingly, economic instruments such as environmentally related taxes and tradable permits.

It has long been argued that economic instruments such as tradable permits or environmentally related taxes are more economically efficient than more direct forms of regulation. While concerns about their environmental effectiveness have often been raised in the past, if well-designed and properly enforced they can be more environmentally effective, particularly tradable permits.[9] Whilst evidence is still being gathered, there is increasing empirical support for this assertion. The US Acid Rain Policy which involves an SO_2 emissions trading programme is, by all measures, a success (Ellerman, 2004). However, design is key. For instance, the widespread use of exemptions from environmentally-related taxes for many low-cost abaters is neither efficient nor effective (OECD, 2006).

Some emerging technologies may reduce environmental pressure by preventing pollution or encouraging more efficient use of resources. However, some technological developments can increase pressures on the environment.

The wider use of more "flexible" policy instruments, such as market-based instruments and performance standards which are not excessively prescriptive, can have far-reaching consequences for how firms address environmental concerns, perhaps resulting in secondary benefits in areas which are not directly targeted by the policy itself. For instance, economic instruments can encourage facilities to adopt an EMS and other environmental management tools. Similarly, more flexible policy measures encourage oversight of environmental matters by senior management and finance/accounting positions (Johnstone, 2007). The "mainstreaming" of environmental concerns in the facility can lead to more pro-active environmental strategies.

Designing environmental policies for small and medium-sized enterprises (SMEs) is an increasing focus of OECD governments. Their relatively small size makes environmental policy a particular burden for SMEs. Sweden and Australia attach particular importance to this issue, and their special Regulatory Impact Analyses focus on the effects of environmental regulations on SMEs. In addition, they are inducing measures to reduce the administrative costs associated with environmental permits.

And finally, the public policy framework is only one influence on firms' environmental practices, with many other stakeholders providing incentives for "corporate environmental responsibility". Recent work in the OECD and elsewhere has shown that other stakeholders, including financial markets, local communities, consumers and employees can also have an important influence (see Johnstone, 2007 for a review of recent work in this area).

Key trends and projections: technology and the environment

Technological change can take on different forms, such as innovations in production processes or the invention of new products, with different potential impacts on the environment. While some innovations help reduce environmental pressures – for example, through reductions in pollution emissions or more efficient use of resources – other innovations can increase environmental pressures. In many cases, the overall effects are ambiguous or uncertain. For example, biofuels and nanotechnologies may have positive effects in one area, but negative effects in another such as increased pressure on land resources. There is a need to better understand the environmental impacts of technologies to make informed policy decisions. Decision-making can be supported by early quantitative environmental assessments of emerging technologies and comparison of the effects of competing technologies. More generally, the development of indicators for environment-related innovation would support efficient policy design.

Several recent key innovations have significantly contributed to environmental protection, and may continue to do so in the future. For instance, carbon capture and storage methods can reduce CO_2 emissions by absorbing the CO_2 emitted from particular production processes. It is estimated that by 2030, the cost of carbon capture and storage technologies could fall below USD 25 per tonne of CO_2 (IEA, 2006b). In wastewater and solid waste treatment, micro-organisms are now being used to transform hazardous material into less dangerous compounds, and to decrease odour and dust generation. Innovation in the fabrication of solar cells (for instance, through the use of nanotechnology) has increased their efficiency significantly. Multi-junction solar cells now provide a 35% increase in available power per solar panel area, compared to existing technologies. In Japan, over 20% of biotechnology applications sold are for industrial-environmental applications. In China, the figure is over 10%; much higher than many OECD countries (Beuzekom and Arundel, 2006). Hybrid vehicle innovations have already resulted in the production of cars using a combined gas-electricity engine, while further research is underway to facilitate the use of hydrogen fuel cells. However, this latter example illustrates the complexity of the environmental assessment of different innovations – while emissions of some local air pollutants may decrease, there may be concerns about end-of-life disposal.

One area which has seen significant innovation in recent years is renewable energy. Following the development of "first-generation" (*e.g.* hydropower, biomass combustion) and "second-generation" (solar heating, wind power, etc.) technologies, some "third-generation" technologies presently being commercially exploited include concentrating solar power, ocean energy, enhanced geothermal systems and integrated bio-energy systems (IEA, 2006c). Partly as a result of such innovations, costs are coming down and the use of renewables is increasing (Figure 1.6). The role of the public sector in providing

Figure 1.6. **Annual average % change in renewable energy production, 1990-2004**

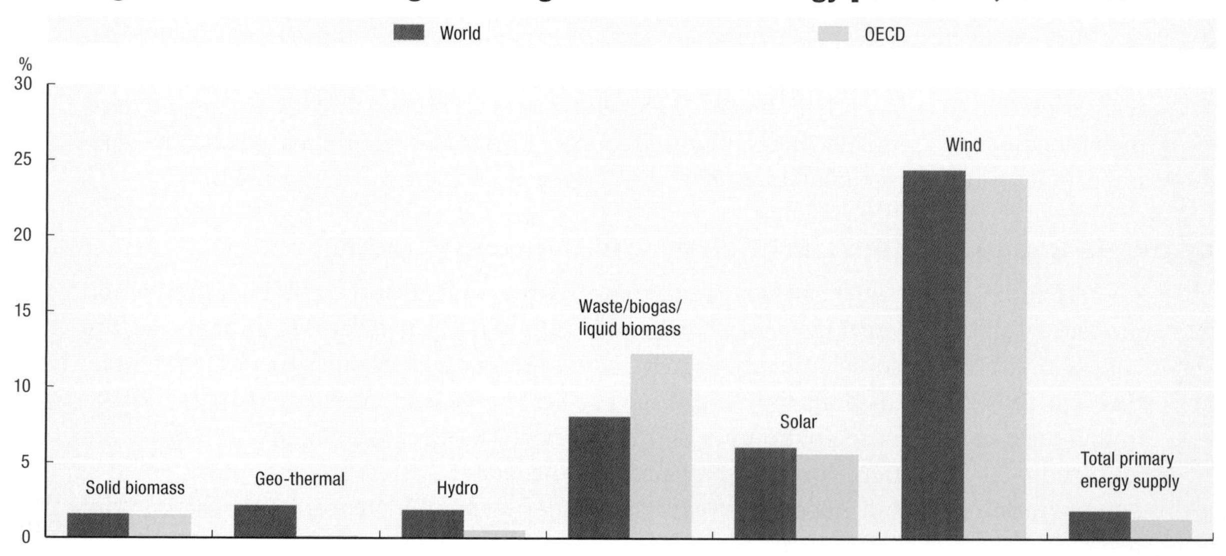

StatLink http://dx.doi.org/10.1787/258704524353

Source: IEA, 2006c.

incentives for such technologies has been significant. Many governments have dedicated research and development (R&D) programmes for this field. Moreover, diffusion of such technologies to developing countries is an important element in a number of multilateral environmental agreements (*e.g.* the Clean Development Mechanism, see Chapter 7 on climate change).

Innovations in information and communications technology (ICT) can affect the cost and quality of monitoring environmental policy in several ways. Innovations in product-tracking technologies are improving tracking of potentially hazardous or recyclable products. Satellite-based mapping technologies reduce the cost of monitoring resource exploitation. Monitoring costs of emissions from large stationary and smaller non-point and mobile sources are falling with innovations in sensors. According to the OECD's Triadic Patent Family (TPF) Database[10] (see Figure 1.8), patents granted for technologies to monitor environmental impacts increased seven-fold in the last two decades. This is significant, since improved monitoring through ICT can increase the environmental effectiveness of policy measures.

While environmental R&D is increasing, its share in total R&D remains small. Environment-related patent activity is also increasing, but no faster than the general rate of patenting.

Many OECD countries have been increasing their investment in environmental R&D to boost technological developments that improve environmental quality. Figure 1.7 shows the evolution over time of the share of environmental R&D in total R&D for several OECD countries.[11] In the OECD's *Science, Technology and Industry Outlook* (2004b), a majority of countries cite environment-related concerns in their science and technology priorities, including: Australia (environmentally sustainable Australia); Austria (environment, energy and sustainability); France (development of renewable energy); Germany (clean processes

Figure 1.7. **Share of environmental R&D in total government R&D, 1981-2005**

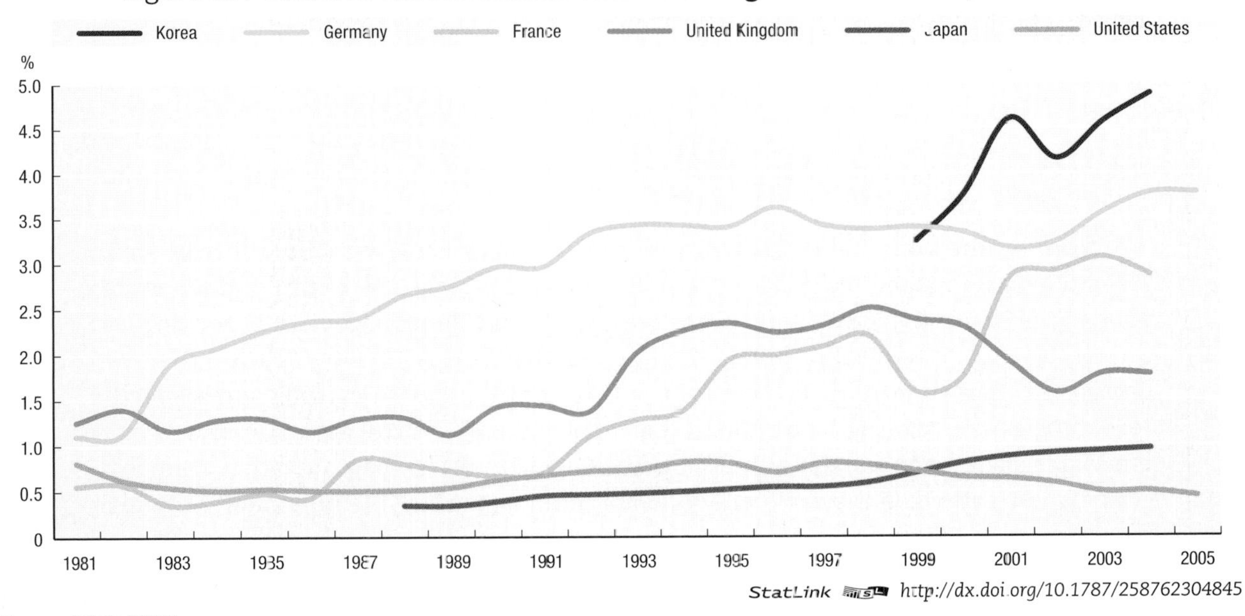

StatLink http://dx.doi.org/10.1787/258762304845

Source: OECD (2005).

and production technologies); Hungary (environmental protection); Norway (energy and environment); United Kingdom (sustainable energy) and the United States (climate, water and hydrogen).

Japan's increase in the share of environmental R&D in total R&D since the 1990s has been significant, but the share is still low compared to some European countries and recent Korean figures. Since the late 1990s, the share has been lowest in the US (although this may be due to discrepancies in data collection). Collecting harmonised data on environmental R&D expenditures would allow a better comparison of the innovation priorities of different countries.

To explore environment-related innovation further, patent data from the OECD's Triadic Patent Family (TPF) Database were extracted.[12] As Figure 1.8 illustrates, there has been continuous growth over recent years (particularly in air and water pollution innovations) except for solid waste and recycling, which peaked in the early 1990s. However, the rate of growth is generally lower than for overall TPF patent activity.

Figure 1.8. **Number of TPF patents in the environmental area, 1978-2002**

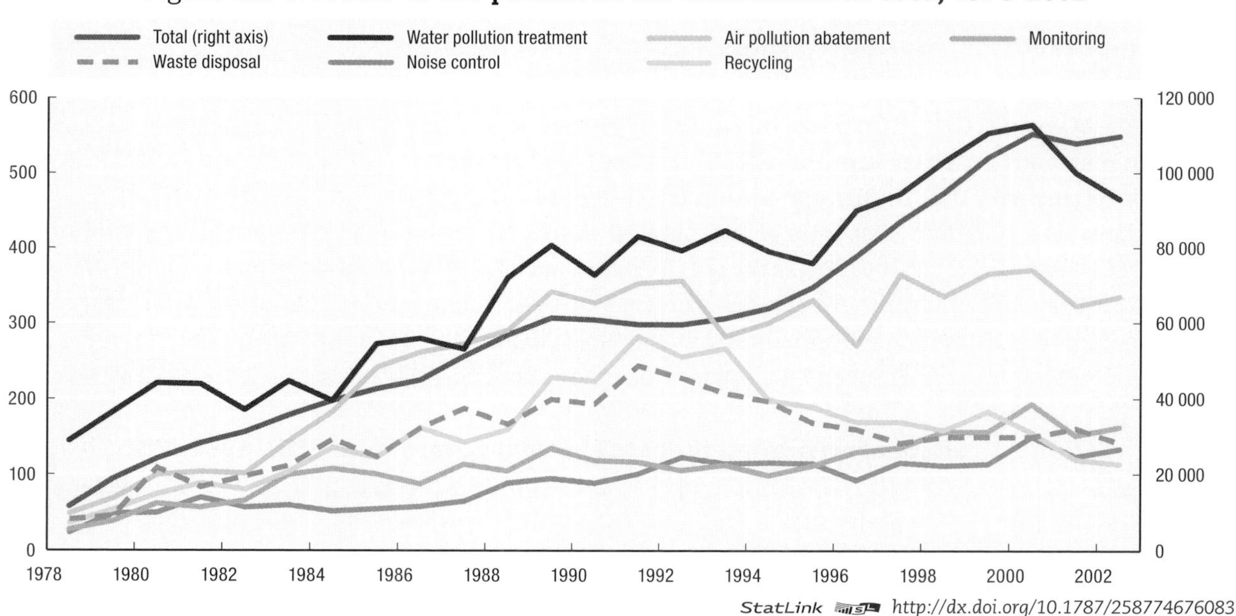

StatLink http://dx.doi.org/10.1787/258774676083

Source: Data drawn from the OECD Project on Environmental Policy and Technological Innovation *www.oecd.org/env/cpe/firms/*.

More recent work has looked more closely at patent activity in a number of areas, including waste-related technologies, motor vehicle emissions abatement and renewable energy (see Figure 1.9).[13] While patent activity in solid waste management is growing less quickly than TPF patenting in general, the growth rate is higher for the other two categories. It is important to note that within these broad categories some specific technologies are growing faster than others. For instance, in the area of motor vehicle emissions abatement, there is a trend towards engine re-design patents rather than post-combustion patents. Renewable energy, solar and (particularly) wind exhibit very high growth rates.

Figure 1.9. **Growth rates in patents in selected environmental areas, 1995-2004**

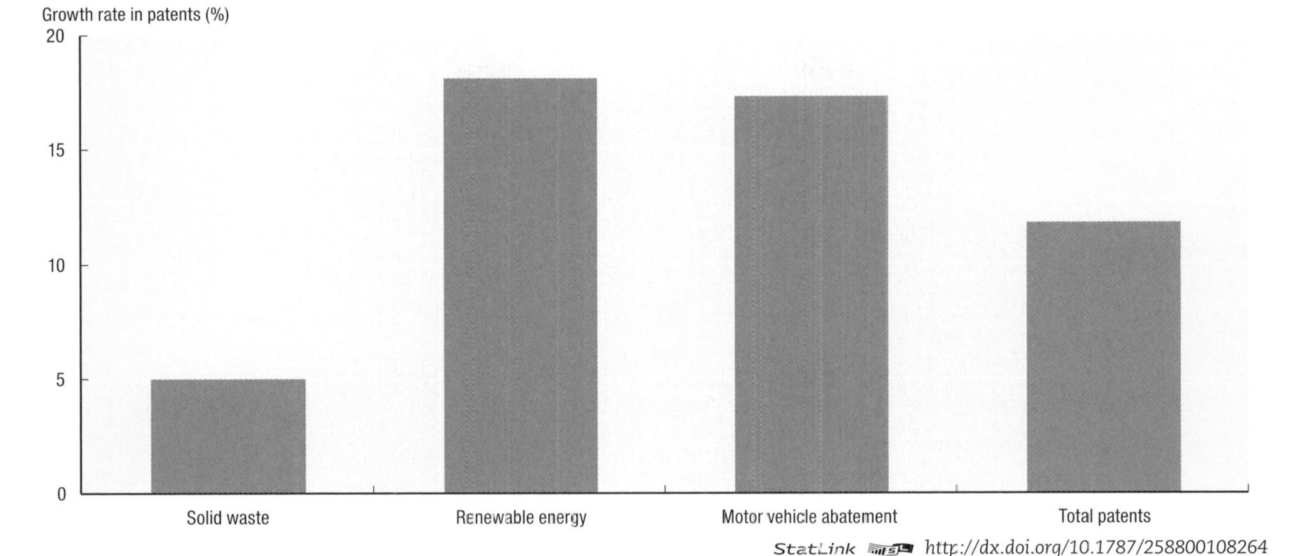

StatLink http://dx.doi.org/10.1787/258800108264

Source: OECD (2007b).

Policy implications

There are two types of externality which public policies can help to internalise in order to increase private returns on innovation and provide for the socially optimal reduction in environmental pollution.[14] The first is the fact that it is often difficult for innovators to capture the benefits of their inventions due to information spillovers. OECD governments introduce a wide variety of policies to internalise such externalities (Jaumotte and Pain, 2005b). The second type of externality involves market failures which mean that users of environmental resources like clean air or fresh water treat them as costless factors of production. Policies are therefore needed to tackle both sources of externality.

However, using a single instrument to internalise both externalities – *e.g.* through subsidies for environment-related R&D or for investments in specific environmental technologies – is unlikely to efficiently achieve both objectives. In general, separate policy instruments will be needed to address each externality. The use of only one instrument requires the regulator to have a very detailed knowledge of the market (development path, technological opportunities, etc.) that is to be regulated. As such, co-ordination between policy-makers in the innovation and environmental areas is key.

While the evidence is still being gathered, there is some support for the hypothesis that market-based instruments can be particularly effective in inducing innovation (Vollebergh, 2007). On the one hand, a market-based instrument gives firms continuous incentives to innovate. In the case of more direct forms of regulation, when the required standard is met the firm has no further incentive to innovate, unless the regulation is made more stringent.[15] This will, however, depend in part upon the nature of the regulation. In practice, it is likely to be administratively easier for the regulatory authority to adjust performance-based standards than technology-based standards through time. Moreover, under technology-based regulations, potential innovation is constrained by the nature of the standard itself. It is likely that not all emission-reducing innovations will be permitted by regulatory authorities. Market-based instruments and well-designed performance standards broaden this potential space for innovation, since any emission-reducing innovation will meet regulatory requirements. Thus, Johnstone and Labonne (2006) find that facilities' investments in environmental R&D increase with the flexibility of the environmental policy instrument. A greater understanding of the drivers of innovation in the environmental sphere is needed.

Notes

1. See *www.un.org/jsummit/html/basic_info/basicinfo.html.*
2. See also Chapters 17 (Energy), 11 (Waste and material flows), and 10 (Freshwater).
3. Residential sector energy use is defined as the energy consumed by households, excluding transport-related energy use.
4. "Decoupling" occurs when the growth rate of an environmental pressure is less than that of its economic driving force (*e.g.* GDP) over a given period.
5. Excluding hydropower.
6. For further information, see *www.oecd.org/env/cpe/firms.* A collection of papers reviewing some of the main findings arising out of the project can be found in Johnstone, N. *Environmental Policy and Corporate Behaviour* (Edward Elgar/OECD, 2006).
7. See: *www.iso.org/iso/iso_catalogue/management_standards/iso_9000_iso_14000.htm.*
8. A sub-set of countries provide data on such expenditures by business.
9. The effectiveness of tradable permits is attributable to the fact that they are unique among policy instruments in setting a cap on total emissions, thus obviating the need for policy adjustments in the face of economic growth or arrival of new firms.
10. This database only includes patents granted by the Japanese Patent Office, the European Patent Office, and the US Patent and Trademark Office (Dernis and Kahn, 2004).
11. While government R&D is only a small proportion of total R&D (and may even crowd out some private R&D), private sector R&D data disaggregated by socio-economic objective are not available. For R&D data, government budget appropriations or outlays for R&D provided by the Government Budget Outlays or Appropriations of R&D (GBOARD) database were used (OECD, 2005).
12. In order to extract the relevant environmental patents from the database, a search filter (Schmoch, 2003) is applied, consisting of a combination of International Patent Classifications (IPC) that are closely related to the environmental sector, as well as keywords to exclude or to include in order to take into account all patents fitting the description. This provides a measure of the number of environmental patents deposited in all three offices in six different environmental areas. Work is on-going at the OECD Environment Directorate to refine the search algorithms.
13. Data drawn from the OECD Project on Environmental Policy and Technological Innovation. A publication summarising initial outputs from the project is forthcoming.
14. See Johnstone and Labonne (2006).
15. To foster innovation, standards can be gradually updated with stricter requirements. The legislation can also specify short-term and long-term limit values.

References

Beuzekom, B. van and A. Arundel (2006), *OECD Biotechnology Statistics,* OECD, Paris.

DEFRA (UK Department of Environment, Food and Rural Affairs) (2006), *The Food Industry Sustainability Strategy,* DEFRA, London, available at: *www.defra.gov.uk/farm/policy/sustain/fiss/pdf/fiss2006.pdf* (retrieved on 10 March 2007).

Department of Energy/Energy Information Agency (2006), *International Energy Outlook*, Department of Energy/Energy Information Agency, Washington DC.

Dernis, H. and M. Kahn (2004), "Triadic Patent Families Methodology", *STI Working Paper 2004/2*, OECD, Paris.

Ellerman, D.A. (2004), "The US SO_2 Cap-And-Trade Programme", in OECD (2004), OECD *Tradable Permits: Policy Evaluation, Design and Reform*, OECD, Paris.

FAO (Food and Agriculture Organisation of the United Nations) (2003), *World Agriculture: Towards 2015/2030 – An FAO Perspective*, Earthscan Publications Ltd , London.

Henriques, I. and P. Sadorsky (2006), "Environmental Management Systems and Practices", in Johnstone, N. (ed.) *Environmental Policy and Corporate Behaviour,* Edward Elgar, Cheltenham and OECD, Paris.

IEA (International Energy Agency) (2006a), *World Energy Outlook*: 2006 Edition, IEA/OECD, Paris.

IEA (2006b), *Energy Technology Perspectives*, IEA/OECD, Paris.

IEA (2006c), *Renewables Information*, IEA/OECD, Paris.

IEA (2007), *Key World Energy Statistics*, IEA/OECD, Paris.

Jaumotte, F. and N. Pain (2005a), "From Ideas to Development: The Determinants of R&D and Patenting", *OECD Economics Department Working Paper* No. 457 [ECO/WKP(2005)44], OECD, Paris.

Jaumotte, F. and N. Pain (2005b) "An Overview of Public Policies to Support Innovation", *OECD Economics Department Working Paper* No. 456 [ECO/WKP(2005)43] OECD, Paris.

Johnstone, N. (2007), *Environmental Policy and Corporate Behaviour*, Edward Elgar, Cheltenham and OECD, Paris.

Johnstone, N. and J. Labonne (2006), "Environmental Policy, Management and R&D", in *OECD Economic Studies*, No. 42/1.

OECD (2002), *Towards Sustainable Household Consumption: Trends and Policies in OECD Countries*, OECD, Paris.

OECD (2003), *Pollution Abatement and Control Expenditures in OECD Countries* [ENV/EPOC/SE(2003)1], OECD, Paris.

OECD (2004a), *OECD Environmental Data Compendium*, OECD, Paris.

OECD (2004b), *Science, Technology and Industry Outlook 2004*, OECD, Paris.

OECD (2005), *Research and Development Statistics*, OECD, Paris.

OECD (2006), *The Political Economy of Environmentally Related Taxes*, OECD, Paris.

OECD (2007a), *OECD Environmental Data: Compendium 2006/2007*, OECD, Paris.

OECD (2007b), *Science, Technology and Industry Scoreboard*, OECD, Paris.

OECD (2007c), *Instrument Mixes for Environmental Policy*, OECD, Paris.

OECD (2007d), *Pollution Abatement and Control Expenditure in OECD Countries* [ENV/EPOC/SE(2007)1], OECD, Paris.

OECD (2007e), *OECD Environmental Performance Review China*, OECD, Paris.

OECD (2008), *Environmental Policy and Household Behaviour: Evidence in the Areas of Energy, Food, Transport, Water and Waste*, OECD, Paris, (forthcoming).

OECD-FAO (2006), *Agricultural Outlook 2006-2015*, OECD, Paris.

Schmoch, U. (2003), *Definition of Patent Search Strategies for Selected Technological Areas: Report to the OECD*, OECD, Paris.

Serret, Y. and N. Johnstone (eds.) (2006), *The Distributional Effects of Environmental Policy,* Edward Elgar, Cheltenham and OECD, Paris.

Vries, De F. (2007), *Environmental Regulation and International Innovation in Automotive Emissions Control Technologies* (ENV/EPOC/WPNEP(2007)2, *www.oecd.org/env*, OECD, Paris.

Vollebergh, H. (2007), *Impacts of Environmental Policy Instruments on Technological Change* [COM/ENV/EPOC/CTPA/CFA(2006)36/FINAL], OECD, Paris.

WBCSD (World Business Council for Sustainable Development) (2004), *Mobility 2030: Meeting the Challenges to Sustainability*, Sustainable Mobility Project Calculations World Business Council for Sustainable Development July 2004. Report available at: *www.wbcsd.org/web/publications/mobility/mobility-full.pdf* .

WWF (World Wildlife Fund) (2006), *Fish Dish – Exposing the Unacceptable Face of Seafood*, WWF, Gland, Switzerland, available at: *www.panda.org/about_wwf/what_we_do/marine/help/seafood_lovers/fish_dishes/fish_chips/cod_issue/index.cfm* (retrieved on 11 April 2007).

ISBN 978-92-64-04048-9
OECD Environmental Outlook to 2030

Chapter 2

Population Dynamics and Demographics

This chapter examines the close relationship between population growth and demographics and the environment. Between 2005 and 2030, world population is expected to grow from 6.5 to 8.2 billion people. The enlarging population, mostly in developing countries, will put more pressure on the environment through increased production and consumption. The demographic features of ageing and migration are particularly relevant from an environmental perspective. Ageing populations have specific consumption patterns, some of which – such as expanded leisure time and income for travel – are associated with increasing environmental impacts. Migration can exacerbate pressures on local environments by increasing density in receiving regions. Environmental conditions will also influence population dynamics, such as through environmental refugees and environment-related disease outbreaks. The number of environmental refugees is expected to grow in the coming decades as a result of the impacts of climate change.

KEY MESSAGES

- Between 2005 and 2030, it is projected that world population will grow from 6.5 to 8.2 billion people. Almost all of the global increase in population will originate in the developing world; the OECD's share of world population will drop from 23% in 1980 to 15% in 2030.
- In addition to the general population growth, two demographic features are particularly relevant from an environmental perspective – ageing and migration:
  - The number of people aged over 60 will increase from 0.7 to 1.9 billion between 2005 and 2050; three out of four of these people will live in the developing world. In 2050, the grey dependency ratio – *i.e* the number of people over 65 years of age that are "dependent" economically on those of working age – will reach 46 to 100 in the USA, 60 in Europe, and 70 in Japan (compared to 20, 27, and 28 in 2005 respectively).
  - Over the same period, 98 million people (net number) will migrate, mainly within regions or from less developed to more developed countries.

World population (millions) 1970-2030

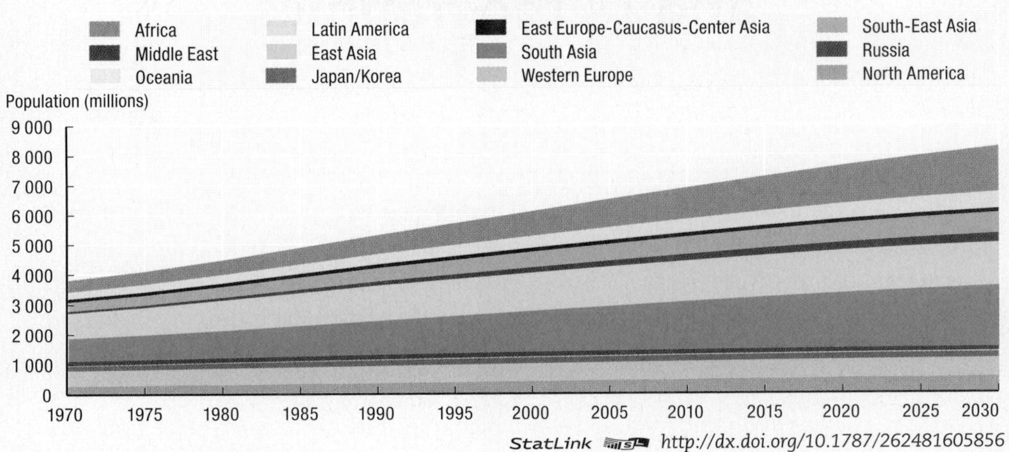

StatLink http://dx.doi.org/10.1787/262481605856

Environmental implications

The growing population will put increasing pressure on the environment, through increased production and consumption.

Ageing populations have specific consumption patterns, some of which – such as increased leisure time and income for travel – are associated with increasing environmental impacts.

Migration, which can also be driven by environmental degradation, can exacerbate pressures on local environments by increasing density in receiving regions and contributing to desertification in sending ones. It can also increase vulnerability to disasters.

Consequences of inaction

Environmental conditions will also influence population dynamics, as is apparent from environmental refugees and environment-related disease outbreaks. The number of environmental refugees* is expected to increase in the coming decades as a result of the impacts of climate change. This might exacerbate security issues.

* Note that the notion is not an official category, which explains why there is no systematic collection of data.

Introduction

Population dynamics are a key driver of environmental change for a number of reasons. People are a driver of economic growth, putting demands on services which have impacts on the environment, and putting direct pressures on the environment by consuming natural resources (including land for food cultivation, housing and infrastructure; energy and wood for fuel; and water) and causing pollution (to air, soil, water, etc.) Population dynamics also affect labour,[1] which is a major driver of growth (in numbers and via labour productivity) in this *OECD Environmental Outlook* Baseline.

Human impacts on the environment vary with changes in levels and modes of consumption and the technologies involved (Prugh and Ayres, 2004). The increasing consumption of the global consumer class, rising population and increasing incomes in developing countries will accelerate environmental pressures from energy, transport, water use and waste production. Chapter 1 on consumption, production and technology sheds some light on expected trends in consumption patterns for households in both OECD and developing countries. It analyses the relationships between consumption patterns and population dynamics, economic development, ageing and changing lifestyles. Income dynamics and income disparities will matter. So will sociological trends: the declining number of people living in each household generates additional per capita levels of consumption of land and energy. At the same time, each urban area has a specific ecological footprint, linked to the efficiency of its use of land, energy and other resources and its capacity to manage housing, develop collective transport systems, collect and treat waste, and secure urban safety.

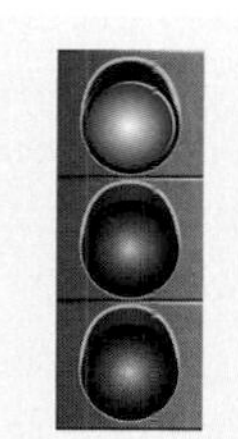

Continuing environmental degradation in some regions will generate additional migrations by the most vulnerable clusters of the population.

In turn, the environment is a driver of population dynamics. Environmentally stressed areas are subject to specific migrations, as testified by the number of environmental refugees (25 million people in 1994, according to UNEP, half of them in Africa). The situation is likely to deteriorate further, as the number of people living in medium to high water-stressed areas is expected to increase by 60% from 2005-2030 under the OECD Baseline (see also Chapter 10 on freshwater). The increasing frequency of extreme weather events, changes in regional food production patterns and, in the longer term, sea level rise, are likely to result in migrations. Environment-related disease outbreaks can also affect population dynamics (see also Chapter 12 on health and environment).

The combination of increasing population density and environmental degradation in many areas worldwide accelerates vulnerability to disasters, for example in the Philippines. Poverty is generally recognised as one of the most important causes of vulnerability to environmental threats (UNEP, 2002).

Keys trends and projections

Population projections

Under the *OECD Environmental Outlook* Baseline, global population is expected to increase from slightly under 6.5 billion in 2005 to 8.2 billion in 2030. The OECD Baseline is based on the medium projection of the United Nations (see United Nations, 2005), in which global population is expected to stabilise at around 9.1 billion inhabitants by the middle of this century. This projection assumes that there will be no demographic catastrophe, and that progress in medical technology will be incremental (see Box 2.1).

Box 2.1. **Assumptions and key uncertainties**

The projections presented in this chapter are based on a number of assumptions:

- The United Nations' medium projections for population are based on the hypothesis that total fertility in all countries converges toward 1.85 children per woman. However, if every second woman in the world has one child more than anticipated, the world population would be 10.6 billion by 2050 instead of 9.1 billion; the population would be 7.7 billion if every second woman in the world has one child less than anticipated. On a country basis, the pace of convergence towards the 1.85 fertility rate may alter the projections by 2030.
- The United Nations' projections on the number of old and very old people only partially incorporate the increases in life-span longevity that have been seen recently (Oeppen and Vaupel, 2002). Additional increases in life expectancy would significantly increase the size of the ageing population, with resulting consequences for consumption patterns, and the social and economic demand for pensions, health-care and other age-related services.
- Hypotheses about labour participation modify economic growth projections, as the contribution of employment to growth is expected to decline, and labour productivity will increasingly become the major factor in economic growth. Should labour participation rates stabilise in OECD countries, macroeconomic projections would not be expected to change significantly (labour would substitute for capital, and production costs in labour-intensive industries would decrease), but the consequences are unclear for migration (*e.g.* would a higher participation rate affect immigration policies and international flows of migrants?) and environmental pressures (*e.g.* what are the environmental consequences of a more or less labour-intensive growth pattern?).
- Migration is an uncertain factor in population and labour force projections.
- In this chapter, countries are considered as single entities. This fails to account for sub-national discrepancies, especially in very large countries.* A disaggregated approach or one focused on ecosystems would provide a more accurate understanding of the environmental consequences of demographic trends.

* See OECD (2003) for an analysis of differences in the structure of population at sub-national level in a number of OECD countries (especially Canada, Portugal, USA, France, Spain, Mexico and Australia).

The fundamental dynamic affecting word population trends is that fertility decreases in a country as the country develops. Economic development, then, is a key factor underlying demographic trends and explaining the contrasted patterns between developed and developing countries, and their convergence over time.

Ninety-five per cent of the global population growth to 2030 will take place in developing countries (Figure 2.1), with the 50 least developed countries experiencing especially rapid population growth. In contrast, the population in OECD countries is expected to stabilise; the share of OECD countries in the world population will drop from 23% in 1980 to 15% in 2030. Note that half the global population growth will come from nine countries only, including India, the USA and China,[2] while in 51 countries (including Germany, Italy, Japan, and the Community of Independent States), population is expected to be lower in 2050 than in 2005.

Figure 2.1. **Population growth by region, 1970-2030**

In billions

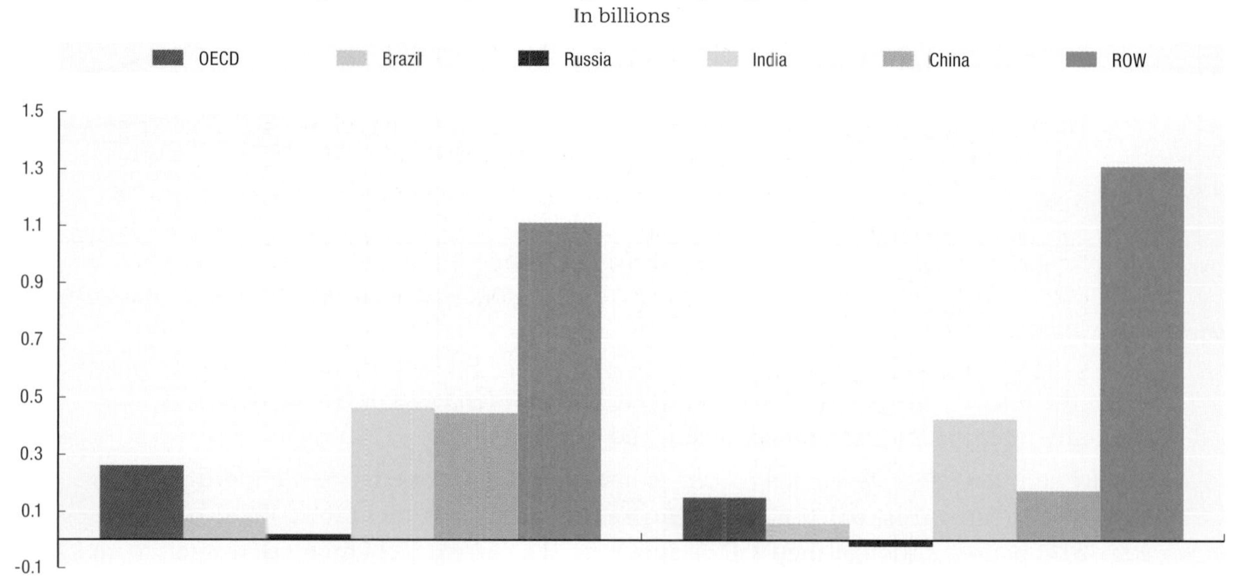

StatLink http://dx.doi.org/10.1787/258802883605

Source: Based on UN, 2004.

The significant population growth in developing countries to 2030 will place additional pressures on the environment, both in growing cities and in rural areas where populations are increasing. Without appropriate infrastructure (housing, energy, transport) and environment-related services, new urban dwellers will generate additional pressures on the environment. In rural areas the poorest people tend to have a high dependence on natural resources. In turn, the increased land and resource pressure is likely to deepen poverty and fuel migration.

Further population growth to 2030 will place additional pressures on the environment unless accompanied by improved environmental policies and infrastructure.

The different dynamics between developed and developing economies result from varied mixes of fertility and mortality trends, which are linked to poverty and to economic growth (Figure 2.2) The link with migrations will be discussed in the following section.

Most industrialised countries already have below-replacement fertility levels, at 1.56 children per woman in 2005. The United Nations expects this will remain so to 2050,

Figure 2.2. **Fertility rates by region, 1970-2040**

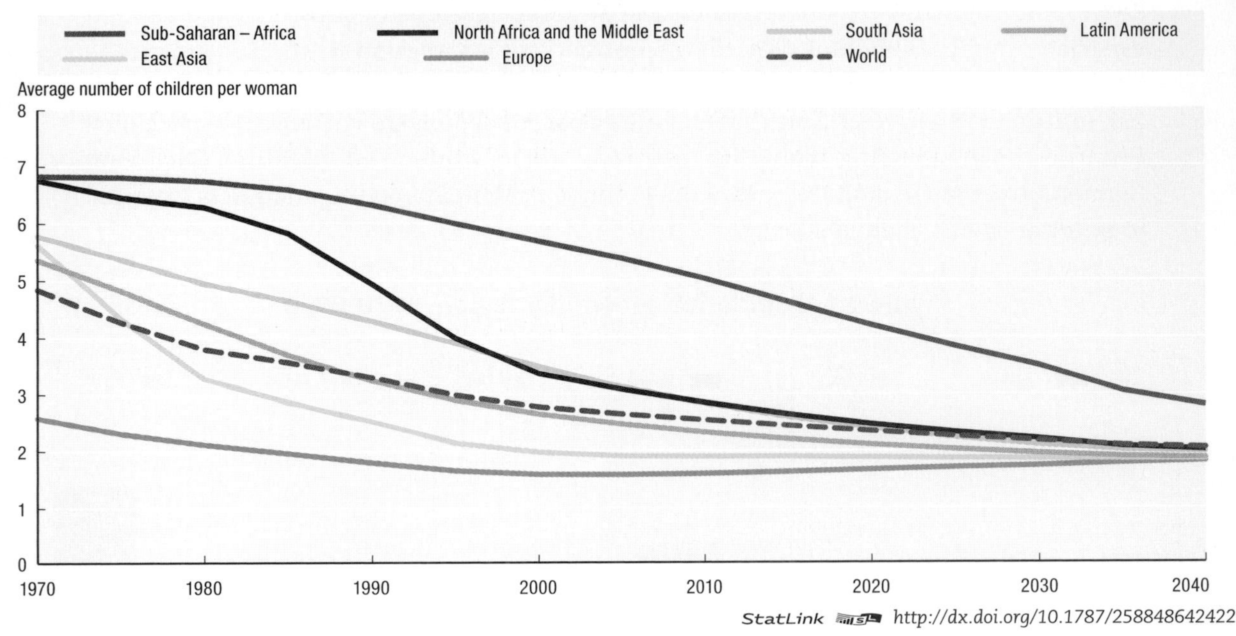

StatLink http://dx.doi.org/10.1787/258848642422

Source: Based on UN, 2004.

when the fertility rate is expected to be about 1.85 children per woman. At the same time, mortality rates are low in these countries, and still decreasing.

By contrast, least developed countries are expected to experience high fertility rates to 2030. These rates will on average remain above replacement level over the 2005-2030 period, although they will decline from the current 5 children per woman to an expected 3.36 children by 2030. In the rest of the developing world, the steady decline in fertility rates which started in the 1960s will continue, and below-replacement levels are likely to be reached in most countries by 2030 (2.01 children per woman, compared to 2.51 in 2005). These countries are also experiencing declining mortality rates, though this trend is being shattered by the HIV/AIDS epidemic in heavily-affected countries.

Countries of the former Soviet Union have a specific profile reflecting the degradation of social and sanitary services which has increased mortality rates. The Russian Federation and the Ukraine in particular are witnessing higher mortality than in the 1960s, and life expectancy in these countries is shorter than it used to be.

Population structure: ageing of populations

Ageing has (favourable and less favourable) consequences on the environment through consumption patterns (housing and land use, transport, tourism, food and drugs, etc.) and sensitivity to environmental constraints (*e.g.* vulnerability to heat-related, illnesses and air pollution effects on respiratory systems). It is associated with population influxes into sunbelts, coastal areas and river valleys, in OECD countries and elsewhere. It has macroeconomic consequences as well, due to public spending and related services – such as pensions, health care, long-term care, education and unemployment transfers – and to age-related trade-offs between current consumption and saving for future generations (ECFIN, 2006). Ageing also affects labour force participation rates, standards of living, urban planning and mobility.

The ageing of the population is a result of the combination of declining fertility and longer life expectancy. It is a dominant trend in OECD countries (see Figure 2.3), especially in North America, Europe, Korea and Japan. UN projections (United Nations, 2005) now show that ageing will occur even faster in the developing world. By 2050, the world is expected to host 1.9 billion people aged over 60 years, 1.2 billion more than it did in 2005. A projected 80% of these over 60-year-olds will live in the developing world. Over the same period, the number of people aged 80 years or more will be multiplied by 4.6: from 86 million in 2005, to 394 million in 2050.

Figure 2.3. **The grey dependency ratio**

Selected countries, 1970-2030

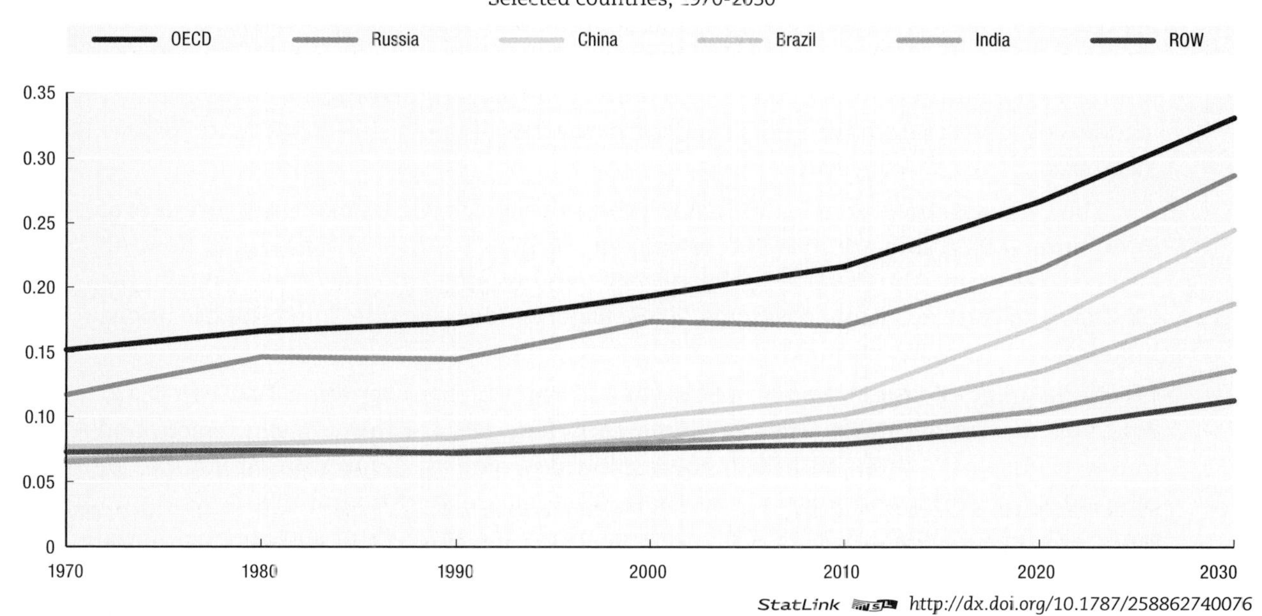

StatLink http://dx.doi.org/10.1787/258862740076

Note: Ratio of people aged 65 years and over to those of working age.

Source: Based on UN, 2004.

One consequence of ageing is a decrease in labour force participation (see also Chapter 3 on economic development). Between 2000 and 2030 the Baseline projects that labour force participation in OECD countries will fall due to a combination of demographic changes and downward pressures from government policies. Thus, by 2030 it projects that labour force participation rates will vary from 49-71% in most OECD regions. However, most countries are likely to employ policies to maintain or increase labour force participation. European economies have set an employment rate target of 70%, which should be reached by 2020.

In this area, policies may influence the decisions of members of the working-age population (particularly women) to participate in the labour force. Indeed, female employment rates, which are very uneven across OECD countries and worldwide, are likely to rise, making them a major driver of change in the workforce. Raising the age of retirement is also being implemented or considered by a number of OECD countries. Migration is yet another option to enlarge the labour force. These policies will have specific environmental consequences. Typically, migration will reallocate people across territories (see below).

Migrations, international and domestic

Migrations change the distribution of population across countries and lands; they can be domestic or international. International migrations directly connect OECD and non-OECD countries. From an environmental perspective, they can add pressures on regions which are already stressed (*e.g.* aggregating people into over-crowded urban areas, or contributing to desertification). They can also be fuelled by environmental pressures. In some circumstances, migrations can exacerbate tensions and security issues.

According to the United Nations, between 2005 and 2050, migrations to more developed countries will more than offset the natural population decline in these countries (United Nations, 2005). Over this period, 98 million migrants will leave the less developed regions (less than 4% of the expected population growth in these regions), and the same amount will reach more developed countries (net figure[3]). However, most migrants to the world's rich countries do not come from among the world's poorest, but from middle-income countries or from the middle and upper reaches of the income distribution of low-income countries (Goldin, 2006).

The United Nations anticipates that the countries which will be major net receivers of international migrants are the USA (which will account for half of the annual flow, on average), Germany (thus reversing the current trend of population decline), Canada, the UK, Italy and Australia (United Nations, 2005). Major senders include China, Mexico, India, Indonesia and the Ukraine.

In a survey of recent trends, OECD (2005) suggests that migration flows to OECD countries are largely stable. They predominantly take place within a given region, and follow traditional routes, although some countries emerge as prominent sources of migrants, *e.g.* China and Russia. The share of labour-related migrations is rising, in particular for qualified migrants. This work confirms the strength of sub-regional flows, typically in sub-Saharan Africa, Latin America and Europe. Central and Eastern Europe tend to receive an increasing number of migrants from neighbouring countries, attracted by the new European Union member states; the region is also a source of migration to nearby OECD countries, in particular Austria, Germany and Italy. In Latin America, migrations within the region remain strong, but flows towards OECD countries keep growing; the USA, obviously, but also Europe (the UK and Italy, in particular), via Spain, are the primary destinations. Sub-Saharan Africa experiences essentially sub-regional flows.

New routes from Asia have changed the picture since the late 1960s. Migrants from Asia constitute a major and growing share of the populations received in OECD countries, typically in the USA (34% of migrants received by the USA originated in this region), Canada and Australia (the share of Asian population amounts to 50% of migrants in these countries), and the UK. Asian migrants form a dominant share of temporary, qualified migrants. An increasing variety of routes lead to migration between countries with cultural and historical similarities and it is expected that such routes will be increasingly crowded under demographic pressure.

Domestic migrations change the distribution of a population across a given territory. Rural-to-rural migration – for example, people moving to forest frontiers or to the coasts for new land and resources – can affect biodiversity through loss of species and genetic material, habitat loss and fragmentation, and disruption of ecosystem processes. Increasing migration to regions that are particularly at risk from natural hazards can increase vulnerability, a challenge that is likely to be exacerbated in the future by the impacts of a changing climate.

The distinction between rural and urban settlements (which is sometimes not clear-cut), and the move from city centres to suburbia, also modify both the pressures on the environment and the opportunities to mitigate them. Major impacts relate to land use (competition between natural habitat, agriculture, and human settlements) and environmental pressures, typically in and around big cities (urban sprawl), mountainous areas, coastal areas and internal seas. These make land and urban planning even more relevant from an environmental perspective (see Chapter 5 on urbanisation).

From 2005 to 2030, the world's urban population is expected to increase by more than 2 billion people. Urban conglomerations and mega-cities affect air pollution, and the demand (and opportunities) for environmental services (water and sanitation, waste management). Local environments are particularly deteriorated in slum areas, where it is estimated that 1 billion people (30% of city dwellers) now live. The United Nations Human Settlements Programme anticipates that this number could double by 2030 (UN-Habitat, 2003), a trend fuelled by migration from rural to urban areas.

Notes

1. Via age structure and participation rates, defined as the share of the adult population that considers itself as part of the labour force.
2. Chinese authorities expect the Chinese population to peak at 1.43 billion in 2020.
3. These anticipations are based on past trends, supplemented by an assessment of the policy stance of countries on international migration flows.

References

Alcamo, J., T. Henrichs and T. Rösch (2000), *World Water in 2025 – Global Modeling and Scenario Analysis for the World Water Commission on Water for the 21st Century*, Center for Environmental Systems Research, University of Kassel, Germany.

ECFIN (2006), "The Impact of Ageing on Public Expenditure: Projections for EU25 Member States on Pensions, Health Care, Long-Term Care, Education and Employment Transfers (2004-2050)", *Special Report* 1/2006, European Commission – DG ECFIN, Brussels.

Goldin, I. (2006), "Globalizing with their Feet: The Opportunities and Costs of International Migration", *World Bank Global Issues Seminar Series*, The World Bank, Washington DC.

OECD (2003), *Territorial Indicators of Socio-Economic Patterns and Dynamics*, OECD, Paris.

OECD (2005), *Trends in International Migrations:2004 Annual Report*, OECD, Paris.

Oeppen, J. and J.W. Vaupel (2002), "Broken Limits to Life Expectancy", *Science*, Vol. 296, pp 1029-1031.

Prugh, T. and E. Ayres (2004), "Population and its Discontents", Editors' introduction, *World Watch Magazine*, p. 13.

UNEP (United Nations Environment Programme) (2002), *GEO-3: Global Environment Outlook, 2002*, UNEP, Nairobi.

UN (United Nations) (2004), *World Population Prospects: the 2004 revision*, United Nations, Department for Economic and Social Information and Policy Analysis, New York.

UN (2005), *World Population Prospects. The 2004 Revision, Highlights*, United Nations, New York.

UN-Habitat (United Nations Human Settlements Programme) (2003), *Global Report on Human Settlements*, Earthscan Publications Ltd (on behalf of UN-Habitat), London.

ISBN 978-92-64-04048-9
OECD Environmental Outlook to 2030

Chapter 3

Economic Development

This chapter highlights key trends and developments in the world economy to 2030 and outlines the consequences of the projected economic growth on the environment. The implications of productivity growth are examined at both the regional and sectoral levels. Given the projected expansion of the global economy to 2030, failure to act on environmental challenges will have even more impact in the future than it does today. Natural resource sectors will find demand increasing for their output as large economies like Brazil, the Russian Federation, India and China (BRIC countries) continue to experience rapid growth. Sectors such as agriculture, energy, fisheries, forestries and minerals will need to have strong policies in place to reduce the environmental impact of this rapid growth.

KEY MESSAGES

- The global economy is projected to grow by 2.8% a year from 2005 to 2030 under the *Outlook* Baseline. The annual average growth rate over this period is projected to be 2.2% for OECD countries, 4.6% for BRIC countries, and 4% for the rest of the world (see figure below).
- Growth is expected to be higher at first (3.4% global growth for 2005-10), then slowing to 2.7% for 2010-20, and 2.5% for 2020-30. This is because the Baseline reflects "no new policies", so assumes that some historical trends – such as trade growth – which contribute to economic growth (but are influenced by government policies) slow over the reference period. As a result, the Baseline is somewhat conservative, given that in 2004 and 2005 global growth was 5.1% and 4.9% respectively.
- Economic growth is affected by labour supply, which in OECD countries will decline in some regions as a result of ageing populations. In other OECD regions, ageing populations will be offset by immigration and sufficiently high birthrates. Aggregate labour productivity growth rates are converging in OECD and non-OECD countries, but this does not necessarily mean a convergence in living standards.

GDP growth rates to 2030

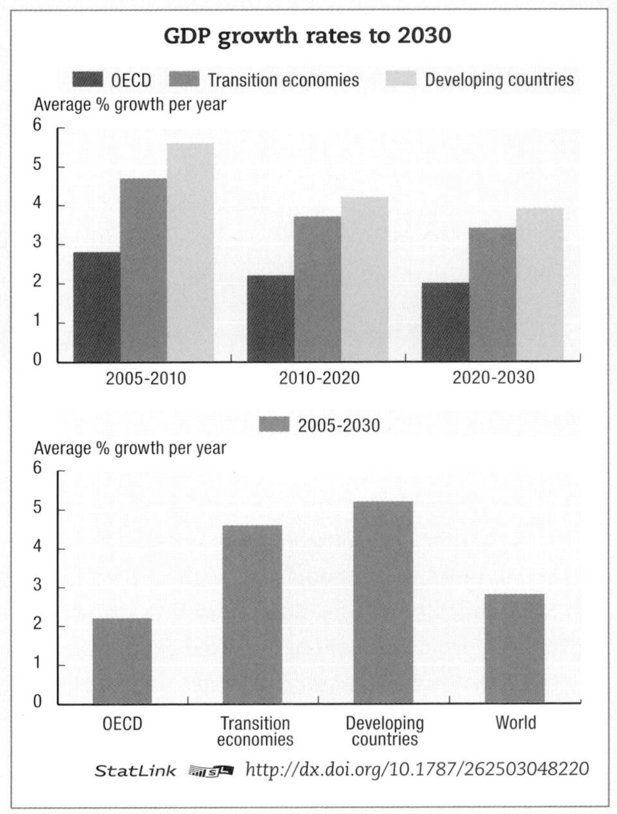

StatLink http://dx.doi.org/10.1787/262503048220

Environmental implications

Increasing aggregate demand and productivity growth will increase demand for material inputs from the environment and increase the amount of by-products that have to be dealt with as waste.

Differences in sectoral growth rates will continue to be manifested as a "decoupling" of economic growth from environmental impacts. This reflects the changing structural composition of economies. The shift towards service-based industries from energy-intensive, polluting industries and agriculture is projected to continue to 2030, reflecting changes in consumer demand.

Technological developments reflected in productivity growth will continue to increase the efficiency of industrial production and reduce levels of pollution and waste per unit of output.

Consequences of inaction

The scale of economic growth anticipated to 2030 under the Baseline is such that failure to act on environmental challenges will have even more impact than it currently does. Natural resource sectors will find demand increasing for their output as large economies like the BRIC countries continue to experience rapid growth. Sectors such as agriculture, energy, fisheries, forestry and minerals will need to have strong policies in place to keep the environmental impact of this rapid growth at an acceptable level. But since all economies will see increasing material wealth, the demand for clean environments will also grow everywhere.

 OECD ENVIRONMENTAL OUTLOOK TO 2030 – ISBN 978-92-64-04048-9 – © OECD 2008

Introduction

Economic growth in OECD regions has been robust for a considerable period of time now and many developing regions have been growing at a rapid pace for at least the past 15 to 25 years. Since the conditions that made that growth possible still prevail (*e.g.* institutional stability, etc.), a fundamental view of the Baseline for this *Outlook* is that the same deep drivers of economic growth will continue into the future – though not at the same intensity as in the recent past (Box 3.1).

Box 3.1. **Sources of assumptions for the modelling framework**

The drivers of economic growth underlying the *Outlook* are largely taken from work by the OECD Economics Department, the International Energy Agency, the OECD Agriculture Directorate, and the UN Food and Agriculture Organization. These drivers include long-term labour productivity growth and labour force participation rates, as well as medium-term developments in trade and the working out of business cycle imbalances. Projections of these economic drivers were constructed to 2030 (2050 for the work related to climate change). They were then transformed into a full economic baseline – both for examining pressures on environmental factors, as well as for looking at mitigating policies – using the ENV-Linkages model.

Other key aspects of the Baseline with respect to economic development include:

- *Energy:* The energy system has been largely calibrated to the 2004 edition of the IEA's *World Energy Outlook* (IEA, 2004; referred to here as WEO2004), although some aspects were updated to the 2006 edition. This essentially means that energy technologies represented in WEO2004 are reproduced for this Baseline. However, even though the technologies are similar, the results obtained for this *Outlook* may be substantially different. The reason for this is that projections for population growth and productivity gain (*i.e.* economic growth) will be different, and have an impact on energy use. Since the *World Energy Outlook* also uses a reference scenario for projecting energy demand, there is a high degree of consistency between that work and the *OECD Environmental Outlook*.
- *Agriculture:* Trends in agricultural productivity will be important to 2030. The trends for yields used in this *Environmental Outlook* were largely adapted from the FAO study *World Agriculture: Towards 2015/2030* (Bruinsma, 2003) where macroeconomic prospects were combined with the views of regional experts. Key agricultural trends emerging from the analysis reported below have been checked for consistency with the *OECD-FAO Agricultural Outlook 2006-2015* (OECD/FAO, 2006). The overall pattern is consistent, so major driving forces are similar between the results of this Baseline and the *Agricultural Outlook*.
- *Technology:* The future is essentially envisioned as a world that is very similar to today's in terms of the role and size of government, policy priorities, taxes, technology diffusion, intellectual property rights, liability rules and resource ownership. It is also similar to today's world in terms of dietal preferences, mobility demand and other consumption habits for given income levels. Since incomes in developing countries will change, there will be some change in consumption patterns, but in a manner that will make them look more like today's developed countries.

Decoupling the environment from economic growth

Although the relationship between the economy and the environment is complex (the Environmental Kuznets Curve[1] illustrates that complexity), inter-relations are strong (Box 3.2). One dimension is illustrated in Figure 3.1, which shows the relationship between GDP and domestic material consumption (DMC). The material input is a straightforward measure of some material flows into the economy (OECD, 2007). The amount of physical inputs into OECD economies rose by roughly 27% between 1980 and 2005, so even in advanced economies there is continued growth in the use of raw materials, highlighting the increased impact that the economy has on the environment.

Figure 3.1. **Domestic material consumption and GDP, 1980-2005**

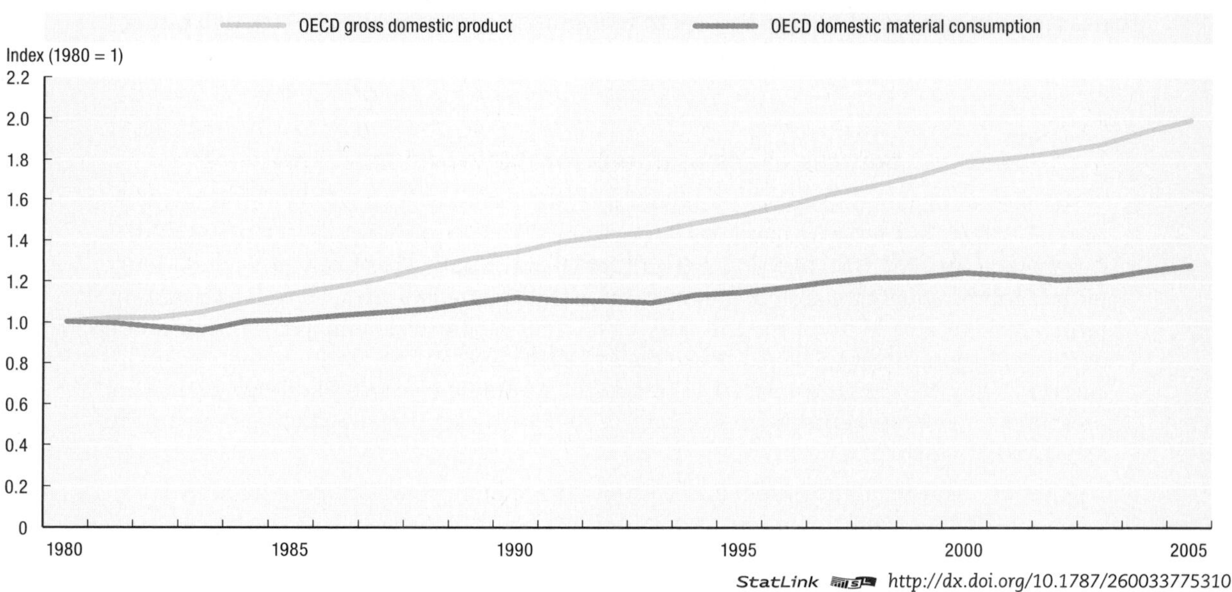

StatLink http://dx.doi.org/10.1787/260033775310

Source: OECD (2007).

An important reason for the divergence between the growth rates of GDP and DMC was that sectors that affect the environment (*e.g.* agriculture, fishing, forestry, minerals, fossil fuels, water, etc.) have been growing more slowly than the rest of the economy. The information and communication technology sectors, the health sectors, the entertainment sectors, etc., are all much bigger today than they were 30 years ago. Even though the sectors that affect the environment are bigger than before, they are a smaller part of the overall economy and thus appear to have "decoupled" from economic growth.

Some relative decoupling of environmental impacts from economic growth can be expected.

Box 3.2. **Interactions between economy and environment**

Another dimension of economy/environment interaction is shown in Figure 3.2. The two lines – economic activity (world GDP) and environmental impact (total ecological footprint, EF) – are represented as indices where both equalled 1 in 1980. The concept of EF (proposed by Rees, 1992, and further developed by the World Wide Fund for Nature, WWF) is controversial, partially because it includes land needed to absorb the CO_2 emitted from burning fossil fuels. As such, we treat it as reflecting overall environmental trends rather than as a detailed measure that policy could target. Moreover, the EF is less problematic when used relative to a base year – many of the constituent parts are better measured as changes than as levels.

While there has been some relative "decoupling" between GDP and EF, they both grew between 1961 and 2003: GDP by more than four times, EF by more than three times. With the scale of economic growth projected under the *Outlook* Baseline, an even stronger divergence between GDP and EF will have to be achieved just to maintain current levels of environmental quality. The divergence shown in the figure is an average for all countries. Some will have stronger divergence than others. If the upward trend in the ecological footprint incorporated all environmental and inter-generational impacts, then the outcome would not necessarily be problematic: there would be no basis for arguing that there are limits to growth. However, conditions for market failure are common when dealing with environmental issues (*i.e.*, externalities, non-rivalry, non-excludability), and hence environmental policies will be needed to correct these failures.

Figure 3.2. **Economy and environment, 1961-2003**

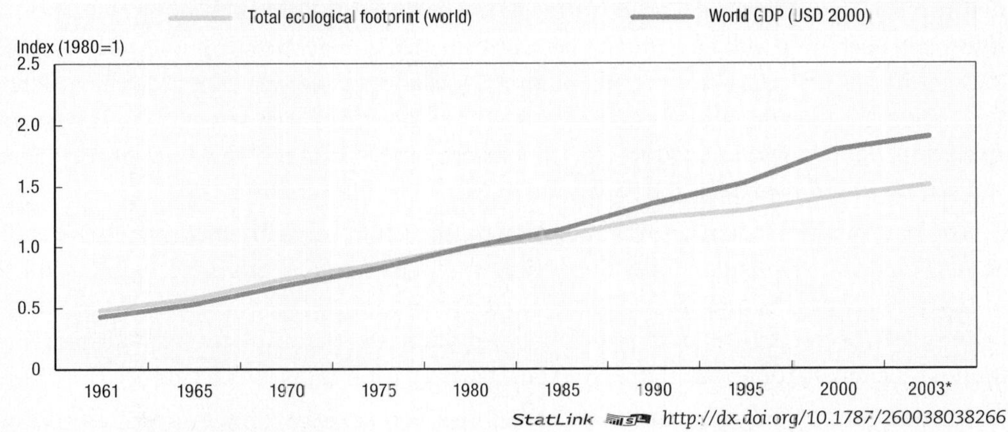

StatLink http://dx.doi.org/10.1787/260038038266

* Last data available (2005 data are not yet available).

Source: WWF (2006); World Bank (2006).

Key trends and projections

Much of the economic growth that will occur to 2030 can be explained using a limited number of primary drivers: labour force growth, labour productivity growth and trade growth. The latter two, productivity and trade growth, require large investments, so they are associated with substantial structural change. Long-term projections of how these drivers will evolve give a good indication of how growth in GDP might develop, including consumption of goods and services that affect the environment. Figure 3.3 outlines trends in these variables from 1980 to 2001. Since data on labour force growth in non-OECD countries are inconsistent, population growth has sometimes been used instead. Each of these three drivers is assessed in more detail below.

Figure 3.3. **Growth trends (average % per year), 1980-2001**

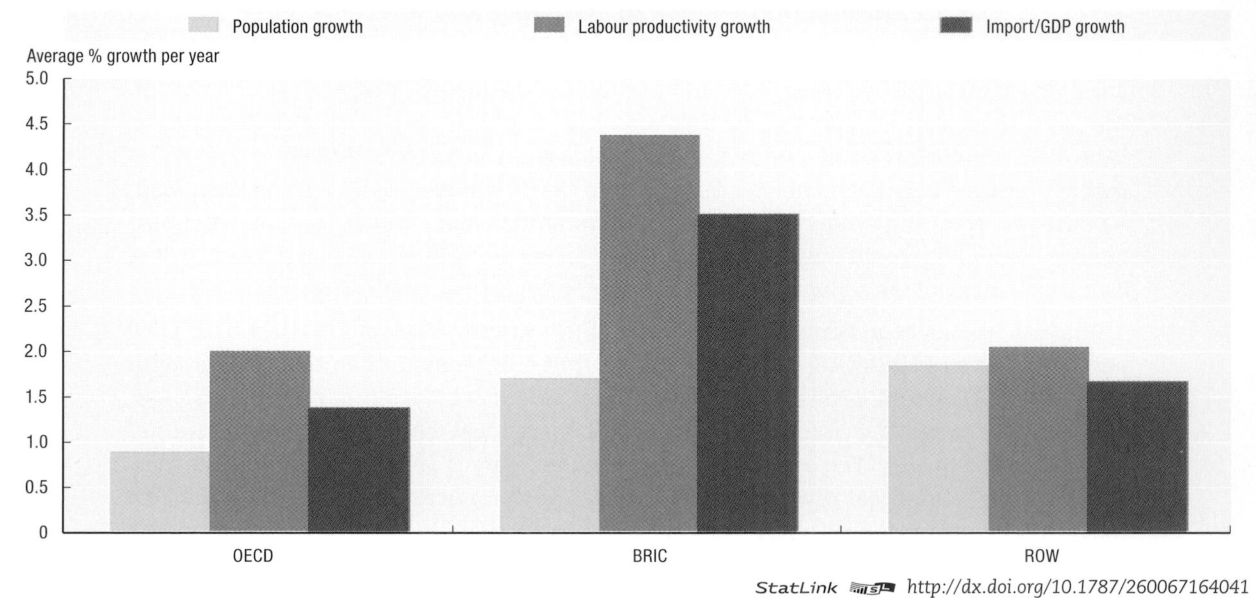

StatLink http://dx.doi.org/10.1787/260067164041

Source: United Nations, 2005; OECD STAN database; World Bank, 2006.

Population is expected to continue to increase (see Chapter 2 on population dynamics and demographics), leading to a larger labour force, and more capacity for production and consumption (see Chapter 1 on consumption, production and technology). Population growth has been, and will continue to be, a strong driver of economic growth even in some OECD countries. The US economy, for example, has had an average GDP growth of just over 3% for the past decade, while labour productivity growth was just over 2%; increases in population (labour force) account for the discrepancy between the two trends. Strong population growth is expected to continue in the US in the Baseline.

The participation of the adult population in the labour force is evolving in the Baseline. The participation rate is generally defined as the percentage of the adult population that considers itself part of the labour force (*i.e.* those who are either working, or looking for work). In OECD countries, government policies are seen as complementing demographic changes by exerting downward pressures on participation rates (OECD, 2003).

For non-OECD countries, the trend of labour force participation is projected to slowly move toward the OECD average. In OECD countries the unweighted average participation rate has been approximately 60% for more than 30 years. A convergence to this 60% average was thus assumed for non-OECD countries (but only 1% of the gap is closed per year).

The OECD region is projected to increase its labour force by 10% between 2005 and 2030, while those in the BRIC countries and the rest of the world increase by 27% and 50%, respectively (Figure 3.4).

Global productivity has been growing steadily, in aggregate, since at least 1980 (for most OECD countries productivity has been increasing for the past two centuries). The importance of productivity growth is that it implies that each person will produce more economic output for each hour worked. This raises living standards, but also increases demand for material inputs from the environment and increases by-products that have to be dealt with.

Figure 3.4. **Labour force projections, 2005-2030**

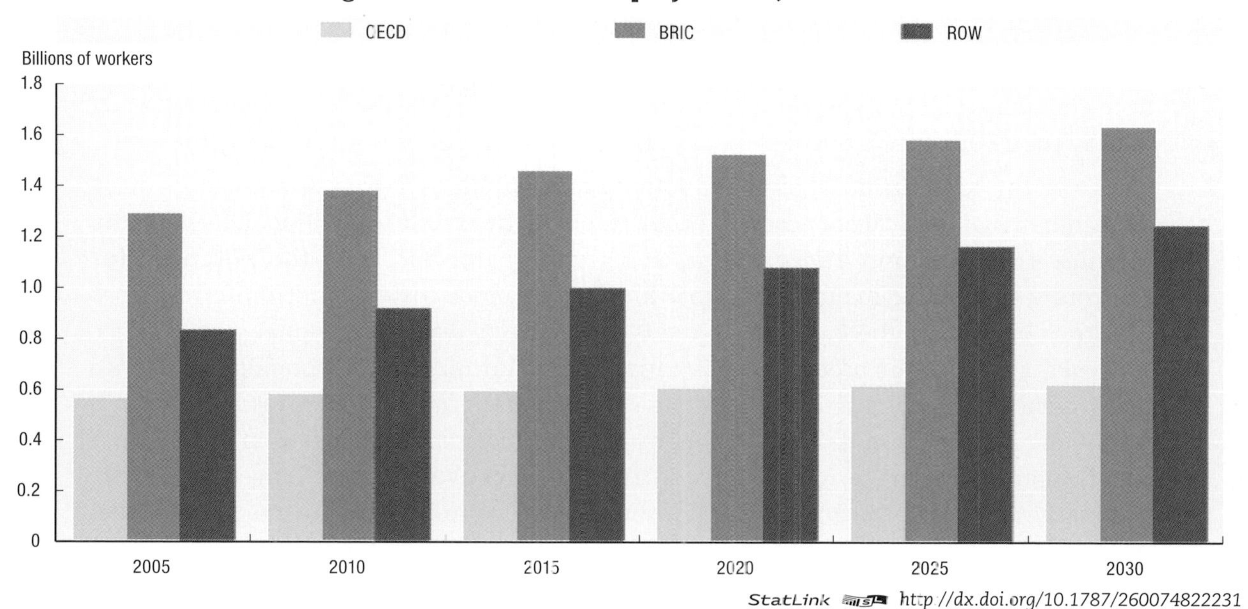

StatLink http://dx.doi.org/10.1787/260074822231

Source: OECD, using United Nations (2005) data.

Long-term productivity growth in the Baseline is set at 1.75% per year. This is a rough (and somewhat understated) historical average for OECD countries, and is generally consistent with some long-term historical trends that are shown in Table 3.1.

In the *Outlook* Baseline, all countries eventually move toward 1.75% per year productivity growth (albeit at a very slow rate). This is applied at a national level (GDP per hour, or day, worked) so it implies that eventually everyone is producing 1.75% more per year of value-added.

Table 3.1. **Productivity in historical perspective for the UK and USA: average annual % rate of change**

UK		USA	
1780-1831	0.4	1800-1855	0.4
1831-73	1.2	1855-90	1.4
1873-1913	0.9	1890-1927	2.0
1913-50	1.6	1929-66	2.5
1950-73	3.1	1966-89	1.2

StatLink http://dx.doi.org/10.1787/256624436857

Source: Crafts, 2003.

Trade is another issue that is crucial for understanding future growth. Trade allows countries to specialise according to their strengths, so that the productive capacity of all countries is increased (see also Chapter 4 on globalisation). Figure 3.3 shows that the import/GDP ratio has been growing; in other words imports grew substantially more rapidly than GDP. In the Baseline, there is a continued growth of trade, but it stabilises relative to GDP (see below for more discussion). This implies that there is an ongoing expansion of trade, but that the proportion of goods and services that are traded internationally does not change.

Aggregate economic growth

The aggregate rate of economic growth projected in the Baseline is shown in Table 3.2. The first five years show a continuation of recent rapid growth rates, though some deceleration is evident (consistent with a return to potential growth rates – a view generally held by the OECD, International Monetary Fund, African Development Bank and others).

Over the long term, labour productivity and population growth are the primary determinants of the scale of economic activity. Productivity and population determine the amount of production and consumption that occurs, and thus the potential for environmental impacts. Since the growth of labour productivity is not uniform across countries or regions, the Baseline for the *Outlook* reflects different regional growth rates. However, since labour productivity is ultimately determined by technologies that are discoverable by all countries, the Baseline assumes that growth rates across regions will ultimately be the same, asymptotically converging towards 1.75% per year, based on a rough historical average for OECD countries.[2] This convergence of long-term labour productivity growth towards 1.75% (although most countries do not reach this level by 2030),[3] implies a steady decline in global GDP growth from a peak that is achieved roughly in 2005. Comparing developing country growth rates with those of developed countries, Table 3.2 shows that developing countries will continue to grow at much higher rates than developed countries.

Table 3.2. **Global annual average GDP growth (%, 2005-2030): Baseline**

	2005-10	2010-20	2020-30	2005-30
OECD	**2.8**	**2.2**	**2.0**	**2.2**
North America	3.5	2.5	2.3	3.1
US and Canada	*3.4*	*2.4*	*2.3*	*2.6*
Mexico	*5.3*	*3.6*	*3.1*	*3.7*
Europe	2.5	2.1	1.8	2.1
Pacific	1.6	1.8	1.3	1.6
Asia	*1.4*	*1.7*	*1.2*	*1.5*
Oceania	*3.5*	*2.5*	*2.2*	*2.6*
Transition economies	**4.7**	**3.7**	**3.4**	**4.6**
Russia	4.7	3.9	3.6	3.9
Other transition economies	4.8	3.5	3.2	4.4
Developing countries	**5.6**	**4.2**	**3.9**	**5.2**
China	7.2	4.9	4.1	5.0
East Asia	5.3	4.3	3.7	4.3
Indonesia	*5.7*	*4.5*	*3.9*	*4.5*
Other East Asia	*5.2*	*4.3*	*3.7*	*4.2*
South Asia	6.5	5.1	4.5	5.1
India	*6.5*	*5.2*	*4.5*	5.2
Other South Asia	*6.5*	*4.8*	*4.4*	5.0
Middle East	4.6	3.6	3.9	3.9
Africa	5.4	4.2	4.4	4.5
Latin America	3.8	2.9	2.8	3.6
Brazil	*3.4*	*2.8*	*2.5*	2.8
Other Latin America	*3.9*	*3.0*	*3.0*	3.2
World	**3.4**	**2.7**	**2.5**	**2.8**

StatLink http://dx.doi.org/10.1787/256624520840

Source: OECD *Environmental Outlook* Baseline.

Technology and productivity

Technology plays an important part in determining productivity since it affects a worker's activity (consider the difference between OECD farmers today and farmers 500 years ago). Since labour productivity is assumed to continue to grow following past trends, there is an implicit assumption that new technologies will continue to be developed. Current trends in information and communication technologies (ICT) are therefore assumed to continue, as are trends in biotechnologies and nanotechnologies. New technologies can be assumed to transform societies over the coming decades to the same extent that they have over past decades, even if the specific areas of technological development have been changing. What is key for the *Outlook*, however, is how these technologies affect the environment. In looking to 2030, productivity growth in the Baseline has been assumed to be environmentally neutral. That is, technology does not, by itself, reduce environmental impacts.

The long-term trend in productivity growth of 1.75% means that there is more value added for each hour worked. Having each worker increasing value added per hour worked means that consumption is increasing. This means that even if population growth were to fall to zero, there would still be increases in the volume of production and consumption. In the goods-producing sectors, the environmental impact results from the fact that as they grow, they will require more material inputs, but not as much as would be implied by the increase in output: future growth will thus be less environmentally damaging *per unit of output*. How much more material input is required will depend on how productivity growth affects production and prices. The relative decoupling that was illustrated in Figure 3.1 will be manifested in the Baseline to 2030 as a result of both productivity growth and growth differences between material-using and other sectors.

Other assumptions can be made that would change the amount of material per unit of value added. That is, the nature of technological change could be changed in the Baseline. However, these other assumptions are more interesting as consequences of policy actions rather than as general characteristics that are embedded in the Baseline.

A number of regions encompassed in Table 3.2 warrant additional comments.

North America

Labour force growth is an important driver of economic growth in North America. Of the growth shown in the table, 1.2% annually is the result of increases in the labour force between 2005 and 2030. An important part of that increase in labour force in the US and Canada will be migration into the region from developing countries (average labour force growth in the US and Canada is 1% between 2005 and 2030).

China

China's labour productivity growth averaged just over 5% per annum from 1980 to 2001. Since GDP growth was significantly higher than that, a substantial part of its past growth clearly originated in increases in the employed labour force (*i.e.* population increases). Indeed, between 1980 and 2001, China's population increased by roughly 30%. In the future, this source of economic growth is not expected to continue as strongly and the existing population will begin to become more dependent on younger cohorts. In spite of these downward pressures, the long-term projection for China is for GDP and productivity growth to both remain over 4% through 2030.

South Africa

South African productivity growth from 1980 to 2001 was actually lower than for many of its sub-Saharan neighbours. Moreover, it is projected to have one of the lowest population growth rates of that region; thus its labour supply will grow more slowly than many of its neighbours and its economy will not perform as strongly as many of the other sub-Saharan economies. It will remain a strong regional power, but some of its neighbours will be catching up.

Central Europe

Productivity growth has been (and will likely remain) robust in Central Europe, but this is combined with a slowing population growth; in many cases the population is declining. Underlying this is a low fertility rate which is aggravated by migration to Western Europe. Central Europe, therefore, will generally show rising standards of living, even while aggregate GDP growth does not reflect the strength of that rise.

Eastern Europe, the Caucasus and Russia

Given the change in economic systems that occurred with the political opening of the 1980s and 90s, these regions have limited data for gauging future growth potential. This is particularly the case for sectoral productivity growth, where the considerable structural change of the post central-planning regimes will contribute to a higher long-term growth path. To address this dearth of data, this region is assumed here to converge to the *relative* sectoral growth of Western Europe. That is, aggregate productivity growth reflects the region's own trend, but sectoral proportions become similar to developed Western trends.

Japan

Japan is assumed to continue to have strong productivity growth, but demographic trends will pull down aggregate GDP growth. While this is similar to the trend in much of Europe, it is expected to be stronger and more keenly felt in Japan.

Middle East

A history-based view of economic growth in the Middle East is necessarily rather pessimistic. The volatility of that region has resulted in uneven periods of growth which, on average, are very low. Even some of the better performers in the region turn out to be less than extraordinary when examined more carefully. For example, while overall growth in Israel has generally been very good, per capita growth is only mediocre. It turns out that much of Israel's economic growth was the result of immigration that expanded the labour force. Since Israel now has no net migration, that channel of growth has been eliminated and is not assumed to re-open during the projection period. Some countries in the region are currently doing well as a result of the upward spike in oil prices. However, an oil-commodity boom is not something that can be counted on for the long-term.

Latin America

Latin America has a long experience with development and with alternative strategies to achieve it. Many of the countries in Latin America have gone through bursts of rapid growth, followed by downturns where much of the gains are lost. The two largest economies in the region, Brazil and Argentina, both have been through numerous cycles of growth and retraction which go back over a 60-year period. One should therefore be rather cautious in projecting optimistic long-term growth for the region.

Sectoral results

The Baseline to 2030 illustrates a changing structure of the economy with productivity growing at different rates in different sectors. In particular, agricultural productivity is generally higher than manufacturing productivity, and manufacturing productivity is generally higher than service-sector productivity. These trends in the future are particularly important because they lead to changes in the composition of output, and the environmental impacts of various areas of the economy. Growth in individual sectors is also an important source of overall economic growth since growth can occur through re-allocation of resources away from low-value (high-productivity) sectors to higher-value (low-productivity) sectors.

Table 3.3 shows the relative sectoral size (measured as the production share of each sector in the gross output of the economy) in the Baseline. Comparing 2001 and 2030 for each sector illustrates the change in the economies' composition. For illustrative purposes, the 26 sectors available in the model have been aggregated to seven sectors (see also Figure 4.5 in Chapter 4, Globalisation).

Table 3.3. **Shares of sectors in 2001 and 2030 (in gross economic output)**

	OECD		BRIC		ROW	
	2001	2030	2001	2030	2001	2030
Agriculture	2%	1%	9%	6%	8%	6%
Forestry and fishing	0%	0%	1%	1%	1%	2%
Energy and mining	3%	2%	6%	4%	9%	7%
Non-durable man.	10%	7%	12%	8%	14%	10%
Durable manufacturing	23%	17%	32%	29%	24%	21%
Trade and transport	18%	18%	15%	16%	16%	16%
Services	44%	54%	25%	37%	28%	38%

StatLink http://dx.doi.org/10.1787/256642330685

Source: *OECD Environmental Outlook* Baseline.

Underlying these changes are projections of different productivity growth across sectors that move workers out of sectors like agriculture and manufacturing. Technological developments mean that goods and services can often be supplied more cheaply, lowering the prices of various goods and the wages paid to produce them. This leads workers to move to other sectors where wages are higher. Thus, underlying changes in composition are technological developments that make some goods and services easier to produce than others (other sources of change include consumer tastes, see also Chapter 1 on consumption, production and technology).

Trade results

In both the short and medium terms, reductions in various transaction and communication costs can be expected to continue and will encourage trade. In both OECD and non-OECD countries, the ratio of imports to GDP will thus continue to increase, as the production of goods and services are gradually rationalised on a global basis. However, the sheer potential size of economies like China and India suggests caution in projecting increases in imports as a percentage of GDP. Large developed economies like the United States and Japan have a low import-to-GDP ratio because the service sectors are much larger than manufacturing, agriculture and other sectors that produce tradable goods. For an economy like China, which is expanding very rapidly, a projection that trade will keep increasing as a share of GDP would imply that manufacturing-driven growth would continue at a pace that seems implausible. Analyses of China's growth already show that it is having a strong impact on its neighbours.[4] A reasonable conjecture is that, at some point, the ratio of imports to GDP will level off (in China's case, even a levelling off may be optimistic, since it will imply a ratio of imports to GDP that is more than three times that of other large economies, such as the US).

Even with imports levelling off in relation to GDP, the Baseline still projects considerable growth in imports to 2030. This is because GDP itself is increasing strongly. Figure 3.5 illustrates the growth of imports in the aggregate regions.

Figure 3.5. **Baseline import growth to 2030**

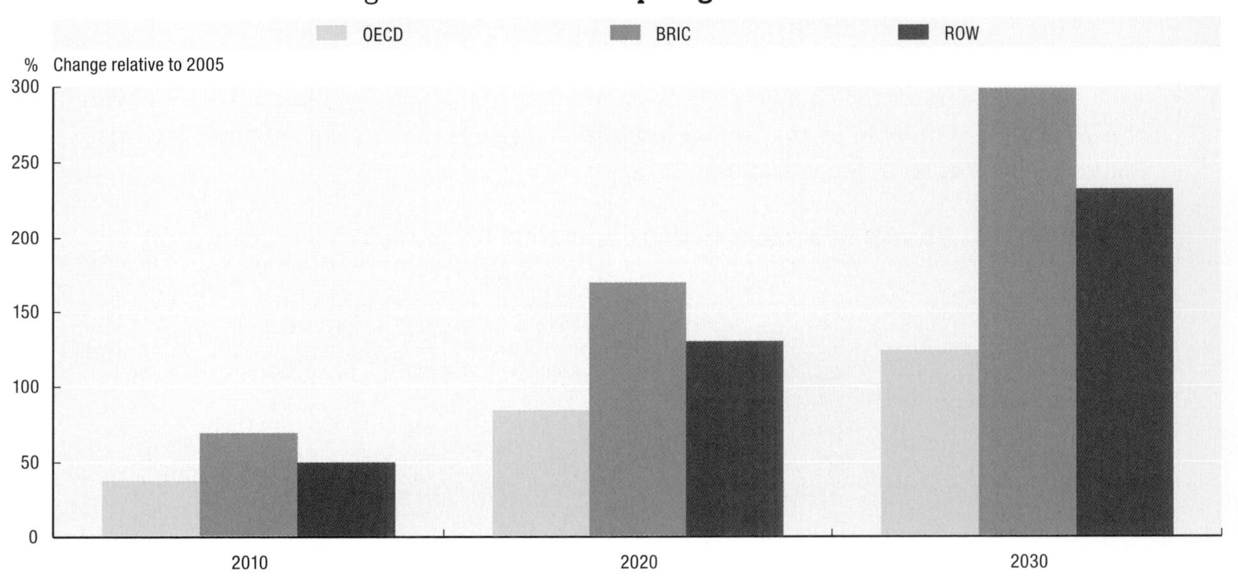

StatLink http://dx.doi.org/10.1787/260144407002

Source: OECD Environmental Outlook Baseline.

Even without increased imports relative to GDP, the amount of trade in absolute terms (imports) is projected to grow considerably by 2030, implying greater environmental impacts from factors such as invasive alien species, CO_2, NO_x and SO_x from fossil fuel use, PM (particulate matter) and ozone, and accidents such as oil spills (see Chapter 4, Globalisation, for further discussion).

Policy implications

The Baseline developed here represents future projections assuming no new government policy. The implications of this Baseline are increasing environmental pressures across the board. These are discussed in detail in the chapters in the second part. However, it is worth noting some trends here that are related to the economic development projected in the Baseline.

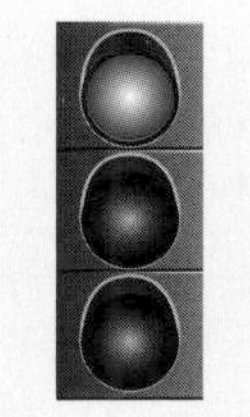

Rapidly growing economies will lead to increasing demand for natural resources to 2030.

Natural resource sectors will see greater demand as large economies like the BRIC countries continue to experience rapid growth (Figure 3.6). Sectors such as agriculture, energy, fisheries, forestry and minerals will need to have strong policies in place that keep the environmental impacts at acceptable levels. All economies are expected to experience increasing material wealth, leading to greater demand for clean environments. But economic growth will occur in the context of a global ecosystem that cannot be easily expanded – so without strong policies, the impact of the economy on the ecosystem is likely to increase. This observation is not to suggest that there are limits to growth, rather that there are inevitable choices to be made between ecosystem and economy if global material well-being is to approach the level that advanced economies currently enjoy.

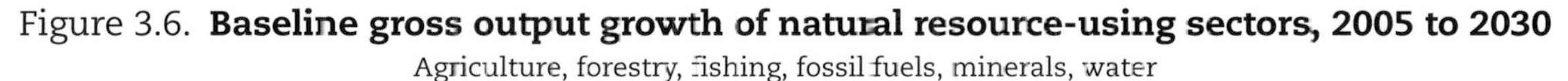

Figure 3.6. **Baseline gross output growth of natural resource-using sectors, 2005 to 2030**

Agriculture, forestry, fishing, fossil fuels, minerals, water

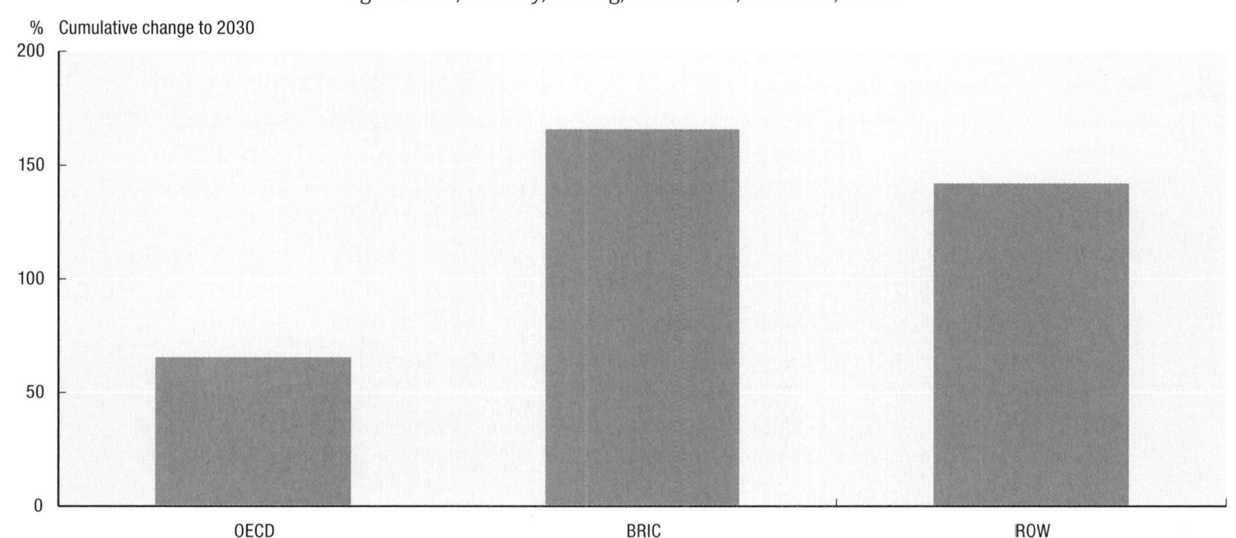

StatLink http://dx.doi.org/10.1787/260248515355

Source: *OECD Environmental Outlook* Baseline.

In this context, failure to act on environmental issues will have even more impact than it does today. The time between first recognition of a problem and national or global consensus to correct it will be much shorter when the world economy is that much more disruptive of ecological systems. Long periods of debate about the need and scale of action on environmental issues may not be possible when the global economy is twice its current size.

Notes

1. Grossman and Krueger (1995).
2. This convergence in growth rates does not imply, however, that income levels will also converge. Convergence in income levels would imply that policies and social preferences would ultimately be identical across countries, whereas convergence in growth rates only implies that countries have access to the same production technologies.
3. Countries slowly converge to that rate by closing the *growth rate gap* by 2% per year (implying that half the gap is closed in about 35 years).The process of moving toward growth convergence occurs in two stages: *i)* moving from current productivity growth rates to the average for 1980-2001 (this is largely completed by 2015) *ii)* then moving to the 1.75% growth target, by closing the growth rate gap by 2% per year. In other words, a country whose productivity is growing at 5% at the beginning of the convergence process will grow at 4.94% in the following year, 4.87% the next year, and so on.
4. McKibbin and Woo (2002) suggest that China's accession to the World Trade Organization was already a strong enough factor to cause its neighbours to potentially de-industrialise.

References

Bergh, van den, J.C.J.M. and H. Verbruggen (1999), "Spatial Sustainability, Trade and Indicators: An Evaluation of the 'Ecological Footprint'", *Ecological Economics*, Vol. 29, No.1, pp. 63-74.

Bruinsma, J. (2003), *World Agriculture: Towards 2015/2030. An FAO Perspective*, United Nations Food and Agriculture Organization, Rome.

Crafts, N.F.R. (2003), "Quantifying the Contribution of Technological Change to Economic Growth in Different Eras: A Review of the Evidence", *London School of Economics Working Paper* 79/03, pp. 31.

FAO/OECD (2006), *Agricultural Outlook 2006-2015*, UN Food and Agriculture Organization and Organisation for Economic Co-operation and Development, Rome/Paris.

Grossman, G.M. and A.B. Krueger (1995), "Economic Growth and the Environment", *Quarterly Journal of Economics*, Vol. 110, pp. 353-378.

IEA (International Energy Agency) (2004), *World Energy Outlook*, International Energy Agency, Paris.

Maddison, A. (2001), *The World Economy: A Millennial Perspective*, Development Centre Study, OECD, Paris.

McKibbin, W. and W. Woo (2002), "The Consequences of China's WTO Accession on its Neighbours", paper presented at the *Columbia University Asian Economics Panel*, New York, October.

OECD (2003), *Labour-force Participation of Groups at the Margin of the Labour Market: Past and Future Trends and Policy Challenges*, [ECO/CPE/WPI(2003)8], OECD, Paris.

OECD (2007), *Measuring Material Flows and Resource Productivity – The OECD Guide*, OECD, Paris.

Rees, W. (1992), "Ecological Footprints and Appropriated Carrying Capacity: What Urban Economics Leaves Out", *Environment and Urbanisation*, Vol. 4, No. 2.

United Nations (2005), *World Population Prospects: The 2004 Revision*, document ESA/P/WP.193, February, Department of Economic and Social Affairs, Population Division, United Nations, New York.

World Bank (2006), *World Development Indicators*, Washington, DC.

WWF (World Wildlife Fund) (2006), Living Planet Report 2006, WWF International, Gland, Switzerland.

ISBN 978-92-64-04048-9
OECD Environmental Outlook to 2030

Chapter 4

Globalisation

Globalisation is one of the key drivers of economic and environmental change. The interactions between globalisation and the environment occur at different levels, and the impacts can be both positive and negative. The quality of environmental governance at all levels is crucial for realising the potential environmental gains from globalisation. However, current environmental policies and institutions are not keeping pace with economic globalisation, especially in developing countries, and need to be reinforced. Better integration of environmental issues with trade and investment policies is needed. Governments have an important role to play in creating a framework that promotes and supports environmental innovation and the dissemination of more environmentally-friendly technologies in global markets.

KEY MESSAGES

- Globalisation is one of the key drivers of economic change. The interactions between globalisation and the environment occur at different levels, and the impacts can be both positive and negative, depending on the assimilative capacity of the environment, natural resource endowments and governments' capacity to put in place and enforce adequate environmental policies.
- An increasing number of bilateral and regional trade agreements deal with environmental issues, offering new opportunities for making trade and environment objectives mutually supportive. However, these are still relatively few, their treatment of the environment varies and governments and companies have to deal with rapidly evolving and increasingly complex sets of rules. Recent investment agreements also tend to deal with a broader set of issues, including concerns related to health, safety, and the environment, and may thus help create a more sustainable framework for foreign investment.
- Multinational enterprises are key vectors of globalisation. While delocalisation of polluting activities to countries with lower environmental standards may occur in some instances, many multinational enterprises apply high environmental standards to their activities worldwide, thus contributing to the globalisation of better corporate practices. However, recent accidents involving large multi-nationals from OECD countries, and the questionable environmental performance of enterprises from emerging economies, underline the need for continued vigilance.

Environmental implications

The quality of environmental governance at all levels is crucial for realising the potential environmental gains from globalisation. However, current environmental policies and institutions are not keeping pace with economic globalisation, especially in developing countries, and need to be reinforced.

Globalisation is changing the patterns of trade and investment activities, with emerging economies playing an increasing role. As the economic weight of emerging economies continues to grow, their contribution to environmental pressures grows as well.

The number of trade and investment agreements which include commitments to co-operate on environmental matters is increasing, although these are still comparatively few.

Governments in emerging economies and developing countries are becoming increasingly aware of the need to improve their domestic investment frameworks in line with sustainable development objectives and some are starting to better integrate environmental concerns into such frameworks.

Globalisation can contribute to the wider use of environmentally-related technologies.

Policy implications

- Support emerging economies to play a role in maximising the positive environmental benefits of globalisation and minimising its negative impacts. This will require new and strengthened approaches to international environmental co-operation and better integration of environmental issues with trade and investment policies.
- Enable domestic co-ordination between ministries of environment and ministries of industry, and other innovation policy-makers to promote a consistent and effective innovation strategy that also allows environmental innovations to be competitive in global markets. Governments have an important role to play in creating a framework that promotes and supports environmental innovation and the dissemination of more environmentally-friendly technologies in global markets.

Introduction

The term "globalisation" has been widely used to describe a process in which the structures of economic markets, technologies and communication patterns become progressively more international over time. Higher levels of investment, deeper liberalisation of international trade regimes, intensified competition and rapid technological change, including in the area of information technologies, are some of the main drivers of this process. While economic integration is a dominant feature of globalisation, social, cultural, political and institutional aspects are also important. Changes in consumption patterns through growing demands and easier access to goods and services, increased transport and energy needs, global access to innovation and knowledge, all play a role in globalisation – and all have an impact on the environment.

This chapter focuses primarily on the economic aspects of globalisation, which relate particularly to a dynamic and multidimensional process of economic integration whereby national resources become more and more internationally mobile while national economies become increasingly interdependent (OECD, 2005a). The chapter describes those aspects of economic globalisation which have the closest links to environment, and which are primarily manifested through increased trade and investment and the growing role of multinational enterprises in contributing to environmental outcomes. Other aspects of globalisation are dealt with in Chapter 1 (Consumption, production and technology), 7 (Climate change), 14 (Agriculture), 16 (Transport), 17 (Energy) and 22 (Global environmental co-operation).

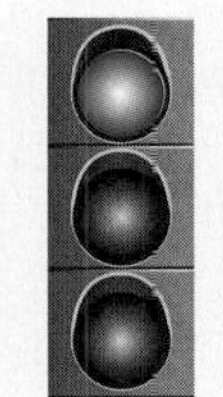

Current environmental policies and institutions are not keeping pace with economic globalisation, especially in developing countries, and need to be reinforced.

The pace and scale of today's globalisation is without precedent. One of its distinctive features is the emergence of large players such as Brazil, Russia, India and China (OECD, 2007a). Another feature is the increasing role of non-state actors, such as multinational enterprises (MNEs) and financial institutions in shaping the global economic agenda. A further aspect of globalisation is that economies become more intertwined and local developments have impacts beyond national boundaries and jurisdictions.

The environment is not confined by national boundaries: there is a single shared atmosphere, ecosystems and watersheds cross national borders, and pollution moves across entire continents and oceans. Countries have recognised that responding to global environmental challenges requires global solutions and international co-operation. Emerging challenges due to economic globalisation, such as the rapidly increasing greenhouse gas emission levels in emerging economies and growing competition for energy and natural resources, as well as the expanding role of non-state actors and increasingly complex interactions between states, present new challenges for environmental governance, including at the global level (Najam *et al.*, 2007; and see Box 4.1).

Box 4.1. **Discussion of globalisation and environment in UNEP**

Environment ministers discussed globalisation and environment at the February 2007 meeting of the United Nations Environment Programme's (UNEP) Governing Council/ Global Ministerial Forum. They recognised that globalisation created and enhanced many opportunities for better promoting sustainable development. At the same time, they agreed that appropriate environmental policies and institutions were required if the opportunities provided by globalisation were to be realised and the risks minimised. There was wide agreement that while the international community had created a variety of bodies to deal with environmental issues, deterioration of natural resources had not been successfully halted or reversed. Unco-ordinated approaches at global, regional and national level, as well as duplication and fragmentation of mandates had exacerbated this situation. Lack of co-ordination was not limited to the UN system, but also involved governments, the private sector and civil society.

The current UN reform process provides an opportunity to discuss how global environmental governance arrangements could be strengthened. However, at this time there is no consensus on how this might be done. Some countries favour the establishment of a "UN Environmental Organisation" to provide better political guidance, legitimacy and effective co-ordination. Others are not convinced that such an organisation is either necessary or desirable, and are instead looking to improve the efficiency and co-ordination of existing arrangements (see Box 22.3 in Chapter 22).

Source: UNEP (2007).

The interactions between globalisation and the environment occur at different levels, and the impacts can be both positive and negative, depending on a variety of factors. These include the assimilative capacity of the environment, natural resource endowments and governments' capacity to put in place and enforce adequate environmental policies. The overall environmental impact of globalisation is difficult to anticipate and will be largely determined by the balance between the efficiency gains and increased pollution and resource consumption associated with more globalised economic activity. The efficiency and effectiveness of environmental and natural resource governance regimes will also be crucial factors.

Globalisation can lead to both positive and negative impacts on the environment.

The impacts of globalisation on the environment will also vary from country to country. For example, increased trade liberalisation can allow for more efficient use of resources in one country, but can also exacerbate resource extraction in other countries. In the case of China, increased imports of timber will relieve pressures on the country's forests; on the other hand, China's huge demand for raw materials is putting more pressure on exporting countries, and can result in overall negative impacts (OECD, 2007b) (see Box 4.2).

Inevitably, the distribution of benefits and environmental pressures will differ, raising issues of equity and social justice. The linkages between globalisation and environment work in both directions: the economic changes brought about by globalisation have an impact on the environment, but changes in environmental conditions and measures also have an impact on the economy.

Box 4.2. Environmental impacts of China's accession to the World Trade Organization

A study conducted by the China Council for International Cooperation on Environment and Development assessed the environmental impacts of China's accession to the WTO in several sectors. For agriculture, the report considered that the impact could be positive if increased trade liberalisation shifted production from products requiring high levels of land, water and chemical inputs to more labour-intensive products. It recommended that this shift should be supported by measures to reduce subsidies for chemical inputs, increase support for advisory services, disseminate information about foreign environmental requirements for agricultural products, and strengthen domestic standards.

Timber imports are projected to increase five-fold from 1995 to 2010, in part to support the production of wood products, notably furniture, for export. While this may have a beneficial impact on Chinese forests, particularly if accompanied by improved forest management, it may also contribute to unsustainable forestry practices in supply countries in Asia and beyond. The report recommended that China should consider reducing escalating tariffs on finished wood products, and strengthen its international co-operation to combat illegal logging and to promote sustainable forestry throughout the entire product chain.

WTO accession has contributed to a sharp rise in aquaculture exports, whose volume currently is roughly equivalent to China's net imports of agricultural products. Environmental problems have been exacerbated by this trend (*e.g.* nutrient and chemical pollution, substrate eutrophication and red tides). However, the report argued that these costs could be outweighed by the economic and environmental benefits if appropriate policies are put in place to ensure high product standards, strengthen control of land-based marine pollution, manage resources effectively to optimise the quality and quantity of products produced, disseminate information, provide technical support, and participate in international activities related to standards for aquaculture.

Source: CCICED (2004).

Globalisation, growth and the environment

Globalisation contributes to accelerated economic growth, particularly through increased trade and investment activity. While this is undoubtedly a positive development, it needs to be accompanied by adequate environmental policies to address the negative impacts of such growth on the environment. Globalisation also stimulates economic development by integrating emerging economies into the global economy. Developed countries have a special responsibility for leadership on environmental and sustainable development issues worldwide, historically and because of the weight that they continue to have in the global economy and the environment. However, as the economic weight of emerging economies continues to grow, their contribution to environmental pressures grows as well, and so does the expectation that they will help to address global environmental challenges.

Globalisation can promote more efficient and less-environmentally damaging patterns of economic development; for example, by helping to concentrate production in countries that have a comparative advantage in energy and natural resource endowments. It can also help to promote the development and diffusion of cleaner technologies. Economic growth and poverty reduction generally also lead to an increased demand for better environmental quality, and the additional wealth can be directed towards environmentally related investments and increasing capacity for environmental protection.

On the other hand, growing economic activity increases overall resource and energy consumption, generates more waste, higher pollution levels, etc. The latter can be due, for example, to extending areas under cultivation for agricultural production destined for export, or increased trade in energy-, materials- or pollution-intensive goods. Subsidies to support such economic activities may reinforce the negative environmental impacts.

Globalisation can also promote structural changes in the patterns of economic activity, including its sectoral distribution. These changes may have positive environmental impacts – *e.g.*, a shift from manufacturing to the service sector – or negative environmental impacts, such as an expansion of energy and material-intensive industries.

Competition and the environment: a race to the bottom or to the top?

A salient feature of globalisation is increased competition. Questions about whether, or how, stringent environmental standards affect the competitiveness of an economy are not new. But globalisation and growing competition with new market entrants have brought the issue again to the fore. At the centre of this debate are the ways in which countries are addressing climate change and how this affects their competitiveness in global markets (see also Chapter 21 on institutions and approaches for policy implementation).

Countries are competing to retain production centres and jobs and attract foreign capital; companies are facing stronger competition both from existing actors and new entrants. Globalisation is also associated with the rapid emergence of global value chains. This is motivated by a number of factors, one of which is to enhance efficiency. The growth of international sourcing has also resulted in the relocation of activities abroad, sometimes involving total or partial closure of production in the home country and the creation of new affiliates abroad (OECD, 2007c; Berger, 2005).

The relocation of industries and globalisation of value chains is often associated with the "pollution haven" hypothesis, according to which industries will relocate to countries where environmental standards are low. A related effect is that of a "regulatory chill" of environmental standards, in order to attract or retain investments or create competitive advantages for exporters. There is a vast literature on the "pollution haven" hypothesis, but there is little actual evidence to support it (OECD, 2002). The prevailing opinion is that the empirical evidence is lacking to support the race to the bottom in response to inter-jurisdictional competition (Porter, 1999).

In fact there are examples of where investment activity has actually helped raise the standards of environmental regulation. This is consistent with the "Porter hypothesis", according to which stronger environmental policies can improve a country's competitiveness by fostering innovation and efficiency (Porter, 1990). One reason is that host governments are becoming more selective about the types of investment they allow, refusing or restricting the relocation of brown industries. Another reason is that many multinational enterprises are applying high environmental standards and management practices to their activities worldwide, and are requiring their sub-contractors to apply similar standards as well (OECD, 2004; 2007d). This also provides a basis for governments to adopt or endorse the standards employed by the "front-runners".

OECD ENVIRONMENTAL OUTLOOK TO 2030 – ISBN 978-92-64-04048-9 – © OECD 2008

Key trends and projections

Trade

International trade is a major driver of growth in the world economy, as trade flows continue to increase. Emerging economies are becoming important players, and their shares in world trade are steadily increasing. Since the early 1990s, South-South trade has expanded at a more rapid rate than either North-North or North-South trade, though starting from a much lower base (OECD, 2006a and b).

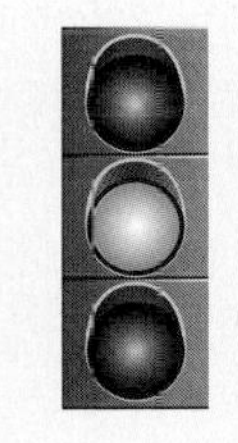

The number of trade and investment agreements which include commitments to co-operate on environmental matters is increasing, although these are still comparatively few.

The United States' economy remains the main engine of global economic growth and international trade, but the growth of world exports of goods and services from China, India and a few other large developing economies such as Brazil is becoming increasingly important (see Figures 4.1 and 4.2). China, for example, absorbed about 6% of world imports in 2005, up from 3.3% in 2000 (WTO Statistics Database, 2007).

Sustained economic development and rising standards of living in China and India have been accompanied by a dramatic increase in Asia's share of world exports and raw material consumption. Russia is likely to continue to benefit from higher prices for oil and other primary commodities such as gas and metals, as well as expansion in domestic demand due to rising real wages and expansionary policies. As the largest and one of the most influential countries in Latin America, Brazil has emerged as a leading voice for developing countries in setting regional and multilateral trade agendas.

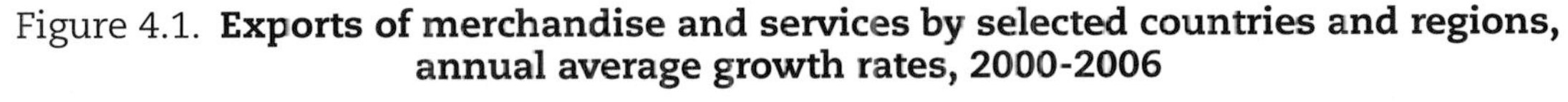

Figure 4.1. **Exports of merchandise and services by selected countries and regions, annual average growth rates, 2000-2006**

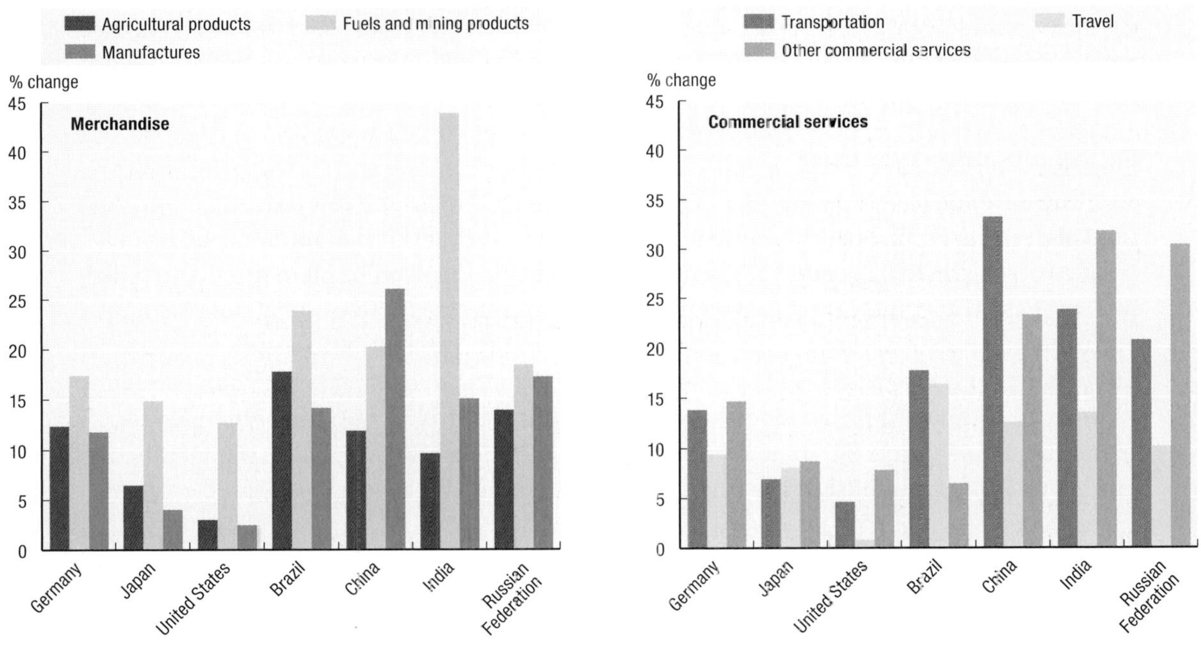

StatLink http://dx.doi.org/10.1787/260262376628

Note: "Other commercial services" include: communication; construction; computer and information services; insurance; financial services; royalties and licence fees; other business services; other personal, cultural and recreational services. Government services are not included.

Source: WTO Statistics Database, 2007.

Figure 4.2. **Total merchandise exports, % of world total per region, 1996 and 2006**

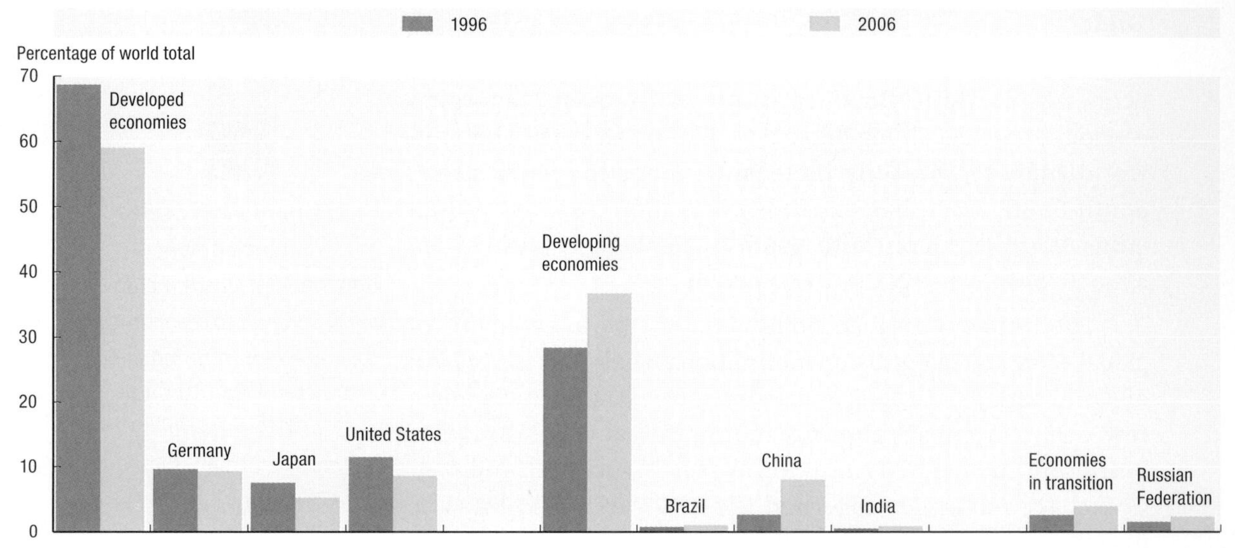

StatLink http://dx.doi.org/10.1787/260283281053

Note: Geographical regions in this figure refer to the UNCTAD classification as follows: Developed economies include OECD countries plus Estonia, Latvia, Lithuania, Malta and Israel. Developing economies: Africa, America (Central and South America and the Caribbean), Asia (eastern, southern, south-eastern and western Asia) and Oceania. Transition countries: Central and Eastern Europe, Caucasus and Central Asia.

Source: UNCTAD Handbook of Statistics on-line, available at *http://stats.unctad.org/*, accessed July 2007.

Since 1980, intra-regional trade has grown in nearly all regions with the exceptions of Central and Eastern Europe, and has steadily accounted for over half of all global trade (UNCTAD, 2007a). The expansion and deeper economic integration of regional trading groups will likely remain a key feature of globalisation towards 2030. The substantial increase in the number of regional (and sub-regional) trade agreements (RTAs) signed over the past 30 years has contributed to intensified trade and has allowed countries to profit from expanding exports. An increasing number of RTAs include environmental provisions (see Box 4.3).

Future projections for trade

The Baseline developed for the *OECD Environmental Outlook* is a reference scenario, and thus it projects recent developments into the future, excluding the adoption of any new policies. As such, policies and agreements that have already been implemented and which will increase trade and investment liberalisation are reflected, but no new policies aimed at further liberalisation are assumed to be adopted in the Baseline. As a result, the Baseline projections for trade to 2030 reflect growth in trade that is increasing faster than economic growth up to about 2015 as existing policies continue to play out, but which levels off thereafter (see Figure 4.3). Thus, without new policies or other trade-inducing factors, the import-to-GDP ratio will stabilise (the ratio is largely unchanged after 2015).

However as this chapter suggests, it is likely that recent trends towards increasing trade and investment will continue in the future, as a result of new or strengthened agreements between countries and liberalisation policies. Chapter 6 presents a key variation on the current Baseline which reflects this continued increase in trade and investment liberalisation – this "globalisation" variation is also shown in Figure 4.3 for comparison.

OECD ENVIRONMENTAL OUTLOOK TO 2030 – ISBN 978-92-64-04048-9 – © OECD 2008

Box 4.3. **Regional trade agreements and the environment**

Multilateral trade rules provide the best guarantee for securing substantive gains from trade liberalisation for all WTO members. Nevertheless, WTO rules also allow the possibility of regional integration and bilateral agreements for members who wish to liberalise at a quicker pace. In this sense, regional trade agreements (RTAs) should be seen as a complement rather than an alternative to multilateral agreements.

Over the last few years, the number of RTAs has significantly increased. While the purpose of many RTAs is to reduce tariffs, a growing number of agreements also deal with other trade-related issues, such as labour and environment. Today, RTAs negotiated by most OECD members include some type of environmental provision.

The scope and depth of environmental provisions in RTAs varies significantly. Among OECD members, Canada, the European Union, New Zealand, and the United States have included the most comprehensive environmental provisions in recent RTAs. The agreements by the United States are unique in that they put trade and environmental issues on an equal footing. Among non-OECD countries, Chile's efforts to include environmental provisions in its trade agreements are particularly noteworthy.

So far, the most ambitious agreements, from an environmental point of view, include a comprehensive environmental chapter, or are accompanied by an environmental side agreement, or both. Some countries consider environmental issues before entering into an agreement, by carrying out a prior assessment of its potential environmental impacts. A few RTAs which did not originally include environmental provisions, have later been complemented by an environmental agreement. This is the case for the MERCOSUR agreement, which has been complemented by a Framework Agreement for Environment.

Environmental elements typically found in many RTAs are environmental co-operation mechanisms. These range from broad arrangements to co-operation in one specific area of special interest to the parties. The areas of co-operation in different RTAs vary significantly, and depend on a range of factors, *e.g.* whether the trade partners have comparable levels of development or not (in which case, co-operation often focuses on capacity building), or whether they have common borders, as is the case between members of the North American Free Trade Agreement (NAFTA).

Environmental standards also figure in a range of agreements, in various forms. The obligation for parties to enforce their own environmental laws is included mainly in agreements involving the United States and Canada. A few RTAs refer more generally to the parties' commitment to maintain high levels of environmental protection. Others, such as those recently negotiated by New Zealand, include references to the inappropriateness of lowering environmental standards. Most RTAs contain clauses reiterating the compatibility between parties' trade obligations and their right to adopt or maintain environmental regulations and standards. Some also include a reference to the compatibility between the agreement and multilateral or regional environmental agreements.

In spite of these developments, the number of RTAs including significant environmental provisions remains small, and some countries, especially developing countries, are reluctant to deal with environmental issues in the context of trade agreements.

Source: Environment and Regional Trade Agreements (OECD, 2007a).

Figure 4.3. **Share of imports in GDP: Baseline and globalisation variation**

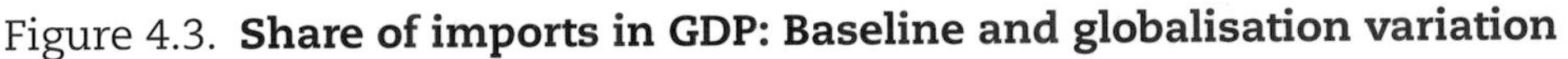

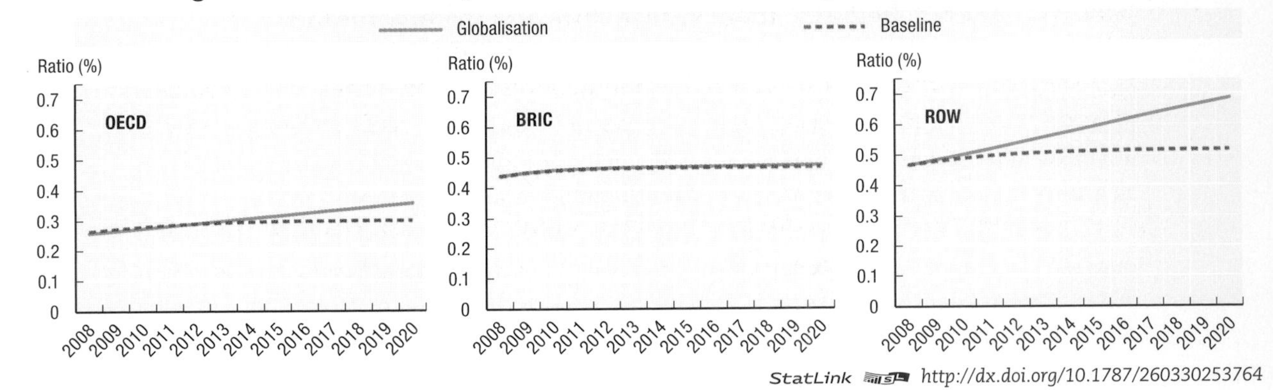

StatLink http://dx.doi.org/10.1787/260330253764

Source: *OECD Environmental Outlook* Baseline and globalisation variation.

Large developed economies like the United States and Japan have a low import-to-GDP ratio because the service sectors are much larger than manufacturing, agriculture and other sectors that produce tradable goods. For an economy like China, which is expanding very rapidly, a projection that trade will keep increasing as a share of GDP would imply that manufacturing-driven growth would continue at an extraordinary pace. For the Baseline, it was conjectured that rapid levelling off will occur in the ratio of imports to GDP for China. Even a levelling off may be optimistic, since it will imply a ratio for China that is more than three times that of other large economies, such as the United States (see also Chapter 3 on economic development).

For the globalisation variation, Figure 4.3 illustrates continued growth in imports in a number of OECD countries, and rapid import growth in the rest of the world (ROW) economies. The very small import growth in BRIC countries reflects the argument that these are large economies that are growing rapidly.

Even with trade agreements that favour particular types of goods, the continued growth of the RoW countries and increasing trade in goods-producing sectors is projected in the Baseline to lead to some movement of polluting industries to those regions. Figure 4.4 shows that nitrogen emissions would be 7% higher in 2030, with similar increases in sulphur and primary energy supply (implying higher CO_2 emissions) in the globalisation variation than under the Baseline.

Figure 4.4. **Environmental implications: Baseline and globalisation variation in 2030**

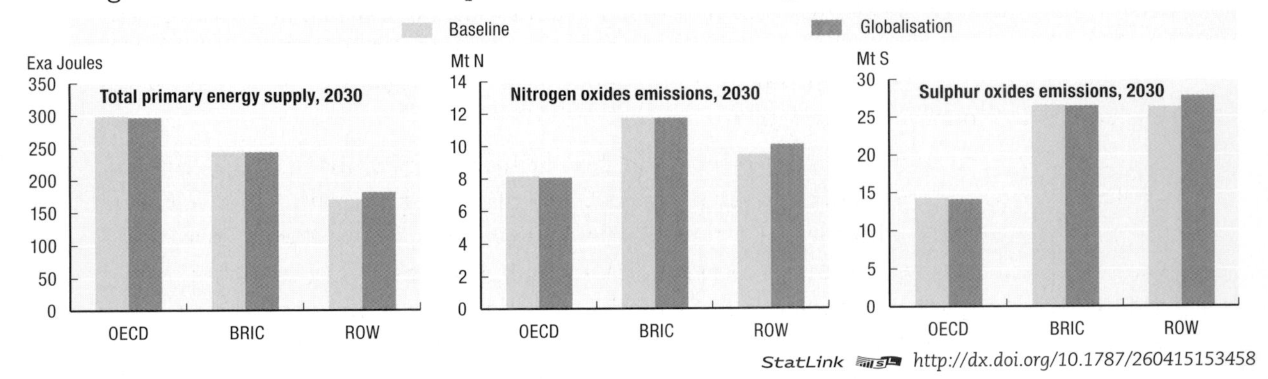

StatLink http://dx.doi.org/10.1787/260415153458

Source: *OECD Environmental Outlook* Baseline and globalisation variation.

Figure 4.5 reflects the projected commercial balance per sector, showing some growth of exports in the manufacturing sector in BRIC countries, and in the energy sector in non-OECD, non-BRIC countries. This confirms the current trend of increased investments in the energy and natural resources sector (mainly oil) in developing countries, described below. Very prominent is the increased exports of services by OECD, with growing imports of manufacturing and energy.

Figure 4.5. **Projected commercial balance by sectors (in million USD), 2005 and 2030**

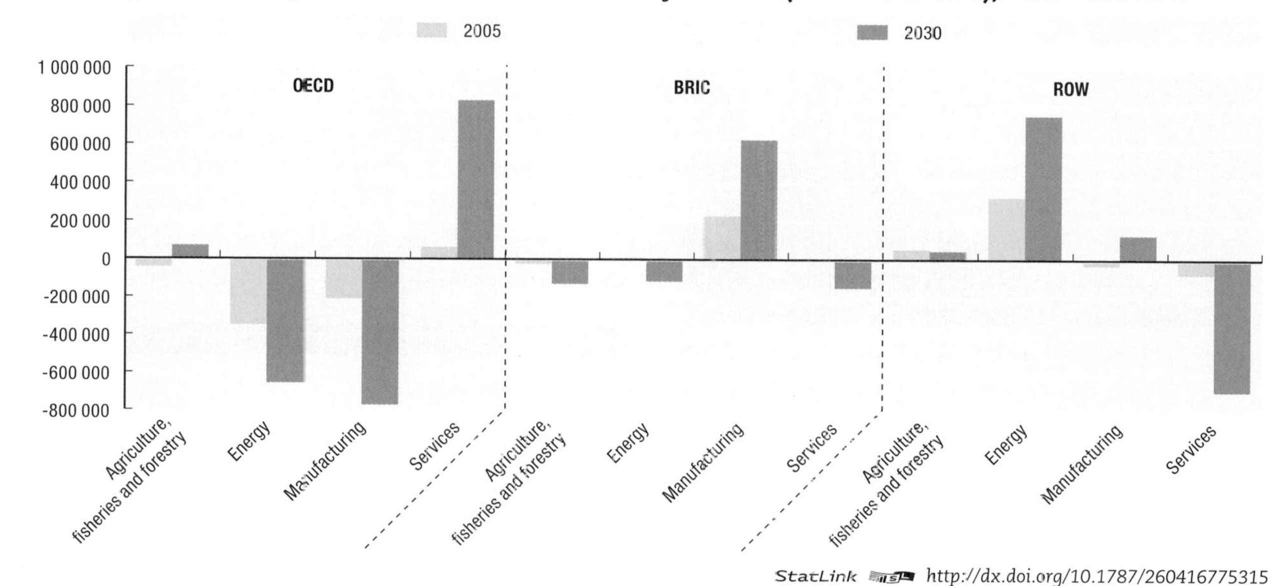

StatLink http://dx.doi.org/10.1787/260416775315

Source: OECD Environmental Outlook Baseline.

International investment

Foreign direct investment (FDI) has been steadily growing. Inflows of FDI grew by 22% in 2006, having already increased by 29% in 2005. Inflows to developed countries in 2006 amounted to USD 801 billion, an increase of 48% over 2005 levels, while those to developing countries reached the highest level ever recorded (for the second time): USD 368 billion. The sharpest rise in FDI was in natural resources, primarily in the petroleum industry (UNCTAD, 2007b).[1]

FDI is increasingly intended to serve global and regional markets, often in the context of international production networks, and the spread of such networks offers, in principle, new possibilities for developing countries and economies in transition to benefit from FDI in the manufacturing sector. In Africa, Latin America and the Caribbean, FDI is still heavily concentrated in the extraction and exploitation of natural resources, with weak linkages to the domestic economy (OECD, 2007a).

Most FDI occurs within the OECD area, and the United States remains the main recipient of FDI, followed by the United Kingdom (OECD, 2007a). FDI remains concentrated in a limited number of countries, with the main non-OECD recipients being China, Russia, Brazil, and India. China has emerged as the largest FDI recipient among all developing countries. South-South FDI has expanded particularly fast over the past 15 years and there has recently been a resurgence of FDI flows to Africa and Latin America, driven by prospects for greater earnings in the extractive industries (UNCTAD, 2007b). Figure 4.6 shows the changes in FDI inflows and outflows in the BRIC and selected OECD countries between 1985 and 2006.

Figure 4.6. **Foreign Direct Investment flows by selected regions and countries, 1985 2006 (in billion USD)**

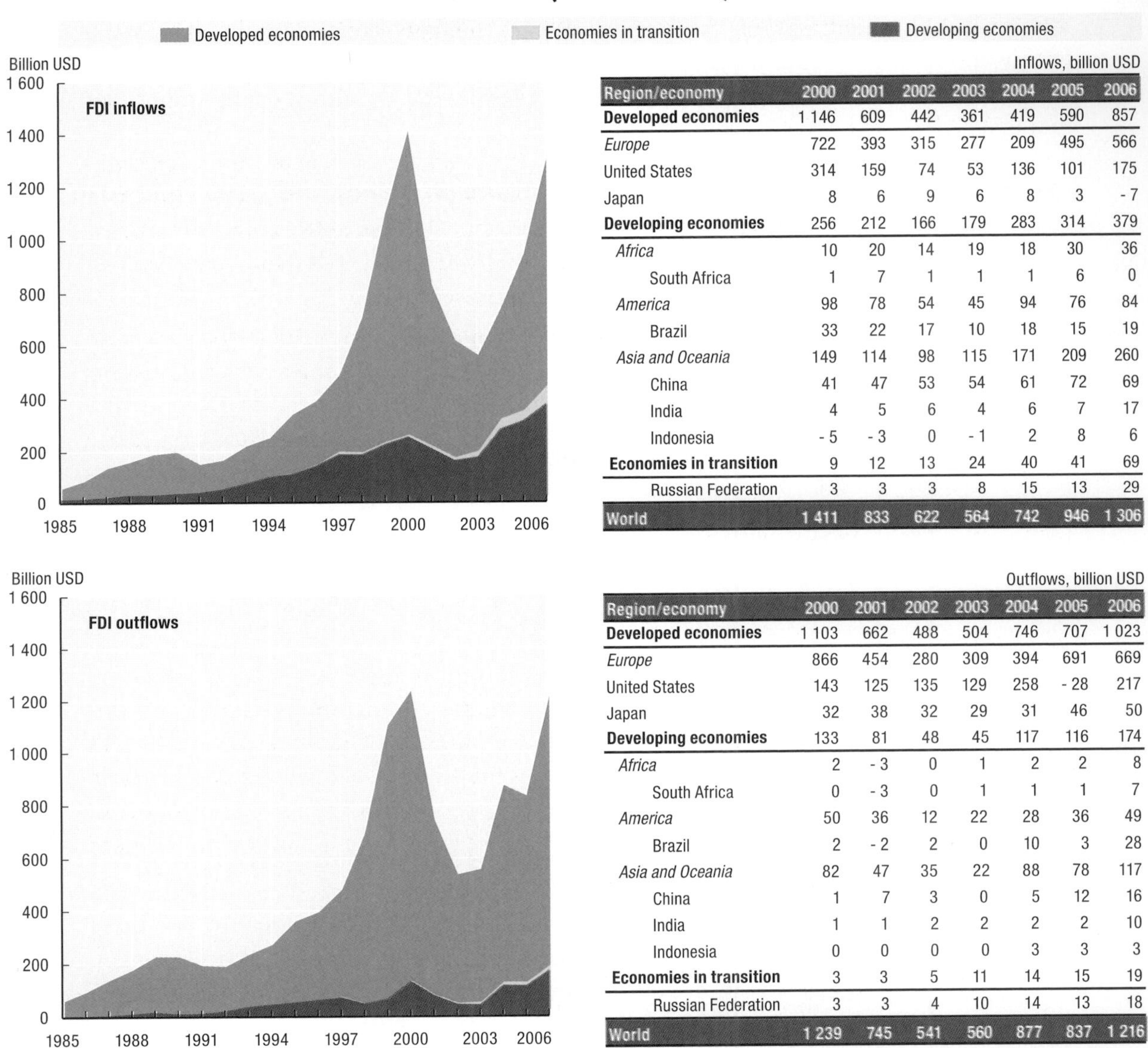

Inflows, billion USD

Region/economy	2000	2001	2002	2003	2004	2005	2006
Developed economies	1 146	609	442	361	419	590	857
Europe	722	393	315	277	209	495	566
United States	314	159	74	53	136	101	175
Japan	8	6	9	6	8	3	- 7
Developing economies	256	212	166	179	283	314	379
Africa	10	20	14	19	18	30	36
South Africa	1	7	1	1	1	6	0
America	98	78	54	45	94	76	84
Brazil	33	22	17	10	18	15	19
Asia and Oceania	149	114	98	115	171	209	260
China	41	47	53	54	61	72	69
India	4	5	6	4	6	7	17
Indonesia	- 5	- 3	0	- 1	2	8	6
Economies in transition	9	12	13	24	40	41	69
Russian Federation	3	3	3	8	15	13	29
World	1 411	833	622	564	742	946	1 306

Outflows, billion USD

Region/economy	2000	2001	2002	2003	2004	2005	2006
Developed economies	1 103	662	488	504	746	707	1 023
Europe	866	454	280	309	394	691	669
United States	143	125	135	129	258	- 28	217
Japan	32	38	32	29	31	46	50
Developing economies	133	81	48	45	117	116	174
Africa	2	- 3	0	1	2	2	8
South Africa	0	- 3	0	1	1	1	7
America	50	36	12	22	28	36	49
Brazil	2	- 2	2	0	10	3	28
Asia and Oceania	82	47	35	22	88	78	117
China	1	7	3	0	5	12	16
India	1	1	2	2	2	2	10
Indonesia	0	0	0	0	3	3	3
Economies in transition	3	3	5	11	14	15	19
Russian Federation	3	3	4	10	14	13	18
World	1 239	745	541	560	877	837	1 216

StatLink http://dx.doi.org/10.1787/260426612155

Source: UNCTAD, FDI online database, 2007, available at *http://stats.unctad.org/*, accessed November 2007.

Very large projected outflows of capital from China to the developing world, particularly Africa, are raising concerns regarding competition for scarce energy resources, and over possible undermining of internationally-recognised standards of corporate conduct (OECD, 2006a). One of the recommendations in the *OECD Environmental Performance Review of China* was that the Chinese government should provide more oversight of the environmental performance of Chinese enterprises, perhaps using the *OECD Guidelines for Multinational Enterprises* (see below) (OECD, 2004).

It is mainly governments' responsibility to ensure that investments contribute to sustainable development by, *inter alia*, ensuring that their adverse environmental effects are adequately addressed and environmental regulations enforced. Recent investment agreements tend to deal with a broader set of issues, including concerns related to health, safety and the environment, and may thus contribute to creating a more sustainable framework for foreign investment. On the other hand, this also means that governments and companies have to deal with a rapidly evolving and increasingly complex set of rules (OECD, 2007f).

The role of multinational enterprises

Multinational enterprises (MNE), both from OECD countries and increasingly also from BRIC countries, have become key actors in the globalisation process. While the capacity of governments to regulate remains broadly within national borders, MNEs operate in many countries of the world. Therefore, corporate environmental behaviour has become increasingly central in the globalisation-environment relationship.

In the past, business tended to regard environmental issues as a challenge or even an obstacle to good economic performance, and in many cases companies preferred risking paying fines for breach of environmental regulations than improving their environmental performance to comply with such regulations. Though this may still occur today, many business leaders perceive good environmental performance as a business opportunity, and increasingly integrate environmental mechanisms into normal management practice. Other factors contributing to this trend are increasingly stringent environmental regulations and enforcement mechanisms, as well as price signals and growing demand from civil society, consumers, shareholders and financial institutions for better environmental performance (OECD, 2004).

Companies are also increasingly taking a pro-active approach to environmental problems, including global problems addressed by multilateral environmental agreements by, for example, engaging in research and development of more energy-efficient production methods, or through market approaches that support biodiversity conservation (OECD, 2005d; 2007f). Leading companies are also recognising the business opportunities of environmental challenges, and are seeing a competitive advantage in moving ahead of changes called for by government regulation and, in some cases, ahead of customer demand. For example, many companies are investing in renewable energy technologies, such as solar and wind energy, and automobile companies are trying to capitalise on the growing demand for more fuel-efficient vehicles through the introduction of hybrid cars (MEA, 2005; OECD, 2007d). Globalisation is also providing new opportunities for innovative companies to access new markets (Box 4.4).

A "green" corporate image and reputation have become key assets for many companies, and many apply the same high environmental standards and practices worldwide in all their plants, thus contributing to globalisation of good environmental corporate practices. Financial institutions, such as development banks, private financial institutions and export credit agencies, as well as rating agencies, increasingly take into account the social and environmental impacts of corporations and of the negative effects of environmental liabilities on stock value (OECD, 2005c). Among the instruments likely to shape international financial activities are the recently revised International Finance Corporation's Performance Standards on Social and Environmental Sustainability, the Equator Principles adopted by a range of banks, and the OECD Recommendation on Environment and Export Credits, adopted in 2003 and revised in 2007. A number of financial indices, such as the FTSE4Good or NASDAQ Clean Edge US Index, have been set up to track the environmental and social performances of publicly traded companies for investors.

Box 4.4. **Environmental innovation and global markets**

An important new development in globalisation is that business R&D strategies are becoming increasingly internationalised. This partly manifests itself through outsourcing and relocation of R&D activities, especially development activities that allow companies to access global talent pools; globalisation of R&D through supply chains and new approaches to partnerships and co-operation.

Growing international markets for environmentally-related technologies provide a further incentive for governments and firms to re-visit their policies in this area. Recent data about the size of this market reveal that large-scale opportunities exist for exporters of environmental goods and technologies. A study by the European Commission estimated the turnover of eco-industries in the EU at EUR 227 billion in 2004, with a growth rate of 7% between 1999 and 2004 (EC, and Ernst and Young, 2006). Globalisation is creating wider markets for environmental technology, and many companies are expanding their operations – including environmentally-related R&D and innovation – to new markets. Much of the expansion of this global market of environmental technologies is expected to occur in emerging countries, especially in China, India and Brazil.

Government policies and regulation continue to be key drivers of environmental innovation, though other factors are gaining importance, including the market opportunities in environment-related sectors. Domestic co-ordination between ministries of environment and ministries of industry and other innovation policy-makers is necessary to promote a consistent and effective innovation strategy that also allows environmental innovations to be competitive in global markets. Some governments are internationalising their national environmental innovation policies in order to scale up the deployment of environmental technologies. Finland, Denmark and Spain, for example, are actively promoting exports of environmental goods and services, and are encouraging, and supporting, domestic firms to become "global exporters".

Adequate enforcement is crucial to create a level playing field in the marketplace: regulatory requirements drive environmental innovation, but they need to apply to all participants. Insufficient enforcement of environmental regulation in one country creates undue advantages for producers and importers who do not comply with the regulation and which have fewer concerns about their reputation. On the other hand, weak enforcement may not provide the incentives that domestic firms need in order to develop internationally-competitive environmentally-related innovations.

Source: OECD (2007e).

In addition, numerous international codes address corporate social responsibility, such as the UN Compact, the Global Reporting Initiative and the OECD's *Guidelines for Multinational Enterprises*, adopted in 1976 and revised in 2000. The OECD guidelines are a set of voluntary recommendations to multinational enterprises in all the major areas of business ethics, including employment and industrial relations; human rights; environment; information disclosure; combating bribery; consumer interests, science and technology; competition; and taxation.The guidelines are global in nature, since they seek to guide companies' behaviour wherever they operate, both in the home country and in host countries.[2]

Governments in emerging economies and developing countries are increasingly starting to integrate environmental concerns into their domestic investment frameworks.

The guidelines' chapter on environment recommends that enterprises establish and maintain an adequate environmental management system, assess and address the foreseeable environmental impacts associated with their products and processes, apply a precautionary approach, and maintain contingency plans for environmental emergencies. They also encourage companies to publish relevant environmental information and engage in adequate communication and consultation with the public and the communities directly affected by their activities. At the national level, many governments have also taken initiatives to promote enhanced environmental performance by companies both at home and abroad, for example, by requiring the publication of annual environmental or sustainability reports (OECD, 2004).

While these trends are encouraging, recent accidents involving large multinationals from OECD countries, and the questionable environmental performance of enterprises from emerging economies, underlines the need for continued vigilance and co-operation between governments and business to strive towards continuous improvement of environmental performance.

Policy implications

Globalisation stimulates economic growth. Ensuring that environmental policies and institutions – at all levels and especially in developing countries – keep pace with economic globalisation and that the benefits of globalisation are equitably distributed are major challenges for governments and society as a whole (OECD, 2005d). The successful conclusion of the Doha Round would be an important step in meeting these challenges (Box 4.5). More efforts are also needed, both in developed and developing countries, to ensure coherence between trade, investment and environment policies in order to take full advantage of growing market opportunities for environmental goods, services and technologies.

Box 4.5. **Ensuring developing countries benefit from trade liberalisation**

Countries have recognised the importance of trade and investment for economic growth in developing countries, and the need to actively support these countries' efforts to access related financial flows. A range of recent OECD studies has confirmed that trade liberalisation has the potential to contribute to improved economic welfare. Implementation of international commitments such as the Monterrey Consensus, the Doha Development Agenda, the World Summit on Sustainable Development's Plan of Implementation and the Millennium Development Goals, which include enhanced market access for developing countries' exports, increased foreign investment in developing countries and emerging economies, and better targeted official development assistance (ODA), will be crucial to prevent a large part of the world from being excluded from the benefits of globalisation.

In the Doha Development Agenda, adopted in 2001, ministers emphasised that international trade can play a major role in the promotion of economic development and the alleviation of poverty. They recognised the need for all countries to benefit from the increased opportunities and welfare gains that the multilateral trading system generates, and noted the particular vulnerability of the least developed countries and the special structural difficulties they face in the global economy. Ministers committed to comprehensive negotiations on agricultural trade aimed at substantial improvements in market access; reductions of, with a view to phasing out, all forms of export subsidies; and substantial reductions in trade-distorting domestic support.

The Doha Development Agenda also provided for an opportunity for negotiation aiming at making development, trade and environment more mutually supportive. Ministers agreed to negotiations that aim to reduce or, as appropriate, eliminate tariffs, as well as non-tariff barriers, in particular on products of export interest to developing countries and on environmental goods and services. Ministers also agreed to consider the effect of environmental measures on market access, especially in relation to developing countries, and those situations in which the elimination or reduction of trade restrictions and distortions would benefit trade, the environment and development.

Source: WTO, 2001; OECD, 2006c; Gurría, 2006.

Countries are actively engaging in bilateral and regional trade and investment agreements. A positive development is that the overall quality of such agreements is improving though the inclusion of environment and sustainable development considerations. Lessons learnt in the negotiation and implementation of those agreements could be used to enhance the multilateral trading system and create sound international investment frameworks which support sustainable development (OECD, 2007e and f).

Energy security and competition for scarce natural resources will be important factors influencing trade and investment patterns in the coming years. These factors pose challenges to governments, not least in terms of international environmental governance, but they also provide opportunities for new technology development and deployment. Globalisation can contribute to the wider use of environmentally-related technologies. Governments have an important role to play in ensuring adequate framework conditions for environmental innovation and their dissemination in global markets. Mechanisms to create and expand markets will help to further promote innovation and deployment of environmentally-related technologies, including those related to renewable energies and energy efficiency. This also involves developing new mechanisms for co-operation between governments and business that provide strong incentives for innovation and continuous improvement of environmental performance (OECD, 2007d).

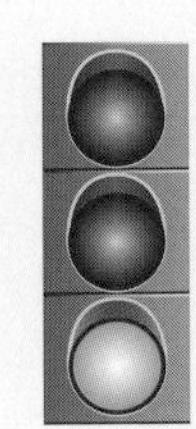

Globalisation can contribute to the development and wider use of environmentally-related technologies.

While markets are becoming increasingly global, environmental requirements are still set at the national or regional level. On the one hand, policy experimentation can help to identify more efficient and effective environmental policies. On the other hand, diverging requirements may create barriers to the development and diffusion of environmentally-related technologies. Finding the right balance between the expansion of trade and investment in global markets, while maintaining countries' sovereign right to set high environmental requirements, will require further efforts, both at national and international levels (OECD, 2005b).

Fair competition requires that the same rules apply to all players, and this is also true in global markets. Governments need to devise appropriate mechanisms to ensure a level playing field, including effective enforcement of relevant environmental regulation and implementation of commitments under multilateral environmental agreements, and of environmental provisions in trade and investment agreements (OECD, 2007d, e and f).

Notes

1. UNCTAD data for 2006 are preliminary estimates. Figure 4.6 only contains data up to 2005.
2. As of July 2007, all 30 OECD members, as well as Argentina, Brazil, Chile, Egypt, Estonia, Israel, Latvia, Lithuania, Romania and Slovenia had adhered to the OECD guidelines.

References

Afsah, S., B. Laplante and D. Wheeler (1996), "Controlling Industrial Pollution: A New Policy Paradigm", *Research Working Paper* No. 1672, World Bank, Washington DC.

Berger, S. (2005), *How We Compete. What Companies Around the World Are Doing to Make it in Today's Global Economy*, MIT Industrial Performance Center, Boston

CCICED (China Council for International Cooperation on Environment and Development) (2004), *An Environmental Impact Assessment of China's WTO Accession: An Analysis of Six Sectors*, CCICED, Beijing.

EC and Ernst and Young (2006), *Eco-Industry, Its Size, Employment Perspectives and Barriers to Growth in an Enlarged EU*, EC, Brussels.

Gurría, A. (2006), "Doha, the Low Hanging Fruit", in *OECD Observer*, 21 August 2006, OECD, Paris.

Jones, T. (2005), "Trade and Investment: Selected Links to Domestic Environmental Policy", in: Wijen, F., Zoetemann, K. and Pieters, J. (eds.), *A Handbook of Globalisation and Environmental Policy*, Edward Elgar Publishing, Cheltenham, UK.

MEA (Millennium Ecosystem Assessment) (2005), *Ecosystems and Human Well-being. Opportunities and Challenges for Business and Industry*, Island Press, Washington DC.

Najam A., D. Runnals and M. Halle (2007), *Environment and Globalization. Five Propositions*, International Institute for Sustainable Development, Winnipeg, Canada.

OECD (1997), *Economic Globalisation and Environment*, OECD, Paris.

OECD (2002), *Environmental Issues in Policy-Based Competition for Investment: A Literature Review*, OECD, Paris.

OECD (2004), *Environment and the OECD Guidelines for Multinational Enterprises*, OECD, Paris.

OECD (2005a), *Measuring Globalisation. Economic Globalisation Indicators*, OECD, Paris.

OECD (2005b), *Handbook on Economic Globalisation Indicators*, OECD, Paris.

OECD (2005c), *Development, Investment and Environment: In Search of Synergies*, OECD, Paris.

OECD (2005d), *Multilateral Environmental Agreements and Private Investment: Business Contribution to Addressing Global Environmental Problems*, OECD, Paris

OECD (2005e), *Trade that Benefits the Environment and Development*, OECD, Paris.

OECD (2006a), *South-South Trade in Goods*, OECD, Paris.

OECD (2006b), *Trends and Recent Developments in Foreign Direct Investment*, OECD, Paris.

OECD (2006c), *Trading Up: Economic Perspectives on Development Issues in the Multilateral Trading System*, OECD, Paris.

OECD (2007a), *Environment and Regional Trade Agreements*, OECD, Paris.

OECD (2007b), *Environmental Innovation in China: Three Case Studies*, OECD, Paris, forthcoming.

OECD (2007c), *International Investment Agreements: Survey of Environment, Labour and Anti-corruption Issues*, OECD, Paris, forthcoming.

OECD (2007d), *Trends and Recent Developments in Foreign Direct Investment*, OECD, Paris.

OECD (2007e), *Environmentally-related Innovation and Global Markets*, OECD, Paris, forthcoming.

OECD (2007f), *Possible Contribution of the Private Sector to MEAs: Suggestions for Further Action*, OECD, Paris, forthcoming.

Porter, M. (1990), *The Competitive Advantage of Nations*, Free Press, New York.

Porter. G. (1999), "Trade Competition and Pollution Standards: 'Race to the Bottom' or 'Stuck at the Bottom'?", 8:2 *Journal of Environment and Development* 133-151.

UNCTAD (United Nations Conference on Trade and Development) (2007a), *Globalization for Development: Opportunities and Challenges*, Report of the Secretary-General of UNCTAD to UNCTAD XII, UNCTAD, Geneva.

UNCTAD (2007b), *World Investment Report 2007*, UNCTAD, Geneva.

UNEP (United Nations Environment Programme) (2007), *President's Summary of Discussions by Ministers and Heads of Delegation at the Twenty-fourth Session of the Governing Council/Global Ministerial Environment Forum, UNEP/GC/24/L5*, UNEP, Nairobi.

WTO (World Trade Organization) (2001), *The Doha Development Agenda*, WTO, Geneva, *www.wto.org/*.

ISBN 978-92-64-04048-9
OECD Environmental Outlook to 2030

Chapter 5

Urbanisation

An estimated 60% of the world's population will live in urban areas in 2030. Urban populations will expand particularly rapidly in developing countries, where the infrastructure needed to support human health and the environment – e.g. water supply, sewage systems, waste collection – is often not in place. A continuing trend towards urban sprawl, particularly in OECD countries, will put pressure on the environment in the coming decades through land use stress, fragmentation of natural habitats, long-term soil degradation and increases in transport-related greenhouse gas and air pollution emissions. A holistic approach is needed to integrate urban design with spatial planning, social objectives, transport policy and other environmental policies (e.g. waste, energy, water). The diversity of urban areas – in terms of history, geography, climate, administrative and legal conditions – calls for urban policies to be locally developed and tailor-made.

KEY MESSAGES

- The urban area expanded by 171% worldwide between 1950 and 2000, and some studies suggest that it may increase by another 150% to 2030.
- Nearly half the world's population now lives in urban areas, and this proportion is expected to grow to 60% by 2030. About 89% of the total projected urban population growth of 1.8 billion people from 2005 to 2030 will occur in non-OECD countries.

Environmental implications

Continuing urban sprawl will put pressure on the environment through land use stress, fragmentation of natural habitats, long-term soil degradation, and increases in greenhouse gas and air pollution emissions.

Developing countries often lack the necessary urban infrastructure to support human health and the environment – such as water supply and sanitation connections, sewerage and sewage treatment, waste collection and management systems, and public transport networks.

Cities also provide opportunities to improve the quality of urban life. From the perspective of sustainable development, compact cities can make more efficient use of natural resources and service provision by concentrating people and economic activities in a limited area. Economies of scale can minimise the adverse effects of consumption and production patterns on the environment.

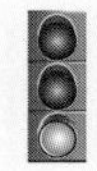

Most OECD cities have made significant progress in reducing their local environmental impacts (*e.g.* urban air and water pollution) through improved wastewater treatment, stricter vehicle emission controls and better public transport provision. Such continuing efforts will be critical to retain the sustainability of city areas.

Policy implications

- Ensure a holistic and long-term approach to integrate urban design with spatial planning, social objectives, transport policy, and other environmental policies (*e.g.* waste, energy, water); better governance and the harmonisation of policy tools will be central for such cross-sectoral integration.
- Implement appropriate financial incentives and building codes to support cost-effective greenhouse gas emission reductions from the building sector. This is particularly important for new building developments, as these buildings may be in place for decades to come.

If the growth in residential building development in China continues at the current rate, about 13 billion m^2 more floor space will be constructed over the next two decades – equivalent to the total building stock currently in place in the EU15 countries. There is an important window of opportunity now to adopt cost-effective energy efficiency measures that will keep the energy demands and greenhouse gas emissions from these new buildings low for their lifetimes.

World population, total, urban and rural, 1950-2030

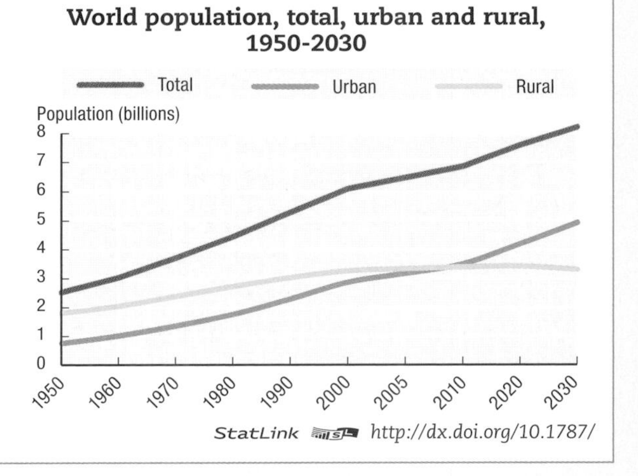

StatLink http://dx.doi.org/10.1787/

Consequences of inaction

Cities concentrate the impacts of human activities – resource use, pollution, and waste – into a small area, and thus often exceed the local capacity of the environment to provide such resources and to absorb the pollution generated. These are not only environmental concerns but also affect the health and well-being of citizens and economic viability. The current unprecedented rate of urbanisation poses formidable environmental, economic and social challenges within individual countries as well as for the world community. Urban environmental problems are now a pivotal issue, and how they are managed has a direct impact on the quality of life for urban dwellers and the achievement of sustainable development locally, regionally and globally.

 OECD ENVIRONMENTAL OUTLOOK TO 2030 – ISBN 978-92-64-04048-9 – © OECD 2008

Introduction

Approximately 49% of the world's population lives in urban areas. It is projected that this will continue to increase in the coming decades to reach about 60% of the population in 2030. For OECD countries, urban populations already exceeded 76% of the total population in 2005, and they are expected to increase to 82% by 2030. Nearly two out of three people globally – and more than four out of five people in OECD countries – will live in city areas by 2030.

Cities[1] provide job opportunities, access to social and environmental services such as education and healthcare and cultural activities. Many cities contribute to a large share of the country's GDP relative to their population and land area (see Table 5.1). Cities also play a key role as transport hubs.

Table 5.1. **Land, population and GDP of selected cities as a share of the country total**

City	Brussels	Budapest	Lisbon	Mexico City	New York	Paris	Seoul	Sydney
Percent of land	2.3	0.8	3.2	0.1	0.1	0.5	0.6	0.02
Percent of population	10.0	25.3	26.3	23.9	7.8	21.2	25.0	24.4
Percent of GDP	44.4	45.6	38.0	26.7	8.5	27.9	48.6	23.5

StatLink http://dx.doi.org/10.1787/256674851347

Note: These data should be interpreted carefully. Due to data availability, data sources for each factor are different. There could be a significant discrepancy between data sources regarding the boundaries of cities, except for Lisbon whose data was provided by the Portuguese National Institute of Statistics (population of 2005, GDP of 2003).
Source: Land: Klein Goldewijk and Van Drecht, 2006; population: UN, 2006; and GDP: OECD, 2006.

Cities can be an efficient living situation from the perspective of sustainable development. The high concentration of people and their related activities can bring economies of scale in providing urban services while minimising some of the adverse effects of consumption and production patterns on the environment and human health. By concentrating people and economic activity in a relatively small area, cities reduce transport distances and often provide more efficient public transport systems – this then reduces transport-related fuel use, air pollution and greenhouse gas emissions. Many cities have a high proportion of the population in relatively compact apartments, without gardens. This can reduce energy and water consumption per person, as well as allowing for the more efficient provision of environmental services such as water and sanitation, and waste collection and recycling.

Most OECD cities have made significant progress in reducing their environmental impacts through improved wastewater treatment, stricter vehicle emission controls and better transport.

On the other hand, dense populations in small areas can simply concentrate some environmental problems such as poor local air quality, high levels of waste generation and pollution emissions, poor quality of urban water bodies, traffic congestion and noise pollution. Because of the concentrated levels of demand for environmental services (*e.g.* water) and the concentrated levels of pollutants, cities may exceed the capacity of the local environment to provide these services or absorb the pollution.

Many OECD countries have made significant progress in dealing with a number of these environmental pressures, for example, increased coverage and level of wastewater treatment, stricter vehicle emission controls, improved public transport, etc. Despite this, OECD countries are still facing tremendous environmental challenges in terms of protection of the natural environment, efficient use of natural resources and the quality of life. Many OECD cities are still suffering poor air quality and some of them are struggling with urban waste (see Chapter 8 on air pollution and Chapter 11 on waste and material flows). In addition, the current trend toward rapid expansion of urban areas, or "urban sprawl", is regarded as one of the major pressures on the urban environment. Sprawling cities consume larger amounts of arable land, require more transport and transport-related infrastructure, and demand more energy. This results in land use stress, fragmentation of natural habitats, increased greenhouse gas emissions, and long-term soil degradation. Infrastructure development is one of the main pressures leading to biodiversity loss, and will be responsible for the largest increase in pressure on biodiversity to 2030 under the *OECD Environmental Outlook* Baseline (see Chapter 9 on biodiversity).

These urban problems are not only environmental concerns, but also raise human health and well-being concerns, such as high levels of vehicle emissions, poor housing and a lack of good quality green space (RCEP, 2007; and see Chapter 12 on health and environment). They also have economic and social impacts, as well as causing economic segregation and undermining social cohesion (Savitch, 2003; Greenberg *et al.*, 2001). Poverty is also closely linked with environmental degradation and environmental justice, although it is not exclusively an urban issue. The growth of cities, particularly in developing countries, has been accompanied by an increase in urban poverty for certain social groups and in particular locations (UNEP, 2002).

Historically, cities have generally evolved in a cycle of transition from urbanisation to suburbanisation and, more recently, to re-urbanisation. In general, most OECD countries are in the phase of suburbanisation or re-urbanisation, while developing countries are mostly in the phase of urbanisation. The focus, intensity and scale of environmental problems that each city faces will vary, in part depending on where they are located within this cycle of urbanisation.

The current unprecedented rate of urbanisation poses formidable environmental, economic and social challenges within individual countries as well as for the world community. Urban environmental problems are now a pivotal issue, and how urban environmental problems are managed has a direct impact on the quality of life for urban dwellers and the achievement of sustainable development locally, regionally and globally.

Key trends and projections

Growing urbanisation

The 20th century saw a tremendous increase in urban population (Figure 5.1). In 2005, there were 3.2 billion urban residents in the world, nearly four times as many as in 1950. World urban population has continued to grow faster than the world population, increasing at an average annual rate of 2.7% between 1950 and 2005, compared to an average annual world population growth rate of 1.7% (see Chapter 2 on population dynamics and demographics).

OECD ENVIRONMENTAL OUTLOOK TO 2030 – ISBN 978-92-64-04048-9 – © OECD 2008

Figure 5.1. **World population – total, urban and rural, 1950-2030**

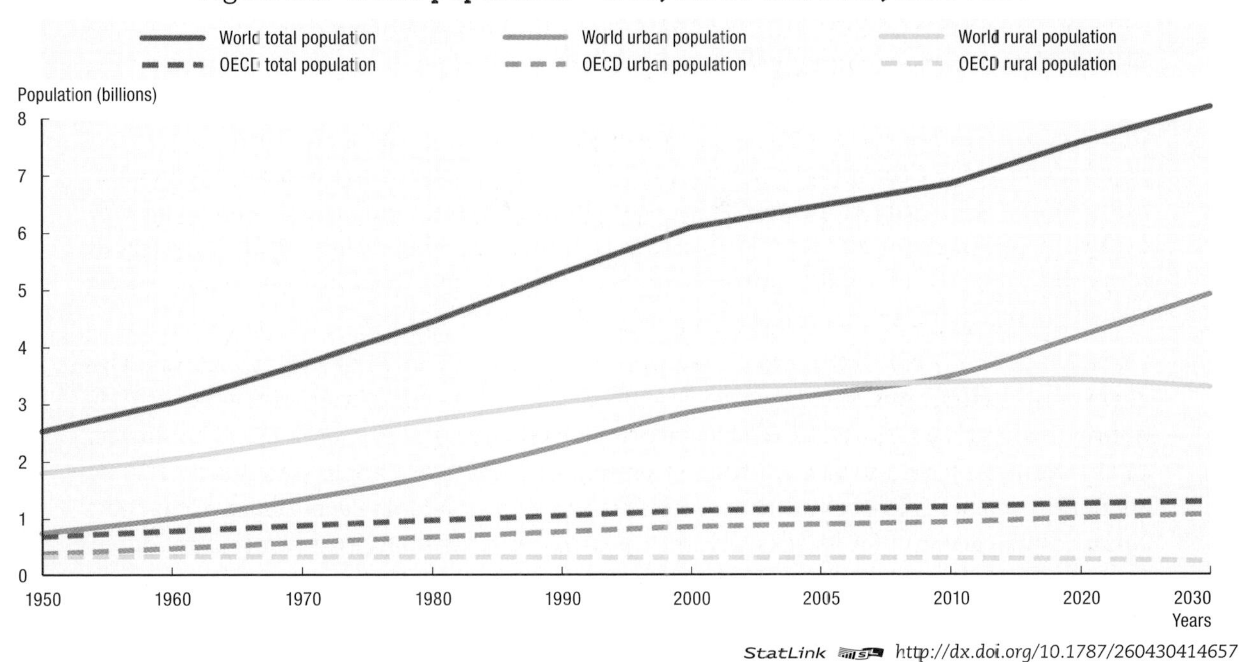

StatLink http://dx.doi.org/10.1787/260430414657

Source: United Nations, 2006

This trend is expected to continue to 2030 (Table 5.2). The majority of the projected population growth will occur in urban areas. World urban population is projected to rise by 1.8 billion between 2005 and 2030, while world population is estimated to grow by 1.7 billion. The absolute growth in total population will be lower than that of the urban population because of the continuing shift in populations from rural to urban areas.

Table 5.2. **World and urban populations, 1950-2030**

	Population (billions)					Average annual rate of change (%)	
County groups	1950	1975	2000	2005	2030	1950-2005	2005-2030
Total population							
World	2.52	4.07	6.09	6.46	8.20	1.73	0.96
OECD	0.68	0.92	1.14	1.17	1.30	0.99	0.40
BRIC	1.07	1.79	2.61	2.75	3.26	1.73	0.68
The ROW	0.77	1.36	2.34	2.54	3.64	2.20	1.45
Urban population							
World	0.73	1.52	2.84	3.15	4.91	2.69	1.79
OECD	0.37	0.62	0.84	0.88	1.07	1.59	0.75
BRIC	0.20	0.45	0.99	1.11	1.77	3.18	1.89
The ROW	0.16	0.45	1.01	1.16	2.07	2.84	2.37

StatLink http://dx.doi.org/10.1787/256727380435

Note: BRIC contains Brazil, China, India and the Russian Federation, and the ROW (Rest of world) indicates all other countries except for OECD and BRIC countries.
Source: United Nations, 2006.

On average, 76% of OECD country populations, or 0.9 billion people, lived in urban areas in 2005. This ranged from 97% in Belgium to 58% in Portugal and 56% in Slovakia. The absolute number of people in urban centres in OECD countries is expected to continue to rise, increasing to 82% of the total population by 2030. But, the overall pace and scale will slow down. Between 2005 and 2030, the average annual growth rate of urban population in OECD countries is expected to be 0.75%, about half of the annual increase rate of 1.59% experienced for the period 1950-2005.

Other than a few exceptions, therefore, the overall urban population increase in most OECD countries between 2005 and 2030 will remain less than 3%. Only the urban populations in the United States, Mexico and Turkey are expected to increase significantly faster at 40%, 16% and 13% respectively. Together, these three countries will represent 69% of future urban population growth of OECD countries. This substantial growth will primarily be fuelled by rural-to-urban migration, regional immigration, and the increasing size of greater metropolitan areas (UN, 2006).

Most of the urban population growth to 2030 will occur in non-OECD countries. The average annual growth rate of urban populations in non-OECD countries will be 2.1% during the period 2005 to 2030, which is more than twice as fast as that in OECD countries. Of the total projected urban population growth of 1.8 billion people from 2005 to 2030, about 89% will occur in non-OECD countries. The rapidly developing economies of the BRIC countries will account for 30% of this urban population growth. By 2030, almost four out of five urban dwellers will be in non-OECD countries.

Most of the world's largest cities will also be located in less developed countries. According to a 2006 UN report, there will be 22 mega-cities with 10 million or more inhabitants by 2015. Only six of these mega-cities will be located in OECD countries.[2]

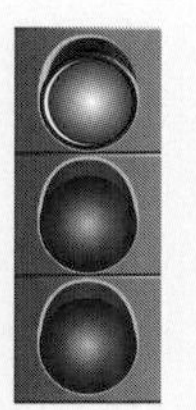

Urban populations are expanding rapidly in developing countries, where the infrastructure needed to support human health and the environment is often not in place.

These demographic changes will have a significant impact environmentally, socially and economically on cities worldwide. Considering that more than 90% of the world's urban growth in the next two decades will be absorbed by cities of developing countries, the impacts are anticipated to be much greater there. In the absence of significantly improved policies, it is likely that a large portion of urban dwellers will be left without access to basic environmental and social services, such as safe and sufficient water, drainage and wastewater treatment, rubbish collection, electricity and heating, and basic health care (UNEP, 2002). As cities have grown in developing countries, so have their slum populations (UN-HABITAT, 2006). In many sub-Saharan African countries, the slum population accounts for over 70% of the urban population, and 51% of the slum population lacks two or more of access to water, access to sanitation, durable housing and sufficient living area.

Urban sprawl

As the population of urban areas has grown, so too has their area (Figure 5.2). Between 1950 and 2000, the total worldwide urban area increased by 171%; 364 065 km^2 of land was converted to urban uses, equivalent to almost the total land mass of Germany (Klein Goldewijk and Van Drecht, 2006). About 50% of this new urbanised area was in OECD countries. The total urban area of the BRIC countries tripled over this period, and those of the rest of the world (ROW) expanded by 4.4 times. The relatively low level of urban land expansion in OECD countries reflects in part their already high level of urbanisation in 1950. As of 2000, OECD cities still made up 58% of total world urban areas. This unprecedented expansion of urban areas not only changes the landscape of the earth but also has significant impacts on our lifestyles.

 OECD ENVIRONMENTAL OUTLOOK TO 2030 – ISBN 978-92-64-04048-9 – © OECD 2008

Figure 5.2. **Trends in urban area expansion, 1950-2000**

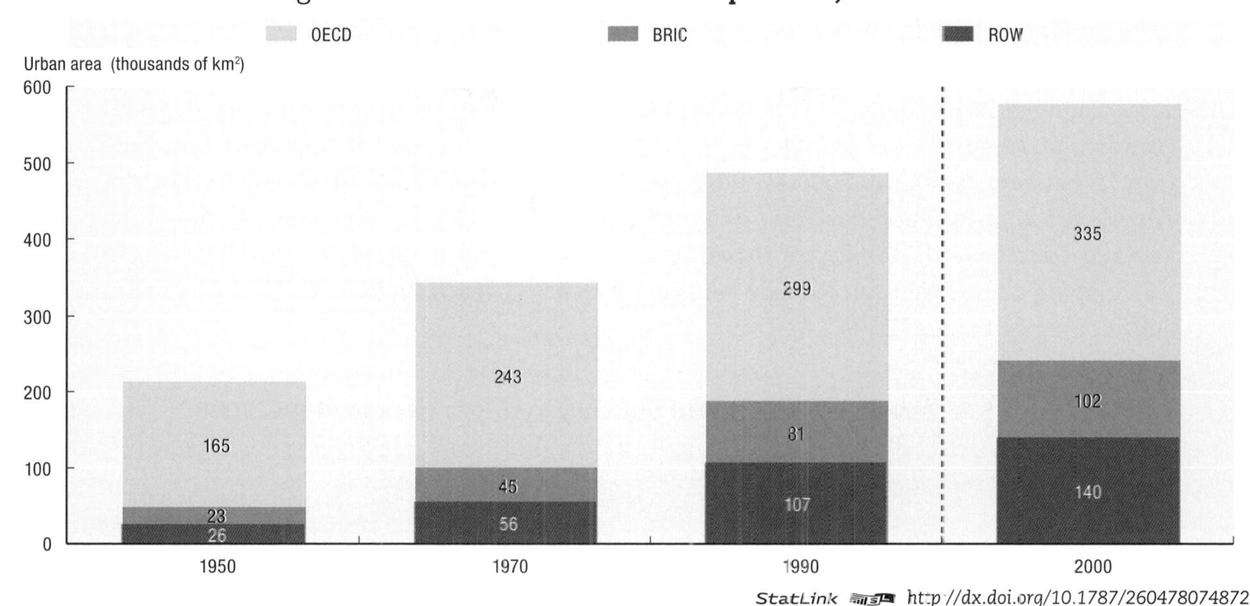

StatLink http://dx.doi.org/10.1787/260478074872

Source: Klein Goldewijk and van Drecht, 2006.

Historically, the physical expansion of cities has been driven by urban population growth, a trend that is seen today in developing countries. Recent urban expansion in OECD countries, on the other hand, is now largely driven by urban sprawl. Urban sprawl can clearly be seen by the fact that urban land expansion has been faster than population growth (Figure 5.3). Urban areas expanded by 171% worldwide between 1950 and 2000, whereas world population grew by only 142%. In particular, the extent of urban areas in OECD countries increased by 104% while the population increased by only 66%. The

Figure 5.3. **Incremental increases to population and urban areas, 1950-2000**

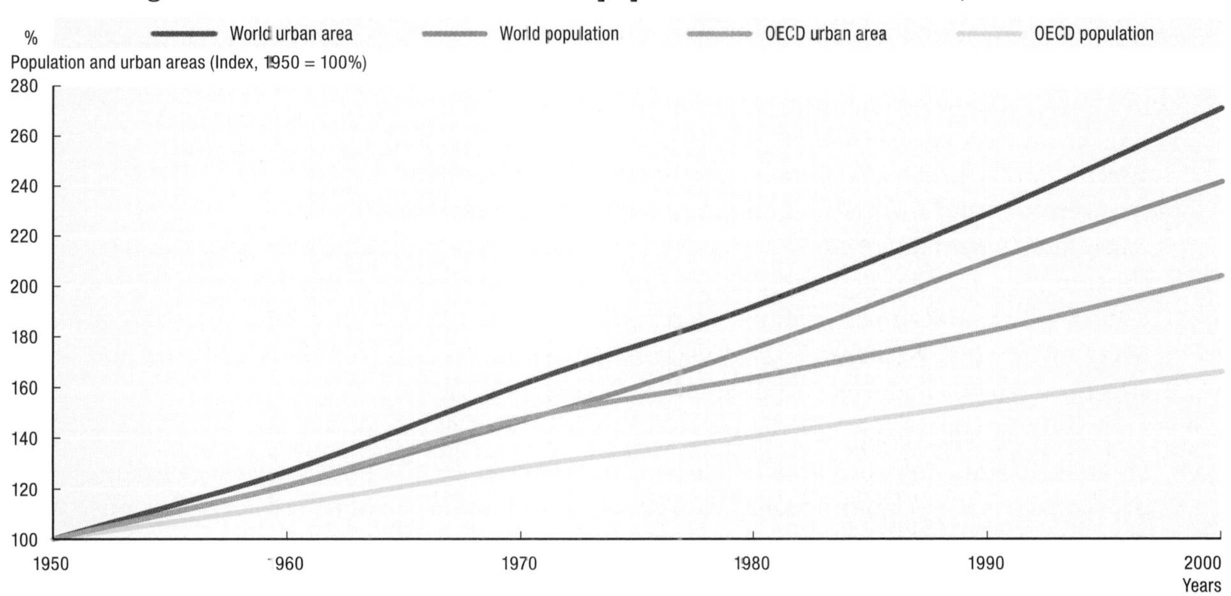

StatLink http://dx.doi.org/10.1787/260510206137

Source: Klein Goldewijk and van Drecht, 2006.

significance of this trend is that, on average, each person consumes more space. The phenomenon of urban sprawl has mainly been seen in North America, but it is becoming a more common phenomenon in other OECD regions as well.

Changes in the average density of urban areas further highlight the current trend toward urban sprawl. An analysis of the average density changes of 90 cities around the world between 1990 and 2000 found decreases in average urban area density both in developed countries and developing countries (Angel *et al.*, 2005). Moreover, the rate of the average decrease in density of urban area was faster in developed countries than in developing countries, even though the average density was already three times higher in developing countries (Table 5.3).

Table 5.3. **Average density and built-up area per person, 1990-2000**

Category	Average urban area density (person per km²)			Average built-up area per person (m²)		
	1990	2000	Annual % change	1990	2000	Annual % change
Developed countries	3 545	2 835	–2.2%	280	355	2.3%
Developing countries	9 560	8 050	–1.7%	105	125	1.7%

StatLink http://dx.doi.org/10.1787/256784002801

Source: Angel *et al.*, 2005.

There are a variety of driving forces behind urban sprawl, including preferences for individual housing, increased mobility, relative abundance of land and land use planning incentives. Urban sprawl involves relatively low-density building on arable and other land outward from the metropolitan core (TRB, 1998; Carruthers, 2003).That is, it takes place at the expense of agricultural land, forest, open space or wetland, with a concomitant loss in the economic, recreational and ecological values that those ecosystems provided. A European Environmental Agency (EEA) study conducted on the land use trends of 23 European countries over 1990-2000 showed that the largest land category replaced by urban development was agricultural land (EEA, 2005). About 48% of the land areas that changed to artificial surfaces during 1990-2000 were originally arable land or permanent crops, and nearly 36% were pasture and mixed farmland. This land consumption has impacts not only within the built-up area but also for considerable distances around it in terms of how land surfaces are reshaped, with valleys and swamps being filled, large volumes of clay and rock being extracted, and sometimes rivers and streams re-channelled. This increases stress on ecosystems and species.

Urban sprawl is a common trend in OECD countries, and can increase transport, pollution, resources use and greenhouse gas emissions.

Urban sprawl not only affects ecosystems, but also the economic and social conditions of cities. It tends to cause population de-concentration in urban centres and generates brownfield sites – abandoned, vacant or under-used former industrial areas (Greenberg *et al.*, 2001; Savitch, 2003). Increasing brownfield generation can lead to insufficient use of established social infrastructures. Furthermore, the segregation of land uses associated with low density and spread-out urban development tends to result in a relatively high level of infrastructure construction – roads, water and sewer systems, schools and privately owned utility systems – that would not be necessary under more compact development (TRB, 2002). Urban sprawl therefore contributes to undermining efficient energy and resource use, thus incurring unnecessary exploitation of natural resources and emission of pollutants.

Nonetheless, without a significant change in our lifestyles or relevant policies, this trend is likely to continue. If average urban densities continue to decline at the rate seen between 1990 and 2000, cities are likely to grow by 150% in area by 2030 worldwide (Angel *et al.*, 2005). Furthermore, if developing countries follow the same path of urban sprawl in the coming decades as more developed countries, the scale of urban expansion could be much greater and the magnitude of its impacts on the environment and human society even more pronounced.

Urban transportation

One of the impacts of urban sprawl is an increasing dependence on the automobile for intra- and inter-metropolitan travel. Urban sprawl entails building extensive transportation systems because houses are increasingly far away from workplaces and commercial centres. This newly constructed transport infrastructure, in return, spurs further urban sprawl – investments made in new motorways or road connections attract new development along the improved transport lines.

It is estimated that transportation networks in OECD countries may take up about 25-30% of land use in urban areas, and almost 10% in rural areas (EEA, 2002). Besides the impacts on land use, this transport infrastructure network also poses a threat to habitats and biodiversity (see Chapter 9 on biodiversity). The fragmentation and degradation of the natural landscape and the isolation of habitats create new barriers to natural migration and the movement of animal populations. In particular, these negative effects significantly increase when urban expansion happens in environmentally sensitive areas, such as on coastlines, floodplains or wetlands.

Increased average trip length and suburb-to-suburb trips also increase fuel consumption and related emissions of air pollutants and greenhouse gases. A strong relationship can be seen between low density cities and high fuel consumption for private transportation, as observed in low density cities such as Sacramento and Houston in the United States (Newman *et al.*, 1999; Kenworthy *et al.*, 2005). Conversely, some Asian cities such as Seoul or Tokyo have relatively high population density and low per capita fuel use for private transportation. This implies that in general there is a significant increase in transport-related fuel consumption in cities as densities fall (Figure 5.4).

Building stock

Buildings occupy a significant volume of urban land and alter the natural urban ecosystem. They also require large amounts of natural resources for their construction, and during their operation consume energy, water and other materials, and emit various kinds of solid, liquid and gas contaminants. At the demolitions stage, they generate vast quantities of wastes into the environment as well. As such, the building sector has significant impacts on the environment and human health.

Figure 5.4. **Energy use per capita in private passenger travel *versus* urban density, selected world cities**

Source: Newman and Kenworthy, Copyright © 1999 by the authors. Reproduced by permission of Island Press, Washington, D.C., 1999.

The construction sector accounts for between one-third and one-half of commodity flows when expressed in terms of weight; this inevitably generates a considerable amount of construction and demolition waste (OECD, 2003).[3] The energy consumed to operate residential, commercial and public service buildings accounts for around 25-40% of final energy consumption in OECD countries. For the UK, it is estimated that the construction, occupation and operation of buildings are responsible for 45% of total UK CO_2 emissions, with 27% of the total coming from domestic buildings (RCEP, 2007). Furthermore, relatively high levels of pollutants arising from building materials and components (*i.e.* finishes, paints and packing materials) can pose various health problems, such as irritation of the eyes, nose and throat, headaches and dizziness.

There is great potential for cost-effective GHG emission reductions from buildings, with the right incentives and building codes.

Buildings can last for decades, even centuries, and it is projected that more than half of existing buildings will still be standing in 2050. As such, the environmental impacts of buildings constructed today will continue for years to come (see Box 5.1 for a Chinese example). With appropriate incentives and building regulations, there is considerable potential to reduce the sector's environmental impacts. There is a lot that can be done to promote more energy efficient buildings in particular, including use of efficient lighting, heating and cooling systems; improved insulation material; passive solar design; greater use of energy-efficient appliances; etc. For example, it is estimated that passive solar heating and passive solar cooling can reduce the heating and cooling load by up to 50% for some buildings at no additional costs, and the efficiency of lighting technologies has improved in recent years such that some estimates show that efficiency gains of 30-60% can be achieved (IEA, 2006). Such improvements could make buildings much more energy efficient, and significantly reduce their contribution to greenhouse gas emissions.

Box 5.1. **Environmental effects of the residential sector in China**

The Chinese Ministry of Construction anticipates that by 2020 an additional 180 million people will reside in China's cities. Residential building floor area has already been dramatically increasing since 2000; according to projections about 13 billion m^2 more residential floor space will be constructed in the next two decades, which is equivalent to the total floor area of all the existing residential buildings in the EU-15 countries.

With such rapid growth forecast for the residential sector, a window of opportunity exists now to significantly improve the energy efficiency of new buildings. The choices made today will determine the efficiency of energy use, and the emissions of air pollution and greenhouse gases from the building sector for many years to come. However, China is likely to face significant challenges in addressing energy efficiency in buildings. Several building energy conservation standards have been set up since the mid-1980s, but only 7% of the existing building stock complies with the regulations. Moreover, due to the relatively high price and lower availability of other energy sources, coal is still the main fuel for heating residential buildings and is likely to be so for some time to come. Currently, energy consumption for space heating is 50% higher than in industrialised countries with a similar climate.

To improve the poor level of energy conservation in buildings, in 2006 the Ministry of Construction set more stringent energy standards. These are first being piloted in Beijing and Shanghai, where the aim is to cut energy consumption in buildings by two-thirds; these standards will apply nationwide by 2010. The government's 11th 5-year plan aims to reduce 89.5 TWh of energy consumption in the building sector by 2010, of which 57 TWh should be saved through new construction and the other 30 TWh from retro-fitting of old, inefficient buildings.

Economic incentives will be needed to complement the regulations, along with institutional reforms to ensure better compliance. The Chinese government – together with key stakeholders such as developers, energy suppliers, households and local government – faces an opportunity and a challenge to significantly improve the energy efficiency of the rapidly expanding new building stock.

Source: Based on information provided by IDDRI (Institut du développement durable et des relations internationales), Paris, 2007.

Policy implications

Over the past few decades, there have been notable environmental improvements in some urban areas. For example, air pollution from transport, especially road transport, is decreasing in OECD countries, thanks to increasingly strict emission standards for transport providers. More innovative policies have been implemented, such as road and congestion pricing (see Box 5.2); environmentally related tax reform; improvement of public transport; and speed control and travel time sanctions for heavy duty vehicles (See Chapter 16 on transport). There are, furthermore, positive examples in practice of the significant expansion of green belt or spaces, decontamination of rivers, development of sewerage and waste management systems and in urban planning (such as brownfield redevelopment in the UK and the USA, and compact cities in Scandinavian countries).

However, there is still much to be done to create environmentally sustainable cities. First of all, integrating related policy tools and objectives will be essential to address urban environmental problems. While each specific policy tool might be valuable in isolation, it can fail to achieve its full potential benefits unless it is adopted in a carefully integrated and cross-sectoral way.

Box 5.2. **Congestion charging**

Congestion charging is primarily intended to address environmental and congestion problems in urban areas with the price incentive levied on vehicle use within the urban zone. There are only a few cases in the world where congestion charging has actually been implemented, the most recent full example being London since 2003. Seoul introduced a partial congestion charging system in 1996. Recently, Stockholm finalised a full-scale experiment. The USA has decided at federal level to carry out a large number of trials using congestion charges, and many other OECD cities, such as Copenhagen, are now discussing the introduction of congestion charging.

In London, the Central London Congestion Charging Scheme involves a daily charge of GBP 8 (2007) for most vehicles in the central core of London during peak periods. Since its introduction in 2003 it has been credited with reducing traffic and traffic-related externalities (congestion, accidents, and air pollution) in London. Since 2001, emissions of NO_x have decreased by 13% and PM emissions by 15%. Approximately half of these reductions has been attributed to changes in vehicle technologies; the other half is likely to be the result of the congestion charges. Congestion charging is also estimated to be responsible for a 16% reduction in CO_2 emissions within the charge zone (TFL, 2006; Beevers and Carslaw, 2005).

The London case illustrates that congestion charging can be an effective congestion reduction strategy and an efficient way to improve mobility and to reduce transport-related pollution and GHG emissions in urban areas. Furthermore, congestion charging can help cities encourage active transportation and ease the operation of businesses, and thus retain their attractiveness.

In addition, many cities, such as Mexico, regulate traffic levels by only allowing cars with number plates ending with either even or odd numbers on certain days. Cities such as Shanghai have banned petrol scooters and only allow liquefied petrol gas (LPG) scooters to be used within the city. In Hong Kong, nearly all taxis run on LPG. In New Delhi, rickshaws which run on compressed natural gas (CNG) are popular.

The integration of spatial planning, transport and environmental policies is particularly crucial because they are so closely related. Land use policies need to take account of travel time, car dependency, greenfield use, access to goods and services, air pollution, noise, greenhouse gas emissions and energy consumption. Spatial policies sometimes influence transport variables much more than transport policy itself does. Integrating land use policy with decoupling objectives in the transport sector is important. Changes in land use regulations may be needed to provide incentives for mixed-use areas with high density.

Furthermore, health and social concerns also need to be integrated into the design and management of urban polices. Urban planning which takes account of urban poverty and health issues will promote the accessibility of the poor to basic environmental services as well as to green space, which will eventually contribute to social cohesion as well. Recent attempts to improve governance and to adopt a more strategic approach to the economic development and social and environmental sustainability of cities are leading to the emergence of what is sometimes termed the "entrepreneurial city". This is a proactive city which aims to mobilise social, political and economic resources in a coherent institutional framework to develop – and sustain long-term support for – a clear social and economic development strategy (OECD, 2001).

Policy integration and successful implementation require a new approach to the governance of urban areas which enables close co-ordination between different policy areas and better co-operation between different levels and orders of governments and local stakeholders. A mechanism for effectively co-ordinating priorities among various levels and orders of governments is particularly critical. One approach would be to develop a comprehensive policy framework which encompasses the essential policy objectives of environmental sustainability, human health and well-being, consulting widely with all relevant stakeholders in its development. The development and implementation of Local Agenda 21 Strategies in a number of cities is a good example of integrated environmental management at the urban level. For example, the Local Agenda 21 Strategy adopted by Copenhagen has led to noticeable improvements in air quality, greenhouse gas emission, energy use, ecological footprint, recycling and the number of buildings constructed using sustainable construction methods and techniques (EC, 2006).

From a sustainable development perspective, cities present formidable challenges, but also provide an opportunity for establishing efficient living environments. There is no single solution that will apply to all cities. The diversity of urban areas in terms of history, geography, climate, administrative and legal conditions calls for urban policies to be locally developed and tailor-made.

Notes

1. "Cities" in this report refer to urban or urbanised areas, including the contiguous territory inhabited at urban levels of residential density and their additional surrounding areas. This is a similar concept to urban agglomeration or metropolitan region.
2. Tokyo (35.5 million people in 2015), Mexico City (21.6), New York (19.9), Los Angeles (13.1), Osaka-Kobe (11.3) and Istanbul (11.2).
3. The OECD (2003) report related to the construction sector for both urban and non-urban development.

References

Angel, S., S.C. Sheppard and D.L. Civco (2005), *The Dynamics of Global Urban Expansion*, World Bank Transport and Urban Development Department, Washington D.C.

Beevers S.D. and D.C. Carslaw (2005), "The Impact of Congestion Charging on Vehicle Emissions in London", *Atmospheric Environment* 39, 1-5.

Carruthers J.I. (2003), "Growth at the Fringe: The Influence of Political Fragmentation in United States Metropolitan Areas", *Regional Science* 82, 475-499.

European Commission (2003), *Achieving Sustainable Transport and Land Use with Integrated Policies*, European Commission Energy, Environment and Sustainable Development, Brussels, *www.transplus.net*.

European Commission (2006), *Communication from the Commission to the Council and the European Parliament on Thematic Strategy on the Urban Environment*, SEC(2006)16, European Commission, Brussels.

EEA (European Environmental Agency) (2002), *Towards an Urban Atlas: Assessment of Spatial Data on 25 European Cities and Urban Areas*, EEA, *http://reports.eea.europa.eu/environmental_issue_report2002_30/en*, European Environmental Agency, Copenhagen.

EEA (2005), *The European Environment: State and Outlook 2005*, EEA, *http://reports.eea.europa.eu/state_of_environment_report_2005_1/en*, pg. 308-311, European Environmental Agency, Copenhagen.

Greenberg M. *et al.* (2001), "Brownfield Redevelopment as a Smart Growth Option in the United States", *The Environmentalist* 21, 129-143.

IEA (International Energy Agency) (2006), *Energy Technology Perspectives – Scenarios and Strategies to 2050*, OECD/IEA, Paris.

Kenworthy, J.R. and F.B. Laube (2005), "An International Comparative Perspective on Sustainable Transport in European Cities", *European Spatial Research and Policy*, Vol. 12, No. 1/2005.

Klein Goldewijk K. and G. Van Drecht (2006), "Hundred Year Database on the Environment (HYDE) 3: Current and Historical Population and Land Cover", in A.F. Bouwman, T. Kram and K. Klein Goldewijk (eds.), *Integrated Modelling of Global Environmental Change. An Overview of IMAGE 2.4*, Netherlands Environmental Assessment Agency (MNP), Bilthoven, The Netherlands.

Newman P. and J. Kenworthy (1999), *Sustainability and Cities*, Island Press, Washington DC.

OECD (2001), *Cities for Citizens: Improving Metropolitan Governance*, OECD, Paris.

OECD (2003), *Environmentally Sustainable Buildings – Challenges and Policies*, OECD, Paris

OECD (2006), *OECD Territorial Reviews: Competitive Cities in the Global Economy*, OECD, Paris.

RCEP (Royal Commission on Environmental Pollution) (2007), *The Urban Environment*, *www.rcep.org.uk/* The Royal Commission on Environmental Pollution, London.

Savitch, H.V. (2003), "How Suburban Sprawl Shapes Human Well-Being", *Journal of Urban Health: Bulletin of the New York Academy of Medicine* Vol. 80, No.4, pp. 590-607.

TFL (Transport For London) (2006), *Central London Congestion Charging: Impacts Monitoring (4th Annual Report)*, Transport for London, London.

TRB (Transportation Research Board)/National Research Council (1998), *The Costs of Sprawl – Revisited*, National Academy Press, Washington DC.

TRB/National Research Council (2002), *Costs of Sprawl – 2000*, National Academy Press, Washington DC.

UN (United Nations) (2006), *World Urbanization Prospects: The 2005 Revision*, United Nations Department of Economic and Social Affairs/Population Division, New York, *www.un.org/esa/population/publications/WUP2005/2005wup.htm*.

UNEP (United Nations Environment Programme) (2002), *Global Environment Outlooks*, Earthscan Publications Ltd., London/Sterling, VA.

UN-HABITAT (2006), *State of the World's Cities 2006/7*, UN-HABITAT, Nairobi.

WCED (World Commission on Environment and Development) (1987), *Our Common Future*, Oxford University Press, Oxford and New York.

ISBN 978-92-64-04048-9
OECD Environmental Outlook to 2030

Chapter 6

Key Variations to the Standard Expectation to 2030

The Outlook Baseline assumes that, without any new policy action, world economic growth and globalisation to 2030 will follow similar trends as seen over the past few decades. This is just an assumption and should not be seen as a forecast of the future: it represents what might happen without any major new events or policies. This chapter explores some of the uncertainties associated with the Baseline, and examines how projections might vary with different assumptions about the productivity growth rate and the rate of globalisation. These variations to the Baseline suggest that higher medium-term growth would amplify impacts on the environment, and increased trade and changing patterns of production would lead to higher energy demands for the world as a whole. These variations illustrate the considerable differences that changes in a few key drivers could make to the nature of the world economy and its pressures on the environment.

KEY MESSAGES

The *Outlook* Baseline assumes that world economic growth and globalisation will follow the same trends to 2030 as seen over the past few decades. This is an analytical tool and should not be seen as a forecast of the future: it represents what might happen without any major new events or policies. But other scenarios are possible, and this chapter explores some of them to: *a)* prepare policy-makers for a range of alternative outcomes, and *b)* gauge how they might affect policy prescriptions:

- Economic growth variations (variations 1-3 below): the five years between 2002 and 2007 witnessed much higher world economic growth rates than previously. Variation 1 projects these recent strong growth rates to 2020 to explore their medium-term impact. Variation 2 assumes countries' labour productivity growth levels off towards 1.25% over the long term instead of 1.75%. This reduced rate of labour productivity growth is more consistent with longer-term (*i.e.* longer than 20 years) historical rates of growth across all countries. Variation 3 assumes that productivity growth levels off to 2.25%. Given recent global growth rates and advances in transportation and communication technology, this is a plausible – if optimistic – long-term outcome.
- Globalisation variation (variation 4): this assumes continued strong increases in trade, *e.g.* as a result of explicit trade policies and/or "autonomous" reductions in the costs of international trade. These factors have been omitted from the *Outlook* Baseline in an effort to clearly distinguish a reference case from a policy case.

Environmental implications

The higher medium-term growth (variation 1) would increase impacts on the environment. If emissions of greenhouse gases from energy were 16% higher in 2030, the impacts would clearly be significant for climate change since an additional 1.7 gigatonnes of CO_2 would be emitted.

Variations in the rates of long-term productivity growth (variations 2 and 3) have less impact on the horizon to 2030, but have larger consequences for the environment in the longer term. Nonetheless, the faster growth represented by the 2.25% rate (variation 3) will mean greater and earlier impact on the environment than growth of 1.25% (variation 2). Though human material well-being will be better off, traditional sources of market failure regarding the environment imply that policy frameworks will need to be reinforced.

The increased trade and changing patterns of production (variation 4) will redistribute polluting activities and cause an overall increase for the world as a whole. While globalisation may not in itself lead to much larger economies, it can have environmental impacts through the much wider dispersion of stages of production (see graph).

For the developing world (ROW), the impact of increased trade on key environmental variables (variation 4, see graph opposite) is expected to be generally negative. This has some implications for policy coherence (*i.e.* achieving development and environmental goals in non-OECD countries). In OECD countries, there is a mild increase projected in total primary energy supply under a globalisation scenario, leading to increased greenhouse gas emissions. There is also a notable decrease in emissions of nitrogen oxides.

Selected environmental impacts of the globalisation variation

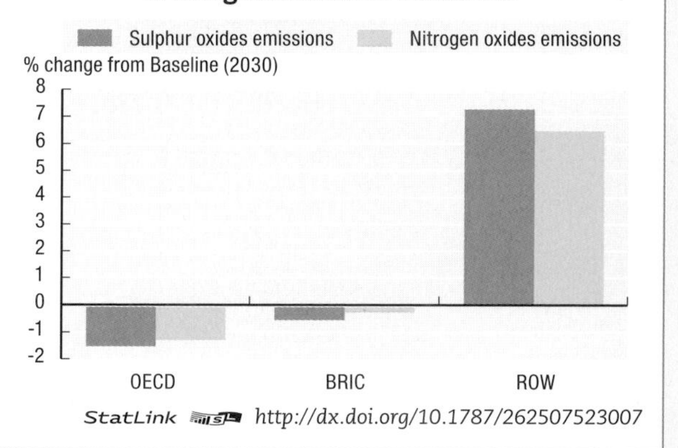

StatLink http://dx.doi.org/10.1787/262507523007

Policy implications

These variations illustrate the considerable differences that changes in a few key drivers can make to the nature of the world economy. Given this level of variability, anchoring the *Outlook* in historical trends for the critical economic and social drivers of environmental change is important – both for putting the Baseline on a firm foundation, as well as for exploring the repercussions of various policy initiatives.

Introduction

The *OECD Environment Outlook* Baseline to 2030 is a reference case to explore sources of future environmental pressures and the impacts of policies on those pressures. It is not an attempt to predict what the world economy will actually look like over the next quarter of a century – it is simply a representation of what the world economy *could* look like if it continued on its present course. The Baseline, as with all quantitative analysis, remains highly uncertain and is primarily useful as an analytical tool.

Other factors besides the choice of the Baseline contribute to uncertainty in this *Outlook*. For example, the way its questions are framed; the models used and how they were combined; and technical assumptions such as resource efficiencies and fuel mixes. Annex B provides an introduction, along with a focus on some specific uncertainties and limitations, of the suite of models used for the *Outlook*.

This chapter explores some uncertainties related to the Baseline, and asks what the likely impacts would be of varying some of the key assumptions in the Baseline (productivity growth rate and a different path towards globalisation).

Figure 6.1 gives an idea of how variable different model results can be. This compares the *Outlook* Baseline projections for CO_2 emissions with some scenarios from the IPCC's (Intergovernmental Panel on Climate Change) Special Report on Emission Scenarios (SRES) programme (IPCC, 2000).[1] The large gap between the dotted lines indicates a number of differences between models, including fundamental elements of model structure, as well

Figure 6.1. **CO_2 energy emissions: OECD and SRES results**

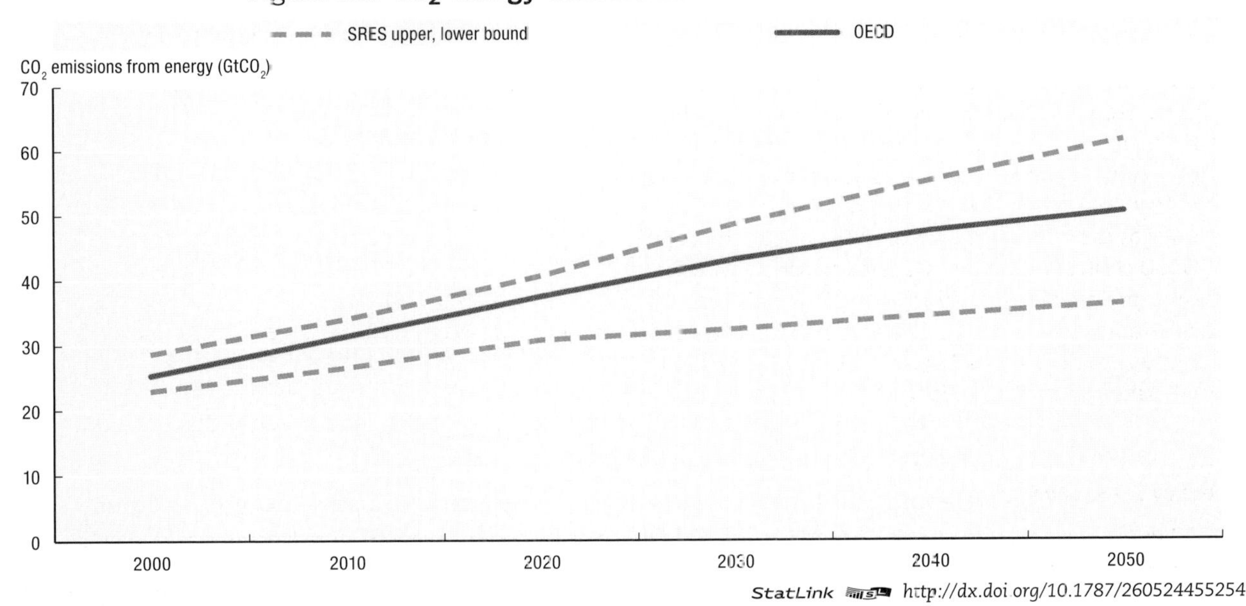

StatLink http://dx.doi.org/10.1787/260524455254

Source: OECD calculations from IPCC Fourth Assessment Report data.

as differences in model parameters. As the figure makes clear, it is important to explore sources of uncertainty in model results so that policy development takes into account a range of possible outcomes.

This chapter is limited to exploring some key variations in assumptions used to develop the Baseline. There are, however, many other areas where assumptions could also be varied, so the results reported here do not constitute a full "sensitivity analysis" *per se*. Nonetheless, the variations studied here illustrate the kinds of impacts that varying assumptions in other areas may have. The variations chosen here also potentially have the widest possible impacts on overall environmental and economic policy. Variations in issues not explicitly included here are potentially important (*e.g.* population, energy, technology, etc.), but have been studied elsewhere. The population projections used in the Baseline, for example, are developed by the United Nations along with high and low variations. This chapter thus acknowledges that the *Outlook* results are conditioned on a particular set of perspectives, and that these provide a useful starting point.

Types of baselines

How a view of future (economic and environmental) outcomes will be used is of prime importance to how a baseline and its variations are developed. Many recent studies, such as the IPCC SRES (IPCC, 2000), the Millennium Ecosystem Assessment (2005), and the United Nations Environment Programme's Global Environment Outlook (UNEP, 2002), have used a series of "storylines" to outline possible evolutions of the world economy. These storylines have the advantage of providing internally consistent baselines that follow from given themes.

In a storyline approach, a narrative is developed describing geopolitical and economic trends. This may be extended to outlining contrasting narratives with major alternatives that provide information on how events may develop. For example, a common approach is to have two major axes on which the narratives are based. Such a case is shown in Table 6.1, where the themes are the degree of globalisation on one axis, *versus* the degree of free-market rule on the other. Each of these quadrants can be further subdivided to give more variation in the possible outcomes facing policy-makers – so *globalising free-markets* can look at different dimensions of economic globalisation, for example.

Table 6.1. **Main axes of variation of narratives**

	Global integration	Regionalism
Economic emphasis	Globalising free-markets	Quasi protectionism
Environmental emphasis	Accounting for global externalities	Local "sustainability"

Table 6.1 implies that different storylines will involve important policy changes for trade, social programmes and the environment. Storylines are thus useful tools that enhance an understanding of potential future outcomes, and can lead to early discussion of what may be needed to avoid undesirable consequences.

A storyline approach, however, cannot easily be used for *policy analysis* without considerable additional detail. Policy analysis requires the careful disentanglement of a new policy from the state-of-the-world without that policy. Specifically, it requires a complete juxtaposition of policy/no-policy alternatives concerning the issue of interest. Mixing policy between the alternatives leads to confusion over the impact of a particular

OECD ENVIRONMENTAL OUTLOOK TO 2030 – ISBN 978-92-64-04048-9 – © OECD 2008

policy agenda. To undertake policy analysis with a storyline, it must be accompanied by a complete quantification of all the drivers behind the storylines. With that information, subsequent analysis can determine what *additional* policies will be necessary to implement social objectives.

Sources of variation in reference scenarios

Broadly speaking, there are at least three important sources of uncertainty in a model-based analysis:

i) *Uncertainty in the model parameters.* Model parameters define unchanging relationships between different parts of the environment/economy. For example, the response of consumer demand to changes in the price of a good or service is often given by a fixed parameter. Moreover, simple models may use parameters to abstract away from behaviour that is complex but not of immediate interest. For example, the relationship between income and savings may be fixed in some models even though people's savings behaviour is actually very complicated. Since parameters are derived from empirical sources, there is statistical uncertainty in the value of the parameter. Dealing with that uncertainty is often done by examining the impact of small changes in parameters on the model's results.

ii) *Uncertainty in the model structure.* There are numerous theories that can be used to underpin a model's structure. If the foundation of the model's structure is wrong, then the results will also be wrong. This source of model uncertainty can be partially dealt with by analysing the model's properties in detail. This can highlight where the model is consistent with "good" analytical/empirical results and where it may be weak. In the general circulation models used for studying future climate change, this area of uncertainty is known as "perturbed physics", where some of the underlying physics of the model are tested for robustness.

iii) *Uncertainty in the drivers* being input into the model to generate results. A model may be developed that does an excellent job of reproducing current economic/environmental outcomes, but it still requires projections of future drivers to underpin it – the issues outlined in Chapter 3. Uncertainty in those drivers translates directly into uncertainty in the model projections.

Given the wide range of results that can be caused by these three sources of uncertainty, how can analysts derive useful policy lessons to aid decision-making? To answer this we discuss each of these areas of uncertainty in turn.

i) *Model parameters:* some of the mathematical equations in the model (*i.e.* those that were estimated or calibrated to obtain model parameters) can be modified to reflect inherent uncertainty. Specifically, they can have a random component introduced to reflect the statistical variability (distribution) of the underlying behaviour. For example, since modelled consumer behaviour is an average over many individuals, there is a good deal of variability in any equation that represents consumer demand – even when it is for a well-specified product such as a car. The random component that is introduced in the equation represents the variability in the underlying behaviour. In fact, this random component would have been an integral part of the equation that was used to estimate the equation's parameters. In the full model where the equation is used, the random part of the equation can then be varied – allowing the model to be studied for uncertainty in the behaviour that the equation represents. By doing a systematic check on all of the model's random parts, a picture can be drawn of the overall randomness of the model. It can then be used to reflect uncertainty of the model in response to various policies. One drawback of this technique is that assumptions have to be made about the statistical properties of the random part of the equation. Without additional analysis to study different distributions, there may be false confidence in the knowledge of the model's uncertainty – all that may really be known is uncertainty as represented by a particular statistical distribution. This first area of uncertainty concerning the ENV-Linkages model is dealt with in more detail in other work done at the OECD. One important lesson from that work is that while the *quantitative* results that come from the model can change with revisions to parameters, the *qualitative* results are much harder to overturn.

ii) *Model structure*. This area of uncertainty is more likely to change qualitative results, but is not dealt with here since it would require changes to the model that are not particularly interesting for the *Outlook*. Overcoming this area of uncertainty is more closely related to choosing between analytical paradigms that distinguish different schools of thought. The computable general equilibrium (CGE) modelling framework used here is a popular analytical tool for understanding economic phenomena. Its use has expanded considerably with the increased interest in quantitative analysis of environmental policy (see Bergman, 2005). This second area of uncertainty suggests that different models will give different results. How do we treat those differences? If each model were randomly drawn from a population of models, such that each draw had a statistically normal distribution, then a sample of model results could be treated as a statistical sample. One could then construct a mean and variance of the results and discuss them using terminology like "statistical significance". If the distribution of models and their results are not known, then constructing a sample mean, and variance around that mean, is a matter of pure aesthetics – it provides little scientific data other than to note certain aspects of the data (policy decisions inferred from the results would be ill-informed). It does, however, provide a basis for informing expert opinion by allowing experts to collect information that they would not otherwise have. Succinctly, those types of results are useful to analysts, not to the non-expert.

iii) *Drivers*. This third area of uncertainty is the focus of this chapter. As discussed above under "Types of baselines", the development of the baseline depends on the ultimate use of the analysis. Similarly, understanding the uncertainty inherent in the baseline depends on the analysis being undertaken with the baseline. The storyline approach outlined above represents an attempt to deal with uncertainty of the future when the range of possible outcomes is itself the key issue. That is, when "future gazing" is a key reason for building baselines, then storylines that span the widest possible futures are imperative. On the other hand, when studying particular policy agendas, the range of uncertainty can be narrowed considerably by focusing attention on key alternatives to a reference case (*i.e.* the baseline) that are most important for the policy issues under consideration. A starting point for looking at those alternatives would be to examine variations in the key drivers of the baseline.

Key variations in the drivers

Environmental outcomes are heavily influenced by the economy. The sheer scale of economic activity can lead to impacts on the environment that accumulate over time and can lead to large scale changes in the quality of the environment. Economic growth is thus an important determinant of the environmental outlook (see also Chapter 3, Economic development).

Long-term economic growth is primarily influenced by a handful of factors, of which the most crucial are the growth of the labour force (population) and the growth of technical knowledge (productivity). Globalisation contributes to growth through gains from comparative advantage (allocative efficiency), but its influence on growth continues only while increasing globalisation is possible. Globalisation, however, is very important for questions concerning the distribution of sources and impacts of environmental drivers (see also Chapter 4, Globalisation).

Changes in population are difficult to predict since economic factors endogenously combine with fertility and longevity to influence growth rates (see also Chapter 2, Population dynamics and demographics). The variability in projections is such that, for 2030, the United Nation's range of projections includes 7% above and below their medium variation. In other words, the annual population growth may be just under 0.3% higher or lower than their central projection. The implications for economic and environmental impacts would be rather significant at both extremes.

Long-term changes in productivity growth of similar magnitudes are also plausible given past trends. The *Outlook* Baseline assumes that all countries move towards a long-term labour productivity growth of 1.75%.[2] This rate is consistent with the longer-term growth experience of economies that have achieved development, and whose productive capacity then essentially grows at the rate of technological advance.

Variations in the aggregate productivity of countries

This section looks at three alternative versions of productivity growth:

i) Variation 1. Whilst the Baseline makes future projections based on the trend in world economic growth between 1980 and 2001, the world economy has actually performed considerably better since 2001 (Figure 6.2). This first variation explores what would happen if this recent high growth – particularly in countries like China – continues in the medium term (to 2020).

Figure 6.2. **World GDP growth (annual), 1980-2008**

Baseline

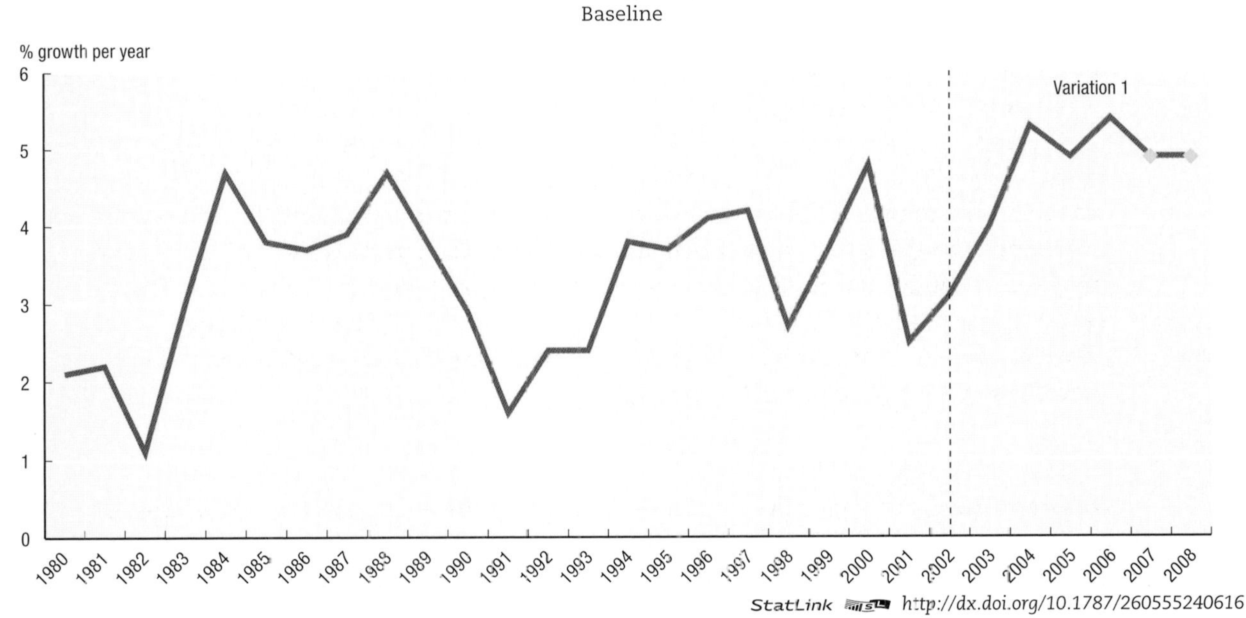

StatLink http://dx.doi.org/10.1787/260555240616

Note: The last two data points (2007 and 2008) are projections by the IMF.

Source: IMF, 2007.

ii) Variation 2. Countries' productivity growth is assumed to go down to 1.25% over the long term[3] instead of 1.75%. This reduced rate of productivity growth is more consistent with longer-term (*e.g.* greater than 20 years) historical rates of growth across all countries. Even this growth rate is high if one looks at the world trend over the past 2000 years (Maddison, 2003).

iii) Variation 3. Productivity growth is assumed to reach 2.25%. While this would be unprecedented over a long period for the world as a whole, given recent global growth rates and advances in transportation and communication technology, this is a plausible – if very optimistic – long-term outcome.

Variation 1: results

Table 6.2 shows how GDP projections change compared with the Baseline when they are derived from post-2000 average growth rates. The results are quite dramatic, especially for the non-OECD regions, reflecting the fact that many of these regions have seen particularly strong economic performance over the past five years.

Table 6.2. **Variation 1: percentage change from Baseline for GDP using recent (5-year) productivity trends**

	2010	2020	2030
OECD	**0.4**	**3.4**	**4.3**
North America	0.2	6.5	8.0
US and Canada	*0.2*	*6.3*	*7.5*
Mexico	*–0.1*	*9.7*	*14.6*
Europe	–0.1	0.7	0.8
Pacific	1.6	–0.3	–0.5
Asia	*1.7*	*–1.1*	*–1.5*
Oceania	*0.6*	*6.6*	*7.8*
Transition economies	**4.4**	**23.5**	**43.3**
Russia	3.4	17.1	30.6
Other EECCA	10.0	54.0	104.6
Other non-OECD Europe	2.4	14.3	25.5
Developing countries	**2.9**	**21.3**	**41.3**
East and SE Asia, Oceania	3.9	29.3	58.7
China	*6.1*	*42.1*	*83.6*
Indonesia	*–1.5*	*2.1*	*5.6*
Other East Asia	*0.5*	*6.0*	*11.1*
South Asia	*2.8*	*19.7*	*36.3*
India	*3.3*	*20.8*	*38.2*
Other South Asia	*1.3*	*15.9*	*29.5*
Middle East	4.3	19.3	30.1
Africa	2.8	19.5	34.2
Northern Africa	*1.2*	*9.4*	*16.8*
Republic of South Africa	*3.8*	*21.9*	*35.1*
Other sub-Saharan Africa	*4.2*	*28.8*	*49.7*
Latin America	–0.2	4.2	7.3
Brazil	*–1.3*	*–0.9*	*–0.6*
Other Latin America	*1.2*	*10.5*	*16.8*
Central and Caribbean	*–2.1*	*–4.7*	*–7.1*
World	**1.0**	**8.4**	**15.9**
European Union	0.0	1.1	1.4
BRIICS	3.9	28.7	57.1
ROW	1.5	12.3	22.6

StatLink http://dx.doi.org/10.1787/256826112658

The growth outlined in Table 6.2 is likely to affect the environment in important ways. For example, if, as in the past, the growth of emissions of greenhouse gases is only partially related to GDP (which Table 6.2 shows to be 16% higher), so the additional growth caused emissions to be only 10% higher, the impacts would clearly be significant for climate change since an additional one gigatonne of CO_2 would be emitted from energy alone.[4] A more aggressive policy would be needed to prevent such emissions. A rough gauge of the additional environmental impacts can be derived by looking at the environmental elements of the System of Environmental and Economic Accounts (United Nations, 2000; *i.e.* the "green" national accounts):[5]

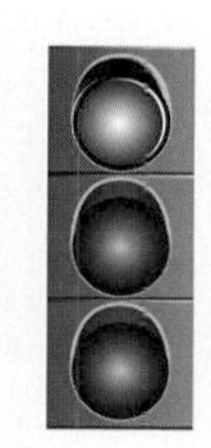

Stronger growth than projected in the Baseline could have significant negative impacts on climate change.

i) *Flow accounts for pollution, energy and materials.* These impacts would be a larger draw on the national accounts, reflecting more environmental damage from the generation of pollutants, solid waste, etc. In developed countries this is less of a problem since standards for clean air, clean water, etc., already exist. In developing countries, however, a political process that is slow to respond to environmental problems would result in greater harm being done (exposures to harmful air particulates and unclean water, for example, are already very high in rapidly developing economies; World Bank, 2007).

ii) *Natural resource asset accounts.* These impacts would also be larger, reflecting greater depletion through changes in stocks of natural resources such as land, fish, forest, water and minerals.

iii) *Valuation of non-market flows and environmentally adjusted aggregates.* The non-market valuation of environmental impacts and adjustment of several macroeconomic aggregates for depletion and degradation costs would reflect the increased activity that enhances existing externalities and market failures.

The very rapid productivity growth illustrated in this variation reflects existing concerns about China's rapid growth (OECD, 2007; World Bank, 2007). When growth is rapid, the ability of the political process to manage it and ensure that policy is able to cope with adverse consequences becomes strained. Policy processes require time to identify issues and build consensus around the need for corrective action. When that time is not available because growth is moving too quickly, then there is a risk that policy will fall significantly behind economic growth and environmental externalities will become much more severe than would otherwise be the case.

Variations 2 and 3: results

In Table 6.3, the long-term growth rate is changed from 1.75% to 2.25% (Variation 3) and compared with the Baseline. The resulting change in economic growth is substantially smaller than the change seen in Table 6.2. A convergence toward 1.25% is also shown in Table 6.3 (Variation 2).

The asymmetry shown in the table between the two growth objectives illustrates how much closer growth rates generally are to 2.25% in the initial years. That is, since the convergence occurs by a slow closing of the gap to 2.25%, there will only be a small increase when the target moves from 1.75% to 2.25% for countries that were already above 1.75%. However, since the initial gap between actual growth and 1.25% will be larger with higher growth rates, the dampening effect of the lower target will be stronger.

Table 6.3. **Change from Baseline in GDP (%) from long-term change in productivity growth**

	V2: target of 1.25%			V3: target of 2.25%		
	2010	2020	2030	2010	2020	2030
OECD	**0.0**	**–3.5**	**–7.3**	**0.0**	**2.7**	**6.4**
North America	0.0	–4.3	–9.0	0.0	3.1	6.4
US and Canada	0.0	–4.2	–8.8	0.0	3.2	6.7
Mexico	0.0	–5.7	–11.6	0.0	1.6	3.5
Europe	0.0	–2.7	–5.0	0.0	1.9	5.8
Pacific	0.0	–3.0	–6.5	0.0	2.9	7.2
Asia	0.0	–3.0	–6.4	0.0	3.0	7.5
Oceania	0.0	–3.0	–6.7	0.0	2.3	4.9
Transition economies	**–0.3**	**–2.1**	**–4.9**	**0.1**	**0.6**	**1.4**
Russia	–0.3	–2.1	–4.8	0.1	0.6	1.4
Other EECCA	–0.3	–2.1	–4.9	0.1	0.6	1.4
Other non-OECD Europe	–0.3	–2.1	–4.9	0.1	0.6	1.4
Developing countries	**–0.3**	**–2.2**	**–5.1**	**0.1**	**0.6**	**1.5**
East and SE Asia, Oceania	–0.2	–2.1	–4.9	0.1	0.6	1.4
China	–0.3	–2.2	–5.2	0.1	0.6	1.5
Indonesia	–0.3	–2.3	–5.2	0.1	0.6	1.5
Other East Asia	–0.2	–1.7	–4.3	0.1	0.5	1.2
South Asia	–0.3	–2.3	–5.2	0.1	0.6	1.5
India	–0.3	–2.3	–5.2	0.1	0.6	1.5
Other South Asia	–0.3	–2.3	–5.2	0.1	0.7	1.5
Middle East	–0.3	–2.3	–5.8	0.1	0.7	1.7
Africa	–0.3	–2.2	–5.3	0.1	0.6	1.6
Northern Africa	–0.3	–2.1	–4.9	0.1	0.6	1.4
Republic of South Africa	–0.3	–2.3	–5.7	0.1	0.7	1.7
Other sub-Saharan Africa	–0.3	–2.2	–5.6	0.1	0.6	1.6
Latin America	–0.3	–2.1	–5.0	0.1	0.6	1.5
Brazil	–0.3	–2.1	–5.0	0.1	0.6	1.4
Other Latin America	–0.3	–2.1	–5.0	0.1	0.6	1.5
Central and Caribbean	–0.3	–2.2	–5.3	0.1	0.6	1.5
World	**–0.1**	**–3.1**	**–6.6**	**0.0**	**2.1**	**4.8**
European Union	0.0	–2.7	–4.9	0.0	1.9	5.8
BRIICS	–0.3	–2.2	–5.1	0.1	0.6	1.5
ROW	–0.2	–2.0	–4.9	0.1	0.6	1.4

StatLink http://dx.doi.org/10.1787/256871785576

The rates of economic growth illustrated in Table 6.3 are less worrying for the environment over the next 25 years than those of Table 6.2. However, in the longer-term, the faster growth represented by the 2.25% rate will mean greater impact on the environment sooner, than with growth of 1.75%. Though human material well-being will be better, the resulting environmental impacts will require more urgent efforts to improve environmental outcomes.

Variations in the patterns of globalisation

Globalisation of trade and production has helped to improve the material well-being of vast numbers of people, but perhaps its greatest impact has been on the spread of knowledge and techniques rather than the pure exchange of goods (see Chapter 4 on globalisation). Studies that have attempted to quantify the benefits of trade in terms of increased GDP growth find that this impact is smaller than the impact on GDP of more dominant factors such as population growth and technical change.

While economic growth is a main determinant of the magnitude of environmental pressures, the geographical distribution of environmental impacts is determined by other factors, some of which governments can influence. This section therefore examines the impact that globalisation has on the location of environmental impact, rather than the impact that larger economies have on overall environmental outcomes.

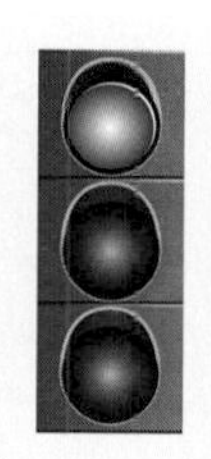

The increased trade and changing patterns of production will increase energy demands significantly for the world as a whole.

Globalisation implies that there is increasing interdependence between countries (*i.e* increasing specialisation of production), so trade is growing by more than any increase in the overall economy. This is in contrast to increasing international commerce that comes about simply from an increase in the size of economies. Globalisation implies an increase in the specialisation of production, and changes in the composition of domestic *versus* foreign-sourced consumption: implying that important structural changes are occurring. So a globalisation variation makes strong predictions about the economic future.

There is good reason to think that current patterns of globalisation are a result of policy initiatives (*e.g.* the successful conclusion and implementation of multilateral trade rounds) and other factors that have promoted trade. However, these implied policies have been omitted from the *Outlook* Baseline in an effort to clearly distinguish a reference case from a policy case. The Baseline also excludes recent reductions in obstacles to trade, such as declines in transportation and communication costs, as well as decreased border delays and other trade impediments. Since the reduction in these factors has been very difficult to quantify, the Baseline assumed that they would level off over the coming decade.

This variation explores what might happen if these past trends were to continue. It assumes continuing declines in:

- Trade margins: the additional revenue received by an exporter when selling on the international market instead of the domestic market. This increases demand by lowering prices in importing countries.
- Invisible costs: the difference between the price at which an exporter sells a good and the price that an importer pays.

For China and India, the changes in globalisation (*i.e.* increasing import/GDP ratio) relative to the Baseline are not implemented since they are already trading large shares of their economies in the Baseline (more than 31% for China, and 21% for India). Large economies, such as the United States and Japan, tend to have lower ratios of imports to GDP than smaller economies, such as Ireland and even Korea.[5] This is because, as economies increase, they tend to focus more on the production of services than the production of goods, and services are generally less traded. Moreover, large economies tend to produce a wider range of intermediate goods domestically, because there is more potential for economies of scale within a large economy. Countries producing a wider range of intermediate goods will show fewer imports relative to GDP (gross output is often several times bigger than value-added – GDP).

Table 6.4 shows the impact on trade of this variation. In many countries, continuing past trends in trade growth will lead to large increases in the import/GDP ratio compared with the Baseline. The changes are important from the perspective of the composition of output (since imports are increasing), but not from the perspective of the overall growth of

Table 6.4. **Percentage change from Baseline of implementing a globalisation variation in 2030**

	Import/GDP % change	GDP % change
OECD	42%	1%
North America	55%	1%
US and Canada	53%	0%
Mexico	65%	5%
Europe	33%	1%
Pacific	44%	1%
Asia	42%	1%
Oceania	54%	1%
Transition economies	**30%**	**1%**
Russia	30%	2%
Other EECCA	25%	0%
Other non-OECD Europe	35%	0%
Developing countries	**31%**	**2%**
East and SE Asia, Oceania	11%	0%
China	0%	–1%
Indonesia	29%	2%
Other East Asia	26%	3%
South Asia	5%	0%
India	0%	–1%
Other South Asia	32%	1%
Middle East	103%	16%
Africa	58%	2%
Northern Africa	82%	2%
Republic of South Africa	42%	1%
Other sub-Saharan Africa	42%	2%
Latin America	70%	1%
Brazil	49%	0%
Other Latin America	96%	1%
Central and Caribbean	39%	0%
World		**1%**

StatLink http://dx.doi.org/10.1787/256886566571

the economy (since GDP is not changing by as much). The increased trade and changing patterns of production will increase energy demands. As the last column shows, this is projected to be substantial in some cases, and significant for the world as a whole (an 8% increase). While globalisation may not in itself lead to much larger economies, it can have environmental impacts through the much wider dispersion of stages of production.

The changes in imports in this globalisation variation, as shown in the table, occur from a somewhat narrow simulation since it does not include tariff reductions from trade agreements such as the Uruguay Round Tariff Agreements. Such agreements usually specify some reduction in tariff levels and are combined with other measures that enhance trade, usually for a particular sector. The results shown below are thus likely to be missing some important details that may be included in future agreements, and thus may misrepresent some of the structural changes that increased trade would actually bring.

Some of the environmental impacts of this variation are illustrated in Figure 6.3. For the ROW regions, the impact is generally negative, and thus has some implications for policy coherence in OECD countries (*i.e.* achieving development and environmental goals in non-OECD countries). In OECD countries, there is a mild decrease in total primary energy supply (TPES). There is a notable decrease in sulphur and nitrogen oxide emissions.

Figure 6.3. **Environmental impacts of the globalisation variation to the Baseline**

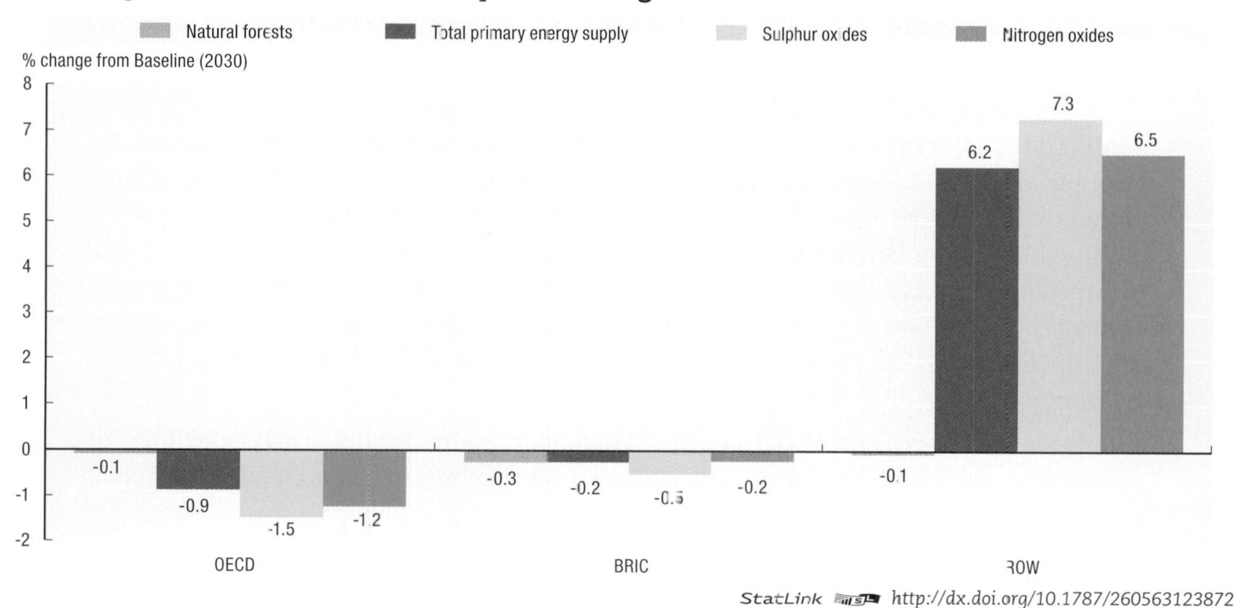

StatLink http://dx.doi.org/10.1787/260563123872

Source: OECD *Environmental Outlook* Baseline and variations.

Alternative scenarios and the Outlook

The methodology applied to obtain the results in the Baseline was specifically designed to be rigorous and give results that would be useful for policy analysis. Other methodologies would give baselines that had important and perhaps even stronger policy implications. Seroa da Motta (2007) notes just how different outcomes can be (Table 6.5).

Table 6.5. **Worldwide growth estimates, 2005-2050 (annual rates)**

Income indicator and country/source	Poncet (2006)	Hawksworth (2006)	O'Neill *et al.* (2005)
GDP			
China	4.7%	3.9%	7.4%
India	4.6%	5.2%	8.3%
Brazil	1.0%	3.9%	5.4%

StatLink http://dx.doi.org/10.1787/257110216826

Source: Seroa da Motta (2007), see references therein.

These scenarios illustrate how changes in important drivers, even without new government policy, can significantly change the nature of the world economy. Given this level of variability, anchoring the *Outlook* in historical trends in the critical drivers is important – both for putting the Baseline on a firm foundation, as well as for exploring the repercussions of various policy initiatives.

Policy implications

Model results are primarily useful to organise and reinforce analytical issues already known from theoretical foundations. Attempting to quantify uncertainty around model results is a necessary exercise in emphasising to analysts and policy-makers just how much information is conveyed in the results. When the uncertainty is conveyed to decision-makers, considerable care has to be exercised in ensuring that outlining the uncertainty does not imply that all is known and there are no sources of additional surprises that may be forthcoming.

Obviously, in addition to the economic variants discussed in this chapter, "technical" assumptions contribute to the uncertainty of the quantitative analyses for this *Outlook*. For example, the Baseline assumes a plausible but still impressive improvement of agricultural productivity. Without it, meeting the demands of the world's population by 2030 would require much more land than projected for this *Outlook*. Similarly, the proportion of coal assumed in the world's energy mix is plausible but by no means a maximum, as has been pointed out by reviewers, especially these from BRIC countries. Thus, also in this respect, the Baseline should not be misinterpreted as the maximum amount of degradation for the environment.

In the remainder of this *Outlook*, the variations presented in this chapter will not be examined further. The intention of the *Outlook* is to explore, at a broad level, issues that policy-makers need to address in the future. For that purpose, having a range of results for each quantitative analysis risks creating greater complexity than is needed. There is always a trade-off between having the maximum amount of good quantitative analysis to draw from (with in-depth analysis of possible variations), and having explanations that are relatively clear and succinct. In this chapter we have illustrated that even without implied changes in government policy, significant variations are possible to the Baseline. For the remainder of the *Outlook*, the focus will be on maintaining clarity in the messages.

Notes

1. More precisely, the dotted lines show the range created by two standard deviations from the median of models results for the SRES scenarios development exercise – though statistical inference is not implied by the range.
2. It also assumes that after 2007 there are two distinct stages that countries go through on their way to the 1.75% growth target: a medium-term process and a longer-term process.
3. A gradual process is imposed that levels off to reach the long-term target. In other words, only a few countries actually reach the target by the end of the *Outlook* horizon.
4. This relationship between emissions growth and GDP growth is generally, but not always, true. China, for example, had average emissions growth of 16% between 2000 and 2005, which was well above average GDP growth.
5. These issues will not necessarily arise with economic growth, but the debate over the Environment Kuznets Curve (Grossman and Krueger, 1995) suggests that there is no reason to assume that economic growth will, by itself, lead to a cleaner environment (Dasgupta *et al.*, 2002; Harbaugh *et al.*, 2002).
6. For the United States and Japan this is 14% and 10%, respectively; while for Ireland and Korea it is 65% and 40%, respectively.

References

Bergman, L. (2005), "CGE Modeling of Environmental Policy and Resource Management", in Maler, K-G. and J. R. Vincent (eds.) *Handbook of Environmental Economics*, Volume 3, Elsevier, Amsterdam.

Dasgupta, S. *et al.* (2002), "Confronting the Environmental Kuznets Curve", *The Journal of Economic Perspectives*, vol. 16, No. 1, pp. 147-68.

Grossman, G.M. and A.B. Krueger (1995), "Economic Growth and the Environment", *Quarterly Journal of Economics*, Vol. 110, 1995, pp. 353-378.

Harbaugh, B., A. Levinson and D. Wilson (2002), "Reexamining the Empirical Evidence for an Environmental Kuznets Curve", *Review of Economics and Statistics*, Vol. 84, No. 3, pp. 541-51.

OECD ENVIRONMENTAL OUTLOOK TO 2030 – ISBN 978-92-64-04048-9 – © OECD 2008

IMF (International Monetary Fund) (2007), *World Economic Outlook: Spillovers and Cycles in the Global Economy*, April, International Monetary Fund, Washington DC.

IPCC (Intergovernmental Panel on Climate Change) (2000), *Special Report on Emissions Scenarios*, Nakicenovic N., *et al.* (eds), Cambridge University Press. Also available at *www.grida.no/climate/ipcc/emission/index.htm*, cited 30 October 2006.

Maddison, A. (2003), *The World Economy: Historical Statistics*, OECD Development Centre, OECD, Paris.

Millennium Ecosystem Assessment (2005), *Ecosystems and Human Well-Being: General Synthesis*, Island Press, Washington, DC.

OECD (2007), *Environmental Performance Review of China*, OECD, Paris.

Seroa da Motta, R. (2007), "Brazil Paper on Environmental Outlook to 2030", paper presented at OECD *Global Forum on Sustainable Development*, 23-24 May 2007, Paris.

UNEP (United Nations Environment Program) (2002) *Global Environment Outlook 3*, Earthscan, London.

United Nations (2000), *United Nations, Integrated Environmental and Economic Accounting – An Operational Manual*, United Nations, New York, sales No. E.00.XVII.17.

World Bank (2007), *The Cost of Pollution in China: Economic Estimates of Physical Damages*, The World Bank, Washington DC.

II. IMPACTS OF ENVIRONMENTAL CHANGE

7. Climate Change
8. Air Pollution
9. Biodiversity
10. Freshwater
11. Waste and Material Flows
12. Health and Environment
13. Cost of Policy Inaction

ISBN 978-92-64-04048-9
OECD Environmental Outlook to 2030

Chapter 7

Climate Change

This chapter examines the projected emissions of greenhouse gases to 2030, by country and sector, and the expected impacts in terms of temperature change and other effects. Without new policies, it is projected that greenhouse gas emissions will increase by about 37% in 2030 compared to 2005 levels, with a wide range of impacts on natural and human systems. The chapter examines the key drivers of increases in greenhouse gas emissions, and explores a range of policy scenarios for reducing these emissions. It finds that early action by all emitters, covering all sectors and all greenhouse gases, can achieve an ambitious emission reduction target at low cost. It highlights the need to share the burden of the cost of mitigation action amongst countries.

KEY MESSAGES

Scientific evidence shows that past emissions of greenhouse gases are already affecting the Earth's climate, with resulting impacts on physical, ecological and social systems (IPCC, 2007a). Global temperatures are about 0.76°C higher than pre-industrial levels. Impacts will become more significant as temperatures and sea levels continue to increase and precipitation patterns shift during the latter part of the century and beyond.

The *Outlook* Baseline projects that current policies and emission trends will lead to a rapidly warming world (see graph and "Consequences of inaction" below). Protecting the climate requires reversing emission trends to reduce global GHG emissions significantly below today's levels by 2050.

Key drivers of emission growth are fossil fuel use (*e.g.* for power and transport) and unsustainable land use policies, including deforestation. Agriculture and waste also contribute to emission growth to 2050.

Recent progress has been made in establishing an international framework for action on climate change. There is also greater policy-making capacity today in many OECD countries to deal with climate change. In non-OECD countries there is also progress, for example to comprehensively monitor and report on emissions, to implement climate change and other relevant policies to reduce greenhouse gas emissions and adapt, and to host Clean Development Mechanism (CDM) projects. This experience will be of value for future climate policies.

Policy options

- Start today to reduce global CO_2 and other emissions in order to stabilise atmospheric concentrations at acceptable levels, and to significantly limit global mean temperature increases, *i.e.* to 2-3°C, rather than the 4 to 6°C projected in the Baseline. This would significantly limit the risk of the worst climate change impacts in the long-term.
- Create conditions for broad participation by all the big emitting countries in mitigation action under a post-2012 framework. This will be essential to achieve these outcomes in a cost-effective manner.
- Develop and strengthen climate-specific policies and measures to put a global price on carbon to stimulate development and deployment of climate-friendly technologies, clean energy systems and provide incentives to change consumer behaviour and business practices.
- Strengthen national frameworks and strategies to better co-ordinate climate change mitigation and adaptation through existing sector policies (*e.g.*, energy, transport, waste, land use and agriculture).
- Expand capacity in national governments to work more effectively with non-governmental actors and organisations, sub-national and city level governments on both mitigation and adaptation.

Cost of mitigation

Emission reductions are not only possible, they are also feasible at limited cost. Simulations in this chapter compare Baseline (no new policy) projections for GHG emissions, global mean temperature and GDP increase with different policy cases of a phased-in carbon tax of USD 25 per tonne of CO_2eq (see graph). Costs of a globally applied tax policy starting in 2008 would decrease GDP by only 1% below its "business as usual" level by 2050. Another more radical scenario involves phasing in a global tax to stabilise atmospheric GHG concentrations at 450 ppm CO_2eq. This policy reduces climate impact substantially (see graph), but has more significant, though manageable, global costs. It is projected to reduce Baseline estimates of GDP by about 0.5% and 2.5% by 2030 and 2050 respectively, amounting to a loss of about 0.1 percentage point a year on average. Aggregate costs of global mitigation (% GDP), with all countries participating, would be lower in the OECD than in the BRIC and ROW countries, underscoring the need for burden-sharing in future agreements.

Impacts of policy scenarios on greenhouse gas emissions

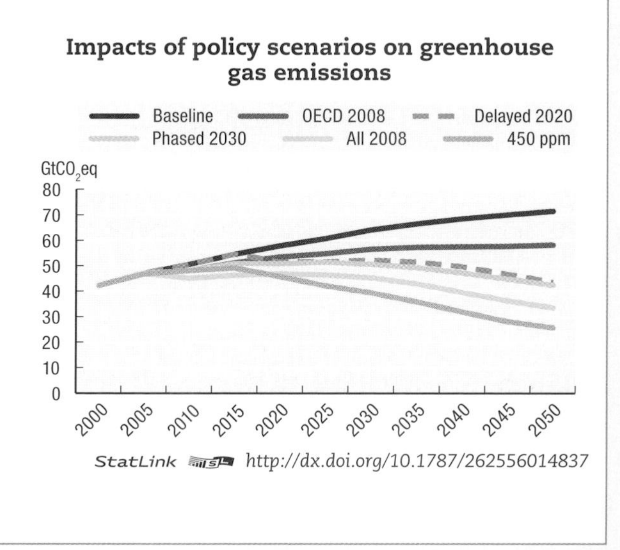

StatLink http://dx.doi.org/10.1787/262556014837

Consequences of inaction

The risks of inaction are high, with unabated emissions in the Baseline leading to about a 37% and 52% increase in global emissions in the 2030 and 2050 timeframe respectively compared to 2005, with a wide range of impacts on natural and human systems. This unabated emission pathway could lead to high levels of global warming, with long-term average temperatures likely to be at least 4 to 6°C higher than pre-industrial temperatures. The costs of even the most stringent mitigation cases are in the range of only a few percent of global GDP in 2050. Thus they are manageable, especially if policies are designed to start early, to be cost-effective and to share the burden of costs across all regions.

Introduction

This chapter presents the *Outlook* results for climate change. It begins with a brief review of the science of climate change to explain the nature of problem. This is followed by a review of historical greenhouse gas (GHG) emission trends and a description of Baseline projections. Next the chapter reviews the nature of the international and national policy challenge to respond to climate change. The chapter closes with a presentation of key results from the *Outlook* policy simulations, comparing the cost and effectiveness of alternative mitigation strategies to limit climate change between now and 2050 (and beyond). Climate change is a "stock pollutant problem" and is thus slow to develop; reductions of greenhouse gas emissions achieved today, and in the decades to come, will affect the climate of future generations. The chapter therefore places the policy challenge of today in the context of long-term climate change outcomes.

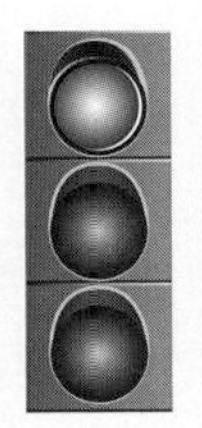

Scientific evidence shows unequivocal warming of the climate system.

Scientific evidence shows unequivocal warming of the climate system (IPCC, 2007a). The global surface temperature increased by 0.76 degrees Celsius from 1850-1899 to 2001-2005. Eleven of the 12 years between 1995 and 2006 rank among the 12 warmest years in the instrumental record since 1850 (IPCC, 2007a; and Figure 7.1). The rate of temperature change has also accelerated, rising to about 0.13°C per decade in the last 50 years, which is about twice the recorded rate of change for the previous 100-year period (IPCC, 2007a); this rate has increased in the last two decades.

The distribution of climate change varies widely by region, with more pronounced warming observed over the interiors of large land masses. Generally regional temperature increases are smaller towards the equator and larger towards the poles. Over the last century, average Arctic temperatures have increased at almost twice the rate of the rest of the world (IPCC, 2007a). Natural factors such as volcanoes and changes in solar radiation cannot explain these phenomena (IPCC, 2007a).

Numerous long-term changes in climate and in natural systems have been observed, many of which are attributable to human activities (IPCC, 2007a). Observed changes include large-scale declines in snow pack and ice cap coverage and glacier retreat in many regions (IPCC, 2007a). Changes have also been observed in many weather extremes since the 1970s, including more intense and longer droughts, particularly in the tropics and subtropics; an increase in the intensity of tropical cyclones (Emanuel, 2005; Webster *et al.*, 2005; IPCC, 2007a); as well as an increase in the frequency of heavy precipitation over most land areas (IPCC, 2007a). The duration and size of wildfires in the western United States are now partially attributed to changes in summer temperatures, precipitation patterns and earlier spring snowmelt (Westerling *et al.*, 2006; IPCC, 2007b). Some evidence of non-linear change is also evident in observed climate change; for example, studies suggest the Atlantic overturning circulation may be 30% slower than between 1957 and 2004 (IPCC, 2007b and c;

Figure 7.1. **Global temperature, sea level and Northern hemisphere snow cover trends, 1850-2000**

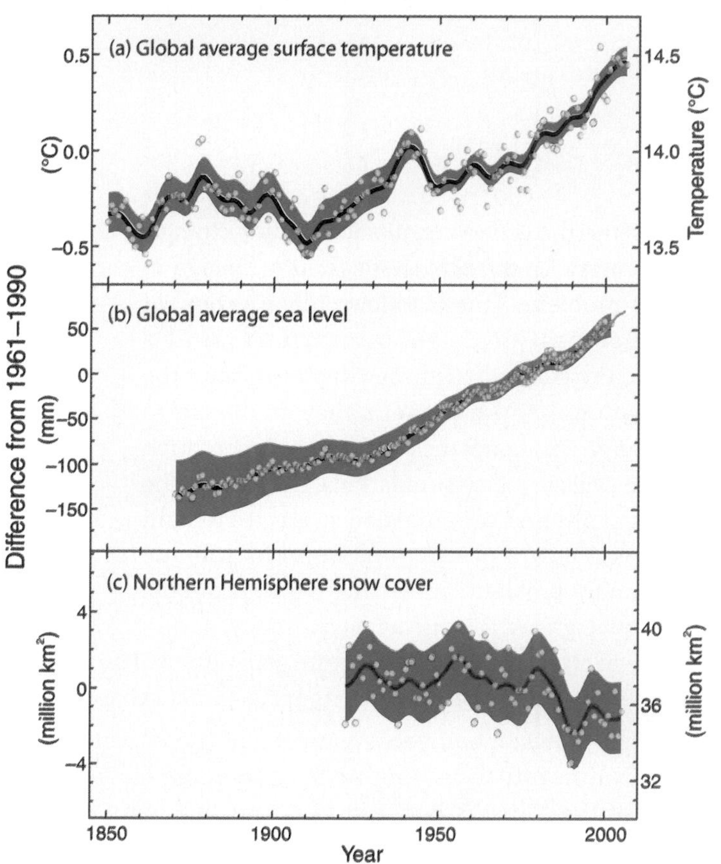

Note: Observed changes in *a)* global average surface temperature; *b)* global average sea level from tide gauge (blue) and satellite (red) data; and *c)* Northern hemisphere snow cover for March-April. All changes are relative to corresponding averages for the period 1961-1990. Smoothed curves represent decadal average values while circles show yearly values. The shaded areas are the uncertainty intervals estimated from a comprehensive analysis of known uncertainties (a and b) and from the time series (c).

Source: Reproduced from IPCC, 2007a, Figure SPM.3.

Bryden *et al.*, 2005). Changes in ocean acidity due to increases in carbon dioxide emissions, reported for the first time in 2004, are altering ocean chemistry and may threaten marine organisms (Feeley *et al.*, 2004; see also Chapter 15 on fisheries and aquaculture). Ecological systems of all types are shifting in elevation and geographical location (IPCC, 2007b; see also Chapter 9 on biodiversity). These observed changes suggest that ecosystems are among the most sensitive of natural and human systems to the pace and the magnitude of climate change, while also the least amenable to managed adaptation.

Most of the observed warming since the mid-20th century is due to changes in greenhouse gas concentrations and can be attributed to human activities (IPCC, 2007a). Climate change is driven by increases in the global population and economic growth, particularly the production and consumption of fossil fuels, the expansion of agriculture and deforestation, all of which have increased GHG emissions (IPCC, 2007a and c).

Atmospheric carbon dioxide and methane (CH_4) concentrations are higher than at any time in the last 650 000 years (Spahni *et al.*, 2005; Siegenthaler *et al.*, 2005; IPCC, 2007a).[1] Increased emissions of CO_2 over the last 100 years increased atmospheric CO_2

concentrations from approximately 280 to 379 parts per million (ppm) in 2005,[2] while methane concentrations increased from 715 to 1 774 parts per billion (ppb) (IPCC, 2007a). Higher concentrations of greenhouse gases in the atmosphere lead to warming, which is offset somewhat by cooling from sulphur aerosols.

As a result of lags in the Earth's systems, particularly the oceans, it is estimated that even if the composition of the atmosphere stabilised today, an additional increase in warming of 0.3-0.9 °C (with a best estimate of 0.6 °C) would still occur over this century (Hansen *et al.*, 2005; IPCC 2007a).[3] Without significant efforts in this century to reduce emissions below current levels, future predictions of climate change suggest it is likely or, in some cases certain, that we will see an acceleration of warming trends, associated climate changes and impacts.

Key trends and projections

Current sources, sinks and historical trends

The principal gases associated with climate change are carbon dioxide (CO_2), methane (CH_4), and nitrous oxide (N_2O), which together accounted for over 99% of anthropogenic GHG emissions in 2005. CO_2 is the dominant greenhouse gas, accounting for 64% of global emissions and about 83% of emissions from OECD countries in 2005, excluding land use and forestry emissions and removals. Including land use change and forestry increases the share of CO_2 in 2005 to 76% globally and does not significantly change the share for the OECD. Hydrofluorocarbons (HFCs), perfluorocarbons (PFCs) and sulphur hexafluoride (SF_6) account for less than 1% of total global anthropogenic GHG emissions, but they are growing quickly. All these greenhouse gases are subject to international obligations under the United Nations Framework Convention on Climate Change (UNFCCC), including national monitoring and reporting of emissions and removals of greenhouse gases.

Fossil fuel combustion is by far the largest global source of CO_2 emissions, accounting for 66% of global GHG emissions in 2005. Of this, fossil fuel combustion in power generation is the most important source, and accounted for about one-quarter of all global GHG emissions in 2005. Electricity-related CO_2 emissions are also a rapidly-growing source of GHGs, particularly in Asia, reflecting both increased electrification rates and the continued predominance of fossil-fired electricity. Global CO_2 emissions from road transport are a significant contributor to global GHG emissions, at 11% of the total in 2005.

Trends in GHG emissions vary widely according to world region. Global anthropogenic GHG emissions (excluding CO_2 emissions or uptake from land use change and forestry and from international bunkers) increased by 28% between 1990 and 2005.[4] This increase was lower in OECD countries (+14%) than in BIC countries (Brazil, India, China), where emissions grew by about 70%. However, emissions in some countries – particularly those in Central and Eastern Europe – fell during the same period. Trends for OECD countries are broadly similar even if emissions or uptake from land use change and forestry are included, in which case OECD countries' emissions increased 10% over the period 1990-2005.[5] BIC countries' emissions also increase even more (nearly 110%) if CO_2 emissions from land use change and forestry are included.[6]

However, between 1990 and 2005 there were also large variations in these trends within different OECD countries. Emissions in nine OECD countries increased by more than 20% in this period,[7] and eight further OECD countries reported smaller increases.[8] However, emissions in several other OECD countries have decreased since 1990, including

Germany, Hungary, Finland, Norway, the Czech Republic and Slovakia, where 2005 emissions were between 67-80% of their 1990 value.

Future projections

There is a large body of literature that assesses future emissions of greenhouse gases (IPCC, 2007c). In almost all such studies, human activities are projected to cause emissions of greenhouse gases to increase for decades or more, unless policies are introduced to alter these trends by providing incentives to limit demand for energy or other emission intensive products, or to change behaviour and technologies in climate-friendly ways.

For the purposes of assessing climate change, the OECD *Outlook* is extended to 2050. Projected GHG emissions trends (including land use change and forestry) by region are shown in Table 7.1 and Figure 7.2. These trends show absolute growth in emissions through 2050 across all regions, with global emissions of all GHGs increasing by about 37% and 52% to 2030 and 2050 respectively. Growth is significantly higher in BRIC and ROW regions compared to the OECD. Accordingly, the share of BRIC and ROW within world emissions increases in this timeframe, growing from 60% in 2005 to 67% in 2050, while the OECD share declines slightly from 40% to 33% in the same period.

Table 7.1 also shows indicators of emission intensity, both per capita and per USD of gross domestic product (GDP). Intensity indicators show that emissions per capita increase in all regions, while emissions per USD of gross domestic product (in 2001 USD) decline across regions. Per capita GHG emissions in BRIC countries were only about one-third of those in OECD countries in 2005 (the equivalent of 5.1 tonnes (T) of CO_2eq per person in BRIC countries compared with 15 T CO_2eq per person for OECD countries)[9] and this pattern continues. The OECD remains the most emission intensive of the regions on a per capita basis, while it is the least emission intensive when measured on a GDP basis.

In the *Outlook* Baseline, CO_2 emissions from energy, industry and land use are also projected to increase from 35.9 GtCO_2 in 2005, to 49.8 GtCO_2 in 2030 and to 55.7 GtCO_2 in 2050, or an increase of 39% and 55% respectively (Figure 7.3).[10] The rapid increase of global energy-related CO_2 emissions is largely as a result of a projected continued expansion in the use of fossil fuel to support growing demand for electricity (Figure 7.3; and see Chapter 17, Energy). Demand for electricity is projected to double between 2000 and 2030, increasing emissions from power generation by 65% to 2030 and by 100% (to 22.2 GtCO_2 compared to nearly 11 GtCO_2 in 2005) to 2050. Global emissions of CO_2 from the transport sector are expected to expand from 6.1 GtCO_2 in 2005, to 9.6 GtCO_2 in 2030 and 12.2 GtCO_2 in 2050, thus roughly doubling by 2050 as the demand for cars increases, particularly in developing countries. Aviation is projected to be the most rapidly growing sub-sector (see also Chapter 16, Transport, and note 6 at the end of this chapter).

The IPCC recently summarised available literature on reference or baseline emission scenarios and established a range of outcomes across these scenarios to 2100. Looking at CO_2 from energy, the IPCC shows an increase ranging from 30-55% between 2005 and 2030, and 50-100% between 2005 and 2050.[11] By comparison, the *OECD Environmental Outlook* projects an increase of about 51% from 2005 to 2030 and 78% to 2050, while the IEA *WEO 2006* shows an increase of about 42% in CO_2 emissions from energy to 2030 from 2005. Both the OECD and the IEA baseline scenarios thus lie in the middle of the full range of emission scenarios available in the literature (Fisher *et al.*, 2007) (see also Chapter 17, Energy).

OECD ENVIRONMENTAL OUTLOOK TO 2030 – ISBN 978-92-64-04048-9 – © OECD 2008

Table 7.1. ***Outlook* Baseline global emissions by region and GHG intensity indicators: 2005, 2030 and 2050**

	2005	2030	2050
		All GHG – Gt CO_2eq	
OECD	18.7	23.0	23.5
BRIC	16.1	23.5	26.2
ROW	12.1	17.6	21.7
World	**46.9**	**64.1**	**71.4**
		Change in GHG, 2030 and 2050	
		% increase	% increase
OECD	Base year	23%	26%
BRIC	–	46%	63%
ROW	–	45%	79%
World	–	37%	52%
		Shares of total GHG by region	
	% share	% share	% share
OECD	40%	36%	33%
BRIC	34%	37%	37%
ROW	26%	27%	30%
		CO_2eq per capita (T/person)	
OECD	15.0	16.8	17.0
BRIC	5.1	6.1	6.4
ROW	5.8	5.9	6.0
World	**7.2**	**7.8**	**7.8**
		CO_2eq per GDP (kg/USD real)	
OECD	0.7	0.5	0.3
BRIC	4.6	2.2	1.3
ROW	2.9	1.6	1.0
World	**1.3**	**0.9**	**0.6**

StatLink http://dx.doi.org/10.1787/257114344671

Note: Figures include land use change and forestry.
Source: *OECD Environmental Outlook* Baseline.

Figure 7.2. **Baseline GHG emissions by regions, 1990 to 2050**

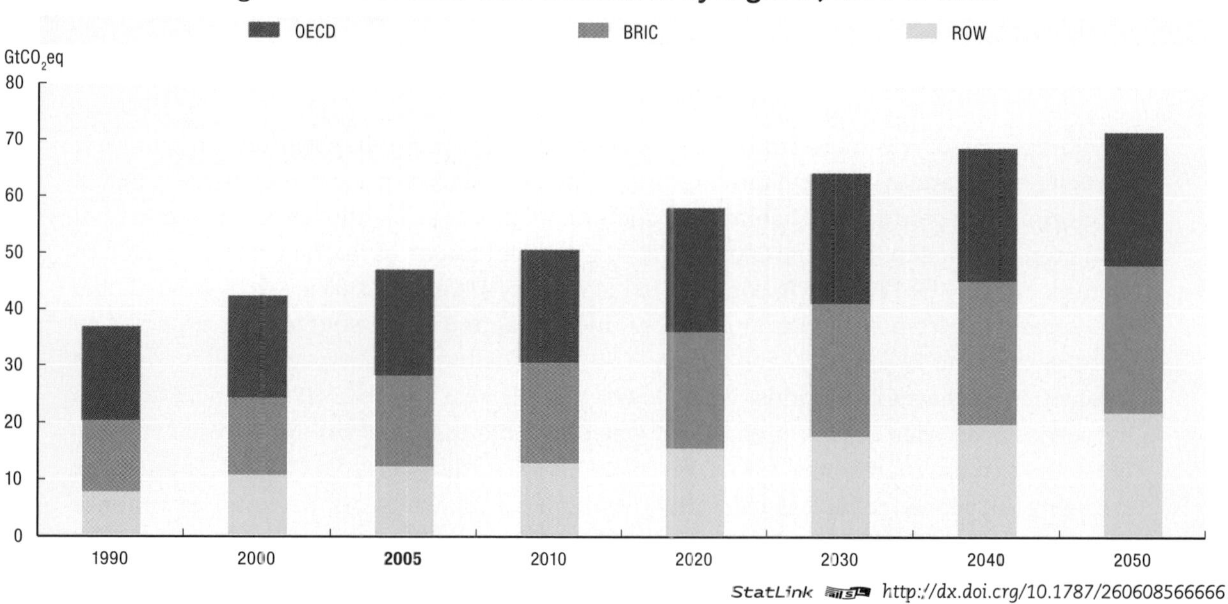

StatLink http://dx.doi.org/10.1787/260608566666

Note: 2005 also included as it is the base year.
Source: OECD Environmental Outlook Baseline.

This *OECD Environmental Outlook* also includes projections of greenhouse gas emissions from non-energy sectors (Figure 7.3). Among the most important of these are CO_2 emissions from global land use change, largely derived from rapid conversion of forest to cropland and grassland in tropical regions. These emissions are estimated to be 5.7 CO_2 Gt per year by 2005, and are projected to decline over the coming decades to 4.1 Gt CO_2 in 2030, and 1.9 Gt CO_2 in 2050. This is due in part to slowing population growth which is likely to reduce pressure on forest areas. Although the quality of inventory data is steadily improving, due to monitoring complexities these projections have large uncertainties, as do the base year estimates.

Figure 7.3. **Total greenhouse gas emissions by gas and CO_2 emissions by source category, 1980-2050**

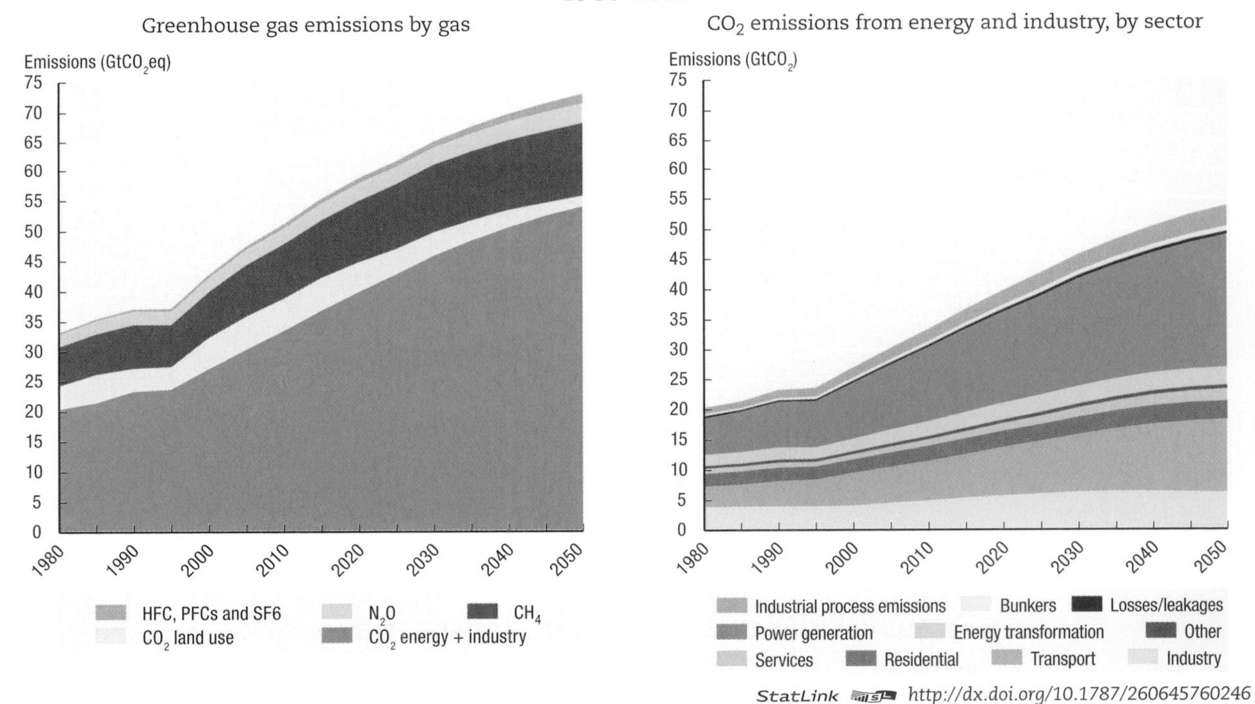

StatLink http://dx.doi.org/10.1787/260645760246

Source: OECD Environmental Outlook Baseline.

Emissions of methane from sources such as solid waste disposal on land, enteric fermentation, natural gas pipelines, rice production, etc. are also projected to increase in line with expanding production of animal products and rice, but at slightly lower rates than total food crop production. Between 2005 and 2030 global emissions of methane are projected to increase roughly by 32%, and to continue to increase to 47% above 2005 levels by 2050. Global N_2O emissions from agricultural practices, industrial and other sources are expected to increase by about 20% by 2030 and 26% by 2050 as agricultural land expands and production intensifies in the next decades, with slower growth nearing 2050. HFCs and PFCs from industrial processes have a high global warming potential and will grow most rapidly, projected to more than double from 2005 to 2030, and nearly quadruple by 2050. These gases are being introduced to replace chlorofluorocarbons (CFCs), which are powerful greenhouse gases and also deplete the ozone layer.[12] By 2050 HFCs and PFCs are projected to contribute roughly 4% of the total change in GHG emissions from 2005.

Policy implications

Successful mitigation of climate change will require an international effort to limit global greenhouse gas emissions significantly below current levels over the long-term (*e.g.* see Figure 7.5). The main international means to address climate change is the United Nations Framework Convention on Climate Change (UNFCCC), which has been ratified by 189 countries. Leadership on the climate change issue has emerged at the highest levels of government in many industrialised countries, and the worldwide prominence of the issue has risen in recent years.

Signatories of the Convention have agreed to work collectively to achieve its ultimate objective (Article 2, UNFCCC), which is: "... stabilisation of greenhouse gas concentrations in the atmosphere at a level that would prevent dangerous anthropogenic interference with the climate system. Such a level should be achieved within a time-frame sufficient to allow ecosystems to adapt naturally to climate change to ensure that food production is not threatened and to enable economic development to proceed in a sustainable manner." By signing the Convention, OECD members and other industrialised nations (or Annex I Parties) agreed to take the lead to achieve this objective, as well as to provide financial and technical assistance to other countries[13] to help them address climate change.

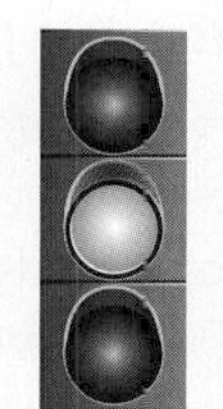

Successful policies to limit GHG emissions will require the participation of all major emitting countries.

In 2005, the Kyoto Protocol entered into force, an event that helped to raise the level of priority attributed to climate change by many governments. The Kyoto Protocol shares the Convention's objectives, but strengthens them through commitments of Annex I Parties (see above) to individual, legally-binding targets to limit or reduce their greenhouse gas emissions. To date 175 countries have ratified the Protocol; 36 of these countries and the EC are required to reduce greenhouse gas emissions below specific levels, a total cut of approximately 5% from 1990 levels by the 2008-2012 period.[14]

When adopting the Kyoto Protocol, governments recognised that it was only a first step in tackling climate change and achieving the Convention's ultimate objective. This has become even clearer today, as the economies and energy demand of some of the developing countries, such as China and India, have grown rapidly in the intervening years, with large increases in emissions (see Figure 7.2). Currently internationally-agreed mitigation targets apply only to industrialised countries and do not extend beyond 2012. At a Conference of the Parties held in Montreal in December 2005, Convention Parties agreed to an on-going dialogue to exchange experiences and analyse strategic approaches for long-term co-operative action to address climate change. This dialogue process will conclude at the Conference of the Parties in December 2007, which is widely expected to agree to launch negotiations for a comprehensive agreement to reduce emissions post-2012.[15] Successfully stabilising atmospheric concentrations to limit emissions and achieve the objectives of the Convention will require the participation of all major emitting countries.

The Convention and the Protocol are not prescriptive, allowing each party the flexibility to decide how to reduce emissions and implement commitments. There is a wide variety of national policies and measures available to governments to mitigate emissions. These include regulations and standards, market-based instruments (emission taxes and charges, tradable permits, and subsidies/financial incentives), voluntary agreements, research and development and information instruments. The environmental effectiveness of policies depends on their stringency and on implementation measures, including monitoring and compliance procedures, whereas the cost-effectiveness will depend to a great extent on how policies are implemented (IPCC, 2007c). Reducing emissions across many sectors and gases requires a portfolio of policies tailored to specific national circumstances. In general, climate change policies will need to be adjusted over time as new knowledge emerges about climate risk as well as about the means to manage climate change and its costs (IPCC, 2007c).

National policy frameworks to address climate change

Governments, corporations, states and cities have recently introduced measures to reduce emissions in the near-term and to promote the development of new GHG-friendly technologies that will be needed in the future. GHG emission trends in industrialised countries suggest that some progress, though still limited, has been made to curb GHG emissions since 1990. Most industrialised nations now have 10-15 years of experience with climate change as a national policy issue, suggesting that it is an opportune time to review and draw lessons from what has been achieved for the future.

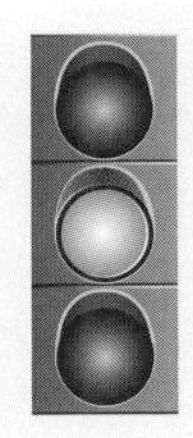

In industrialised countries some progress in curbing GHG emissions has been made, but efforts are insufficient given national and international goals to limit climate change.

There is also growing evidence of more significant policy-making capacity to deal with climate change in many countries compared to earlier years. A look at progress to date in efforts to mitigate emissions highlights several important issues. First is the emergence of climate change specific policies, or those that are truly new and designed to target GHG emission reductions. Such policies are often cross-sectoral, are comprehensive in their coverage of GHGs and are more stringent than early mitigation policies. Examples include emission trading schemes, CO_2 and green energy taxes, voluntary measures with industry to address GHG emissions, targeted regulation (*e.g.* for CH_4 emissions), collaborative research and development programmes.

Second, there is progress in many countries to develop "whole-of-government" efforts to integrate climate change into pre-existing sector policy frameworks. Examples include measures to accelerate investment in energy efficiency through energy policy and to promote mass transport options through transportation policy frameworks. In non-energy sectors, waste minimisation, landfill gas recovery and agriculture fertiliser management are examples of pre-existing measures that have been reinforced due to concern about greenhouse gas emissions. All of these low-cost measures have multiple environmental and economic benefits (*e.g.* see Table 7.2). Importantly, there are numerous local and national co-benefits of taking steps to reduce CO_2 and other GHG emissions other than avoiding climate change, such as reduced air pollution and improved energy security. And at the global level, action to limit HFCs and CFCs will benefit both climate and ozone protection efforts (Velders *et al.*, 2007). In addition, land use planning, agriculture and infrastructure design are increasingly taking into account climate change risk at the local scale, flagging the early development of adaptation (see below).

The third area of progress is the emergence of multilevel governance on climate change issues, both vertically (from local to national) and horizontally (across both governmental and non-governmental actors). Leadership and experimentation by cities and other

OECD ENVIRONMENTAL OUTLOOK TO 2030 – ISBN 978-92-64-04048-9 – © OECD 2008

Table 7.2. **Related aims and co-benefits of sector policies to reduce GHGs**

Sector	Climate policy aims and benefits	Other (non-climate change) benefits
Electricity production and industrial energy use	Encourage fuel switching from coal and oil to low or no-emission energy sources, such as renewable energy and energy efficiency, to reduce CO_2 emissions.	Raise regional and urban air quality and limit SO_x and NO_x air pollution, preserve water quality, protect forests and ecosystems; increase energy security.
Residential – buildings and appliances	Lower energy use requirements of housing and household services, reduce CO_2 emissions.	Lower investment costs for energy suppliers and possibly smooth load; lower operating costs for consumers and avoid pollution from (unnecessary) electricity and/or heat generation; improve comfort and affordability; raise energy security.
Industry – manufacturing	Stimulate investments in energy and materials efficiency, reduce CO_2 and other GHG emissions.	Improve resource efficiency of industrial operations; short- and long-term financial savings; lower energy consumption (and costs); raise profits and energy security.
Transport	Raise the efficiency and emission performance of vehicles and manage demand, reduce CO_2 and possibly other GHG emissions.	Lower congestion in cities and limit harm to human health from urban air pollution; lower dependency on oil imports to raise energy security, gain in technology leadership. However dis-benefits may also exist *e.g.* increased diesel fuel use lowers CO_2 but increases particulates, which have human health risks; also catalytic converters lower NO_x emissions but raise N_2O and CO_2 emissions.
Agriculture	Minimise nitrogen fertiliser use, reduce N_2O emissions.	Lower nitrogen run-off from agriculture and improve water quality; improve sustainability performance.
Waste	Minimise waste, encourage recycling and material efficiency in production and packaging, reduce CH_4 emissions.	Limit needs for costly and unsightly landfilling; improve economic performance.

sub-national governmental authorities are increasingly shaping mitigation strategies. Sweden, the UK and the US, among others, have city governments which have taken the lead on mitigation. Australia, Canada and the US provide examples of proactive state or provincial governments. In the private sector, some companies have also begun to target and regulate GHG emissions. Sub-national regions and cities may also play an essential role in adaptation planning, as seen in emerging efforts in Denmark, Canada, the UK and the US.

Integrating adaptation responses into sector and natural resource management policies is expected to be a key way forward to limit the socio-economic risks of climate change (Agrawala, 2005; Levina and Adams, 2006; McKenzie-Hedger and Corfee-Morlot, 2006). However, much less progress has been made on adaptation compared to mitigation. Adaptation includes coastal zone and water resource management policies as well as disaster prevention and planning policies (*e.g.* to anticipate more frequent flooding, drought, heat waves or fire, depending on the region). Other benefits of such measures include reinforcing sustainability and creating a greater capacity for sectors to respond to climate variability as well as climate change over the longer-term. Table 7.3 highlights the coverage of impacts and adaptation in national reports on progress under the UNFCCC.

In addition to national action on adaptation, the EU is taking steps to advance the adaptation agenda as a priority across its member states. In 2007, the European Commission adopted its first policy document on adaptation highlighting the need for early action where there is sufficient knowledge, using EU research to fill knowledge gaps and integrating global adaptation into external relations policy (CEC, 2007). The OECD Development Assistance and Environment Policy Committees also recently issued a declaration on adaptation, calling for greater co-operation and attention in development assistance and national planning for development (OECD, 2006).

Table 7.3. **Coverage of impacts and adaptation in National Communications under the UNFCCC (including NC2, NC3, and NC4)**

		Climate change impact assessments			Adaptation options and policy responses				
		Historical climatic trends	Climate change scenarios	Impact assessments	Identification of adaptation options	Mention of policies synergistic with adaptation	Establishment of institutional mechanisms for adaptation responses	Formulation of adaptation policies/ modification of existing policies	Explicit incorporation of adaptation in projects
Early to advanced stages of impact assessment	Iceland	■	✖	■					
	Hungary		■	✖					
	Portugal	✖	■	■					
	Estonia	■	■	■					
	Latvia	■		✖	✖				
	Russia	■	✖	■	✖				
Advanced impacts assessment, but slow development of policy responses	Japan		■	■	✖				
	Romania	■	■	■	✖				
	Denmark	■	■	■	✖				■
	Korea	■	■	■	✖				
	Slovenia		■	■	■				
	Ukraine*		✖	■	■				
	Belarus		■	■	■				
	Bulgaria	■	■	■	■				
	Croatia	■	■	■	■				
	Mexico	■	■	■	■				
	Slovak Republic	■	■	■	■				
	Norway		●	■	✖	✖			
	Czech Republic		■	■	■	✖			
	Liechtenstein		✖		✖	■			
	Germany		✖	■	✖	■			✖
	Austria		■	■		■			
	Lithuania	■	■	■	■	■			
	Greece	■	■	■	✖	■			
	Italy*	■		■	■	■			
Moving towards implementing adaptation	Spain		■	■	■	■	●		
	Ireland	■	■	■		■	■		
	Finland	■	■	■	■		■		
	Poland		■	■	✖	■		■	
	Switzerland	■	■	■	✖	■		■	
	Sweden	■	■	■		■		■	
	United States	■	■	■	■	■		■	
	Canada		✖	●	■			■	■
	New Zealand		■	■	■	●	✖	●	
	Belgium		■	■	✖	■	■	■	
	Australia		■	■	✖	■	■	■	
	France		■	■	■	■	■	■	
	Netherlands			■		■	■	■	■
	United Kingdom		■	●	●	■	■	●	■

* NC2/NC3 only.

Coverage:

- Extensive discussion (dark shading)
- Some mention/limited discussion (light shading)
- No mention or discussion (no shading)

Quality of discussion:

- ■ Discussed in detail, *i.e.* for more than one sector or ecosystem, and/or providing examples of policies implemented, and/or is based on sectoral/national scenarios.
- ✖ Discussed in generic terms, *i.e.* based on IPCC or regional assessments, and/or providing limited details/no examples/only examples of planned measures as opposed to measures implemented.
- ● Limited information in NCs, but references to comprehensive national studies.

Source: Gagnon-Lebrun and Agrawala, 2008.

Market-based instruments

A large number of market-based instruments are used in a variety of ways by countries to mitigate GHG emissions. These include emission charges and taxes, product charges, tax differentiation and subsidies.[16] Several OECD countries have implemented modest CO_2 emission taxes or "green" energy taxes intending to limit emissions. For example, in Denmark, Finland, the Netherlands, Norway and Sweden, CO_2 or "green" energy taxes have been in place since the early 1990s. In the Netherlands and Sweden significant energy taxes or rebate/refund systems encourage investments in energy efficiency and the use of renewables. The Swiss government also implemented a CO_2 tax in 2006 (UNFCCC, 2006b).

GHG emission trading is another prominent form of market-based instrument for climate change mitigation. The Kyoto Protocol allows industrialised countries to achieve their emission targets through the use of a number of international market-based instruments that are flexible about where emission reductions take place.[17] These include international emissions trading (Box 7.1), the Clean Development Mechanism (CDM) and Joint Implementation (JI). These flexible approaches help to lower the costs of compliance below what they would be if each country worked alone.

Emission trading is being implemented or considered by a number of national governments, for example the EU, Norway, Japan,[18] Australia and New Zealand, and by sub-national entities such as the states in the US and provinces in Canada. The EU Emission Trading Scheme (ETS) is by far the largest of these and is enabling more than 25 countries to test and gain practical experience with this instrument, including design and competitiveness issues. Implementation of the ETS has included extensive discussions about efficient and politically feasible design options and, more generally, the applicability of a cap and trade approach to GHG emission sources (and sinks). This has also prompted a large number of studies on efficiency and equity issues associated with the distribution of permits, the implications of economy-wide *versus* sectoral programmes, mechanisms for handling price uncertainties, different forms of targets, and compliance and enforcement issues.

Two other "flexibility mechanisms" under the Kyoto Protocol will also generate tradable credits. The Clean Development Mechanism (CDM) allows Annex I Parties to implement project activities that reduce emissions by non-Annex I Parties, in return for certified emission reductions (CERs). The CERs generated by such project activities can be used by Annex I Parties to help meet their emissions targets under the Kyoto Protocol, provided that the projects help developing countries achieve sustainable development.[19] The CDM is growing fast and is currently expected to generate 2.1 billion credits by 2012 (UNEP/RISO *www.cdmpipeline.org*) which is already a significant proportion of the expected gap between mitigation targets and national emissions under current policies.

The second of these "flexible mechanisms" is Joint Implementation, where Annex I Parties may implement an emission-reducing project in the territory of another Annex I Party and generate emission reduction units (ERUs) towards meeting its own Kyoto target. It is likely that many countries will have to implement additional policies and/or take more advantage of these flexibility mechanisms to achieve their Kyoto Protocol emission targets.

Box 7.1. **The European Union Emission Trading System (EU ETS)**

The launch of the EU ETS is one of the most significant recent policy developments aimed at reducing GHG emissions in industrialised countries under the Kyoto Protocol. It is a so-called "cap and trade" system where participants agree to work together through a market to achieve fixed emission reduction targets. Its first, pilot, phase ran from 2005-2007. Its second phase runs from 2008-2012, and its third phase will start in 2013. The EU ETS extends to all EU member states (25 in the pilot phase, and 27 in the second phase). In March 2007, the European Council endorsed an energy and climate package, making an independent commitment to reduce greenhouse emissions by at least 20% by 2020 and concluding that the reduction target would be increased to 30% in the context of an international agreement that includes other industrialised countries. A key challenge for the EU will be delivering on these political commitments. Before the end of 2007, the Commission will present a proposal to amend the Emission Trading Directive as well as a Burden Sharing Decision to achieve the agreed greenhouse gas reduction target.

The EU ETS is significant in all EU countries in terms of the scope of emissions covered under the system, which includes approximately half of gross EU CO_2 emissions from almost 11 500 installations during 2005-2007. The share of CO_2 emissions covered in individual countries varies widely, from approximately 22% in Luxembourg to 78% in Finland. Coverage of the EU ETS will expand during the second phase in terms of numbers of installations, the type of GHG emission covered (with some countries choosing to include industrial N_2O emissions), and potentially also the emission sources covered (*e.g.* aviation).

In the pilot phase of the EU ETS, national allocation plans (including reserves for new entrants) allowed for a slight increase in emissions from the covered facilities above baseline emission levels. Actual emissions were below allocation levels by approximately 8% in 2005 and 2% in 2006, indicating that the allocations in the pilot phase did not effectively constrain emissions below what they would have been otherwise. Allocation for the second phase of the EU ETS is much tighter, with the proposed cap for EU25 member countries lower than their EU ETS emissions in 2005, even though the coverage of phase two is larger than phase one.

A number of factors have affected allowance prices in the EU ETS, including the overall size of the allocation, relative fuel prices, weather and the availability of Certified Emission Reductions (CERs) from the CDM. The market has grown enormously, with over one billion tonnes CO_2eq of allowances, corresponding to over USD 24 billion, traded in the EU ETS during 2006 (Capoor and Ambrosi, 2007). The EU ETS has experienced significant price volatility during its pilot phase, with prices rising to over EUR 30 per tonne CO_2, but then dropping dramatically in April 2006 when emissions data from member states were released showing that they had emitted less than anticipated. By late 2007, prices for phase one allowances were lower than EUR 0.1 per tonne. However, prices for phase two allowances are much higher (EUR 21-23/tonne in October 2007) due in part to the much more stringent allocations in this phase.

From 2013, there may be significant changes in the coverage of the EU ETS and in its links to other schemes – as well as increased harmonisation of the cap-setting, allocation, monitoring, reporting and compliance provisions. The Commission's recommendations for such changes will be made in its review at the end of 2007, and should be finalised during 2008-2009.

Regulations and standards

Regulations and standards specify abatement technologies (technology standard) or minimum requirements for pollution output (performance standard) to reduce emissions. Because performance standards require specific emission levels but often allow firms some discretion in how to meet those requirements, they are regarded as more cost effective than technology standards. Regulations and standards are often most applicable to sectors where consumers do not respond to price signals or where the price elasticity of demand is low (*e.g.* electricity, gas). Relatively few regulatory standards have been adopted solely to reduce greenhouse gases, although standards have been adopted that reduce these gases as a co-benefit. For example, there has been extensive use of standards to increase energy efficiency, including fuel economy standards for automobiles, appliance standards and building codes. Standards to reduce methane and other emissions from solid waste landfills have also been adopted in Europe, the United States, China and other countries. Such standards are often driven by multiple policy objectives, including reducing other pollutants (*e.g.* volatile organic compound emissions), improving safety by reducing the potential for explosions and reducing odours for local communities.

Voluntary agreements

Voluntary agreements and measures (VAs) are agreements between governments and one or more private parties to achieve environmental objectives or to improve environmental performance.[20] They are a common GHG policy in OECD countries (see Box 7.2). It is difficult to compare the "stringency" of agreements in different countries since they use different units, timeframes and/or boundaries. More fundamentally it is difficult to determine the effectiveness of voluntary agreements in reducing GHG emissions below business-as-usual levels (OECD, 2003). However, the benefits of voluntary agreements for individual companies may be significant. Firms may enjoy lower legal costs, enhance their reputation and improve their relationships with shareholders. Negotiations to develop VAs on climate change can help to raise awareness of climate change issues and the potential for mitigation within industry, and help to move industries towards best practices.

Technology research and development

Research and development (R&D) policies may include direct government spending and investment on mitigation technologies and tax credits to improve their performance and lower their costs. Examples of international initiatives that aim to develop and advance cost-effective technologies include the International Partnership for a Hydrogen Economy, the Carbon Sequestration Leadership Forum, and the Asia-Pacific Partnership on Clean Development and Climate. Countries pursue technological R&D in national policy for a number of reasons, such as to foster innovation, induce investments by industry and to help domestic industries to be competitive. Investments in R&D can however be misdirected to the wrong technologies or can result in the "locking in" of inefficient technology paths, and the results may not be seen for decades. While R&D programmes play an essential role, they will need to be supplemented with other policies, for example economic instruments and other incentives such as feed-in tariffs,[21] to promote deployment and diffusion of low carbon technologies and to ensure reductions in GHG emissions.

Box 7.2. **Examples of voluntary agreements in OECD countries**

- Australia's "Greenhouse Challenge Plus" programme: An agreement between the government and an enterprise/industry association to reduce GHG emissions (see *www.greenhouse.gov.au/challenge*).
- Japanese Keidaren Voluntary Action Plan: Voluntary measures taken by 35 industrial and energy converting sectors to reduce GHG emissions, which are followed up by government review. The relationship between the government and industry in Japan, as well as the unique societal norm, make this voluntary programme unique; in other words there is *de facto* enforcement (see *www.keidanren.or.jp*).
- Netherlands Voluntary Agreement on Energy Efficiency: A series of legally binding long-term agreements based on annual improvement targets and benchmarking covenants between 30 industrial sectors and the government to improve energy efficiency.
- United States Climate Leaders: This partnership encourages individual companies to develop corporation-wide GHG inventories, set aggressive reduction goals, report inventory data annually, and document progress towards their goals, reporting annually to the US Environmental Protection Agency. Since 2002 the programme has grown to include 118 corporations (see *www.epa.gov/climateleaders*).

Policy simulations

Model simulations undertaken for the *Outlook* provide insights into several key policy questions (Box 7.3). This section investigates:

- How climate change impacts compare across alternative mitigation strategies, *e.g.* early action compared to phased or delayed action.
- How modest or phased mitigation achieved through a harmonised, global carbon tax compares to atmospheric stabilisation pathways for mitigation (*e.g.* stabilising atmospheric concentrations at about 450 ppm CO_2eq and above).
- The costs and effectiveness of full *versus* more partial participation in global mitigation strategies.

The rest of this chapter focuses on two main sets of policy simulations: *i)* the implementation of a harmonised global "carbon" tax; and *ii)* implementation of a stabilisation objective, in this case, 450 ppm CO_2eq. Both are projected to lead to significant emission reductions and to alter climate change in the next 50 years. The analysis compares the environmental and economic effects of these different policy choices with the *Outlook* Baseline to 2050. It considers changes in GHG emissions (compared to 2000 emission levels) across regions, sectors and sources, as well as the effects on atmospheric concentrations of GHG and global and regional temperature changes. Ancillary or co-benefits of mitigation are also briefly analysed here focusing on three areas: air pollution, biodiversity and security. Economic effects are described as changes in global and regional economic growth – using GDP – comparing the policy cases to Baseline outcomes in a given year. Finally, sectoral economic effects of the different mitigation cases are considered by comparing changes in value added by sector and region against Baseline developments. The key assumptions and uncertainties associated with such projections and simulations are listed in Box 7.4.

OECD ENVIRONMENTAL OUTLOOK TO 2030 – ISBN 978-92-64-04048-9 – © OECD 2008

Box 7.3. Description of Baseline and policy simulations

Baseline assumptions: The *Outlook* Baseline uses the UN forecast of population growth to 2050 and estimates that global economic growth will be 2.4% per year (expressed in terms of purchasing power parity or PPP) on average to 2050. Productivity growth rates and economic growth, labour force growth rates and population growth are outlined in Chapters 2 and 3.

Policy case 1. Global GHG taxes:

Four cases are considered based on the implementation of a USD 25 tax per tonne of CO_2eq.* As the social costs of carbon** grow over time, the tax is increased in real terms by 2.4% per year. The level of CO_2eq tax used in three of these policy simulations escalates over time (Figure 7.4). The tax applies to the main greenhouse gas (*i.e.* CO_2, CH_4 and N_2O) emission sources across all economic activities, although the timing and countries participating in its application vary by scenario as follows (from least to most environmentally aggressive):

i) **OECD 2008:** OECD countries immediately implement the USD 25 tax on all greenhouse gases and sources.

ii) **Delayed 2020:** all countries impose the tax on greenhouse gas emissions, but the timing is delayed until 2020.

iii) **Phased 2030:** the global tax on greenhouse gas emissions is phased in, beginning with the OECD from 2008; Brazil, Russia, India and China from 2020 and then the rest of the world (ROW) from 2030 onwards.

iv) **All 2008:** in a more aggressive effort to mitigate global GHG emissions, all countries implement the USD 25 tax on CO_2 and other GHG emissions from 2008.

Figure 7.4. **CO_2eq tax by policy case, 2010 to 2050: USD per tonne CO_2 (2001 USD, constant)**

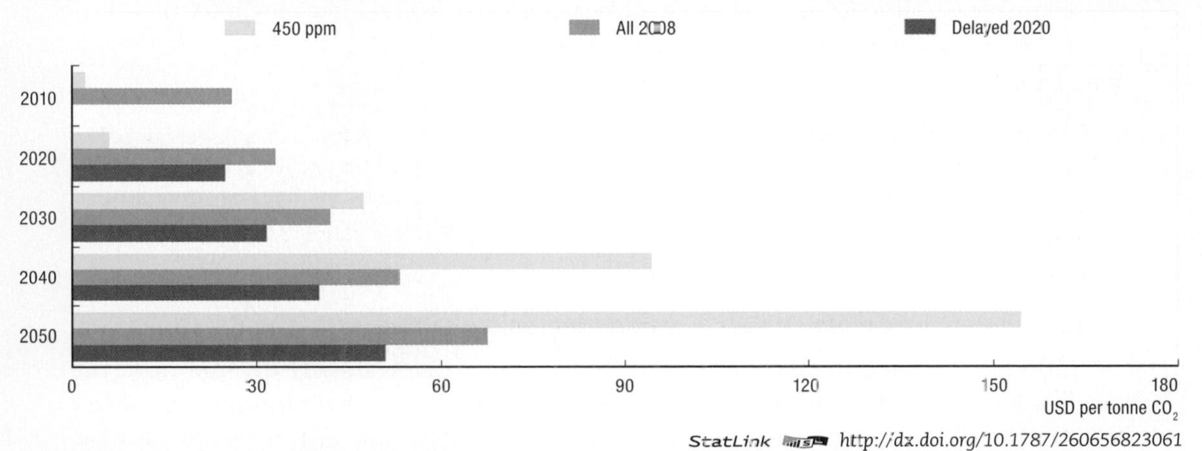

StatLink http://dx.doi.org/10.1787/260656823061

Source: OECD Environmental Outlook Baseline and policy simulations.

Policy case 2. 450 CO_2eq ppm stabilisation:

This policy simulation is chosen to demonstrate the level of effort required to stabilise atmospheric concentrations of GHG at 450 ppm CO_2eq (referred to below as 450PPM) and limit global mean temperature change to near 2°C over the long-term. It provides insights into possible mitigation costs for this aggressive mitigation pathway. It simulates an emission reduction pathway across all world regions in a "least-cost" manner across all sources (and sinks) of greenhouse gas emissions. In addition to cost and effectiveness, the simulation also reviews the technologies needed to achieve this aggressive stabilisation target (see Chapter 17). This allows us to understand what technologies and sources of greenhouse gases are expected to offer the most cost-effective means of reducing emissions significantly over the coming decades. The tax that was applied for this simulation increases from USD 2.4 per tonne of CO_2eq in 2010 to USD 155 in 2050 (in 2001 USD, constant).

A variation on this case is also presented to explore burden-sharing, using a cap and trade approach to implementation.

* Note a comparable tax is assessed as part of the policy packages exercise. See Chapter 20.

** The "social cost of carbon" (SCC) refers to the marginal damage costs of carbon emissions, or the incremental damage cost of emitting one additional tonne of carbon (in the form of CO_2) into the atmosphere. This is the key measure of benefits of mitigation within a cost-benefit analysis approach of policy assessment. See Pitinni and Rahman (2004) for a brief explanation of how integrated assessment models typically estimate SCC.

Box 7.4. Key uncertainties and assumptions

Projections of climate change depend on a number of parameters, all of which are associated with uncertainty in the future, including:

- Estimates of future population, economic growth and technology change: predictions of GHG emissions are influenced by population and economic growth and assumptions about technological changes. While most emission scenarios vary little to 2030, beyond that period GHG emissions could vary significantly if population, labour force participation, productivity, technological progress and economic growth differ from the assumptions in the Baseline.
- Climate sensitivity: this parameter characterises how global temperatures respond to a doubling of CO_2 concentrations. The IPCC in its 2007 report noted that climate sensitivity is likely to be in the range of 2° to 4.5°C with a "best estimate" of 3.0°C. It is very unlikely to be below 1.5°C and values substantially higher than 4.5°C cannot be excluded.
- Abrupt changes and surprises: the *Outlook* Baseline assumes a linear response to increasing concentrations of GHGs. There is however evidence from the paleo-climatic record that the Earth's systems have undergone rapid changes in the past and that these could occur in the future.
- Probability of outcomes, risks assessment: given these, and other, uncertainties, probabilistic assessment is increasingly used to give policy-makers an idea of the likelihood of achieving identified targets (Jones, 2004; Yohe *et al.*, 2004; Mastrandrea and Schneider 2004). For example, Meinshausen (2006) considers the case of a 2°C target, estimating that a 650 ppm CO_2eq concentration level would offer only a 0% to 18% probability of success. This presents climate change in a risk assessment and management framework.
- Adaptation: human systems are likely to respond to climate change through adaptation, while ecological systems are likely to find it more difficult to adapt. The faster global warming occurs, the more difficult and limited adaptation will be. Most current studies of climate change impacts recognise the need to consider adaptation, but few modelling studies integrate adaptation comprehensively into quantitative analyses.

Climate change and global impacts: mitigation policy compared to the Baseline

Climate change outcomes for the different policy cases already diverge from the Baseline by 2050 and this difference will grow over time. In the nearer-term the *Outlook* Baseline projections suggest that without new climate change and environmental policies, GHG emissions will grow at a pace that raises CO_2eq concentrations significantly to approximately 465 ppm by 2030 and further to 540 ppm by 2050, which is predicted to increase global mean temperature by 1.9°C in 2050 (above the pre-industrial level, within a range of 1.7 to 2.4°C; see Table 7.4c).[22] By 2030 the *Outlook* projects that temperature under the Baseline will be increasing rapidly, by about 0.28°C per decade, up from about 0.18°C per decade today, and will continue at this pace until 2050. Factors like reduced sea-ice cover, which would change the regional albedo (reflectivity of the Earth's surface), and enhanced methane emissions from melting permafrost soils may accelerate unmitigated climate change beyond these levels.

Table 7.4 shows growth in GHG and CO_2 emissions for the Baseline and policy cases compared to 2000 emission levels. All of the policy cases, except the OECD 2008 tax, lead to significant emission reductions compared to 2000, with the 450 PPM case showing the greatest reductions in global GHG emissions (–39%), whereas the All 2008 tax case delivers about two-thirds of this emission reduction by 2050. Interestingly the Phased 2030 and Delayed 2020 tax cases significantly reduce emissions from the Baseline but do not deliver

OECD ENVIRONMENTAL OUTLOOK TO 2030 – ISBN 978-92-64-04048-9 – © OECD 2008

Table 7.4. **Policy scenarios compared to Baseline: GHG emissions, CO_2 emissions and global temperature change, 2000-2050**

a. % Change in GHG emissions relative to 2000

Region	Baseline		OECD 2008		Delayed		Phased		All 2008		450 ppm	
	2030	2050	2030	2050	2030	2050	2030	2050	2030	2050	2030	2050
World	52	69	34	38	23	3	20	0	7	–21	–7	–39
OECD	28	31	–14	–43	2	–22	–14	–42	–14	–42	–23	–55
BRIC	72	92	72	92	36	14	36	16	16	–13	4	–34
ROW	65	104	66	103	44	31	55	51	30	5	6	–19

b. % Change in CO_2 emissions relative to 2000

Region	Baseline		OECD 2008		Delayed		Phased		All 2008		450 ppm	
	2030	2050	2030	2050	2030	2050	2030	2050	2030	2050	2030	2050
World	54	72	36	38	31	7	26	3	11	–21	–3	–41
OECD	31	34	–9	–42	8	–18	–9	–41	–9	–41	–18	–55
BRIC	81	106	81	107	50	24	36	16	24	–11	13	–34
ROW	65	104	66	103	50	32	55	51	33	3	7	–25

c. Atmospheric GHG concentrations, global mean temperature, rate of temperature change

Region	Baseline		OECD 2008		Delayed		Phased		All 2008		450 ppm	
	2030	2050	2030	2050	2030	2050	2030	2050	2030	2050	2030	2050
CO_2 Concentration (ppmv)	465	543	458	518	458	507	455	501	448	481	443	463
GMT range (°C)[a]	1.2-1.6	1.7-2.4	1.2-1.5	1.6-2.2	1.2-1.5	1.5-2.1	1.1-1.4	1.5-2.0	1.1-1.4	1.4-1.9	1.1-1.4	1.3-1.8
Rate of GMT chg (°C/decade)	0.28	0.28	0.25	0.23	0.22	0.19	0.22	0.18	0.21	0.15	0.16	0.10

StatLink http://dx.doi.org/10.1787/257115140846

a) The range in global mean temperature change is based on MAGICC model calculations as performed by van Vuuren *et al.* (forthcoming). The MAGICC range originates from emulation of different climate models, here showing the impact of climate sensitivity with a range corresponding to a climate sensitivity of 2.0-4.9 °C. The overall range in transient 21st century climate change was used relative to the IMAGE model outcomes to account for differences in the scenarios.

Source: OECD Environmental Outlook Baseline and policy simulations.

absolute emission reductions in 2050. The OECD 2008 tax shows significant reductions in OECD regions (–43%) yet the global emissions still grow by 38% compared to 2000 emission levels (Table 7.4). The spread of outcomes among these cases demonstrates the importance of full participation by all major emitters and early mitigation efforts if substantial emission reductions are to be achieved by 2050.

Figure 7.5 compares the Baseline and policy cases' GHG emission pathways with longer term stabilisation pathways (*i.e.* for 650, 550 and 450 ppm CO_2eq as well as alternative baseline scenarios). A comparison with the IPCC summary of long-term emission scenarios, in Table 7.5, shows that the *Outlook* Baseline clearly is outside of the range of a stabilisation pathway for 750 ppm CO_2eq, with emissions likely to grow throughout the 2100 period. A baseline of this type would be expected to lead to a global mean temperature increase range of 4-6°C (above pre-industrial, equilibrium).[23]

Compared with the Baseline trajectory, the early and more aggressive policy cases deliver significantly lower concentrations and thus lower temperatures and slower rates of change (*i.e.* as illustrated in the 450 PPM and All 2008 cases). The global tax (All 2008) falls within the 550 ppm CO_2eq target by 2050. Delaying mitigation efforts to 2020 (Delayed 2020), or phasing in participation by large emitters outside of the OECD much more slowly (Phased 2030) raises emissions sufficiently to shift global emissions from a

Figure 7.5. **Global GHG emission pathways: Baseline and mitigation cases to 2050 compared to 2100 stabilisation pathways**

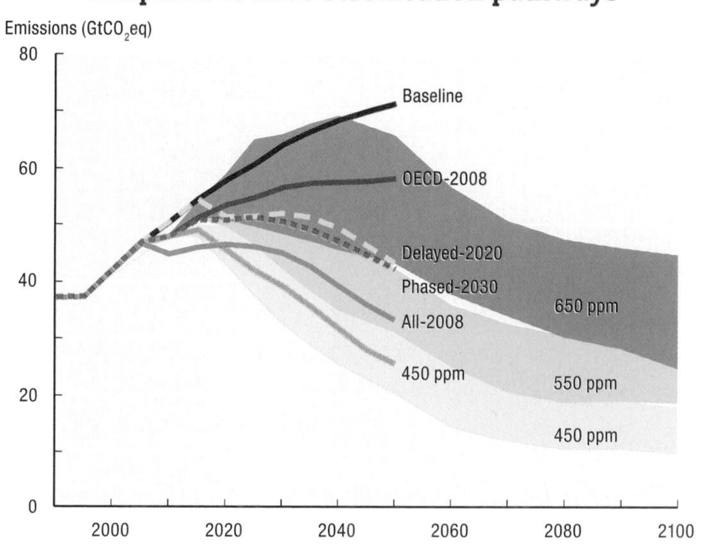

Source: OECD Environmental Outlook Baseline and policy simulations; and van Vuuren *et al.*, 2007.

550 CO_2eq to a 650 ppm pathway. By contrast, the OECD-only tax from 2008 (OECD, 2008) starts to bring global emissions into the pathway early for 650 ppm CO_2eq stabilisation, but by the end of 2050 overshoots this because of the limited participation in mitigation efforts.

Table 7.5 shows quite different climate change outcomes at equilibrium for stabilisation pathways; the *Outlook* Baseline and policy simulations can be considered in this longer-term context. The more comprehensive (in terms of participation) and more stringent policy cases – *i.e.* All 2008 and 450 PPM – are likely to avoid roughly 1-3°C of global mean temperature increase (or more) already in the 2080 timeframe compared with scenarios falling at the high end of stabilisation such as the Category V and VI scenarios in Table 7.5.[24] Similarly decadal rates of temperature change differ significantly among the cases. By 2050, the All 2008 and 450 PPM cases slash the rate of change by half and two-thirds respectively compared to the Baseline, demonstrating a strong climate change response to early and more comprehensive action (Figure 7.6c).

The costs of inaction or delayed action are therefore potentially significant (see also Chapter 13, Cost of policy inaction). The latest IPCC report (2007) suggests greater risks than previously for even relatively low levels of temperature increases (*e.g.* 1-3°C above pre-industrial levels) (Schneider *et al.* 2007; IPCC 2007d). Delay in reducing emissions could have serious consequences for the environment and could be costly, especially if society eventually decides that it is prudent to opt for stringent mitigation targets in the long-term. This is demonstrated by the clear differences in climate change outcomes by 2050 associated with the case of a 10-year delay in policy action (Delayed 2020) compared to cases with earlier mitigation action (450 PPM; All 2008) (Figure 7.6). Other literature also explores these risks (Kallbekken and Rive, 2006; Shalizi, 2006). For example, Kallbekken and Rive (2006) show that immediate emission reductions lower the rate at which global emissions need to be reduced for a given climate target; they show that to achieve a given temperature after a delay of 20 years would require emissions to be reduced at a rate that is 3-9 times greater than if emissions were reduced immediately.

OECD ENVIRONMENTAL OUTLOOK TO 2030 – ISBN 978-92-64-04048-9 – © OECD 2008

Table 7.5. **Characteristics of post TAR stabilisation scenarios and resulting long-term equilibrium global average temperature and the sea level rise component from thermal expansion only**[a]

Category	CO_2 concentration at stabilisation (2005 = 379 ppm)[b]	CO_2-equivalent concentration at stabilisation including GHGs and aerosols (2005 = 375 ppm)[b]	Peaking year for CO_2 emissions[a, c]	Change in global CO_2 emissions in 2050 (% of 2000 emissions)[a, c]	Global average temperature increase above pre-industrial at equilibrium, using "best estimate" climate sensitivity[d, e]	Global average sea level rise above pre-industrial at equilibrium from thermal expansion only[f]	Number of assessed scenarios
	ppm	ppm	Year	Percent	°C	metres	
I	350-400	445-490	2000-2015	−85 to −50	2.0-2.4	0.4-1.4	6
II	400-440	490-535	2000-2020	−60 to −30	2.4-2.8	0.5-1.7	18
III	440-485	535-590	2010-2030	−30 to +5	2.8-3.2	0.6-1.9	21
IV	485-570	590-710	2020-2060	+10 to +60	3.2-4.0	0.6-2.4	118
V	570-660	710-855	2050-2080	+25 to +85	4.0-4.9	0.8-2.9	9
VI	660-790	855-1130	2060-2090	+90 to +140	4.9-6.1	1.0-3.7	5

StatLink http://dx.doi.org/10.1787/257132076082

a) The emission reductions to meet a particular stabilisation level reported in the mitigation studies assessed here might be underestimated due to missing carbon cycle feedbacks (see also Topic 2.3).*

b) Atmospheric CO_2 concentrations were 379 ppm in 2005. The best estimate of total CO_2eq concentration in 2005 for all long-lived GHGs is about 455 ppm, while the corresponding value including the net effect of all anthropogenic forcing agents is 375 ppm CO_2eq.

c) Ranges correspond to the 15th to 85th percentile of the post-TAR scenario distribution. CO_2 emissions are shown so multi-gas scenarios can be compared with CO_2-only scenarios (see Figure SPM.3).*

d) The best estimate of climate sensitivity is 3°C.

e) Note that global average temperature at equilibrium is different from expected global average temperature at the time of stabilisation of GHG concentrations due to the inertia of the climate system. For the majority of scenarios assessed, stabilisation of GHG concentrations occurs between 2100 and 2150 (see also footnote 21).*

f) Equilibrium sea level rise is for the contribution from ocean thermal expansion only and does not reach equilibrium for at least many centuries. These values have been estimated using relatively simple climate models (one low resolution AOGCM and several EMICs based on the best estimate of 3°C climate sensitivity) and do not include contributions from melting ice sheets, glaciers and ice caps. Long-term thermal expansion is projected to result in 0.2 to 0.6 m per degree Celsius of global average warming above pre-industrial. (AOGCM refers to Atmosphere Ocean General Circulation Models and EMICs to Earth System Models of Intermediate Complexity.)

* These are cross-references in the original report. The report is also available on the Internet, see: *www.ipcc.ch*.

Source: Table SPM.6, IPCC (2007d), *Climate Change: Synthesis Report. The Fourth Assessment Report*, Cambridge University Press, Cambridge, UK (reproduced here with the full set of original notes).

Regional effects of mitigation policy compared to the Baseline

The regional distribution of climate change is projected to vary significantly, with many heavily populated regions of the world experiencing temperature changes that are higher than the projected average (see Figure 7.7a for Baseline temperature patterns). With higher temperatures, the hydrological cycle is also projected to intensify under the Baseline case as more water evaporates and on the whole more precipitation results. As with the temperature pattern though, the effect is very unevenly distributed and many areas may even become drier, while adjacent areas receive more precipitation. In already water-stressed areas such as southern Europe and India, the negative impact on agriculture and human settlements would be substantial. The risk of drought-related problems will be highest in areas where the future drop in surplus is large relative to the current level. These areas are likely to include parts of Africa as well as southern Europe, large parts of Australia and New Zealand. Areas with substantial increases over already high levels in 2000 are more susceptible to encounter water drainage or flooding problems. In general, all areas facing considerable changes in surplus will have to adapt to cope with these changes, including through adjustments in water management practices and/or infrastructure.

Figure 7.6. **Change in global emissions, GHG atmospheric concentrations, global mean temperature: Baseline and mitigation cases**

A. Changes in global GHG emissions in 2050 relative to 2000 by policy case

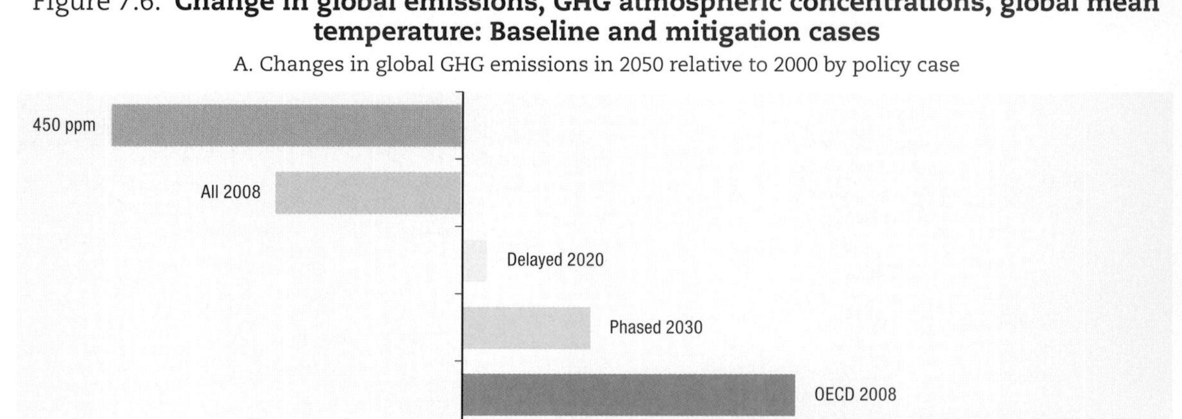

B. Changes in CO_2 concentrations over time by case, 2000 to 2050

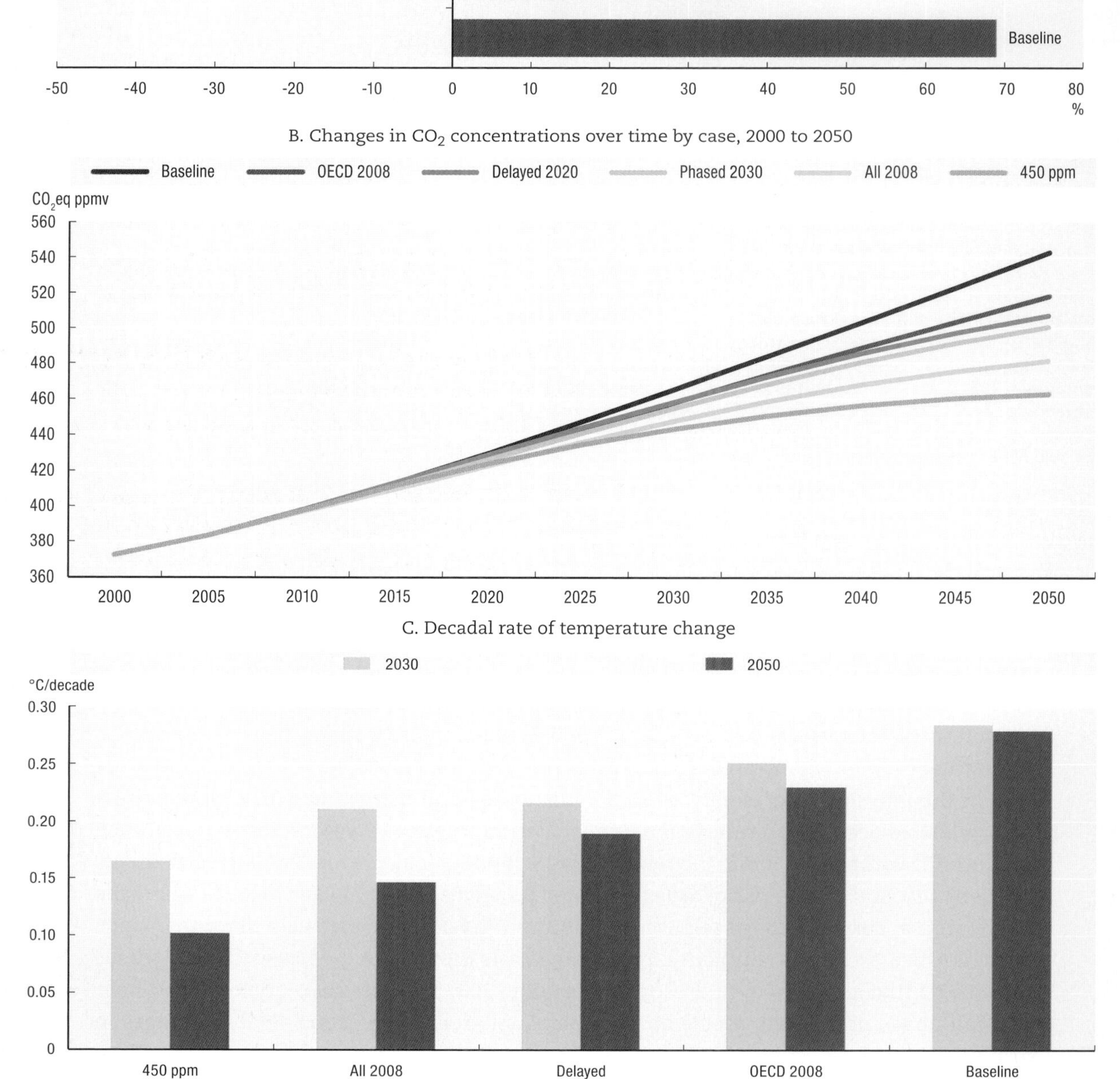

StatLink http://dx.doi.org/10.1787/260741607702

Source: OECD Environmental Outlook Baseline and policy simulations.

Figure 7.7. **Change in mean annual temperature levels in 2050 relative to 1990 (degrees C)**

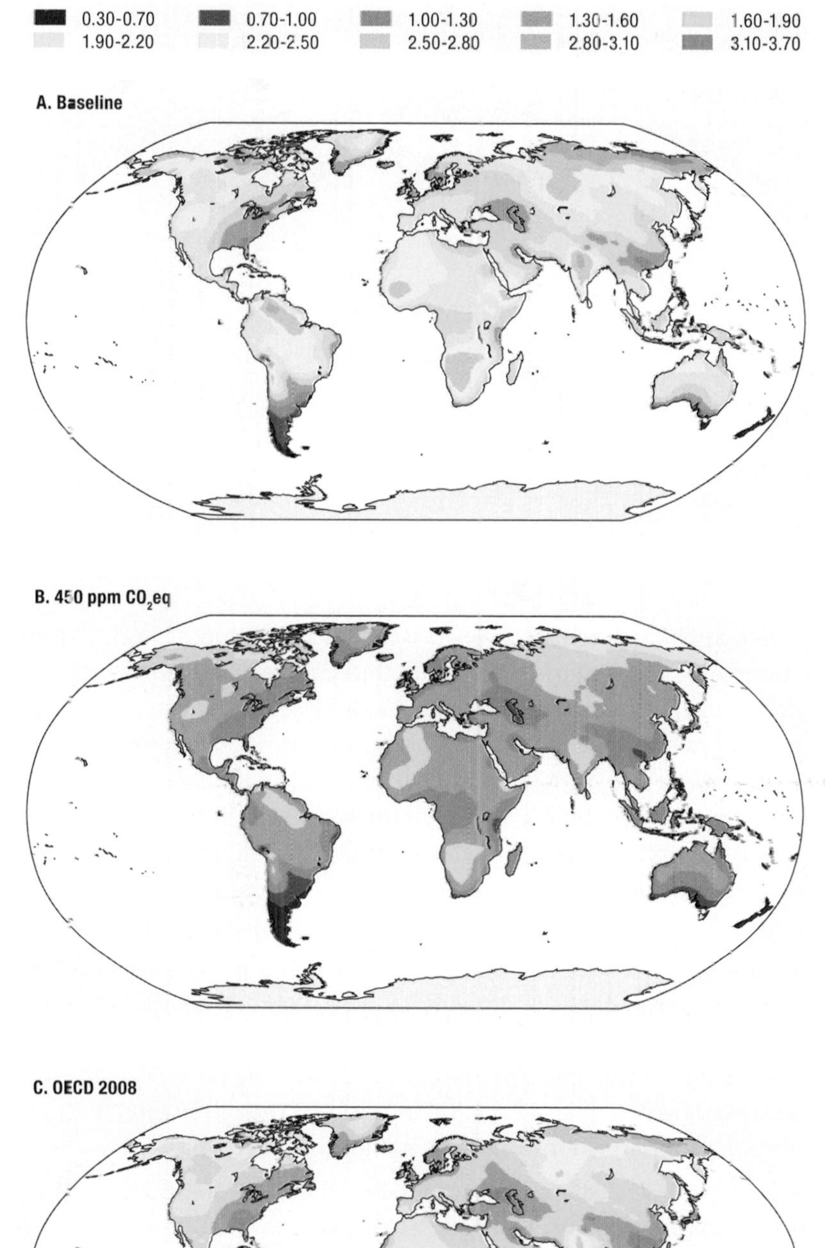

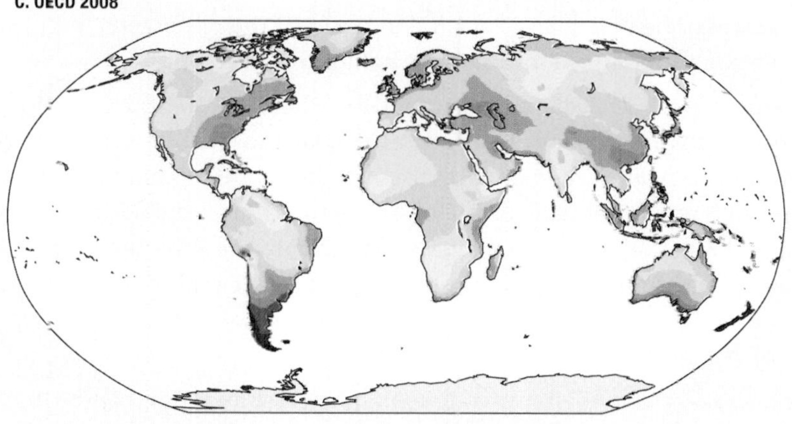

Source: *OECD Environmental Outlook* Baseline and policy simulations.

Figure 7.7. **Change in mean annual temperature levels in 2050 relative to 1990 (degrees C)**(*cont.*)

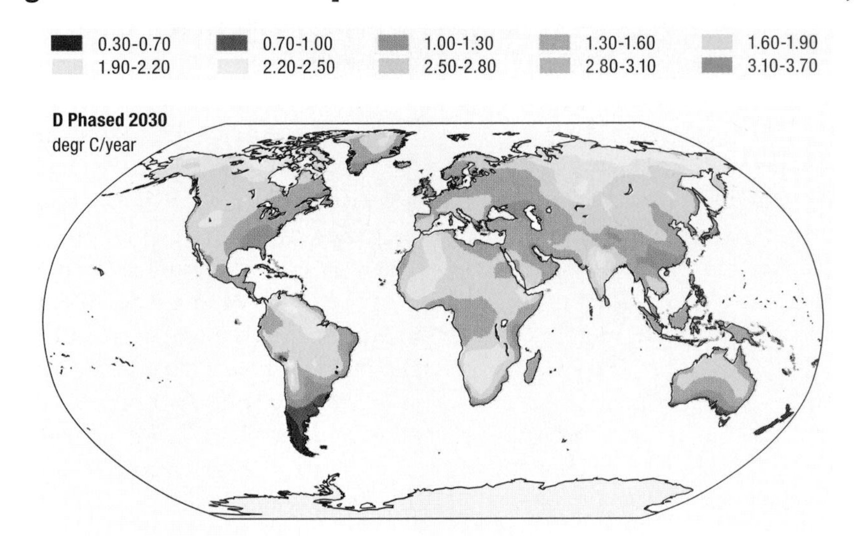

Source: *OECD Environmental Outlook* Baseline and policy simulations.

Climate change is expected to affect productivity, commodity prices and the spatial allocation of the various crop types. Under the *Outlook* Baseline, temperate crops are likely to tend to "move north" as growing conditions nearer to the equator become less suitable to 2030 and beyond, while growing conditions may improve at higher latitudes. There is a great deal of uncertainty associated with the potential for irrigation, the availability of fertilisers and changes in pests. For tropical crops like rice, changes in precipitation may affect large areas. Though still uncertain and relatively small in the 2030 timeframe, these changes are accounted for in the estimates of future agricultural productivity for all crop types in this *Outlook* (see Chapters 10 on freshwater and 14 on agriculture).

Mitigation policy will affect the pattern of regional climate change and the distribution and magnitude of regional impacts. Already by 2050, regional temperature patterns show much less dramatic changes under the more aggressive and early action mitigation scenarios compared to the case of inaction (Baseline) (see Figures 7.7a-d). These differences between the policy and Baseline in terms of the predicted climate changes will become even more pronounced into the last half of the 21st century.

Contrasting the OECD 2008 tax case with the Phased 2030 and 450 PPM cases shows that the more stringent and more comprehensive the mitigation effort (in terms of participation) in the next decades the more likely it will be possible to limit temperature changes over large regions of the world. The 450 ppm CO_2eq stabilisation case significantly limits global and regional warming by 2050 compared to the Baseline pattern of warming. As noted above, this difference is projected to widen by the end of the century.

Co-benefits of mitigation[25]

As noted above, the co-benefits of greenhouse gas mitigation can be significant, and could include cost-reductions in the achievement of air pollution policy objectives (see Box 7.5) as well as the direct improvement of human health, urban environments or

Box 7.5. **Co-benefits and the cost-effectiveness of climate and air pollution policy**

Accounting for the co-benefits of reduced air pollution and reduced greenhouse gas emissions can have significant impacts on the cost effectiveness of climate and air pollution policy. The co-benefit relationship suggests that co-ordination of policy efforts in these areas could deliver important cost savings. For example, van Harmelen *et al.* (2002) found that to comply with agreed or future policies to reduce regional air pollution in Europe, mitigation costs are implied, but these are reduced by 50-70% for SO_2 and around 50% for NO_x when combined with GHG policies. Similarly, in the shorter-term, van Vuuren *et al.* (2006) found that for the Kyoto Protocol, about half the costs of climate policy might be recovered from reduced air pollution control costs. The exact benefits, however, critically depend on how climate change policies are implemented and on the baseline policies that are used for comparison (Morgenstern, 2000). Most available studies do not treat co-benefits comprehensively in terms of reduction costs and the related health and climate impacts in the long-term, thus indicating the need for more research in this area (OECD, 2000; IPCC, 2007a).

national security. We focus here on the ancillary benefits that accompany GHG mitigation policy in three different areas – air pollution, biodiversity and security – drawing on *Outlook* simulations in the first two areas to illustrate the magnitude of benefits.

Air pollution and biodiversity co-benefits: Outlook *results*

Stabilising concentrations of GHG in the atmosphere at relatively low levels requires reversing trends so that emissions decline in the coming decades. For example, in the 450 ppm case, global CO_2 emissions peak in 2015 and decline thereafter by about 40% relative to 2000 emission levels. Reducing CO_2 emissions by this degree would require a major transformation in the energy sector with energy efficiency, renewable or nuclear energy playing a larger role than in the past (see Chapter 17, Energy). In addition to limiting the scale and the pace of climate change, a transition to clean energy systems and away from fossil fuel combustion will yield a range of environmental benefits including in the area of air pollution and human health. Figure 7.8 shows that the 450 ppm case leads to reductions by 2030 in the range of 20-30% for sulphur oxides (SO_x) and 30-40% for nitrogen oxides (NO_x). SO_x and NO_x cause acid rain, damaging freshwater ecosystems, forest ecosystems and agricultural productivity on a regional scale. NO_x is also a local pollutant and in urban areas is a precursor to ozone formation which is harmful to human health. Urban ozone episodes affect respiratory and lung systems and aggravate asthma and allergies to pollen. In this example, the largest air pollution co-benefits would be found in some of the most rapidly developing and urbanising areas of South Asia (SOA including India), Indonesia and the rest of South Asia (OAS), China (CHN), and eastern Europe and central Asia (ECA). There is also a large relative benefit in North America (NAM – *i.e.* Canada, Mexico and the US) in moving from the Baseline to the 450 PPM case.

As biodiversity will vary with levels of climate change and with approaches to greenhouse gas mitigation policy, ancillary benefits of mitigation policies are also possible in the 2050 timeframe. Using the mean species abundance (MSA) indicator (see Chapter 9, Biodiversity), Figure 7.9 compares the 450 ppm case to the Baseline. These results depend upon the avoided climate change impacts, as discussed above, and the mitigation approaches embedded in the 450 ppm case, where large scale production of second generation biofuels are an important

Figure 7.8. **Air pollution co-benefits of GHG mitigation: reduction in NO_x and SO_x emissions – 450 ppm case and Baseline, 2030**

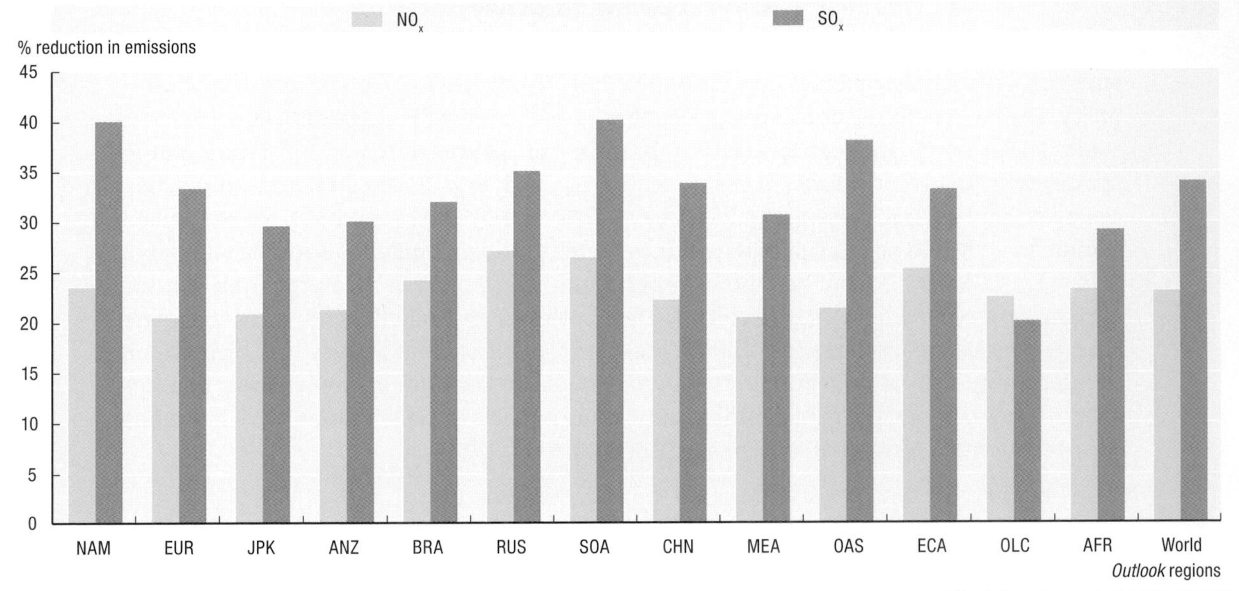

StatLink http://dx.doi.org/10.1787/260800071717

Note: Regional country groupings are as follows: NAM: North America (United States, Canada and Mexico); EUR (western and central Europe and Turkey); JPK: Japan and Korea region; ANZ: Oceania (New Zealand and Australia); BRA: Brazil; RUS: Russian and Caucasus; SOA: South Asia; CHN: China region; MEA: Middle East; OAS: Indonesia and the rest of South Asia; ECA: eastern Europe and central Asia; OLC: other Latin America; AFR: Africa.

Source: OECD Environmental Outlook Baseline and policy simulations.

part of the policy portfolio. This biofuel production will affect land use and biodiversity in various ways. Although the 450 ppm case leads to less climate change than in the Baseline, increased land use for biofuel production causes substantial additional biodiversity loss. However, the net balance for the 450 ppm case between avoided and additional losses is slightly positive: 1% less decrease in mean species abundance than in the Baseline by the middle of the century. This reflects the assumption that greenhouse gas mitigation policies also provide incentives to reduce deforestation and thus develop more compact agricultural activity than would otherwise be the case, which in turn is essential to reach the climate target. However, concrete policy instruments to promote this would need to be developed. The benefits from the reduction in the total amount of land conversion from forest to agriculture under the 450 ppm case compared to the Baseline partly compensate for losses from biofuel production (Figure 7.9). It should also be noted that the recent IPCC assessment presents new evidence that suggests that biodiversity might be more sensitive to climate change than previously believed (IPCC, 2007b and d).

National security

In addition to sector policy co-benefits that are mainly local in scale, there are also national and international co-benefits of climate change mitigation and adaptation in the form of reduced security risks. Climate change will affect world regions unevenly, with the greatest costs likely to fall on the poorest regions (IPCC, 2007b; IPCC, 2007d). The uneven distribution of climate change impacts is due in part to high vulnerability of poor nations, where the ability to cope with climate change is low. It follows that climate change has implications for foreign policy and national security, for example by increasing the flood

Figure 7.9. **Biodiversity effects of the 450 ppm case by 2050**

Mean species abundance: percentage points relative to Baseline

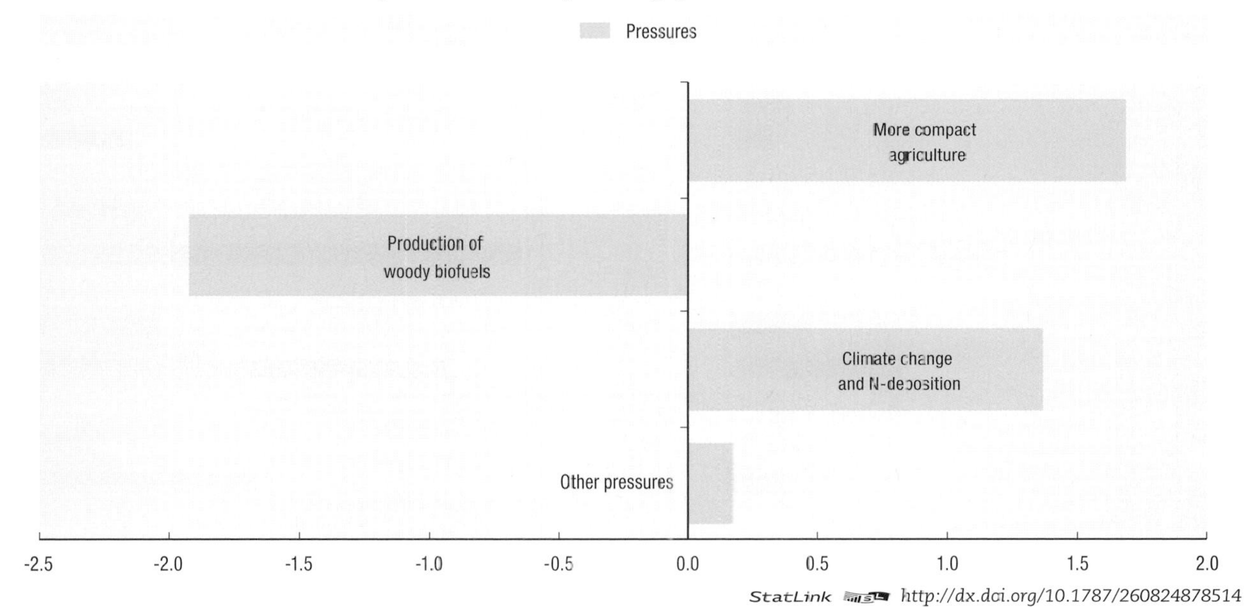

StatLink http://dx.doi.org/10.1787/260824878514

Note: MSA effects are presented as change from the Baseline scenario. Avoided loss in mean species abundance (MSA) is presented as a positive value, and additional loss as negative. The figure shows the effect of each individual pressure factor as well as the total effect of all factors. The MSA biodiversity indicator is further explained in Chapter 9, Biodiversity; see also Alkemade *et al.*, 2006; CBD and MNP, 2007.

Source: *OECD Environmental Outlook* Baseline.

risk and exposure to other extremes in poor and heavily populated regions, in addition to increased competition for resources in already water scarce regions of the world (Brauch, 2002; Barnett, 2003; Campbell *et al.*, 2007). Thus a co-benefit of global mitigation policies is to limit "cascading consequences" and national security risks from otherwise unchecked climate change (Campbell *et al.*, 2007; Oberthuer *et al.*, 2002).

Costs of mitigation and implications for innovation

Figure 7.10a, and b and Table 7.6 compare the economic costs of the different policy cases with the Baseline economic projections for 2030 and 2050. These model simulations assume perfect cost-effective implementation pathways of each mitigation policy case, and therefore could be said to underestimate the true implementation costs. However, the model also assumes there are no opportunities for negative or no-cost mitigation and does not explicitly account for co-benefits as an offset to costs even though these may be significant (*e.g.* see discussion in IPCC, 2007c and above). These limitations might therefore be said to overestimate the costs of mitigation.

Figure 7.10. **Economic cost of mitigation policy cases by major country group**

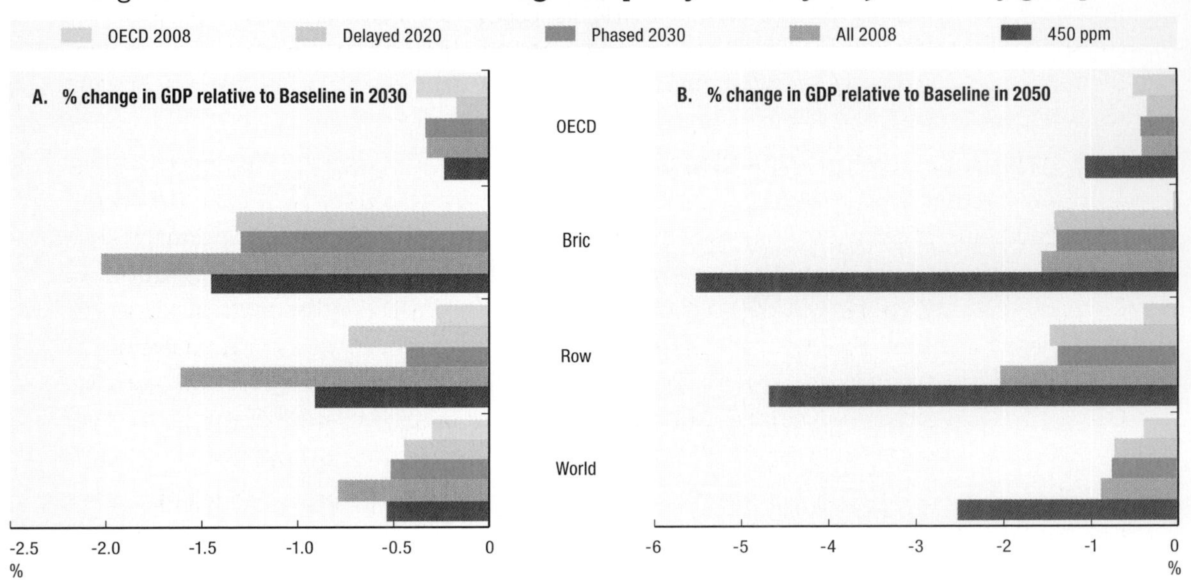

StatLink http://dx.doi.org/10.1787/260827814045

Note: Scales differ.

Source: *OECD Environmental Outlook* Baseline and policy simulations.

Table 7.6. **Change (%) in GDP relative to Baseline of different scenarios, 2030 and 2050**

Case Region	450 ppm		All 2008		Phased 2030		Delayed 2020		OECD 2008	
	2030	2050	2030	2050	2030	2050	2030	2050	2030	2050
OECD	–0.2	–1.1	–0.3	–0.4	–0.3	–0.4	–0.2	–0.3	–0.4	–0.5
BRIC	–1.4	–5.5	–2.0	–1.6	–1.3	–1.4	–1.3	–1.4	0.0	0.0
ROW	–0.9	–4.7	–1.6	–2.0	–0.4	–1.4	–0.7	–1.5	–0.3	–0.4
WORLD	–0.5	–2.5	–0.8	–0.9	–0.5	–0.8	–0.4	–0.7	–0.3	–0.4
BIC	–1.1	–4.7	–1.6	–1.0	–1.0	–0.9	–1.0	–0.9	0.0	0.0
MEA/Russia	–2.9	–10.6	–4.5	–6.0	–2.3	–4.3	–2.4	–4.2	–0.7	–0.8

StatLink http://dx.doi.org/10.1787/257133737368

Source: *OECD Environmental Outlook* Baseline and policy simulations.

The results show that even for the most aggressive mitigation case – stabilising concentrations at 450 ppm CO_2eq – global costs of mitigation are positive, but manageable. Total loss of GDP (relative to the Baseline) is projected to be roughly 0.5% by 2030, rising to about 2.5% by 2050. This is equivalent to slowing annual growth rates in GDP over the 2005 to 2050 timeframe by about 0.1 percentage point. The regional distribution of costs, however, for this stabilisation case differs broadly in the 2030 and 2050 timeframe. OECD costs are projected to be the lowest, at 0.2% and 1% below the Baseline GDP in 2030 and 2050 respectively. The costs in Brazil, Russia, India and China (BRIC) are roughly five times this level and those in the rest of the world (ROW) about four times as high. For the other tax policy cases, the costs are significantly lower in the

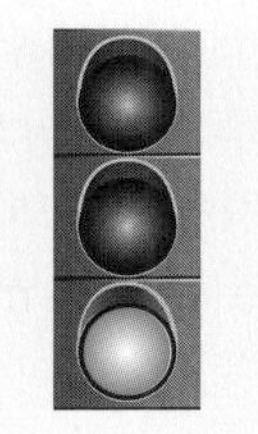

Global costs of mitigation, even for the most stringent mitigation case – stabilising concentrations at 450 ppm CO_2eq – are significant but manageable.

 OECD ENVIRONMENTAL OUTLOOK TO 2030 – ISBN 978-92-64-04048-9 – © OECD 2008

2050 timeframe; however, given the timing of the stabilisation case, the costs in 2030 are sometimes lower under stabilisation than for the USD 25 tax case (see Table 7.6). As noted below, this large regional difference in cost could be addressed through a variety of different burden-sharing mechanisms including, for example, differential target setting in a cap and trade policy scenario.

An important analytical question is the impact of GHG mitigation policy on industrial competitiveness and, possibly, business decisions about where to locate industrial production. Another interesting result from this analysis is that these simulations, with a particularly rich representation of trade, do not show much leakage (or migration) of industrial activity, energy use and CO_2 emissions from the OECD to other parts of the world. This is evident from Table 7.4a, which shows no increase in emissions in other parts of the world under the OECD 2008 tax case, where a tax is imposed in the OECD region alone. Also OECD emission reductions compared to the Baseline (or base year) are comparable across the OECD 2008 tax case and All 2008, or the case where a global tax is imposed.

Oil and natural gas producing countries (including Russia) are projected to experience the greatest change in GDP from mitigation efforts (across all policy cases) because of their economic vulnerability to taxation on the carbon content of fossil fuels (*i.e.* oil and oil products). These countries' export markets for fossil fuels are likely to be affected. Their domestic economies will also be affected significantly since fuel prices are kept low, either through subsidies or exceptionally low energy taxation, which in turn boosts domestic consumption, dependence on fossil fuels and GHG intensity of economic production. This vulnerability might be ameliorated by diversifying the economies of oil-producing countries *and* raising the price of domestic energy to its opportunity cost (*i.e.* the world price, plus whatever taxes are applied to other commodities). While cheap fossil fuels should be a natural comparative advantage for energy-producing economies, they can become liabilities in a carbon-constrained world.[26]

Under the policy simulation of an immediate adoption of a USD 25 tax on CO_2 by all countries (All 2008), annual GDP in the oil-producing countries is estimated to be about 4% and 5% lower than the Baseline in 2030 and 2050 respectively (Table 7.6). Phasing in the tax is projected to roughly halve the economic losses for this oil-producing group of countries, whereas if OECD countries act alone the economic losses associated with the tax fall to about one-tenth of the All 2008 scenario. Of course, as noted above, the environmental effectiveness of the tax in reducing global GHG emissions would also drop significantly if participation is more limited or implementation delayed.

More generally, the high costs of aggressive mitigation (*e.g.* 450 ppm) in non-OECD regions are driven by several factors:

- The large potential for relatively low-cost mitigation in non-OECD regions compared to the OECD becomes especially important under the most stringent mitigation cases.
- The growth in emissions from non-OECD countries is higher than for OECD countries, which means that these countries will need to reduce a relatively larger share of emissions under the stringent mitigation scenario.
- As noted above, the relatively high levels and broad scope of energy subsidisation in some key regions (*e.g.* Russia, newly independent states and many oil-producing regions) raise the cost of mitigation, especially in the energy and the energy-intensive sectors.

Figure 7.11 shows changes in value added[27] by type of industry across major country groupings relative to the Baseline for the 450 ppm stabilisation case in 2030. This

Figure 7.11. **Change in value added: 450 ppm CO_2eq stabilisation case relative to Baseline, 2030**

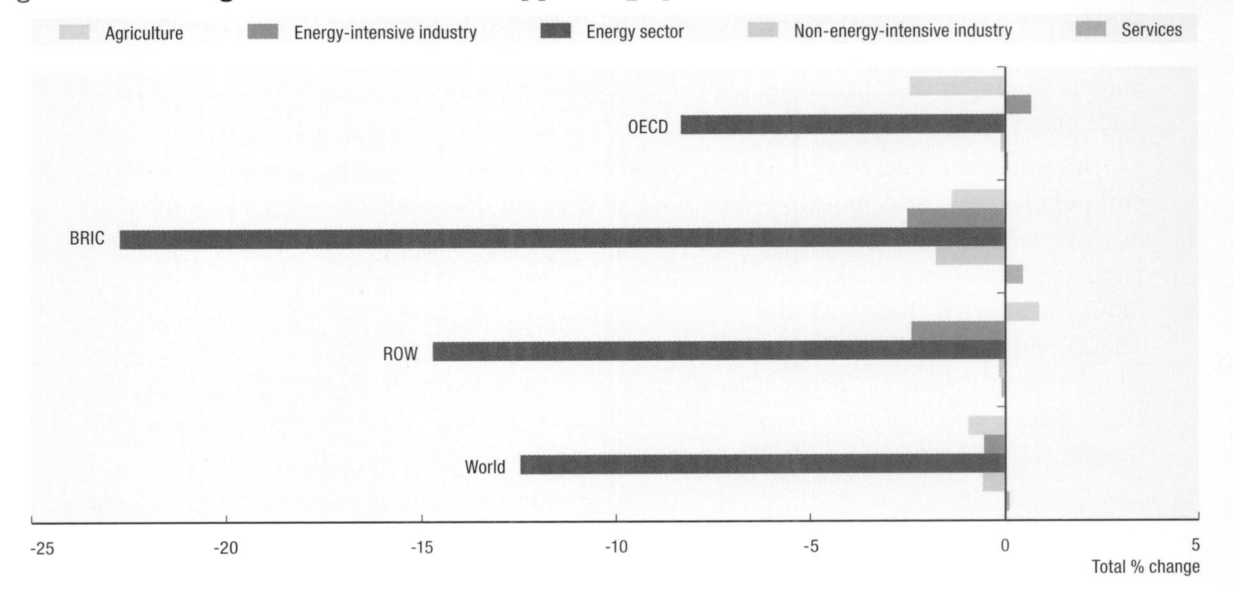

StatLink http://dx.doi.org/10.1787/260838735053

Source: OECD Environmental Outlook Baseline and policy simulations.

demonstrates that the energy sector is a principal source of mitigation; changes in this sector dominate in all the country groups in 2030 and this result continues to 2050 (not shown here). Other sectors show mixed results. Two main factors contribute to the varied outcomes. First, when the cost of energy goes up, firms switch to other inputs. If those other inputs consist of labour and capital, value-added will increase. In general, this should not be enough to completely offset the impact of energy price increase, so the net impact should be negative. However, when there are differences between regions in fossil-fuel intensity of sectoral production, then some sectors in some regions may, in fact, show a net gain. In other words, the heterogeneity of sectoral results illustrated in the figure reflects regional differences in sectoral fossil-fuel intensity.

Burden sharing

These policy simulations suggest a need for a burden-sharing mechanism in any future international collaboration to reduce global emissions. The burden could be shared through a variety of ways, but one that is often discussed is the use of permit allocation under an emission trading system (see Box 7.1 for an example of how this is done in the EU ETS). Another approach would involve allowing each country/region to set its own local price for abating CO_2 emissions. While this may be workable, it may also be vulnerable to the free-rider problem in allocating emission reductions.[28] In a global trading system, it would be possible to allocate permits in a way that allows OECD countries to carry a relatively greater financial responsibility for emission reduction than non-OECD regions. In addition, a global mitigation effort combined with a burden-sharing scheme could be easier (although still difficult) to agree than internationally harmonised carbon taxes. It is generally recognised that creating harmonised taxes will be very difficult, whereas negotiating a system of tradable permits frames the problem of climate change as one of both challenges and opportunities, and brings mutual benefits from co-operation.

All of the tax cases show lower economic costs (see Table 7.6 and Figures 7.10a and b) but are also less effective in avoiding climate change than the 450 PPM case. The 450 ppm case, however, requires policy to be aggressive in mitigating emissions across all regions. Achievement of this stabilisation target through a harmonised tax results in a global GDP loss of about 2.5% by 2050. An emissions trading policy – aiming to achieve the same target – would keep the GDP loss at similar levels. Alternative policies could increase global costs substantially if they do not encourage least cost abatement in a similar manner.

The regional costs of climate policy strongly depend on how international climate policy is implemented. As an alternative to an international carbon tax (explored above), mitigation may be achieved through a so-called cap and trade system, which has a centrepiece agreement on emission reduction targets or caps, and on how these are to be allocated across regions in combination with international emissions trading. In such a system, international trade still allows all countries to benefit from low-cost reductions worldwide (depending on the extent of participation). Figures 7.12a and b show an illustrative example of what could happen to regional emissions and the regional distribution of direct mitigation costs in striving to stabilise greenhouse gas concentrations at 450 ppm CO_2eq through a global trading system.[29]

Under this simulation, part of the emission rights would be traded internationally. Rather than using a uniform global carbon tax to stabilise greenhouse gas concentrations at 450 ppm CO_2eq (see Box 7.3), this example assumes an annual cap on emissions to achieve the same target. The allocation of emission rights in this example is based on gradual per capita convergence worldwide by 2050. Alternative convergence criteria are conceivable (*e.g.* emissions per GDP, or emission thresholds) as well as alternative convergence years. The model simulation assumes that countries trade emission rights in order to minimise their overall cost of abatement. Thus, assuming full trade, full market access and full information, the simulation determines what proportion of emission rights would be traded and how that would affect regional costs of abatement.

Figure 7.12a. **Greenhouse gas emissions by regions in 2050: Baseline and 450 ppm cap and trade regime**[a]

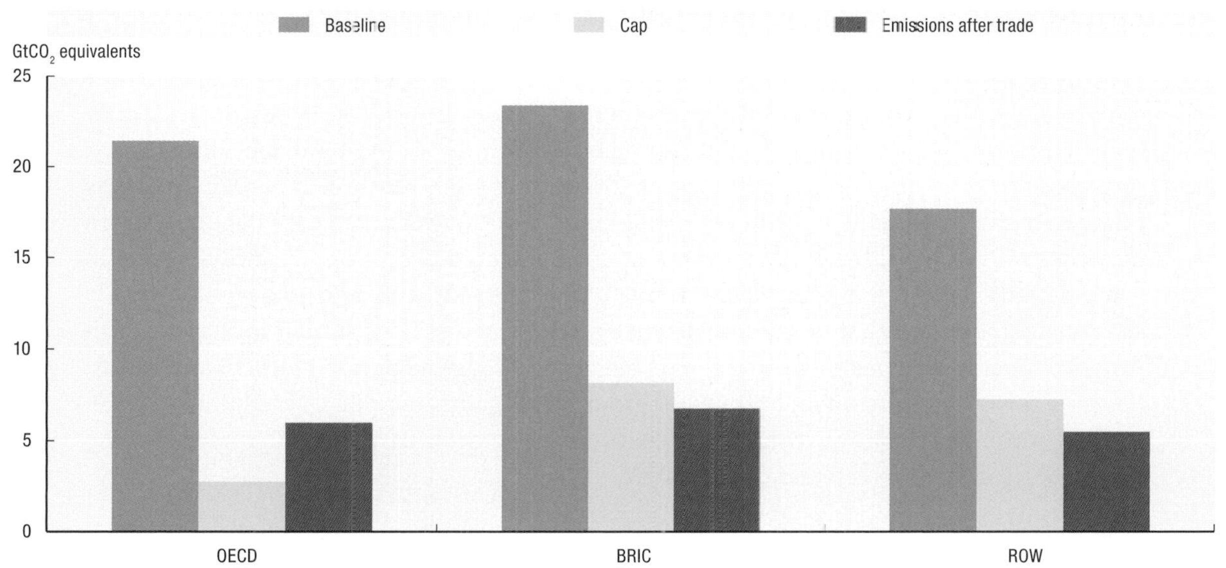

StatLink http://dx.doi.org/10.1787/260866744606

a) Excluding greenhouse gas emissions from land use and forestry.

Source: FAIR model (*www.mnp.nl/fair/introduction*): see note 29 at the end of this chapter.

Figure 7.12b. **Regional direct cost of greenhouse gas abatement under different mitigation regimes, 2050**

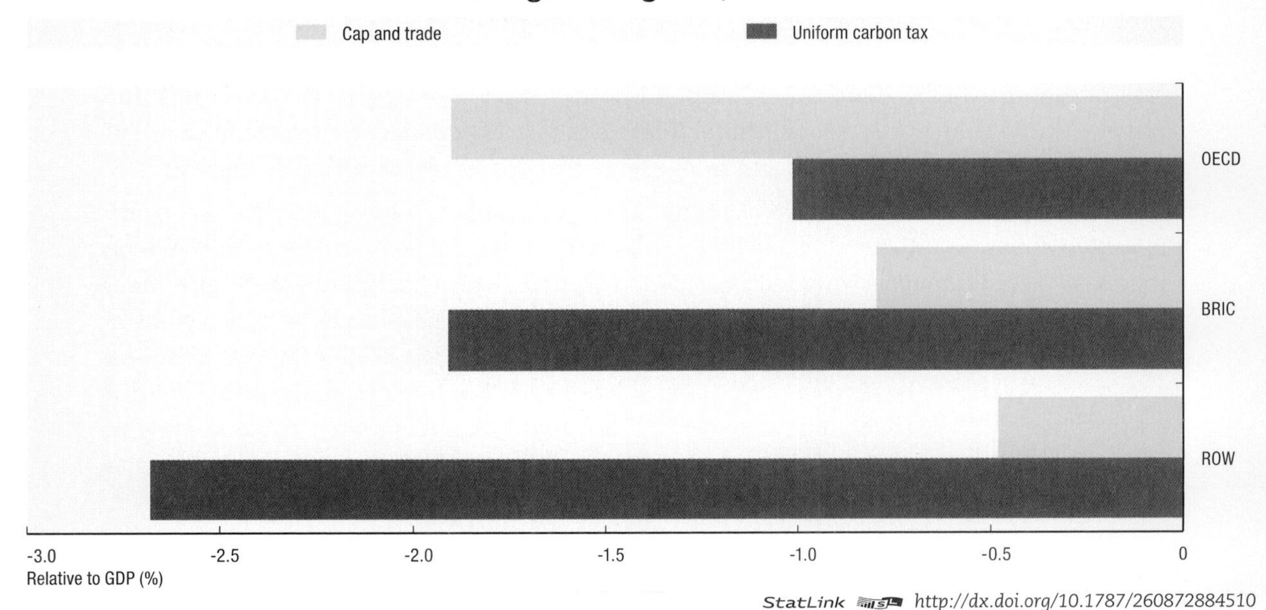

StatLink http://dx.doi.org/10.1787/260872884510

Source: FAIR model (*www.mnp.nl/fair/introduction*): see note 29 at the end of this chapter.

In Figure 7.12a, the difference between the bars representing the Baseline (left) and emissions cap (middle) is the amount of emissions to be cut in each regional grouping to stabilise greenhouse gas concentrations at 450 ppm CO_2eq without trading. In this example, OECD countries would be required to cut emissions by 18.7 Gt CO_2eq by 2050 compared to the Baseline. The difference between these emission caps without trade (middle bars) and emissions after trade (right-hand bars) reflects the emission rights that would be bought or sold between regional groups. In this example of trading, OECD countries buy 3.3 Gt CO_2eq of emission rights by 2050.

The cap and trade system changes the global distribution of direct abatement costs compared with the uniform global tax case (see Figure 7.12b). The costs to OECD countries of achieving 450 ppm CO_2eq stabilisation are more than in the global tax case because they are assigned more ambitious emission reduction targets. These OECD targets are partly met by trading, which brings costs down below what they would be if met unilaterally. Importantly, this simulation limits the imposition of high costs in non-OECD regions relative to their GDP, which would otherwise emerge in the global tax case (Figure 7.10). In moving towards 2050, the ROW group of countries would even see net annual gains in some periods under the trading case (*i.e.* in 2025). In the BRIC group, Russia would initially see considerable gains before coming down, by 2050, to a cost level similar to that in North America. Costs are expected during the whole of the simulated period in Brazil and China; however these costs are offset in the BRIC grouping by gains in India. Overall the emission trading simulation shows the direct costs of mitigation in the BRIC region falling significantly under the cap and trade system.

Summary

The unique challenges of climate change mitigation include balancing concerns about its inter-generational consequences, as there is a lag between when action is taken and when results are reaped (*i.e.* in the form of avoided climate change impacts). The consequences of climate change, and vulnerability to these, are also distributed across regions and countries unevenly, with the greatest risk of relative impacts expected to be in regions and countries where emissions are lowest. Mitigation potential and climate change risk also differ widely within a single country, across locations and actors. Distributional considerations are inevitably an important consideration for policy decision-making across all scales of governance. In addition, there are important questions about how much mitigation is desirable and how fast, and how to act in a cost-effective, economically sustainable and equitable manner.

The *Outlook* on climate change leads to a number of important conclusions for policy:

i) The risks of inaction are high, with unabated emissions in the Baseline leading to about a 37% and 52% increase in global emissions in the 2030 and 2050 timeframe respectively, with a wide range of impacts on natural and human systems. This unabated emission pathway could lead to high levels of global warming, with long-term temperature rises likely in the range of 4 to 6°C (equilibrium).

ii) Starting early with mitigation policies that stabilise atmospheric concentrations will limit temperature increases and rates of change significantly by mid-century and could limit long-term temperature increases to 2-3°C.

iii) Broad participation by all the big emitting countries in the coming decades will be required to achieve these outcomes.

iv) The costs of even the most stringent mitigation cases are in the range of only a few percent of global GDP in 2050. Thus they are manageable, especially if policies are designed to start early, to be cost-effective and to share the burden of costs across all regions.

Notes

1. Though from 1990 to 2004, total CH_4 emissions decreased across all OECD countries by roughly 8%, with the largest absolute decreases occurring in Germany, Poland, the United Kingdom and the United States (UNFCCC GHG emission database: *http://GHG.unfccc.int/tables/queries.html*). N_2O emissions have followed a similar trend.

2. CO_2 concentrations are currently increasing at a rate of approximately 1.9 ppm per year (IPCC, 2007a).

3. This warming estimate is relative to 1980-1999; comparing it to pre-industrial temperature adds 0.5 °C for warming of 1.1 °C (best estimate) for a range of 0.8-1.4 °C of warming.

4. Note this period is relevant to accounting for emissions from countries listed in Annex I – or industrialised countries – under the UNFCCC and the Kyoto Protocol. For comparison, emission data are also reported here for non-OECD countries, *i.e.* Brazil, India and China or the BIC group of countries, leaving Russia aside as it has very different patterns of emission growth to the other large non-OECD economies noted here.

5. Data for 2005 are used for all OECD countries except where they were not available: *i.e.*, Greece (2004), Turkey (2004), Mexico (2002) and Korea (2001).

6. Accounting for national emissions according to the Kyoto Protocol separates emissions from land use, land use change and forestry as well as international bunker fuels for aviation and marine activities. The former are accounted for and managed separately by individual nations under the rules for "Kyoto forests", while international bunkers (international aviation and marine fuel use) are to be managed through the agreements under the UN International Civil Aviation Organisation (ICAO) and the International Maritime Organisation (IMO). To date no agreement has been achieved. International bunker fuels were estimated to be about 3% of world CO_2 emissions in 2005 and are growing rapidly (IEA, 2006).

7. In descending order: Korea (1990-2001), Spain, Canada, Portugal, Turkey, Greece, Ireland, Mexico and New Zealand.

8. In descending order: the United States, Austria, Italy, Japan, Switzerland, Australia (where the increase since 1990 has been below the increase allowed under the Kyoto Protocol), Luxembourg and Iceland (both of which have had increases of less than 1% since 1990).
9. Note CO_2eq is used in two ways in this chapter. First it is a "unit" of measurement of aggregate emissions across greenhouse gases. This is based on a reporting convention adopted by the IPCC – global warming potentials – which refer to the integrated radiative forcing of each gas in comparison to that of CO_2 in a given timeframe. Similarly CO_2eq concentrations combine the concentrations of different greenhouse gases into a single metric, accounting for the different radiative forcings of each. See IPCC 2007a, p. 133 for a full description.
10. The *OECD Environmental Outlook* Baseline for CO_2 emissions from energy has been calibrated to that developed by the International Energy Agency (2006) in their *World Energy Outlook* (WEO), which looks in-depth at world energy developments to 2030.
11. The upper and lower bounds of the baseline scenarios represent one standard deviation around the median of the entire distribution of emission pathways within the baseline.
12. CFCs contribute much more to radiative forcing than HFC/PFCs do today or in future predictions, so their reduction is significant to climate change.
13. These countries are also referred to as Annex II countries or Parties (where they are ratified Parties to the Convention or the Protocol).
14. The US signed the Kyoto Protocol but have not ratified it.
15. A number of parallel processes are also proceeding towards a similar end, *e.g.* the Gleneagles dialogue initiated in 2005, among others.
16. See for example: OECD/EEA database on instruments used for environmental policy and natural resources management: *www2.oecd.org/ecoinst/queries/index.htm* [last accessed 17 July 2007].
17. This is because as GHGs are a global pollutant, the impacts are not related to the source or location of the emission.
18. This is a voluntary system.
19. As of 7 Feb. 2007, approximately 112 million CERs are expected to be generated through registered projects.
20. Voluntary agreements and measures are a subset of a larger set of "voluntary approaches" that may include unilateral actions by industry and other stakeholders.
21. The regulated price per unit of electricity that a utility or supplier has to pay for renewables-based electricity from private generators.
22. Unless otherwise noted, this *Outlook* assumes a climate sensitivity of 2.5°C per doubling of carbon dioxide concentrations in the atmosphere, which is lower and more conservative than the IPCC AR4 (IPCC, 2007a). Using the IPCC "best estimate" of climate sensitivity (*i.e.* 3°C) would raise the central estimate of temperature change associated with the Baseline emission pathway.
23. The trajectory of Baseline emissions beyond 2050 is not clearly defined. Based on the emission trajectory to 2050, it is unlikely that the Baseline would lead to stabilisation of greenhouse gas concentrations at a level below the IPCC "category V" and "VI" scenarios (see Table 7.5). This suggests that an indicative value of minimum equilibrium temperature change under the *Outlook* Baseline would be 4-6°C.
24. The case cited here is for 450 ppm CO_2, which is roughly equivalent to 550 ppm CO_2eq taking into account the concentrations of all GHGs in the atmosphere. The data for temperature change in the 2080s associated with stabilisation pathways are cited in Carter *et al.*, 2007. Baseline temperature estimates for the 2080s are taken from van Vuuren *et al.*'s 2007 "modified B2" scenario, which is similar to our Baseline to 2050. Tim Carter and Detlef van Vuuren provided the data for this calculation.
25. See Chapter 8, Air pollution, for a discussion of air pollution policies in their own right. See Chapter 12 for the related benefits in terms of human health. Typically these are more ambitious policies than the co-benefits of climate change policies.
26. Though not shown, Norway fares better than Russia in response to mitigation policy because its domestic energy prices are closer to those of its competitors.
27. Value added is the contribution to GDP of any particular industrial activity, sub-sector or sector.
28. The free-rider problem refers to a situation where parties in a negotiation have an incentive to let others do most of the work.

29. This simulation was conducted using the FAIR model (*www.mnp.nl/fair/introduction*). Unlike Figure 7.10, costs estimated by this simulation and presented in Figure 7.12b are the direct costs of mitigation; that is, they do not represent change in GDP growth as a result of shifts induced in the wider economy. Although the metric for measuring economic effects is slightly different than the one described for the ENV-Linkages simulations above, the relative change from the Baseline to policy simulations – or between policy cases – is indicative of the results that would be obtained using ENV-Linkages. The simulation has been done at the level of 26 global regions, although Figures 7.12a and b aggregate the results to three regional groups. This aggregation masks some of the more detailed results which show that intra-regional trading also occurs to lower the overall costs of mitigation.

References

Agrawala, S. (ed.), (2005), *Bridge Over Troubled Waters: Linking Climate Change and Development*, OECD, Paris.

Alkemade, J.R.M. *et al.* (2006), "GLOBIO 3: Framework for the Assessment of Global Terrestrial Biodiversity", in A.F. Bouwman, T. Kram and K. Klein Goldewijk (eds) *Integrated Modelling of Global Environmental Change. An Overview of IMAGE 2.4*, Netherlands Environmental Assessment Agency (MNP), Bilthoven, The Netherlands.

Barnett, J. (2003) "Security and Climate Change", *Global Environmental Change* 13:7-17.

Brauch, H.G. (2002), "Climate Change, Environmental Stress and Conflict", in G.F.M.f.t. Environment (eds) *Climate Change and Conflict*, Part II, German Federal Ministry for the Environment, Berlin.

Bryden, H.L., H R. Longworth and S.A. Cunningham (2005), "Slowing of the Atlantic Meridional Overturning Circulation at 25 degrees North", *Nature*, Vol. 438, 1 December 2005.

Campbell, K.M., J. *et al.* (2007), *The Age of Consequences: The Foreign Policy and National Security Implications of Global Climate Policy*, Center for Strategic and International Studies; Center for a New American Security, Washington DC.

Capoor, K. and P. Ambrosi (2007), *State and Trends of the Carbon Market*, The World Bank, Washington DC.

Carter, T.R. *et al.* (2007), "New Assessment Methods and the Characterisation of Future Conditions", in M.L. Parry *et al.* (eds.), *Climate Change 2007: Impacts, Adaptation and Vulnerability. Contribution of Working Group II to the Fourth Assessment Report of the Intergovernmental Panel on Climate Change*, Cambridge University Press, Cambridge, UK.

CBD (Convention on Biological Diversity) and MNP (Netherlands Environmental Assessment Agency) (2007), "Cross-roads of Life on Earth: Exploring Means to Meet the 2010 Biodiversity Target; Solution-oriented Scenarios for Global Biodiversity Outlook 2", *CBD Technical Series No. 31*/MNP report nr. 555050001, Secretariat of the Convention on Biological Diversity (sCBD) and Netherlands Environmental Assessment Agency (MNP), Montreal and Bilthoven.

CEC (Commission of the European Communities) (2007), *Adapting to Climate Change in Europe: Options For EU Action*, Green Paper from the Commission to the Council, the European Parliament, the European Economic and Social Committee and the Committee of the Regions; COM(2007)354 final, Commission of the European Communities, Brussels.

Emanuel, K. (2005), "Increasing Destructiveness of Tropical Cyclones over the Past 30 Years", *Nature*, Vol. 436: 686-688, 4 August 2005.

Feeley, R.A. *et al.* (2004), "Impact of Anthropogenic CO_2 on the $CaCo_3$ System in the Oceans", *Science*, Vol. 305, 362-366.

Fisher, B.S. *et al.* (2007), "Issues Related to Mitigation in the Long Term Context", in O.R.D.B. Metz *et al.*, (eds.), *Climate Change 2007. Mitigation. Contribution of Working Group III to the Fourth Assessment Report of the Inter-governmental Panel on Climate Change*, Cambridge University Press, Cambridge.

Gagnon-Lebrun, F. and S. Agrawala (2008), "Implementing Adaptation in Developed Countries: An Analysis of Broad Trends", *Climate Policy*, in press.

Hansen, J. *et al.* (2005), "Earth's Energy Imbalance: Confirmation and Implications", *Science*, Vol. 308. No. 5727, pp. 3008, 3 June 2005.

Harmelen, T. van *et al.* (2002), "Long-term reductions in costs of controlling regional air pollution in Europe due to climate policy", *Environmental Science and Policy*, 5(4), pp. 349-365.

IEA (International Energy Agency) (2006), *World Energy Outlook 2006*, OECD, Paris.

IEA (2007a), *CO_2 Emissions from Fuel Combustion 2007*, OECD, Paris.

IEA (2007b), *World Energy Outlook 2007*, OECD, Paris.

IPCC (Intergovernmental Panel on Climate Change) (2007a), "Summary for Policymakers", in, S. Solomon *et al.* (eds.), *Climate Change 2007: The Physical Science Basis. Contribution of Working Group I to the Fourth Assessment Report of the Intergovernmental Panel on Climate Change*, Cambridge University Press, Cambridge, United Kingdom and New York.

IPCC (2007b), "Summary for Policymakers", in M.L. Parry *et al.* (eds.), *Climate Change 2007: Impacts, Adaptation and Vulnerability. Contribution of Working Group II to the Fourth Assessment Report of the Intergovernmental Panel on Climate Change*, Cambridge University Press, Cambridge, United Kingdom and New York.

IPCC (2007c), "Summary for Policymakers", in B. Metz *et al.* (eds.), *Climate Change 2007: Mitigation. Contribution of Working Group III to the Fourth Assessment Report of the Intergovernmental Panel on Climate Change*, Cambridge University Press, Cambridge, United Kingdom and New York.

IPCC (2007d), "Summary for Policymakers", in B. Metz *et al.* (eds.), *Climate Change 2007: Synthesis Report, Fourth Assessment Report of the Intergovernmental Panel on Climate Change*, Cambridge University Press, Cambridge, United Kingdom and New York.

Jones, R. (2004), "Managing Climate Change Risks", in Corfee-Morlot, J. and S. Agrawala (eds.), *The Benefits of Climate Change Policies: Analytical and Framework Issues*, OECD, Paris.

Kallbekken, S. and N. Rive (2006), "Why Delaying Emission Cuts is a Gamble", in Schellnhuber *et al.* (eds.) *Avoiding Dangerous Climate Change*, Cambridge University Press, Cambridge.

Levina, E. and H. Adams (2006), *Domestic Policy Frameworks for Adaptation – Part I: Annex I Countries*, in ENV/EPOC/IEA/SLT(2006)2, OECD, Paris.

Mastrandrea, M., and S. Schneider (2004), "Probabilistic Integrated Assessment of 'Dangerous' Climate Change", *Science* 304:571-575.

McKenzie-Hedger, M. and J. Corfee-Morlot, (eds.) (2006), *Adaptation to Climate Change: What Needs to Happen Next?*, report of a workshop in the UK EU Presidency, UK Environment Agency and DEFRA, London.

Meinshausen, M. (2006), "What Does a 2 degree C Target Mean for Greenhouse Gas Concentrations?", in Schellenhubner J. *et al.*, (eds.) *Avoiding Dangerous Climate Change*, Cambridge University Press, Cambridge.

Morgenstern, R. (2000), "Baseline Issues in the Estimation of Ancillary Benefits of Greenhouse Gas Mitigation Policies", in *Ancillary Benefits and Costs of Greenhouse Gas Mitigation, OECD Proceedings of an IPCC Co-Sponsored Workshop*, 27-29 March 2000, in Washington DC, OECD, Paris.

Oberthuer, S., D. Taenzler and A. Carius (2002), "Climate Change and Conflict Prevention: The Relevance for the International Process on Climate Change", in *Climate Change and Conflict*, Part III, German Federal Ministry for the Environment, Berlin.

OECD (2000) *Ancillary Benefits and Costs of Greenhouse Gas Mitigation, OECD Proceedings of an IPCC Co-Sponsored Workshop*, 27-29 March 2000, in Washington DC, OECD, Paris.

OECD (2003), *Voluntary Approaches for Environmental Policy: Effectiveness, Efficiency and Usage in Policy Mixes*, OECD, Paris.

OECD (2006), *Declaration on Integrating Climate Change Adaptation into Development Co-operation*, adopted by Development and Environment Ministers of OECD Member Countries on 4 April 2006, COM/ENV/EPOC/DCD/DAC(2005)8/FINAL, OECD, Paris.

Pittini, M., and M. Rahman. (2004), "Social Costs of Carbon", in J. Corfee-Morlot and S. Agrawala (eds.) *The Benefits of Climate Change Policy: Analytical and Framework Issues*, OECD, Paris.

Schneider, S.H. *et al.* (2007), "Assessing Key Vulnerabilities and the Risk from Climate Change", in M.L. Parry *et al.*, (eds.), *Climate Change 2007: Impacts, Adaptation and Vulnerability.Contribution of Working Group II to the Fourth Assessment Report of the Intergovernmental Panel on Climate Change*, Cambridge University Press, Cambridge, UK.

Shalizi, Z. (2007), "Energy and Emissions: Local and Global Effects of the Giant's Rise", in A.L. Winters and S. Yusuf (eds.), *Dancing with Giants: China, India and the Global Economy*, World Bank and Institute of Policy Studies: Washington DC and Singapore.

Siegenthaler, U. *et al.* (2005), "Stable Carbon Cycle-Climate Relationship during the late Pleistocene", *Science*, Vol. 310: 1313-1317, 25 November 2005.

Spahni, R. *et al.* (2005), "Atmospheric Methane and Nitrous Oxide of the Late Pleistocene from Antarctic Ice Cores", *Science*, Vol. 310: 1317-1321, 25 November 2005.

Tol, R. (2005), "The Marginal Damage Costs of Climate Change: An Assessment of the Uncertainties", *Energy Policy* 33:2064-2074.

UNFCCC (United Nations Framework Convention on Climate Change) (2006a), *National Greenhouse Gas Inventory Data for the Period 1990-2004 and Status of Reporting*, FCCC/SBI/2006/26, 19 October, UNFCCC, Bonn.

UNFCCC (2006b), *Synthesis of Reports Demonstrating Progress in Accordance with Article 3, Paragraph 2, of the Kyoto Protocol* FCCC/SBI/2006/INF.2, 9 May, UNFCCC, Bonn.

Velders, G. *et al.* (2007), "The Importance of the Montreal Protocol in Protecting Climate", *Proceedings of the National Academy of Science* 104: 4814-4819.

Vuuren, D. van *et al.* (2006), "Exploring the Ancillary Benefits of the Kyoto Protocol for Air Pollution in Europe", *Energy Policy*, 34, pp. 444-60.

Vuuren, D. van *et al.* (2007), "Stabilizing Greenhouse Gas Concentrations at Low Levels: an Assessment of Reduction Strategies and Costs", *Climatic Change* 81 (2):119.

Vuuren, D. van *et al.* (Submitted), *Temperature Increase of 21st Century Stabilization Scenarios*, submitted to PNAS.

Webster, P.J. *et al.* (2005), "Changes in Tropical Cyclone Number, Duration, Intensity in a Warming Environment", *Science* Vol. 309: 1844-1846, 16 September 2005.

Westerling, A.L. *et al.* (2006), "Warming and Earlier Spring Increases Western US Forest Wildfire Activity", *Science Express*, 6 July 2006.

Yohe, G., N. Andronova and M. Schlesinger (2004), "To Hedge or not to Against an Uncertain Climate Future", *Science* Vol. 306: 416-417.

ISBN 978-92-64-04048-9
OECD Environmental Outlook to 2030

Chapter 8

Air Pollution

This chapter focuses on projected developments in outdoor air pollution, especially particulate matter and ozone, and on urban air quality. It outlines projections of concentrations between 2000 and 2030, and summarises the impact of three policy simulations on air pollution emissions. Most OECD countries have reduced air pollution in recent decades, decoupling it from economic growth. However, pollution from other countries is increasingly undermining local urban air quality management, effectively making it an international issue. Paying greater attention to marine shipping, dealing with precursors of ground-level air pollution (such as methane) and taking into account the transport of air pollution from one continent to another in domestic air quality policies are all important in combating air pollution.

KEY MESSAGES

The *Outlook* Baseline projects a further deterioration of urban air quality to 2030, especially in non-OECD countries. The health related target levels of particulate matter (PM_{10}) are already being exceeded in most regions.

In OECD countries the main sources of man-made air pollution remain motorised transport and other uses of fossil fuels. In many developing countries wood burning is also a major source of air pollution. Abatement of air pollution from aviation and shipping is lagging behind abatement from road vehicles. Under current policies to abate sulphur dioxide emissions, emissions from shipping will overtake land-based emissions in OECD countries by 2020.

In China, air pollution abatement policies are emerging, including stringent vehicle standards. However, these do not seem to be sufficient yet to decouple air pollution developments from economic growth. Consequently, it is projected that in China and other parts of continental Asia emissions will continue to grow to 2030.

In the Northern hemisphere, pollution from other countries is increasingly undermining local urban air quality management, effectively making it an international issue.

Most OECD countries have reduced air pollution in recent decades, decoupling it from continued economic growth. Measures at local, national and international scales to combat air pollution in OECD countries have been particularly effective. The Baseline assumes this trend will continue as a result of existing policies, but additional measures will be required to further decouple particular pollutants.

Policy options

- Pay greater attention to marine shipping as a growing source of air pollution. This sector has cost-effective potential for abating emissions.
- Increase the impacts of policies by targeting those with important synergies between air pollution and climate change. For example, dealing with precursors of ground-level air pollution (such as methane), is a cost-effective way of addressing both issues.
- Take into account the transboundary movement of air pollution from one continent to another in domestic air quality policies.

Three *Environmental Outlook* (EO) policy packages were simulated with a varying degree of participation by BRIC and other non-OECD countries (see Chapter 20 for more details). Putting in place ambitious but realistic enhanced air pollution policies in OECD countries would significantly reduce by 2030 the urban levels of particulate matter (see graph) and related health impacts (see Chapter 12). Reducing emissions through a global policy package could bring large improvements in urban air quality by 2050.

Annual mean PM_{10} concentration in urban agglomerations, 2030

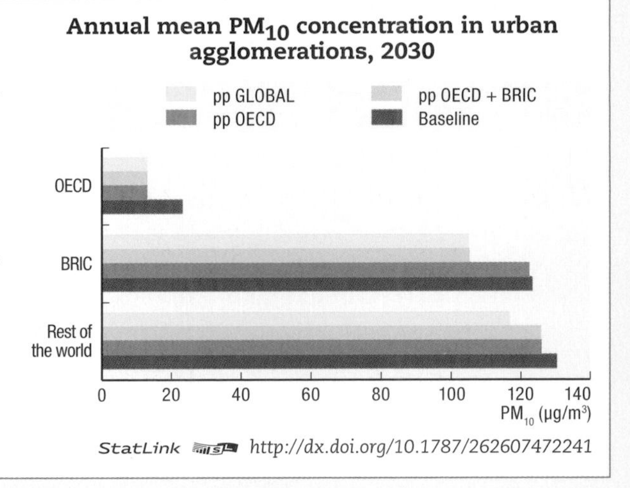

StatLink http://dx.doi.org/10.1787/262607472241

Consequences of inaction

With no new policies to tackle air pollution, urban dwellers, mostly in developing countries, will be exposed to air pollution above health related target levels. In combination with ongoing urbanisation and ageing, an increase in adverse health effects of air pollution is expected between 2000 and 2030.

 OECD ENVIRONMENTAL OUTLOOK TO 2030 – ISBN 978-92-64-04048-9 – © OECD 2008

Introduction

Many epidemiological and toxicological investigations have shown that exposure to air pollution damages health and can lead to hospital admission or premature death (see also Chapter 12 on health and environment). It also affects natural ecosystems. This chapter focuses on outdoor air pollution, especially particulate matter and ozone. Moreover, the chapter deals mainly with urban air quality. It outlines projections of concentrations between 2000 and 2030, modelled as part of the *Outlook* Baseline. The chapter also summarises the impact of three policy simulations on air pollution emissions. The health impacts of these projections are presented in Chapter 12.

Air pollution sources and movement patterns

Particulate matter (PM) is either directly emitted into the atmosphere, or formed in the atmosphere from precursor gases (sulphur dioxide, nitrogen oxides, ammonia and, to a lesser extent, volatile organic compounds).[1] The use of coal and wood for domestic heating and cooking are important sources of PM; natural sources are volcanic emissions, soil resuspension (Sahara dust) and sea spray. Exposure may be exacerbated by the use of open stoves, leading to high indoor concentrations (Box 8.1).

In the Northern hemisphere, pollution from other countries is increasingly undermining local urban air quality management.

Ozone is not emitted into the atmosphere; instead it is formed in photochemical processes. The most important precursors of ozone are nitrogen oxides and organic compounds. Volatile organic compounds are largely emitted by the transport sector and by the use of solvents. Another important precursor is methane; anthropogenic methane emissions mainly come from rice paddies, waste and wastewater treatment, gas and oil mining and animal husbandry.

Although local and regional emissions of air pollutants determine the levels of air pollution and human exposure, air pollutants such as ozone, particulate matter and other long-lived pollutants can be transported over very great distances. Hemispheric transport

Box 8.1. **Indoor air pollution**

Studies have revealed that the number of premature deaths arising from indoor air pollution in developing countries is comparable to the global number arising from ambient air pollution (see for example Smith *et al.*, 2004). Indoor air pollution requires viable, cost-effective interventions that can reduce exposure and improve health. Although awareness has been growing, indoor air pollution from household traditional biomass use has not been a major issue on the global agenda in terms of international, bilateral, or national development assistance. Measures to address indoor air pollution from traditional biomass use are not included in this *Outlook*. But from a global perspective, the potentially large health benefits from tackling indoor air pollution should be a policy priority.

of air pollution, from one continent to another, is an emerging international complication for domestic air quality policies (see Box 8.2 and 8.3). Global atmospheric dispersion models show that the increasing air pollution emissions in Southeast Asia may result in increasing background levels in North America and Europe. This increasing hemispheric background makes an important contribution to the trends in concentrations measured at the more local scale. It may also frustrate abatement policies at the local scale (see Box 8.2). Initially the focus of the ozone problem was on smog episodes, short periods with peak concentrations over 80-100 parts per billion (ppb). Emissions of ozone precursors at the regional to continental scale are responsible for these episodes. Satellite measurements clearly show the regional aspect of tropospheric ozone pollution in northeastern India, eastern United States, Europe, eastern China and west and southern Africa (Fishman *et al.*, 2003). Over China, an increase in nitrogen dioxide concentration is observed by satellites; at the same time a substantial reduction in nitrogen dioxide concentrations over some areas in Europe and USA is found (Richter *et al.*, 2005). These trends in nitrogen dioxide concentrations are in line with trends in emissions of nitrogen oxides: emissions are slowly decreasing in Europe and the USA, but increasing in China.

Impacts of air pollution on health

The most severe health effects of air pollution are from exposure to particulate matter and ozone. It is suggested that there is no safe level for either pollutant: they may even pose a health risk at concentrations below current air quality guidelines (see WHO, 2006 and references therein; and Chapter 12, in this *Outlook* report).[2]

Exposure to particulate matter (PM_{10} or $PM_{2.5}$, small particles with a diameter less than 10 microns (μm) or 2.5 μm, respectively) is one of the greatest human health risks from air pollution. Effects include the risk of respiratory death in infants under one year, as well as increasing deaths from cardiovascular and respiratory diseases and lung cancer. The epidemiological evidence shows adverse effects of particulate matter after both short-term and long-term exposure. Health effects of particulate matter are initiated by their inhalation and penetration in the lungs. Both chemical and physical interaction with lung tissues can irritate or damage the respiratory tracks. Current understanding is that the mortality effects of PM are mainly associated with the smaller particles, those with a diameter of 2.5 μm or smaller. However, effects are also observed with larger particles with diameters in the range of 2.5 to 10 μm.[3]

Toxic and carcinogenic pollutants like heavy metals or polycyclic aromatic hydrocarbon (PAH) are frequently bound to particles. For Europe the proportion of lung cancer attributable to urban air pollution, especially fine particles, can be as high as 10.7%, corresponding to 27 000 cases annually (Boffetta, 2006).

Ground level ozone is a strong photochemical oxidant; it is the main pollutant during summer smog episodes. Ozone impairs pulmonary function, causes lung inflammation and lung permeability and can lead to respiratory problems, increased medication usage, illness and death. Long-term exposure to relatively low levels is of concern. With even low ozone concentrations affecting both health and ecosystems (see below), formation and transport of ozone at the hemispheric scale becomes more important. In the troposphere, ozone also acts as a greenhouse gas (see Chapter 7 on climate change).

Impacts of air pollution on the environment

Ozone also affects vegetation by damaging leaves and reducing growth. Total exposure during the growing season, including at low levels, can have ecosystem-wide impacts. In

 OECD ENVIRONMENTAL OUTLOOK TO 2030 – ISBN 978-92-64-04048-9 – © OECD 2008

Box 8.2. **Travel distances and residence times of various air pollutants**

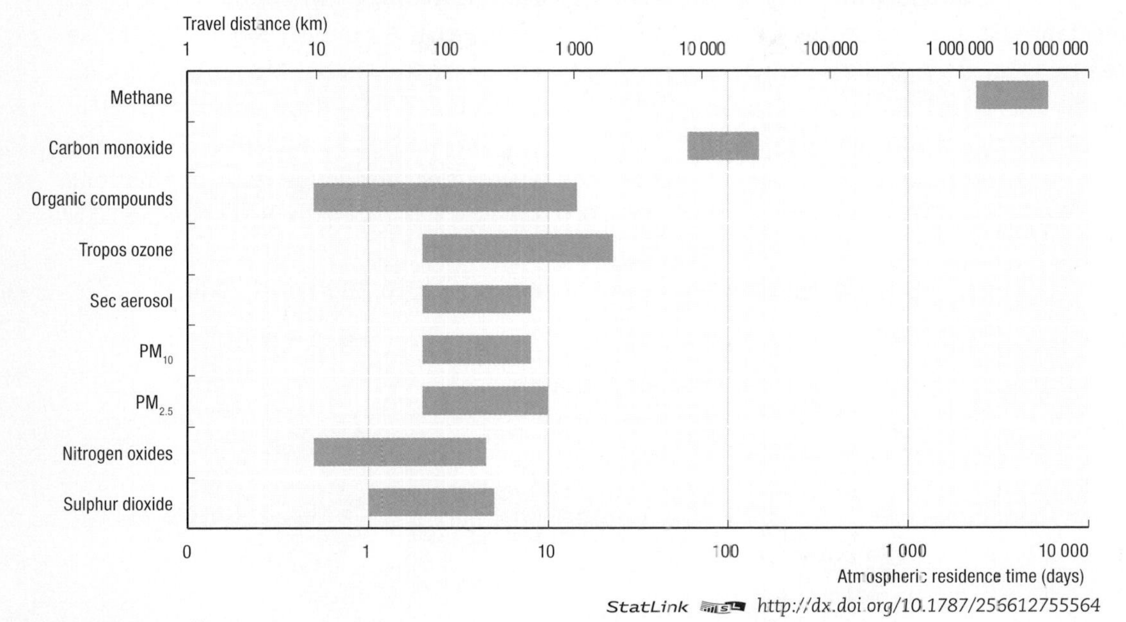

StatLink http://dx.doi.org/10.1787/256612755564

Note: Scales are not linear.

Air pollution is a problem at various different scales: sources (emissions) and impacts are seen at scales ranging from local to global. Pollutants with a very short residence time affect local air quality. Pollutants with a residence time of days or weeks affect air quality at a scale from local to continental. Particulate matter has a residence time of between several days and a week, giving it time to travel across an entire continent; smaller particles travel further than coarser particles. The gaseous precursors of aerosol have generally shorter lifetimes but can travel distances of several hundreds to a thousand kilometres. Ozone at higher altitudes may be transported across an entire hemisphere. The transport distances of the ozone precursors show a wide range. Volatile organic compounds (VOC) are reactive and travel at a continental scale. The reactive mixture of VOC and nitrogen oxides may lead to photochemical smog episodes with high ozone levels at the continental scale. Carbon monoxide and methane are long living (3 months and 8-10 years, respectively) and are transported at the hemispheric and global scale. The increased concentrations of these precursors have doubled the hemispheric ozone background concentration since industrialisation started. Persistent organic pollutants represent a global problem, although some have low residence time in the atmosphere they can be revolatilised and can migrate long distances and persist in different parts of the environment.

Europe, the economic costs of ozone exposure by agricultural crops are estimated to be high (EUR 2.8 billion in the European Union in 2000).

Gaseous pollutants like sulphur dioxide, nitrogen oxides and ammonia have various adverse impacts on vegetation, water bodies and materials. The deposition of nitrogen dioxide, sulphur dioxide and ammonia acidifies terrestrial and freshwater ecosystems ("acid rain"). Acidifying pollutants may also damage building structures and monuments. Eutrophication is a consequence of excess input of nitrogen nutrients (ammonia, nitrogen oxides); the atmospheric input of other nutrients is negligible. Eutrophication disturbs the structure and function of ecosystems, *e.g.* causing excessive algae blooming in surface waters or the loss of biodiversity. Forest decline in Europe, North America, and likely also in other parts of the world has been attributed to acidifying or eutrophying deposition. The deposition of toxic or persistent pollutants may result in an accumulation of these pollutants in the soil and biota.

Key trends and projections

For the *Outlook*, air quality, in particular annual average PM concentrations in residential areas, has been estimated using the Global Urban Air quality Model (GUAM; de Leeuw *et al.* forthcoming) which is a modified version of the Global Model of Ambient Particulates (GMAPS; Pandey *et al.*, 2006). The model is based on urban-specific meteorological data, and emissions and demographic data at the national level. The model is used to estimate the PM levels in more than 3 000 cities worldwide with populations greater than 100 000 (reference period 1995-2000, see Box 8.3 and Figure 8.1) between 2000

Figure 8.1. **Cities included in the assessments, in 2000 and 2030**

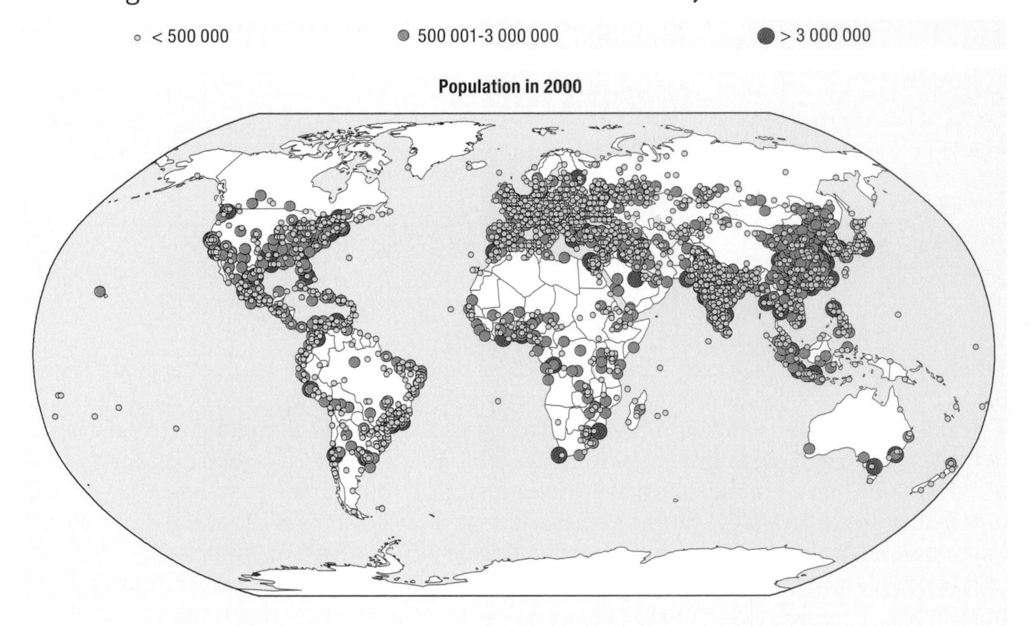

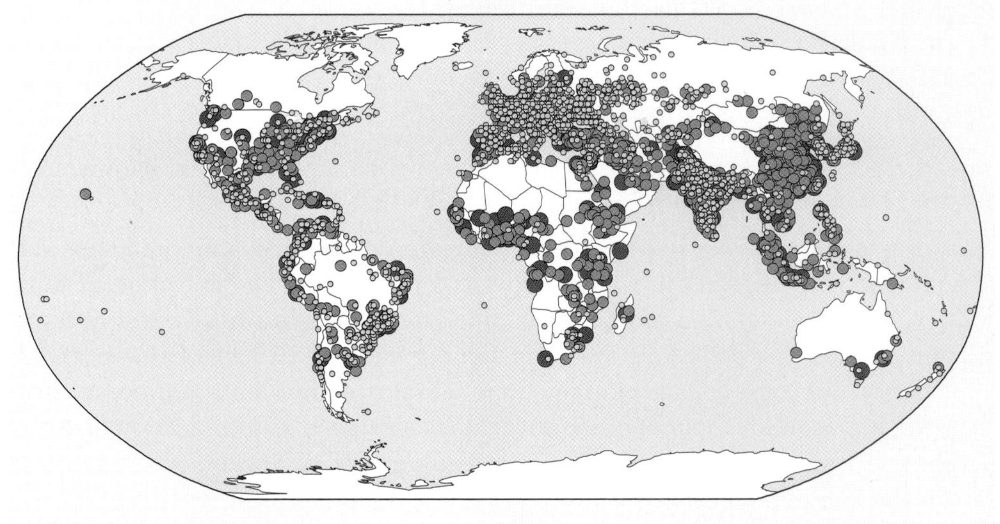

Source: Based on 2000 data taken from Pandey *et al.*, 2006.

OECD ENVIRONMENTAL OUTLOOK TO 2030 – ISBN 978-92-64-04048-9 – © OECD 2008

and 2030. In 2000 the population living in the modelled agglomerations varied between 18-70% of the total population in a region; in total 34% (2 062 million) of the global population lived in the modelled cities. In the period 2000-2030 the urban population is projected to grow both in absolute number (to 3 558 million) and as a proportion of the world population (43% in 2030). Figure 8.1 reflects the projected growth of urban populations, which is particularly strong in Africa and Asia.

Box 8.3. **Key uncertainties and assumptions**

This analysis describes the urban air quality in world cities with a population over 100 000. The list of urban agglomerations is taken from a database* prepared by the World Bank and refers to the situation in the period 1995-2000. Cities smaller than 100 000 inhabitants and fast growing agglomerations which cross the threshold in the period 2000-2030 have not been included. Therefore only a fraction of the total urban population is covered in the analysis; "urban population" in the text refers to the total population in the modelled cities only.

Air quality estimates have been prepared for urban agglomerations only, not for rural areas. Health impacts (presented in Chapter 12) have been assessed only for exposure to ambient air pollution and only for the population in the modelled urban agglomeration, not for those in rural or smaller urban areas. However, health effects from air pollution are also likely even from the lower levels of air pollution in these areas. Indoor air pollution from the use of solid fuels causes serious health effects. Smith *et al.* (2004) have estimated that twice as many premature deaths are attributable to indoor pollutants than to outdoor pollution (Box 8.1). Consequently, the quantitative results which refer to the impact of the exposure to outdoor air pollution on the modelled urban population presented here will actually be an underestimation.

Assumptions made in the development of the Baseline will affect the emissions estimated. Total and regionalised emissions depend on assumptions made about fuel mix, energy efficiency, growth in transport demand *etc.* For example, if more domestic coal is used in eastern Europe this will have a negative impact on urban air quality.

* See: *http://siteresources.worldbank.org/INTRES/Resources/AirPollutionConcentrationData2.xls.*

Particulate matter

For this *Outlook*, the annual average concentration of PM_{10} in cities of more than 100 000 people was modelled for 2000 and 2030. The population-weighted results for 13 regional clusters are presented in Figure 8.2. The World Health Organization (WHO, 2006) recommends three interim targets (decreasing from 70 to 50 to 30 $\mu g/m^3$) and a guideline of 20 $\mu g/m^3$ for PM_{10}. Figure 8.2[4] shows large differences in population-weighted concentrations for each region. In the most polluted areas (Middle East, Africa, Asia except Japan) concentrations are substantially above the WHO Interim Target One of 70 $\mu g/m^3$; these levels are associated with about 15% higher long-term mortality than at the guideline level. OECD-Pacific is the only region with average concentrations below the WHO air quality guideline (AQG). Within each region there are large differences between cities (Figure 8.3). In the Middle East and most of Asia, 70-90% of the urban population is exposed to concentrations above the highest WHO interim target. The

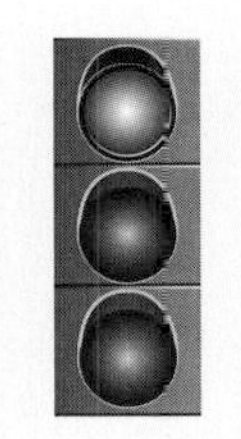

The health related target levels of particulate matter (PM10) are being exceeded in most regions.

Figure 8.2. **Annual mean PM_{10} concentrations, Baseline**

Regional annual mean PM_{10} concentration (population weighted)

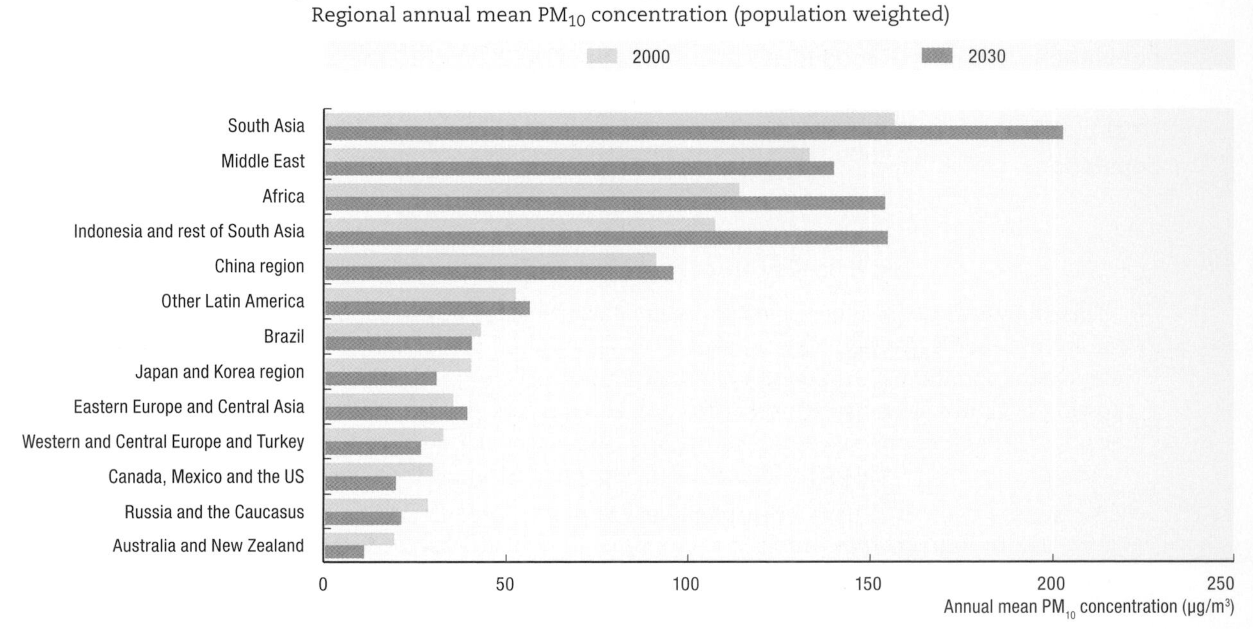

StatLink http://dx.doi.org/10.1787/260874156488

Source: *OECD Environmental Outlook* Baseline.

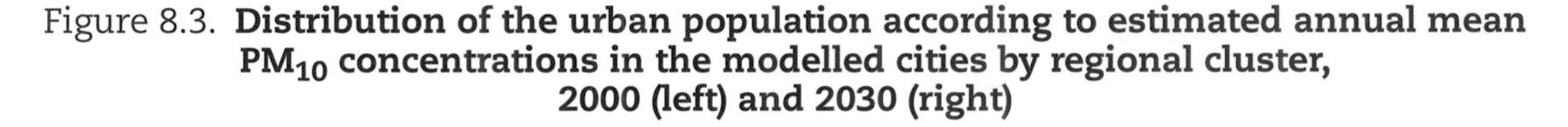

Figure 8.3. **Distribution of the urban population according to estimated annual mean PM_{10} concentrations in the modelled cities by regional cluster, 2000 (left) and 2030 (right)**

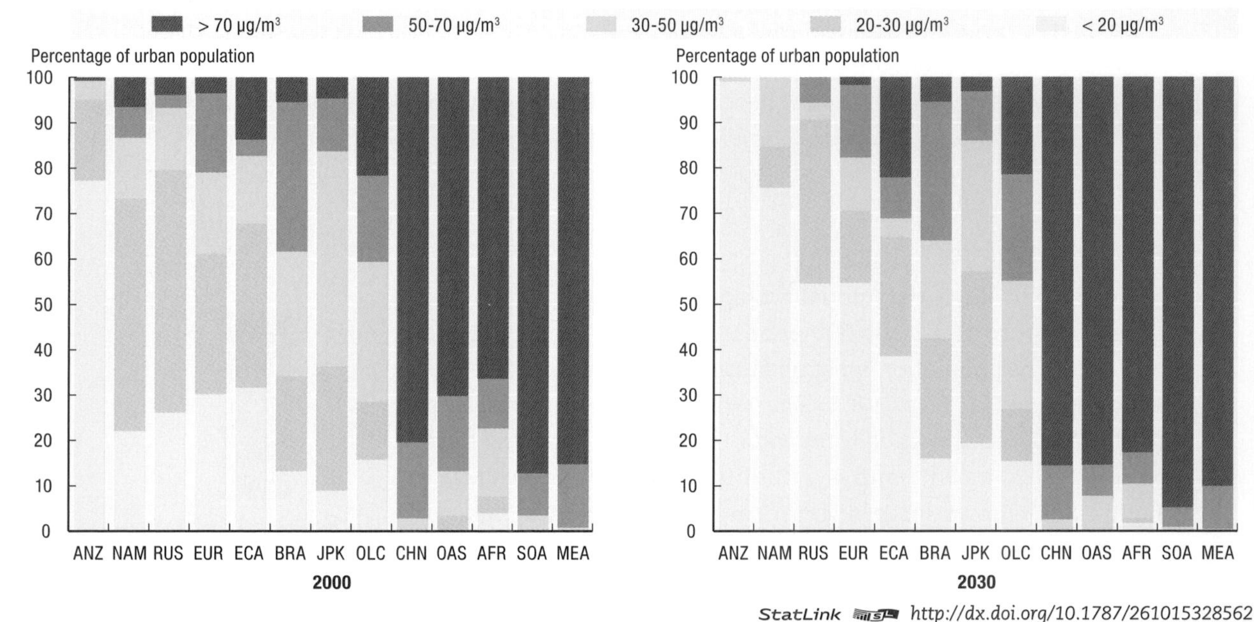

StatLink http://dx.doi.org/10.1787/261015328562

Note: Regional country groupings are as follows: NAM: North America (United States, Canada and Mexico); EUR (western and central Europe and Turkey); JPK: Japan and Korea region; ANZ: Oceania (New Zealand and Australia); BRA: Brazil; RUS: Russian and Caucasus; SOA: South Asia; CHN: China region; MEA: Middle East; OAS: Indonesia and the rest of South Asia; ECA: eastern Europe and central Asia; OLC: other Latin America; AFR: Africa.

Source: *OECD Environmental Outlook* Baseline.

high levels in South Asia are modelled mainly for cities in Bangladesh and Pakistan. Whilst the model may have overestimated the levels here, there is observational evidence that urban levels in Pakistan are frequently above 200 μg/m³ (Ghauri *et al.*, 2007).

Under the Baseline, urban air quality is projected to deteriorate by 2030 in seven of the 13 regional clusters. In the five most polluted regions, 50-90% of the urban population would be exposed to concentrations above the first WHO interim target of 70 μg/m³. An increase in related health impacts is expected (see Chapter 12).

The Baseline emission projection of air pollutants for Russia assumes a doubling of the use of natural gas in the domestic market (from 14 to 20 exajoules primary energy use between 2000 and 2030), while the use of coal remains modest (from 6 to 10 to 8 exajoules primary energy use in 2000, 2020 and 2030, respectively). In combination with stepped-up desulphurisation at power plants, partly to meet obligations under the Convention on Long-Range Transboundary Air Pollution, this is projected to decrease emissions of sulphur dioxide from electricity production (from 3.0 TgS in 2000 to 1.1 in 2030, far less than the 9 TgS per year that were emitted in the mid 1980s and 1990s).[5]

Policy simulations reported later in this chapter typically see energy use in Russia increase less than in the Baseline, with coal use even decreasing (to 2 exajoules), and thus emissions of sulphur dioxide decreasing to 0.4 TgS per year by 2030.

However, developments less favourable than assumed in the Baseline are conceivable. In particular, a far larger share of natural gas production could be destined for export, while coal rather than nuclear power could be used to fill the gap in domestic energy supply (not unlikely if the electricity market were to be liberalised.) Under such developments, air quality would strongly deteriorate rather than improve, unless more ambitious desulphurisation targets for coal, heavy fuel and process emissions are implemented. Therefore, Baseline concentrations in Russia and the Caucasus might be underestimates.

Ground-level ozone

Traditionally the focus on ozone problems was on peak episodes (*e.g.* the Los Angeles smog) where concentrations exceed 60-120 ppb caused by regional emissions of volatile organic compounds and nitrogen oxides. However, recent WHO findings that ozone can cause health problems at low concentration levels has drawn attention to the hemispheric background ozone, where emissions of its longer living precursors (methane, carbon monoxide) are of importance. The increased emissions of these pollutants have led to a steady increase of the hemispheric ozone background concentrations since the beginning of industrialisation (Volz and Kley, 1988).

Figure 8.4 shows the annual average ground level ozone concentrations for 2000 and 2030 for Baseline conditions (Dentener *et al.*, 2005). The maximum concentrations found in the Himalaya region are mainly from a natural origin; these are due to the high altitude of this region and to the strong mixing with ozone-rich stratospheric air. High man-made levels are found over the Arabian Peninsula, over the Mediterranean and the eastern coast of the USA. Observational data in Europe indicate that, despite the decrease in the European emissions of ozone precursors, ozone concentrations are expected to increase, especially in urban areas, because of the interaction with local emissions of nitrogen oxides (EEA, 2006; ETC/ACC, 2007). Under the *Outlook* Baseline, the area with annual mean concentrations exceeding 45 parts per billion is projected to greatly increase by 2030 to become one large continuous area from Spain to Japan, along with two additional areas over coastal USA (Figure 8.4).

Figure 8.4. **Current (2000, top map) and future (2030, bottom map) ozone concentrations at ground level**

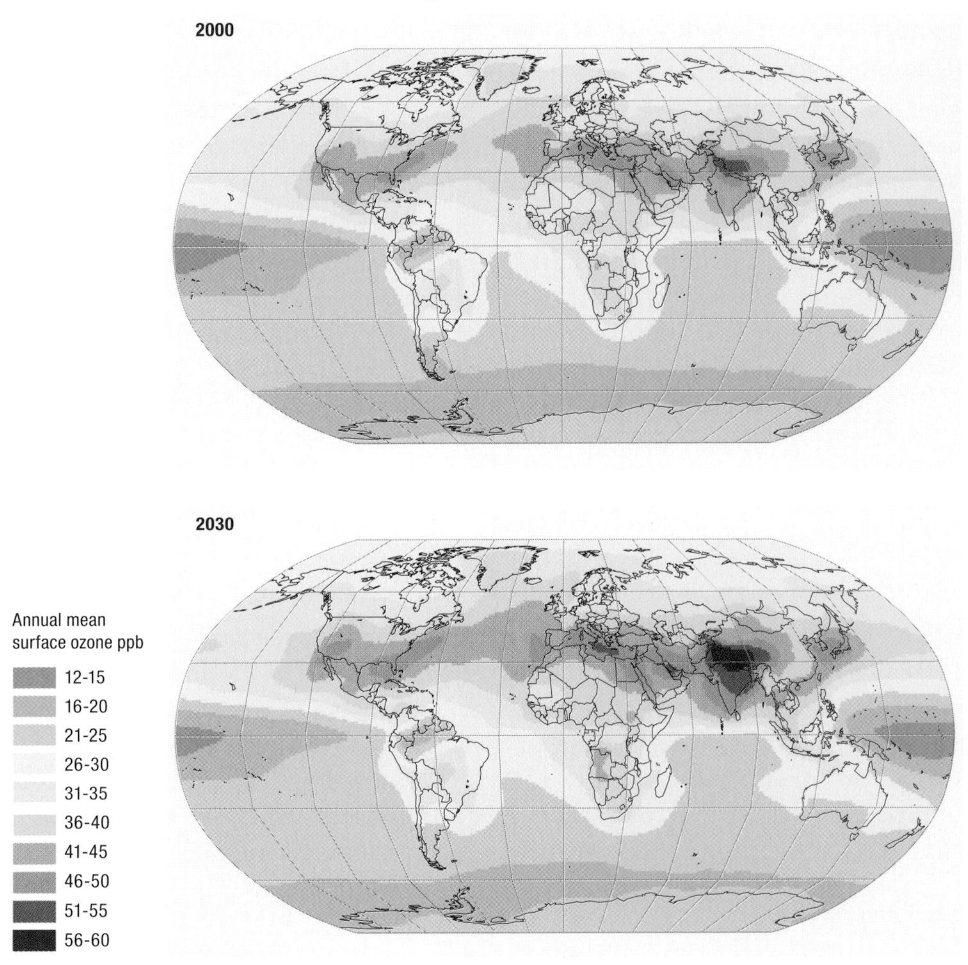

Source: Dentener *et al.*, 2005.

The potential exposure of (modelled) urban population to ozone concentrations is presented in Figure 8.5. Following the recommendations of the WHO, the ozone exposure is expressed as SOMO35[6] as this is the most descriptive parameter for health impacts Ozone exposure is seen to increase globally in this projection. Worldwide, a 25% increase is expected to 2030, but this varies between regions from less than 5% to more than 55%. The implications of these changes in ozone levels on health effects are discussed in Chapter 12.

Policy implications

In the past, the major instrument to address air pollution has been direct government regulation. Major examples are standards for fuel quality used by industry and transport, for emissions from cars and industry, as well as air quality standards and goals for protecting health and vegetation. These "command and control" measures have been very successful and have the advantage that their environmental effect is ensured. However, sometimes these measures can be partly undermined by other developments: increasing car ownership may reduce the positive effect of improved fuel quality, for example. A more

OECD ENVIRONMENTAL OUTLOOK TO 2030 – ISBN 978-92-64-04048-9 – © OECD 2008

Figure 8.5. **Potential exposure of urban population to ozone, 2000 and 2030**

Ozone in urban agglomerations

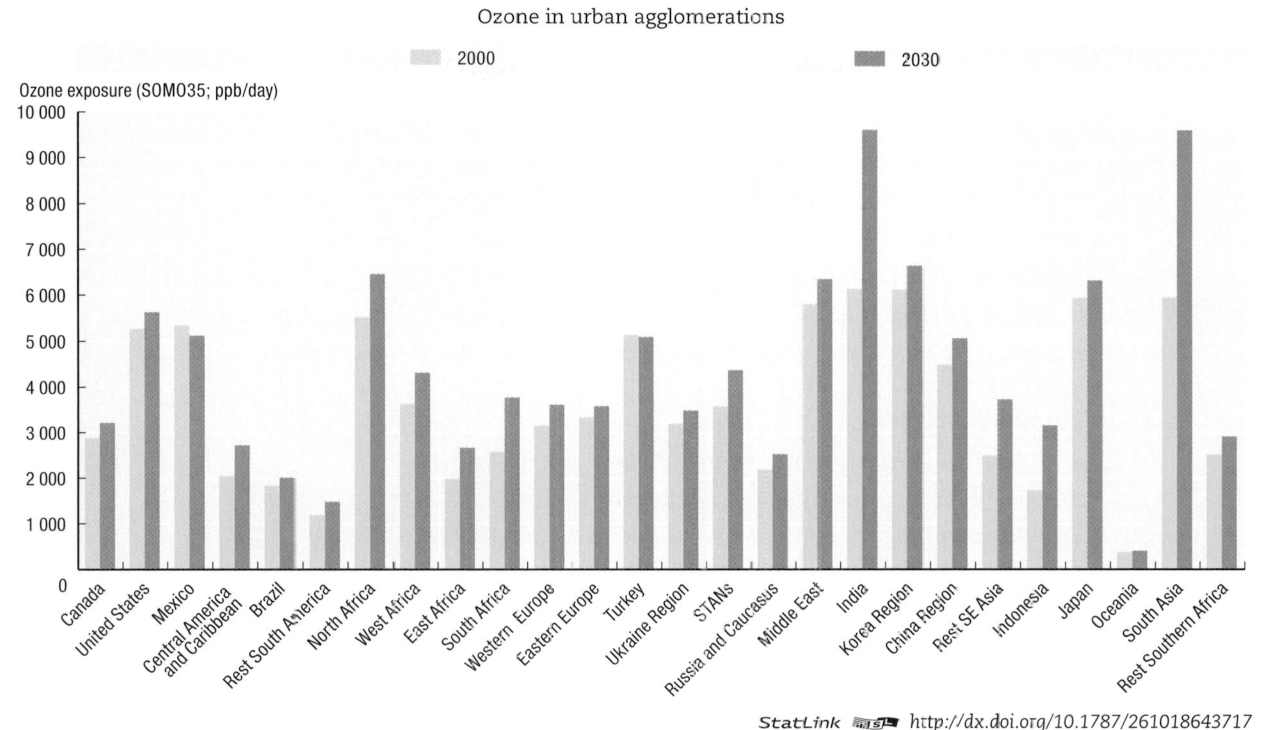

StatLink http://dx.doi.org/10.1787/261018643717

Source: OECD Environmental Outlook Baseline.

extensive discussion on sectoral policies to address air pollution is presented in Chapters 16 (Transport); 1 (Consumption, production and technology); 5 (Urbanisation) and 14 (Agriculture).

Economic instruments

Despite the success of regulation, economic instruments such as taxation and emissions trading have become increasingly popular. They can be more cost-effective than regulation because they give an incentive to the market (industry, transport sector) to take measures which cost the least.

The use of environmental taxes is still limited but growing in many countries. A database operated by the OECD and the European Environment Agency (EEA) lists about 375 environmentally-related taxes in OECD countries, not including some 250 other measures like environmentally related fees and charges. The database includes the energy and transport sectors. About 90% of the tax revenues stem from taxes on motor vehicle fuels and motor vehicles (OECD, 2007).

Most OECD countries have reduced air pollution in recent decades, decoupling it from continued economic growth. But additional measures are needed.

Subsidies, support schemes and green purchasing have proved invaluable for the development and uptake/diffusion of clean technologies like renewable energy and catalytic converters in cars. However, subsidies can have adverse environmental effects; for example, electricity generation from fossil fuels still receives much higher subsidies than renewables. Reforming these sorts of subsidies could improve air quality.

A well-known example of emissions trading is the trading system set up in the US for sulphur dioxide emissions by electricity generating facilities.[7] Regional schemes also exist in the US for emissions trading of nitrogen oxides. In China a pilot scheme is being started to trade emission credits for sulphur dioxide. In the Seoul metropolitan area an emission trading scheme for sulphur dioxide, nitrogen oxides and particulate matter has been initiated. Successful emissions trading systems depend on a formal legal structure, including an effective compliance system with real consequences for non-compliance.

Possible barriers to the use of economic instruments are their relatively high administration costs, the need for complex control technology (*e.g.*, in the case of pollution-dependent road pricing) and an inequitable distribution of costs. Trading schemes may not work very well if emission allowances are not scarce enough, thereby undermining their value.

Voluntary agreements

Voluntary agreements can in principle play a role in addressing air pollution, but there are few recent examples. The closest related example of a significant voluntary agreement is the agreement between the European Automobile Manufacturing Association (ACEA) and the EU and this relates to greenhouse gas emissions rather than conventional air pollution. As with all voluntary agreements, they have to be backed-up by regulation. In fact, performance under the ACEA agreement is falling short of expectations, in particular for Japanese manufacturers (DLR, 2004; IEEP, 2005; Fontaras and Samaras, 2007).

Abating methane emissions can help to improve air quality, as well as reduce greenhouse gas emissions.

In markets with few suppliers, supplier-driven standardisation can be a significant driver for the adoption of cleaner equipment. Power generation turbines are a case in point. This can be particularly important in times of fast expansion of economic activities, such as currently in China, when many installations are being built or renewed (UNEP/RIVM, 1999). Another example, from the late 1990s, was the consultations in Europe between car fuel producers and car manufacturers to co-ordinate the development of cleaner and/or more efficient car engines.

In order to advance new, clean, but high-investment technologies such as hydrogen use in transport, public-private partnerships do have a role. One example is the Clean Energy Partnership for Berlin (CEP-Berlin, 2006) and related demonstration projects co-funded by industry and the European Commission's framework programme for research, technology and development (European Commission, 2006). But these, too, are not specific to the policy area of air pollution.

Synergy and trade-offs with other areas of environmental policy

Policies to decrease air pollution may have a variety of conflicts and synergies with other policy objectives. As an example of conflict, some forms of air pollution, such as sulphur particulates, can provide regional cooling or shading. A reduction in these emissions while greenhouse gas emissions continue is likely to cause a small increase in global warming (see Chapter 7 on climate change). Energy savings and the introduction of renewable energy (wind, solar) are examples of synergy: both greenhouse gases and air pollutant emissions will be reduced. The abatement of ozone is another good example of how synergies can be achieved. Ozone is the third most important greenhouse gas. One of

OECD ENVIRONMENTAL OUTLOOK TO 2030 – ISBN 978-92-64-04048-9 – © OECD 2008

its main precursors is methane, the second most important greenhouse gas. Reducing methane emissions will be an efficient way to reduce emissions of primary and secondary greenhouse gases and will also abate ground level ozone.

There are three categories of synergy across environmental policy areas:

i) Decreasing the volume of an activity (energy use, transport) or limiting its increase will almost surely decrease all the ensuing environmental pressures – greenhouse gas emissions, air pollution, noise and so on.

ii) Clean energy can reduce air pollution and bring other environmental benefits too. But the balance can be positive or negative, depending on the specifics (see also Chapter 17 on energy). Examples are biofuels, hydrogen-powered transport and wind energy. Uncontrolled use of biofuels, although potentially beneficial for reducing CO_2 emissions, should be avoided as they are a source of black aerosol, a particulate which has a serious impact on health and which also contributes to global warming.[8]

iii) End-of-pipe measures and similar technical changes can conflict with other goals. For example, modern diesel engines in cars can lower greenhouse gas emissions, but make it more difficult to decrease emissions of nitrogen oxides. One reason for this is the existence of different regulatory tracks for these issues. Consolidating or at least harmonising these should eventually reassure manufacturers and local governments that timely consideration has been given to such trade-offs.

Overall, policies which address the driving forces more directly tend to have a better chance of enhancing synergies between air pollution reduction and tackling other problems. The schematic view of urban air quality in Box 8.4 offers a useful approach for prioritising the most cost-effective method for tackling poor urban air quality.

Policy simulations: urban air quality

To analyse the potential impact of some of the policy measures described above, three policy options have been simulated and their effect on emissions compared with the Baseline:

i) Enhanced air pollution measures to reduce emissions of sulphur dioxides, nitrogen oxides, volatile organic compounds and carbon monoxide in OECD countries (ppOECD or EO Policy Package).

ii) BRIC countries move to a similar ambition level in air pollution policy (ppBRIC + OECD).

iii) The remaining countries eventually move to the same ambition level (ppglobal).[9]

The effect on emissions is simulated on the basis of what can be achieved with existing technology – even if achieving these levels in reality in some countries is a long way off. The analysis assumes that the air pollution policies are part of a broader movement to boost environmental policies, with either OECD countries, or OECD and BRIC, or all countries stepping up their ambition in environmental policies (see Chapter 20 on environmental policy packages). In this manner, some indication of trade-offs and synergies can be gleaned from the modelling results.

The policy simulations model development towards – but not quite reaching – maximum feasible reduction of air pollutants (as defined by the International Institute for Applied Systems Analysis). To keep the policies realistic, albeit ambitious, the model assumes that eventual emission levels for each country remain 3 to 14% above what could be achieved with Maximum Feasible Reduction (MFR). For example, compared with the costs for fully implementing the MFR options in the European Union, this reduces the additional costs by more than 60% (Amman *et al.*, 2005).

Box 8.4. **Urban air quality**

A schematic view of the various contributions to urban air pollution

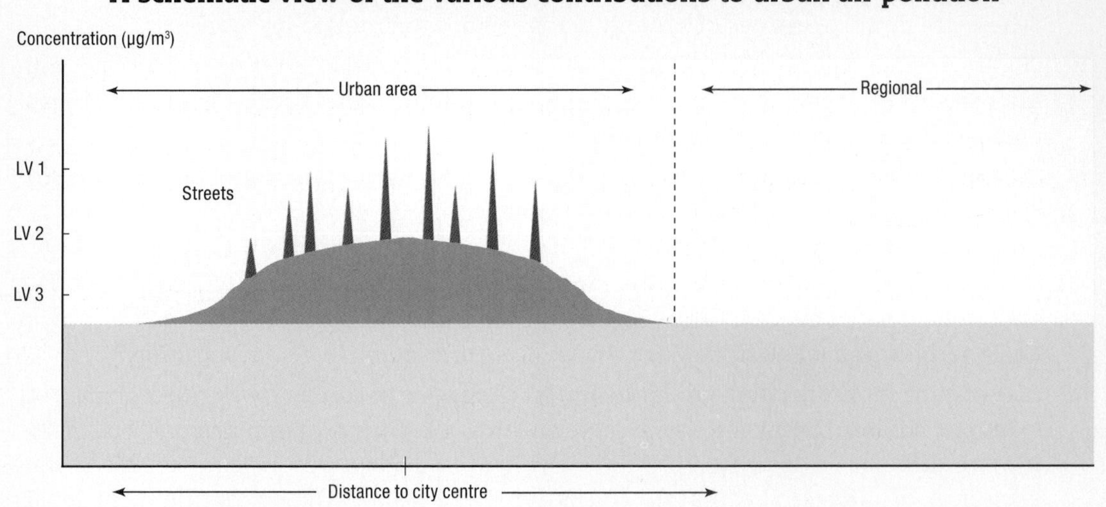

Air quality may vary within a city depending on traffic intensity, population density, physical topography and the weather conditions. At traffic (or industry) hot spots, high concentrations of pollutants result from local emissions. These concentrations are additional to the urban background contribution which is caused by diffuse emissions within the city itself. The regional background concentrations occur outside the city and are derived from urbanised areas, even those some distance away, rural emissions and emissions on a hemispheric level.

The magnitude of various contributions will vary from place to place and from time to time. Nevertheless, a schematic view like this may guide a first analysis of abatement policies. Consider, for example, an air pollution limit value LV1. This level is exceeded in a number of hot spots; the most cost-efficient way to reduce the pollution level will be to introduce local abatement levels. Limit value LV2 is exceeded at nearly all hot spots and in parts of the urban background area. Here abatement should focus both on sources within the city as well as on local sources. For limit value LV3, attainment can not be realised by reducing emissions in the city alone; reductions at the regional or hemispheric level are also needed.

A further refinement of the three policy packages assumes that countries will only start implementing the air pollution policies beyond the Baseline after their GDP at purchasing power parity (ppp) per capita reaches a certain income level. The speed of introduction of air pollution policies beyond the Baseline is also assumed to be dependent on GDP per capita. For example, in the BRIC policy package, India would start implementing these policies somewhat later than China.

The speed of implementation of emission control options is assumed to range between 15 and 30 years. The implementation is assumed to take at least 15 years; large point sources and transport will see enhanced emission controls introduced first, with other diffuse sources being addressed typically a decade later.

Abatement of emissions from shipping is included in the policy packages, as this becomes a cost-effective option in regions that have brought a good part of the land-based emissions under control. Emissions of sulphur dioxides from sea shipping affect air quality large distances downwind, typically thousand of kilometres or more away from the source and possibly in the middle of a continent. For example, it is expected that if emissions from

 OECD ENVIRONMENTAL OUTLOOK TO 2030 – ISBN 978-92-64-04048-9 – © OECD 2008

sea ships are not further restricted, the increase in shipping will negate land-based emission control efforts in Western Europe by 2020 (Cofala *et al.*, 2007).

Nitrogen oxides and sulphur dioxide emissions

Figure 8.6 gives the simulated emission profiles for sulphur dioxide and nitrogen oxides (the pollutants most relevant for PM). By 2030, worldwide emissions of nitrogen oxides in the case of enhanced environmental policies worldwide (ppglobal) are projected to be 31% lower than the Baseline emissions; and 37% lower for sulphur dioxides. By 2050, worldwide emissions of sulphur dioxides would be 84% lower than the Baseline; and 63% for nitrogen oxides.

Figure 8.6. **Emissions of sulphur dioxides and nitrogen oxides: Baseline and policy cases**

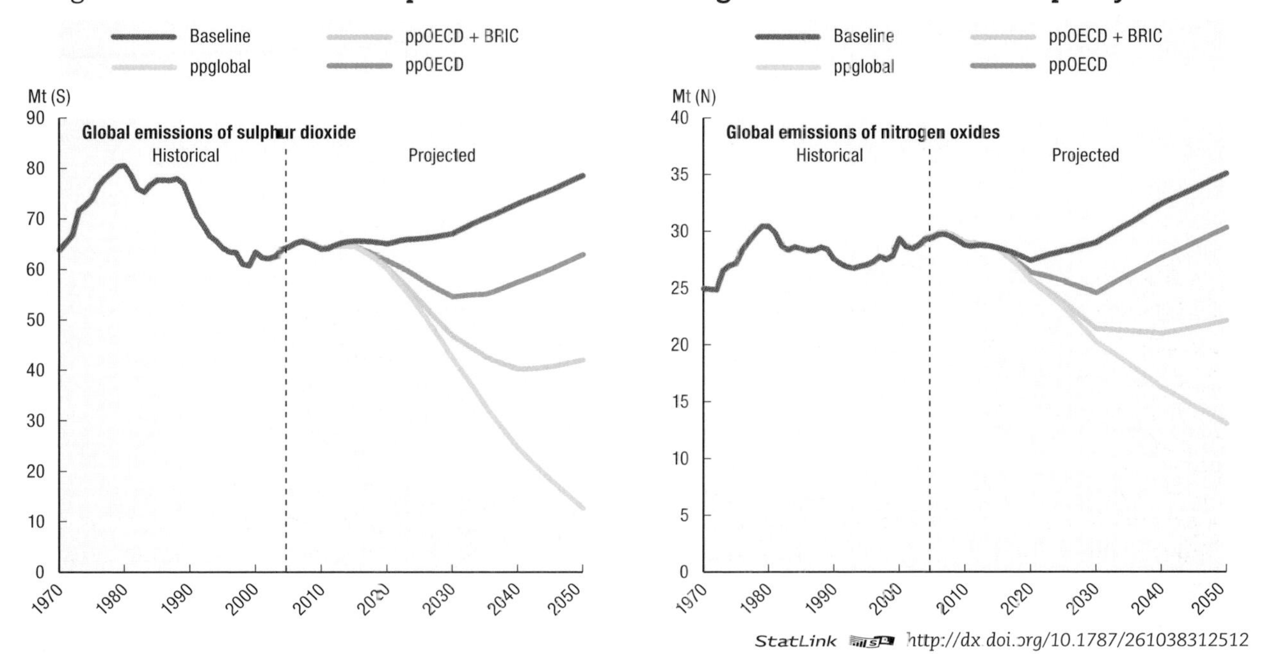

StatLink http://dx.doi.org/10.1787/261038312512

Source: *OECD Environmental Outlook* Baseline and policy simulations.

For sulphur dioxides, the largest contribution in 2030 to the decrease in emissions comes from OECD countries (10 million tonnes sulphur per year less), closely followed by the BRIC countries (–8 Mt S). By 2050, in the case of enhanced environmental policies worldwide, the rest of the world delivers the largest reduction (–32 Mt S) compared to the Baseline, while the OECD cluster stabilises at –11 Mt S and the BRIC countries decrease further to –23 Mt S per year.

Policy-induced decreases in emissions of nitrogen oxides are less steep, as it is technically more difficult and thus more costly to achieve deep emission cuts. By 2030 and 2050, the decrease in OECD countries is projected to be –4 and –5 Mt nitrogen per year respectively. The BRIC cluster achieves a lesser decrease by 2030 (–3 Mt N/year) but more than double that (–8) by 2050. The rest of the world sees a small decrease in nitrogen oxides emissions by 2030 (–1 Mt N/year), but steady decreases after that, passing –9 Mt/N year by 2050.

Figure 8.7 shows the development of sulphur dioxide emissions for the Baseline and policy cases for each of the regional clusters: OECD, BRIC and the rest of the world. Note that on balance, the rest of the world's sulphur dioxide emissions begin to deviate (in the ppglobal case) from the Baseline almost at the same time as those from BRIC countries. This reflects the relative weight of South Africa and marine shipping.

Figure 8.7. **Sulphur dioxide emissions, 1970-2050**

Sulphur dioxide emissions, Baseline and policy cases, by regional group

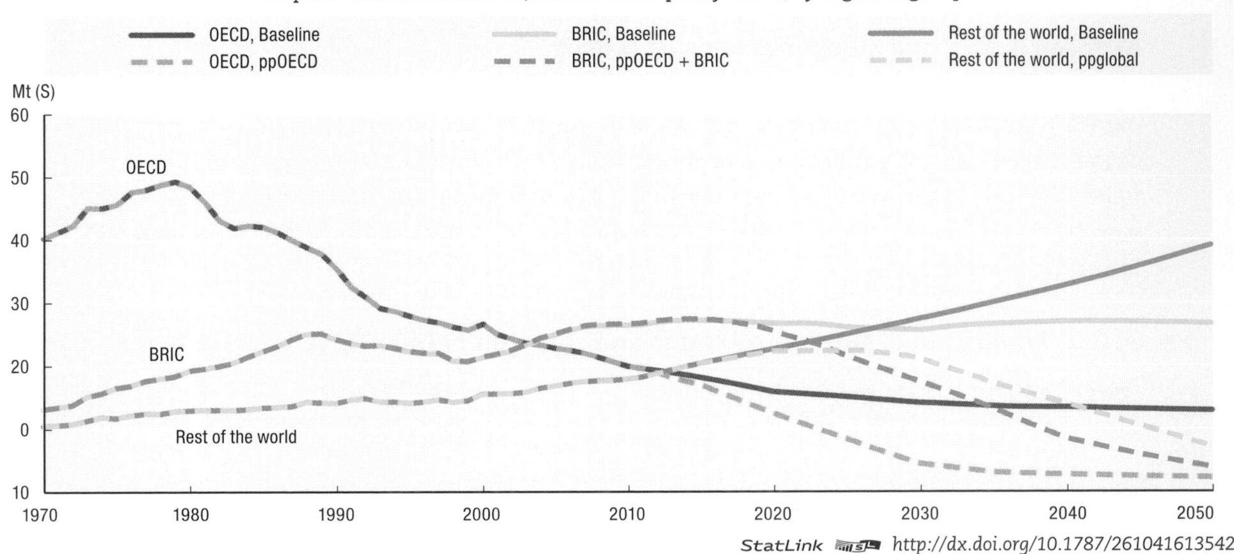

StatLink http://dx.doi.org/10.1787/261041613542

Source: *OECD Environmental Outlook* Baseline and policy simulations.

Particulate emissions

Urban PM_{10} concentrations have also been estimated for each of the three policy cases (Figure 8.8). The case of ppOECD leads to a 35-45% reduction in PM_{10} compared to the Baseline and an estimated reduction of 5% or less in BRIC and the rest of the world. In the ppBRIC+OECD case a concentration reduction of about 25% is estimated, although South

Figure 8.8. **Annual mean PM_{10} concentrations (μg/m^3) for the 13 regional clusters, 2030, Baseline and three policy cases**

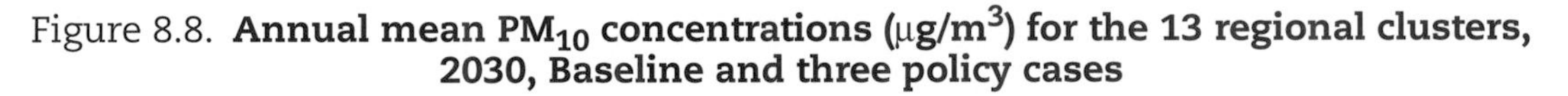

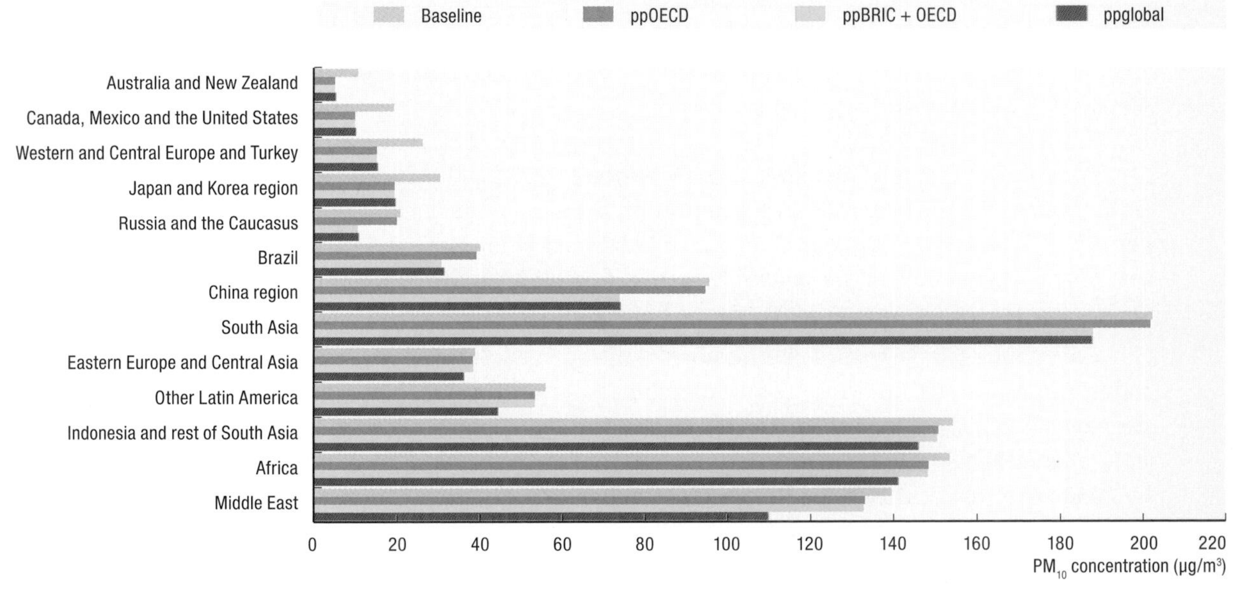

StatLink http://dx.doi.org/10.1787/261062032532

Source: *OECD Environmental Outlook* Baseline and policy simulations.

Asia lags behind with a reduction of 8%. The case of ppglobal results in a small reduction (5-8%) in the rest of the world. For the period 2030-2050, projections suggest that emissions of the PM-precursor gases sulphur dioxide and nitrogen oxides will not substantially decrease in the OECD under this simulation of enhanced environmental policies worldwide; PM concentrations will remain more or less constant here. In the BRIC countries and the rest of the world, the measures result in strong emission reductions starting around 2020 and 2030, respectively. A further reduction in concentrations is expected here.

Population exposures for 2030 in the Baseline and the most stringent policy case (ppglobal) are compared in Figure 8.9. Under the ppglobal case the situation is projected to improve, but large proportions of the urban population are still expected to be living in cities with annual mean PM_{10} concentrations exceeding the WHO interim target 1 of 70 $\mu g/m^3$. The health impact assessment of these policy cases is discussed in Chapter 12 on health and environment.

Figure 8.9. **Distribution of the urban population according to estimated annual mean PM_{10} concentrations in the modelled cities, 2030, Baseline (left) compared to policy case ppglobal (right)**

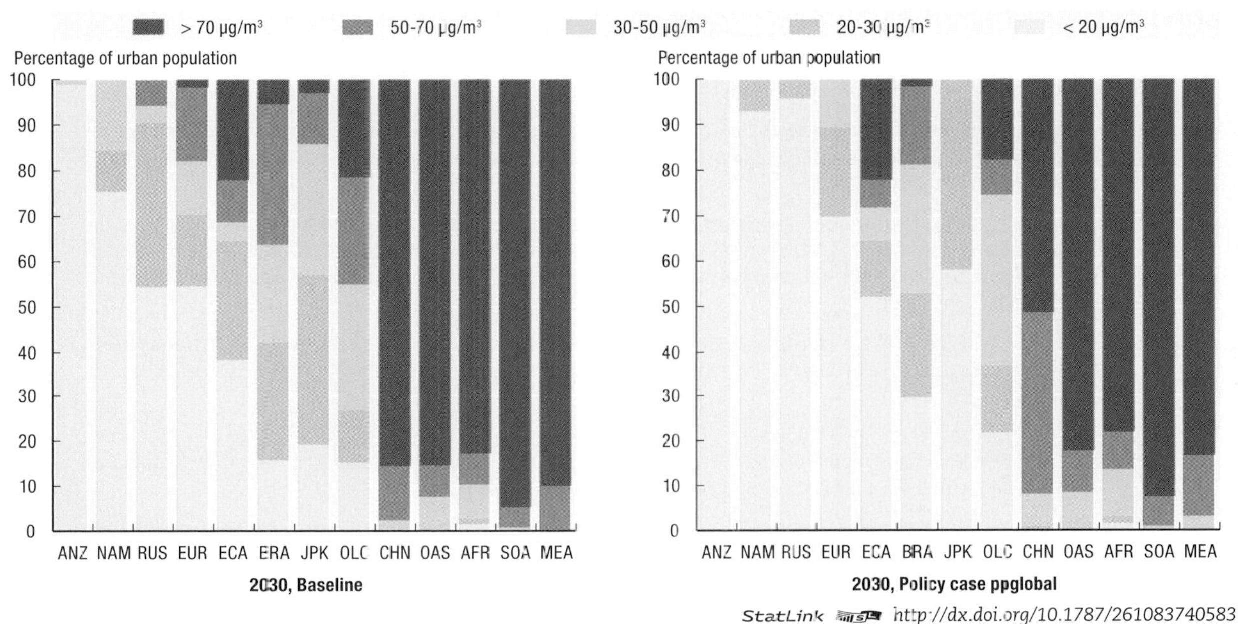

StatLink http://dx.doi.org/10.1787/261083740583

Note: Regional country groupings are as follows: NAM: North America (United States, Canada and Mexico); EUR (western and central Europe and Turkey); JPK: Japan and Korea region; ANZ: Oceania (New Zealand and Australia); BRA: Brazil; RUS: Russian and Caucasus; SOA: South Asia; CHN: China region; MEA: Middle East; OAS: Indonesia and the rest of South Asia; ECA: eastern Europe and central Asia; OLC: other Latin America; AFR: Africa.

Source: OECD Environmental Outlook Baseline and policy simulations.

Notes

1. Nitrogen oxides are a mixture of nitrogen monoxide and nitrogen dioxide.
2. The World Health Organization has published air quality guidelines for reducing health impacts of air pollution (WHO, 2006). These are regularly reviewed based on expert evaluation of current scientific evidence.
3. Information on ambient concentrations is much more widely available for coarser particles (PM_{10} or total suspended particles) than for the fine particles ($PM_{2.5}$). The analysis presented in this chapter first models PM_{10} concentrations in urban areas. Next, in the health impact assessment

(see Chapter 12) $PM_{2.5}$ concentrations are estimated from the modelled PM_{10} concentrations using an observed $PM_{2.5}/PM_{10}$ concentration ratio.

4. In this chapter the 13 regional clusters are abbreviated as: NAM: North America; EUR: OECD Europe; JPK: OECD Asia; ANZ: OECD Pacific; BRA: Brazil; RUS: Russia and Caucasus; SOA: South Asia; CHN: China region; MEA: Middle East; OAS: Other Asia; ECA: Eastern Europe and Central Asia; OLC: other Latin America and Caribbean; AFR: Africa.
5. One TgS (teragramme) corresponds to one billion kilogrammes of sulphur.
6. That is, the sum of excess of daily maximum 8-h means over the cut-off of 35 ppb calculated for all days in a year; SOMO35 is expressed in ppb/day.
7. See: *www.epa.gov/airmarkets/*.
8. Aerosol plays a double role in climate change: "white" aerosol (*e.g.* sulphate formed from burning sulphur-containing coal) has a cooling effect but "black" aerosol (soot) may cause warming.
9. Referred to as the *Environmental Outlook* (or EO) policy package in Chapter 20.

References

Amann M. *et al.* (2005), *CAFE Scenario Analysis Report Nr. 4: Target Setting Approaches for Cost-effective Reductions of Population Exposure to Fine Particulate Matter in Europe*, Background paper for the CAFE Steering Group, International Institute for Applied Systems Analysis (IIASA), 13 February 2005, Luxembourg, Austria.

Boffetta, P. (2006), "Human Cancer from Environmental Pollutants: The Epidemiological Evidence", *Mutation Research* 608, 157-162.

CEP-Berlin, (2006), *Berlin on the Way to Becoming the Hydrogen Metropolis – CEP Lays Another Cornerstone for Emission-Free Mobility*, Clean Energy Partnership press release, March 2006, *www.cep-berlin.de/presse/total/EN_CEP_PM_060314.pdf*.

Cofala, J. *et al.* (2007), *Analysis of Policy Measures to Reduce Ship Emissions in the Context of the Revision of the National Emissions Ceilings Directive*, Contract No 070501/2005/419589/MAR/C1. International Institute for Applied Systems Analysis (IIASA), Laxenburg, Austria, *http://ec.europa.eu/environment/air/pdf/06107_final.pdf*.

Dentener F. *et al.* (2005), "The Impact of Air Pollutant and Methane Emission Controls on Tropospheric Ozone and Radiative Forcing: CTM calculations for the Period 1990-2030", *Atmos. Chem. Phys.*, 5, 1731-1755.

DLR (2004), *Preparation of the 2003 Review of the Commitment of Car Manufacturers to Reduce CO_2 Emissions from M1 Vehicles, final report of Task A: Identifying and Assessing the Reasons for the CO_2 Reductions Achieved Between 1995 and 2003*, DLR, Berlin, December 2004.

EEA (European Environment Agency) (2006), "Air Pollution by Ozone in Europe in Summer 2005", *Technical Report No 3/2006*, European Environment Agency, Copenhagen.

ETC/ACC (European Topic Centre on Air and Climate Change) (2007), "European Exchange of Monitoring Information and State of the Air Quality in 2005", *European Topic Centre on Air and Climate Change Technical Paper* 2007/x, forthcoming, ETC/ACC, Bilthoven, The Netherlands.

European Commission (2006), *Hydrogen for Clean Urban Transport in Europe*, project summary, European Commission, Brussels. *http://ec.europa.eu/energy/res/fp6_projects/doc/hydrogen/factsheets/hyfleet_cute.pdfhttp://ec.europa.eu/energy/res/fp6_projects/doc/hydrogen/presentations/hydrogen_for_transport.pdfhttp://ec.europa.eu/energy/res/fp6_projects/hydrogen_en.htm*

Fishman J., A.E. Wozniak and J.K. Creilson (2003), "Global Distribution of Tropospheric Ozone from Satellite Measurements Using the Empirically Corrected Tropospheric Ozone Residual Technique: Identification of the Regional Aspects of Air Pollution", *Atmos. Chem. Phys.* 3, 893-907.

Fontaras, G. and Z. Samaras (2007), "A Quantitative Analysis of the European Automakers' Voluntary Commitment to Reduce CO_2 Emissions from New Passenger Cars Based on Independent Experimental Data", *Energy Policy* 35 (2007) 2239-2248.

Ghauri, B., A. Lodhi and M. Mansha, (2007), "Development of Baseline (air quality) Data in Pakistan", *Environmental Monitoring Assessment*, 127, 237-252.

IEEP (Institute for European Environmental Policy) (2005), *Service Contract to Carry Out Economic Analysis and Business Impact Assessment of CO_2 Emissions Reduction Measures in the Automotive Sector*, Institute for European Environmental Policy, Brussels, June 2005.

Leeuw de F., H. Eerens, R. Koelemeijer and J. Bakkes, (2008), *Estimations of the Health Impacts of Urban Air Pollution in World Cities in 2000 and 2030*, OECD, Paris.

OECD (2007), *Policy Brief: Environmentally Related Taxes: Issues and Strategies*, OECD, Paris, *www.oecd.org/dataoecd/39/18/2674642.pdf*

Pandey, K.D. *et al.* (2006), *Ambient Particulate Matter Concentrations in Residential and Pollution Hotspot Areas of World Cities: New Estimates based on the Global Model of Ambient Particulates (GMAPS)*, The World Bank Development Economics Research Group and the Environment Department Working Paper, The World Bank, Washington DC.

Richter, A. *et al.* (2005), "Increase in Tropospheric Nitrogen Dioxide over China Observed from Space", *Nature* 437, 129-132.

Smith K.R., S. Mehta and M. Maeusezahl-Feuz (2004), "Indoor Air Pollution from Household Use of Solid Fuels", in M. Ezzatti, A.D. Lopez, A. Rodgers, and C.U.J.L. Murray, (eds.) *Comparative Quantification of Health Risks: Global and Regional Burden of Disease due to Selected Major Risk Factors*, Vol. 2, pp. 1435-1493, World Health Organization, Geneva.

UNEP (United Nations Environment Programme) (2006), *GEO Yearbook 2006: An Overview of Our Changing Environment*, United Nations Environment Programme, Nairobi.

UNEP/RIVM (1999), *GEO-2000 Alternative Policy Study for Europe and Central Asia: Energy-Related Environmental Impacts of Policy Scenarios*, UNEP/DEWA&EW/TR.99-4 and RIVM 402001019, UNEP, Nairobi.

Volz, A and D. Kley (1988), "Evaluation of the Montsouris Series of Ozone Measurements Made in the Nineteenth Century", *Nature* 332, 240-242.

WHO (2006), *Air Quality Guidelines; Global Update 2005: Particulate Matter, Ozone, Nitrogen Dioxide and Sulphur Dioxide*, World Health Organization Regional Office for Europe, Copenhagen, Denmark.

ISBN 978-92-64-04048-9
OECD Environmental Outlook to 2030

Chapter 9

Biodiversity

Biodiversity loss is expected to continue to 2030, particularly in Asia and Africa. This chapter examines the sources of this loss – land use changes, unsustainable use of natural resources, invasive alien species, global climate change and pollution – and explores policy responses to halt further damage. Protected areas, which have grown significantly in number during the past few decades, will become increasingly important in the preservation effort as agricultural and urban land use expands. While many of the biodiversity "hotspots" worldwide are situated in developing countries, OECD countries have a role to play in helping to support their conservation and sustainable use through global and regional agreements, as well as through working together to address market and information failures.

KEY MESSAGES

The *Outlook* Baseline projects continued biodiversity loss to 2030 (as measured by human interference in biomes), with particularly significant losses expected in Asia and Africa.

Continued population and economic growth will put pressure on biodiversity through land use changes, unsustainable use of natural resources and pollution. Climate change will also put pressure on biodiversity in the coming decades.

Agriculture will continue to have major impacts on biodiversity. It is projected from 2005 levels that, in order to meet increasing demands for food and biofuels, world agricultural land use will need to expand by about 10% to 2030 – for crops and livestock together.

Although protected areas have expanded rapidly during the past few decades, the biomes represented in that coverage are uneven. Marine areas are thought to be under-represented in all categories of protected areas.

Many policy instruments are available to governments to mitigate the impact of economic growth on biodiversity. Since studies generally show that biodiversity has considerable direct and indirect value – and markets often fail to fully capture that value – additional pro-biodiversity policies are needed, for which governments have the necessary tools at their disposal.

The number and extent of protected areas have been increasing rapidly worldwide in recent decades; they now cover almost 12% of global land area.

Policy options

- Work toward sustainable use of biodiversity in the long term, but expand the biomes covered by some level of protection so as to ensure that the widest possible range of biodiversity is being preserved.
- Improve existing policy frameworks to minimise impacts of further economic growth on biodiversity.
- Expand policies (market-based approaches) so that current values of biodiversity are reflected in market activities.
- Enhance programmes to combat the spread of invasive alien species.
- Help support the conservation and sustainable use of biodiversity "hotspots" in developing countries through global and regional agreements, as well as through working together to address market and information failures.
- Ensure that trade liberalisation is not harmful to biodiversity in countries expected to expand output.

Consequences of inaction

- The loss of biodiversity through continued policy inaction is expected to be significant both in measurable economic loss and difficult-to-measure non-marketed terms.
- Inaction to halt biodiversity loss can lead to further losses in essential ecosystem services – such as carbon sequestration, water purification, protection from meteorological events, and the provision of genetic material.

Introduction

Biodiversity worldwide is being lost, and in some areas at an accelerating rate (Pimm *et al.* 1995). According to the Millennium Ecosystem Assessment (MEA 2005a), the main sources of biodiversity loss are land use changes (usually associated directly or indirectly with increasing populations, *e.g.* conversion to agriculture); unsustainable use and exploitation of natural resources (especially fisheries and forestry); invasive alien species; global climate change; and pollution (*e.g.* nutrient loading). While these are the immediate sources of the loss of biodiversity, the underlying problem is that biodiversity is usually not fully accounted for by consumers in the market place – there is often no distinction between biodiversity-friendly goods and those that damage biodiversity. Without government intervention, the market place has difficulty making that distinction. That so few policies have been enacted to mitigate biodiversity loss is an indicator of the strength of the underlying market failure, especially since there is considerable evidence for direct and indirect values of biodiversity that are not reflected in the market (*e.g.* OECD, 2002).

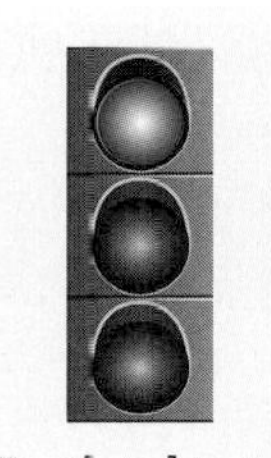

Further losses in biodiversity and ecosystem services are expected to 2030.

Looking forward, many factors will affect biodiversity in ways that will either harm or help it. Nowhere is this potential for changes in biodiversity greater than in two areas: *i)* the increase and extension of agricultural activity, which often results in biodiversity loss; and *ii)* the creation and sustainable use of protected areas, which mitigate further biodiversity loss. Agriculture has historically had the largest impact on biodiversity, and it is expected to continue to be a major factor in the future. Protected areas are a fairly recent phenomenon, but their importance for biodiversity in the future will become key. Over longer time horizons, a source of biodiversity loss whose potential looms very large is climate change. However, the uncertainty around its impact is also large at this stage and its impact within the time frame under consideration here may be small compared with other sources (see also Chapter 13, Cost of policy inaction).

Future pressures on biodiversity are closely linked to increases in economic activity, with associated changes in consumption and production patterns. Under the *OECD Environmental Outlook* Baseline, world population is expected to be 30% higher in 2030 and, when coupled with increasing material well-being (the world economy may be twice as big in 2030 as it was in 2005), this is likely to exacerbate current pressures on ecosystems. Ensuring that economic development is sustainable will require satisfying human needs and wants in such a way that valuable biodiversity and ecosystem functions are not lost, in particular as many of these ecosystem functions – including carbon sequestration, water purification, and the provision of genetic material – directly support economic and social well-being. While many of the biodiversity "hotspots" worldwide are situated in developing countries, OECD countries have a role to play in helping to support their conservation and sustainable use through global and regional agreements, as well as through working together to address market and information failures.

Key trends and projections

A rough measure of biodiversity loss can be obtained using a relatively simple indicator called mean species abundance.[1] Figure 9.1 compares biodiversity (MSA) in 2000 and 2050 with a hypothetical level chosen to reflect low human interference. The results for 2000 are based on data available in the IMAGE model, while those for 2050 are based on the combined results of ENV-Linkages and IMAGE. The MSA on a global basis is projected to decline by 10% between 2000 and 2030 (7 percentage points).

Figure 9.1. **Historical and projected future changes indicated by mean species abundance, 2000-2050**

Potential = 100%

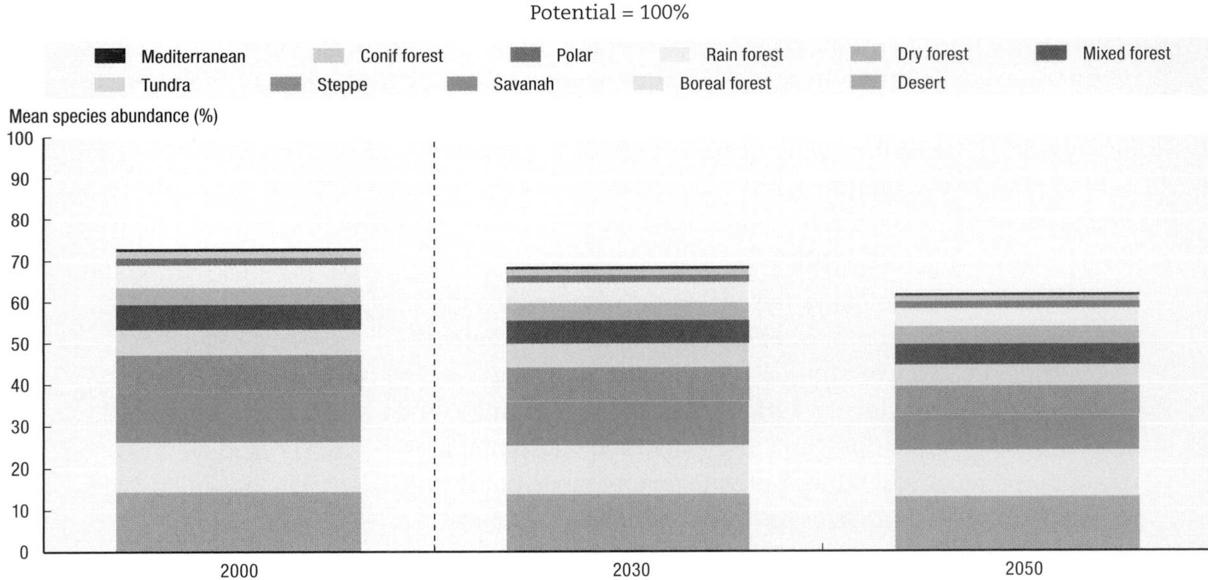

StatLink http://dx.doi.org/10.1787/261146122600

Note (with indicated change): Boreal forest (–5%); Desert (–6%); Tundra (–7%); Polar (–2%); Conif forest: temperate coniferous forest (–8%); Mixed forest: temperate broadleaf and mixed forest (–12%); Mediterranean: Mediterranean forest, woodland and shrub (–10%); Dry forest: tropical dry forest (0%); Rain forest: tropical rain forest (–14%); Steppe: temperate grassland and steppe (–15%); Savannah: tropical grassland and savannah (–20%).

Source: OECD Environmental Outlook Baseline.

In April 2002 the Conference of the Parties to the Convention on Biological Diversity adopted a strategic plan. This committed parties to significantly reduce the current rate of biodiversity loss (by "mainstreaming" biodiversity concerns) at the global, regional and national level by 2010 (Decision VI/26). This objective was subsequently endorsed by the World Summit on Sustainable Development, and was reinforced by G8 environment ministers following their meeting in Potsdam in March 2007. That target would certainly change the trend outlined in Figure 9.1, but has not been reflected in the Baseline because the specific policies that would be needed to achieve it are not yet in place.

Figure 9.2 shows that according to the Baseline, future biodiversity loss to 2030 (as measured by MSA) is likely to mainly come from pressures from agriculture (32%) and infrastructure (38%). Infrastructure development includes urbanisation, transportation networks and other elements of human settlement. The significant loss to infrastructure is

Figure 9.2. **Sources of losses in mean species abundance to 2030**

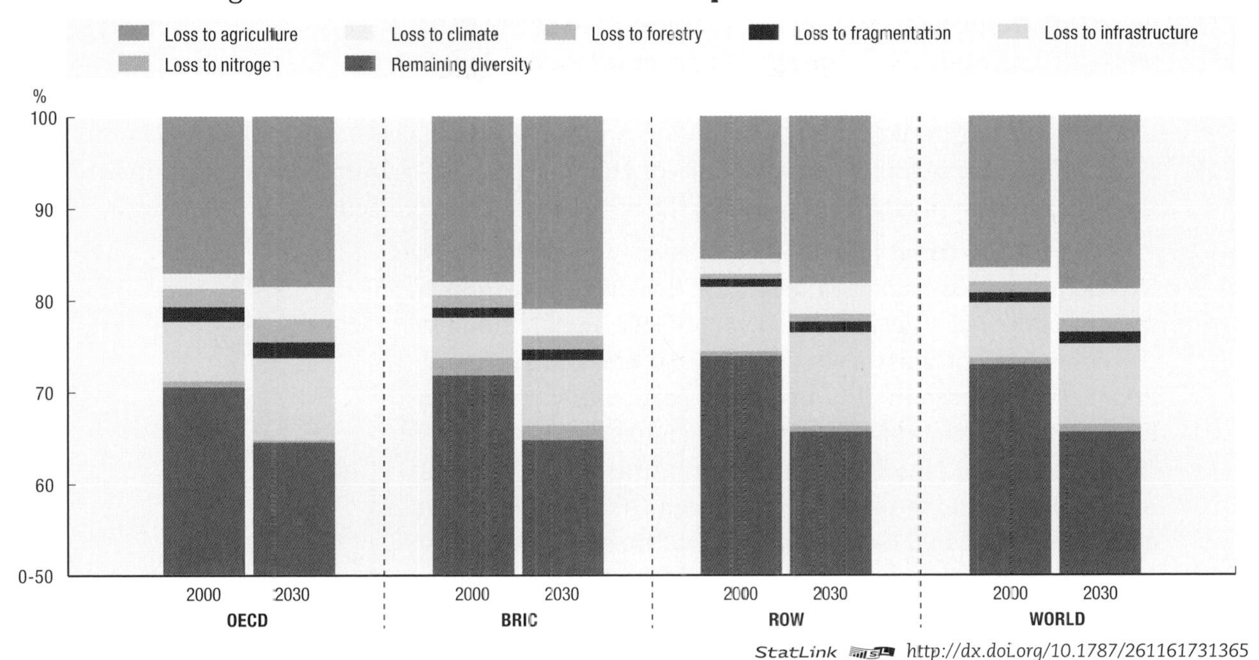

StatLink http://dx.doi.org/10.1787/261161731365

Source: OECD Environmental Outlook Baseline.

an indication that increased population with increased wealth will lead to a spreading out of people that will affect natural areas more heavily.

To 2030, growth in agricultural production is expected to lead to further pressures on biodiversity through land use changes in the vast natural areas of North America and Australia/New Zealand. In the densely populated regions of Western Europe and Japan we are already seeing high levels of human encroachment on nature. All OECD regions, however, show further decline due to expanding infrastructure and other influences.

The Russian and other former Soviet Union economies featured a relatively high MSA biodiversity score in 2000 (roughly 83% of pristine state) with only limited further losses (down to roughly 78% of pristine state) projected by 2030. This is mainly because of the vast natural and sparsely populated areas of this region. By contrast, from an already low starting point, biodiversity in OECD Europe (48%) is projected to deteriorate further to 40% in 2030. Expansion of agricultural land in new EU member states and infrastructure are the main drivers of this downward trend.

Significant differences in both levels and trends for biodiversity are also found between different developing regions. In East Asia agricultural areas are projected to decrease, but quickly expanding infrastructure, high levels of nitrogen deposition and some mild early impacts of climate change more than offset that effect. In both South and Southeast Asia, biodiversity declines (as measured by MSA) of at least 10 percentage points are anticipated. In South Asia, expanding agriculture is the main cause, while in densely populated Southeast Asia infrastructure expansion and fragmentation play a bigger role. In all developing regions climate change, notably changes in precipitation, are also expected to affect biodiversity.

Land use changes

Conversion of land away from biodiversity-rich natural conditions is perhaps the greatest pressure on ecosystems and biodiversity. The 2005 *Millennium Ecosystem*

Assessment suggests that "Most changes to ecosystems have been made to meet a dramatic growth in the demand for food, water, timber, fibre and fuel" (MEA, 2005a). Forestry activity and agriculture have been the primary drivers of this biodiversity loss. The MEA found that more land was converted to agriculture in the 30 years following 1950 than during the 150 year period between 1700 and 1850. Similarly, the *Global Biodiversity Outlook* 2 (SCBD, 2006) also identifies habitat loss – or land use change – arising from agriculture as the leading cause of biodiversity loss in the past, as well as in projections for the future.

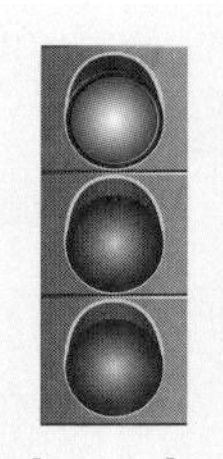

Land use change for agriculture is the main source of biodiversity loss worldwide.

The further increase in food crop lands worldwide of 16% to 2030 (from 2005) expected under the Baseline will continue to be an important factor in biodiversity loss, mostly through the conversion of grasslands and forested areas to farmland. Projected increases in crop lands are particularly notable in Russia, South Asia, developing Africa and some (but not all) OECD countries (see Figure 9.3). Agricultural land area is expected to decrease to 2030 in the Asian OECD region (Japan and Korea). It should be emphasised that these results reflect minimal changes in policy and technology. Changing those assumptions could result in large changes in some of these trends. For example, the location of these increases is driven in part by continuing tariffs and other agricultural policy measures. A policy simulation was undertaken with ENV-Linkages to reflect the gradual removal of agricultural tariffs, and the impacts of this on land use examined (Box 9.1).

Furthermore, Heilig *et al.* (2000) use FAO/IIASA data to show that by applying existing technologies already in use elsewhere, China could feed itself in 2025 using less land than it did at the turn of the century. However, many of those technologies are unlikely to be implemented while labour costs are low and government policy does not encourage high-productivity farm production.

Figure 9.3. **Change in food crop area, 1980-2030**

1980 = 100%

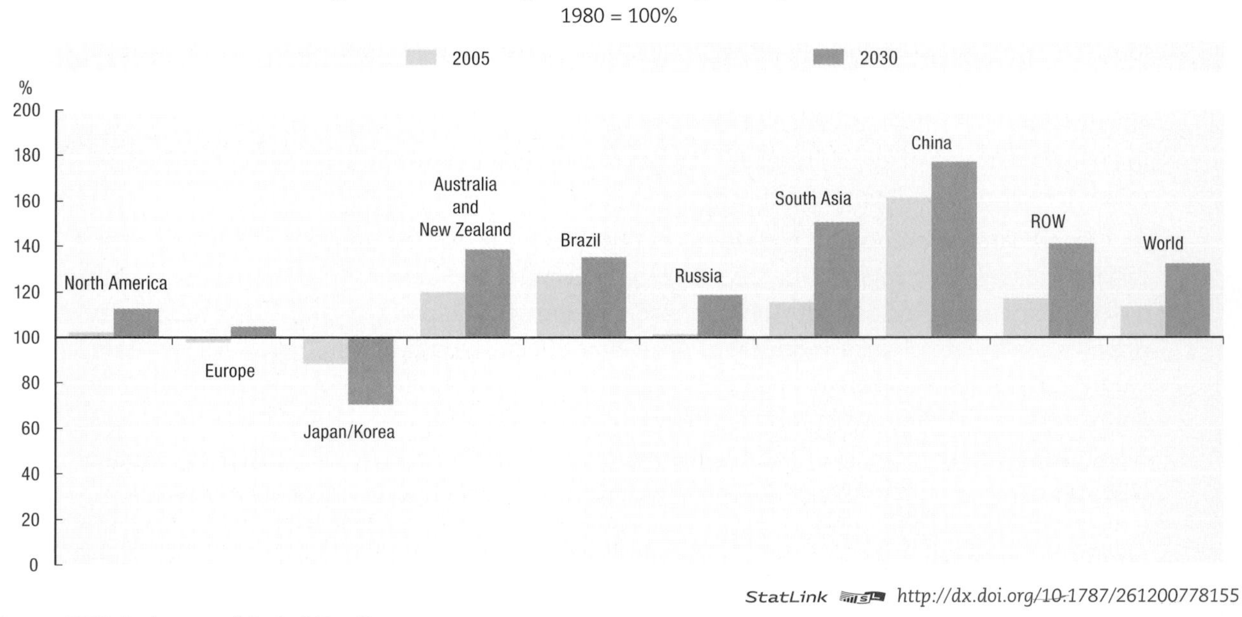

StatLink http://dx.doi.org/10.1787/261200778155

Source: OECD Environmental Outlook Baseline.

Box 9.1. **Modelling the impact of agricultural tariff reductions**

Under the Baseline for the *Outlook*, it is expected that increasing demand for food (and biofuels) will lead to more than a 10% increase in all agricultural lands worldwide (16% increase for food crops, 6% increase for grass and fodder, and 242% increase for biofuels). The location of these increases is driven in part by continuing tariffs and other agricultural policy measures. A policy simulation was undertaken with ENV-Linkages to reflect the gradual reduction of agricultural tariffs, and the impacts of this on land use examined. These results are primarily useful in drawing attention to areas where biodiversity policy may need reinforcing: though measuring changes in land use for agriculture can be indicative of changes in pressure on biodiversity, a thorough analysis of impacts on biodiversity would have to account for some counteracting factors.

In the simulation, all countries are postulated to lower their tariffs by 50% by 2030, thus significantly affecting agriculture in a number of sectors in countries where tariffs are high – the simulation reduced only direct tariffs as they existed in 2001.

Total agricultural land use under this simulation of tariff reform would be increased by around 1.8% compared to the Baseline in 2030. This implies that instead of agricultural land increasing by 10%, it would increase by 11.8%. This is combined with the economic benefits that the reforms would bring, and other environmental benefits of more efficient markets and rational land use. While the global trend is upwards, this masks some regional variation, such as increases in some areas (especially Brazil and parts of Southern Africa) and decreases in others (especially those OECD countries where tariffs are high). The decrease shown for Japan in response to this policy would be in addition to the roughly one-third decrease in agricultural land use that occurred between 1980 and 2000.

Whether the increase in agricultural land in Brazil *versus* the reduction elsewhere represents a net loss of biodiversity is not easily answered. Some studies show that Brazil can significantly expand agricultural lands without losing additional rainforest because the expansion is likely to occur instead in the Cerrado region. But the Cerrado region of Brazil also has its own unique biodiversity and does not currently have sufficient protected areas to ensure that biodiversity will not be lost. Adequate protection of the Cerrado and enforcement of the existing policies protecting the rainforest could accompany such agricultural trade liberalisation to ensure sustainable use of biodiversity-related resources even with expanded agriculture. Such a strategy could lead to gains both in worldwide agricultural efficiency, as well as more sustainable use of biodiversity. SCBD (2007) obtained the result that global biodiversity would be damaged by trade liberalisation, mainly as a result of the impacts in Brazil.

Table 9.1 outlines the types of agricultural land use changes that might be associated with tariff reductions in regions with the largest impact – 10 of the models' 34 regions are shown. The changes are relative to the Baseline, meaning that they should be compared to a world which is using 10% more land for agriculture than today.

Table 9.1. **Impact on land types in 2030 of agricultural tariff reform (compared to Baseline)**

Country/region	Change in livestock	Change in crops	Comment
Iceland\Norway\Switzerland	–8.7%	–13.0%	Gain in forested areas, some loss of semi-natural grassland
Japan	2.6%	–21.6%	Gain in forested areas
Korea	0.3%	–14.5%	Switch in crop composition, gain in forested areas
Turkey	–1.3%	–2.4%	Some gain in forested areas, natural pastures
Mexico	0.1%	–3.3%	Less pressure on rainforest
..	..	..	..
USA	0.0%	2.4%	Increased use of marginal cropland
EU members non-OECD	2.8%	1.3%	Loss of forested areas
Australia and New Zealand	4.3%	1.4%	Some loss of forested areas and natural pastureland
Rest of South Africa	6.0%	0.6%	Some loss of forested areas and natural pastureland
Brazil	10.0%	0.0%	Loss of natural pastureland; potential loss of rainforest

StatLink http://dx.doi.org/10.1787/257177550380

Source: OECD Environmental Outlook Baseline and policy simulations.

While biofuel expansion is included in the Baseline, it plays a small role in land use change to 2030. This is in part because the price of oil in the Baseline is assumed to return to levels that do not encourage heavy use of biofuels for transport. Should governments continue to increase support for biofuels, or should oil prices remain significantly above USD 60 indefinitely, there is very large potential for significant shifts of land use to agriculture for biofuel production (see Chapter 14 on agriculture).[2]

While agriculture has had predominantly negative impacts on biodiversity, this is not a universal outcome in all circumstances. The Mediterranean basin, for example, is considered a biodiversity hotspot largely because the conditions that agriculture has created have been conducive to maximising diversity. Alpine meadows are another example of how farming activity can sustain biodiversity. Organic agriculture can also be more biodiversity-friendly than other forms of agriculture because of the lower levels of homogenisation of plant and animal life in and around the farm. However, at very large scales it is not clear whether these benefits can be maintained (Hole *et al.,* 2005). Similar observations can also be made in many regions, both within and outside OECD countries. While they do not change the overall observation that clearing land for agricultural use is generally detrimental to biodiversity, they do call for a more nuanced view in some cases.

It should also be noted that biodiversity can be considerably enhanced through the "greening of agriculture". For example, recent trends in OECD countries towards payments for environmental services to farmers hold out the prospect of achieving increases in biodiversity while simultaneously maintaining or increasing agricultural output (see also Chapter 14 on agriculture).

Unsustainable use and exploitation of natural resources

Over-harvesting of species (especially when it is illegal) reduces biodiversity by decimating specific plant or animal species, as well as by affecting habitats and species' interdependence. For example, over-harvesting of cod in the North Atlantic has led to cascading impacts on the overall food chain in the ecosystem, with resulting impacts on other fish stocks (Frank *et al.*, 2005). Over-harvesting of trees has led to the loss of significant sources of biodiversity in rainforests in both South America and Asia. In the past, over-harvesting of particular species has led to their extinction.

Marine biodiversity is experiencing pressure from both fishing activity and non-fishing sources (see Chapter 15 on fisheries and aquaculture). Given the growth in demand for fish products, increases in pollution and eutrophication of marine environments, alteration of physical habitat, exotic species invasion, and effects of other human activities, the pressure on marine biodiversity from anthropogenic sources will continue to increase to 2030 (see Committee on Biological Diversity in Marine Systems, 1995, for more detail on how each of these sources affects biodiversity). There are also early signs of climate change affecting marine biodiversity, and this is likely to intensify, *e.g.* through increased acidification of oceans (Gattuso *et al.*, 1998).

Roughly 40% of forest area has been lost during the industrial era, and forests continue to be lost in many regions. Between 2005 and 2030, a further 13% of naturally forested area is expected to be lost worldwide under the Baseline, with the greatest rates of deforestation occurring in South Asia and Africa (excluding recent regrowth). This reflects the increasing demand for forest products, with global timber production having increased by 60% in the last four decades (see Box 9.2). However, forests have been recovering in some temperate

 OECD ENVIRONMENTAL OUTLOOK TO 2030 – ISBN 978-92-64-04048-9 – © OECD 2008

Box 9.2. **Environmental impacts of forestry**

Forests are the most biodiversity-rich terrestrial ecosystem. They provide a wide range of values to humans, varying from timber, pulp and rubber, to environmental services. At the global level, forests play a crucial role in regulating the climate and represent a significant carbon reservoir. However forest biodiversity is threatened by deforestation, degradation and fragmentation. The main factors driving biodiversity depletion in forests include pressures from increasing land use for farming and livestock grazing, unsustainable forest management, introduction of invasive alien species, mining and infrastructure development. For the most part, industrial logging and the development of tree plantations are not direct causes of deforestation, but major contributors to forest degradation and fragmentation, which in turn can increase the risk of deforestation.

Demand for wood production

In 2005, about half the world forest area was designated for production of wood and non-wood forest products. Rapidly increasing demands for wood, notably from paper and pulp industries due to growing paper consumption, and from the energy generation sector to supply biofuels, is expected to put further pressures on forest resources and survival. Global roundwood production in 2005 amounted to over 3.5 billion m^2. Industrial roundwood accounted for about half of the total roundwood production, and increased by about 18% between 1980 and 2005. Of all industrial roundwood products, paper and paperboard production grew most rapidly – doubling between 1980 to 2005 as a result of surging demand for paper in developing countries (see also Chapter 19 on selected industries: pulp and paper). Over half of the world's roundwood is used as fuel wood or charcoal, supplying about 10% of the world's energy. Woodfuels are also used as modern biofuels to generate electricity, gases and transportation fuel. Demand for biofuels as primary inputs for electricity is expected to increase by 19% to 2030.

Environmental effects of forestry on forest areas

Forest area and deforestation

Global forest area accounted for about 4 billion hectares or 30% of total land area in 2005. The *OECD Environmental Outlook* Baseline projects that natural forest areas will decrease by a further 13% worldwide from 2005 to 2030, with the greatest rates of deforestation occurring in South Asia and Africa. Primary forests were lost or modified to other forest types at an average rate of 6 million ha per year over the past 15 years, and the rate of loss is increasing.

There are three major forest types according to latitude: boreal/taiga (found throughout the high northern latitudes), temperate and tropical forests. Temperate forests, mostly secondary and plantation forests, have been slightly increasing over a long period due to natural reforestation and forest plantations on abandoned agricultural land. Tropical and boreal forests, however, are under pressure from deforestation and forest degradation in primary forests. With some exceptions, most of the logging in the topical and boreal regions involves "cut-and-go" operations in primary forests, *i.e.*, short-term exploitation of industrial wood products without caring for the long-term regeneration of the forest. Severe degradation of forests can occur due to impacts of felling damage and residual wastes on water, soil, nutrient cycles and species richness. In the tropics, most logging is followed by subsequent transition to other land uses, such as crop production and livestock grazing.

Increasing plantation forests

The increasing development of intensive forest plantations for wood production is another threat to forest biodiversity. Productive forest plantations covered 109 million hectares in 2005, having increased annually by about 2 million hectares between 2000 and 2005. Although the total extent of productive plantation areas is relatively small, they provide 22% of world industrial wood supply (FAO, 2006). The area of productive plantation is expected to increase over the coming decades to meet the growing demand for wood products.

Box 9.2. **Environmental impacts of forestry** *(cont.)*

Forest biodiversity in plantation forests is much less than in natural forests. Plantation forests can affect the soil structure, chemical composition, regional hydrological cycle (and regional ecosystems), and cause significant water depletion in the basin. Other environmental issues in monoculture plantations include genetic impoverishment and increased risk of spread of insects and disease. However, it has been argued that increasing wood production from plantations can reduce the pressures on natural forests for industrial wood extraction. Sustainably managed plantation forests can also play a vital role in conservation of biodiversity by acting as buffer zones for fragmented remaining forests.

Illegal and unauthorised industrial wood production and trade

Illegal logging continues to threaten forest biodiversity, with as much as 8 to 10% of global industrial roundwood production estimated to be sourced illegally (Seneca Creek Associates and World Resources International, 2004). Illegal logging takes place in both developed and developing countries. Illegal logging can have serious environmental, social and economic costs and jeopardise international and national efforts to achieve sustainable forest management. Some cases of illegal logging have been reported as taking place in forest protected areas. The economic costs of illegal logging are tremendous: global market losses of USD 10 billion annually, and government losses may amount to USD5 billion in lost revenues (World Bank, 2006a).

Direct driving forces of illegal logging are the higher profits obtainable than for legal logging, coupled with often low risk of apprehension and/or low penalty costs. These are exacerbated by weak forest legislation. The pressures behind illegal logging are the increasing international demand for wood products and a highly developed international supply chain. At the supply end, it is surprisingly easy for consumers to buy illegally logged products as the origin of most wood products is unverifiable.

Policy responses

Meeting increasing demands for forest resources while maintaining forest coverage and ecosystem quality is a major policy challenge, especially in tropical and boreal regions. There have been considerable international efforts to promote and ensure sustainability in forest management and to tackle illegal logging. Policies that address problems in forestry are particularly beneficial for the environment since this is one area where all three environment-related conventions interact (climate change, biodiversity and desertification).

In order to encourage sustainable forest management further and reduce illegal logging, forest legislation and associated policy systems urgently need to improve. A range of regulatory instruments can be used, including allocating concession rights; regulating inputs and processes such as the use of chemical fertilisers and water; setting standards for intensity and species of harvesting and logging; and the obligatory implementation of environmental impact assessments. It is important that the regulations are based on the best available scientific knowledge on the forest quality and possible impacts of forest activities, and that they are followed by close monitoring of changes in forest quality. Whilst a number of OECD countries have long adopted reduced-impact techniques for wood production, such sustainable practices have not been widely introduced in tropical and boreal forests due to the associated increased production costs and need for investments in training and planning.

Economic instruments – including fees or charges for harvesting and trading of industrial roundwood, charges or non-compliance fees related to certain types of forestry activities, taxation on the conversion of forest land to other uses, and subsidies for afforestation – can be used to encourage more sustainable forest management. At the same time, it is essential to remove or reform existing subsidies which promote excessive logging and access to natural forests, such as subsidies for establishing plantation forests or agricultural fields on natural forested land.

Eco-certification is another important instrument for reducing consumers' demand for wood products from unsustainably managed forests. Various certification schemes have been developed by the forest industry, environmental NGOs and the EU. It is important to develop a clear set of indicators to ensure sustainability of the forests managed under each of the certification schemes.

countries in recent decades, with much of this in forest plantations. Plantations are providing an increasing proportion of harvested roundwood, amounting to 22% of the global harvest in 2000. However, plantation forests are often monocultures, and so exhibit much less biological diversity and richness of ecosystems than natural forests. Demand for forest products is expected to continue to rise in coming years, in particular for emerging economies such as China and India, and with it the pressures of illegal logging and a continuing trend toward plantation forests.

Invasive alien species

Invasive alien species are a human-induced problem that is thought to rank high as a contributor to past biodiversity loss (see Wilson, 2002) and which is unlikely to abate by 2030. Many of the human vectors that have contributed to species migration are strengthening with increased economic wealth. For example, trade and travel are both expected to grow strongly in the future, and both have been prominent as agents for moving species outside their natural ranges (ballast water used by ships, and seeds or animals carried on vehicles are classic examples). Historically, many species have also been deliberately introduced for economic benefit: it is estimated that some 98% of the world's agricultural production results from sources that are not native to the areas where they are currently grown or raised. This includes crops and animal species. The combination of purposeful and accidental transplants of species that are in some cases harmful has led to a large human-induced impact on species distribution.

Invasive species can have an impact on biodiversity both within an ecosystem, by disturbing the balance of species in the ecosystem, and globally, by making the worldwide distribution of species more monolithic. This is particularly evident on the island of Hawaii, where only one-quarter of the original (pre-European contact) bird species remain, and where almost one-half of the free-living flowering plants are aliens introduced since European contact (Wilson, 2002). These new species make Hawaii look similar to many other tropical areas, whereas its isolation had once made it unique.

Table 9.2 illustrates the magnitude of environmental impacts of a small sample of invasive alien species. A few estimates put the number of alien species in the tens of thousands for just a handful of countries (Atkinson and Cameron, 1993; Perrings *et al.*, 2000; Pimentel *et al.*, 1999).

Table 9.2. **Environmental impact of invasive alien species**

Invasive species	Some impacts
Crazy ant (*Anoplolepis gracilipes*)	Forms multi-queen super-colonies in rainforests in Pacific Islands. Kill arthropods, reptiles, birds and mammals on the forest floor and canopy. Eats leaves of trees and farms sap-sucking insects.
Brown tree snake (*Boiga irregularis*)	Arrival in Guam caused the near-total extinction of native forest birds.
Avian malaria (*Plasmodium relictum*)	Arrival and spread through mosquitoes has contributed to the extinction of at least 10 native bird species in Hawaii and threatens many more.
Miconia (*Miconia calvescens*)	Spread in Pacific has led to its taking over of large areas, displacing native vegetation, and increasing landslides due to its superficial root structure.
Water hyacinth (*Eichhornia crassipes*)	Now found in more than 50 countries on five continents. Its shading and crowding of native aquatic plants dramatically reduces biological diversity in aquatic ecosystems.

Source: ISSG, 2000.

Table 9.3 shows some of the economic costs associated with the disruption caused by invasive alien species. While this table gives only some of the associated costs, it is clear that they can be very large. These economic impacts also do not account for many aspects of invasive species that are known to be important but were not measured in the studies; for example, the irreversible impacts of invasive species on local ecosystems.

Table 9.3. **Sample economic impact of invasive species**

Species	Economic variable	Economic impact
Introduced disease organisms	Annual cost to human, plant, animal health in USA	USD 41 billion per year
A sample of alien species of plants and animals	Economic costs of damage in USA	USD 137 billion per year
Salt cedar (*Tamarix* spp.)	Value of ecosystem services lost in western USA	USD 7-16 billion over 55 years
Knapweed (*Centaurea* spp), and leafy spurge (*Euphorbia esula*)	Impact on economy in three US states	USD 40.5 million per year direct costs USD 89 million indirect
Zebra mussel (*Dreissena polymorpha*)	Damages to US and European industrial plants	Cumulative costs 1988-2000 = USD 750 million to 1 billion
Most serious invasive alien plant species	Costs 1983-92 of herbicide control in the UK	USD 344 million/year for 12 species
Six weed species	Costs in Australian agro-ecosystems	USD 105 million/year
Pinus, hakeas and *acacia* spp.	Costs to restore South African floral kingdom to pristine state	USD 2 billion
Water hyacinth (*Eichhornia crassipes*)	Costs in 7 African countries	USD 20-50 million/year
Rabbits	Costs in Australia	USD 373 million/year (agricultural losses)
Varroa mite	Economic cost to beekeeping in New Zealand	USD 267-602 million

Source: GISP (2001), and references therein.

Global climate change

The Intergovernmental Panel on Climate Change (IPCC) notes that numerous long-term changes in climate have already been observed (IPCC, 2007). Further changes in climate are expected in the coming decades, driven in part by past emissions, but also by the impossibility of reducing emissions immediately to zero (see Chapter 7, Climate change). These changes to climate have direct impacts on ecosystems and individual species.

Small-scale studies linking changes in climate to biodiversity are growing in number (Parmesan, 2005), but most look at particular species and focus on population changes within a particular ecosystem or biome.[3] A few of those studies link climatic changes and biodiversity through changes in the geographical distribution of species. Species are generally limited by climate to areas where either they – or their food-source – can survive. Small increases in temperature have generally (though not always) been found to cause migration either northwards in latitude, or higher in altitude (Parmesan, 1996). These changes will cause some ecosystems to shrink and others to expand. For example, most ecosystem models predict that tundra will shrink with warming as boreal forests push up from the south. Species dependent on the tundra ecosystem will experience a shrinking habitat and their populations will decline. The northern migration is caused by changes in both maximum daytime temperatures, and minimum night time temperatures. The maximum temperature can determine whether a species is able to find suitable habitat during the feeding and breeding season, whereas minimum temperature can determine whether a species survives the winter chill.

Changing temperatures will also cause mountain ecosystems to change. Warming would put pressure on species to move to higher altitudes. An analysis of ecosystems in California reveals that alpine forests will likely shrink in future climate scenarios (Lenihan *et al.*, 2003). Species dependent on these forests will be at risk. Aquatic ecosystems can also

OECD ENVIRONMENTAL OUTLOOK TO 2030 – ISBN 978-92-64-04048-9 – © OECD 2008

be affected by climate change since some have been shown to be sensitive to small changes in temperature. Cod, for example, can only tolerate a small temperature change before their ability to reproduce is compromised because spawning is triggered by a narrow range of water temperatures. Strong impacts have been observed in coral reef systems that are thought to be linked to the limited climate change that has occurred over the past few decades (Hughes *et al.*, 2003).

The threat of climate change also raises concerns for conservation efforts. Current conservation efforts are geographically static, tending to protect an area rather than a geographically mobile ecosystem. However, if there is a threat from climate change, it may be important to anticipate where future habitat should be, not just where current habitat exists. Conservation efforts may have to consider dynamic strategies to either adjust to moving habitats over time, or create buffer zones and ecological corridors. Given current and evolving land use around many protected areas, leaving enough space for biodiversity to adapt to changes in climate will clearly be difficult. Mitchell *et al.* (2007) identify a number of measures for enhancing adaptation in the UK so that future climate change does not compromise the government's ability to achieve its biodiversity goals. Resilient natural systems will not only benefit biodiversity, but will preserve the "services" that ecosystems provide and could be costly to replace: soil conservation, clean air and water, agricultural productivity, and other less direct economic and social benefits, such as leisure activity (see Chapter 13 for further discussion).

Current model analyses suggest that sufficient warming may occur over the coming decades to put pressure on many species (IPCC, 2007). The impact on biodiversity will depend on the ecosystem. But climate change pressure will be in addition to existing impacts on species and ecosystems from factors such as land use change, invasive alien species, habitat fragmentation from infrastructure development, and nitrogen deposition or other wide-dispersion pollutants.

Industrial and agricultural pollution

Since the 1950s, nutrient loading – *i.e.* anthropogenic increases in nitrogen, phosphorus, sulphur, and other nutrient-associated pollutants – has emerged as a potentially important driver of ecosystem change in terrestrial, freshwater and coastal ecosystems. Moreover, it is projected to increase substantially in the future (see also Chapter 10 on freshwater). Synthetic production of nitrogen fertiliser has been a key driver of the remarkable increase in food production during the last 50 years, but this and other smaller anthropogenic sources of nitrogen now produce more reactive (biologically available) nitrogen than is produced by all natural pathways combined. The damage done by these fertilisers (and other pollutants) has been documented, as has the increasing numbers of marine "dead zones" that are associated with eutrophication (*e.g.* Diaz *et al.*, 2003; Howarth *et al.*, 1996). Some of these impacts are permanent and require substantial human intervention to reverse. The acidification of lakes is known to diminish (though slowly) once sources of acid rain are removed, but the restoration of pre-impact species can only be approximated by restocking efforts (Keller *et al.*, 1999).

While total OECD nitrogen surpluses entering the environment (*i.e.* total nitrogen inputs from fertilisers, manure and atmospheric deposition *less* uptake by agriculture) declined between 1990 and 2002, they have increased in some, mainly non-European, OECD countries. Developing countries showed a decrease in the efficiency of fertiliser use between 1970 and 1995. In some cases this may simply reflect diminishing returns, but in others more of it ended up in the environment rather than being taken up by crops (*e.g.* in China). Nonetheless, some developing countries show a nitrogen deficit balance (particularly Africa), which can translate into a loss of soil productivity through depletion of soil nitrogen and phosphorous pools.

The *Outlook* Baseline projects that nitrogen surpluses will continue to increase for the world as a whole to 2030 as agricultural production expands (and intensifies), and as a result of pressures from untreated wastewater discharges in rapidly growing urban areas. The largest increases in nitrogen surpluses are expected in the Asian region. The impact of other pollutants has been decreasing in North America and Europe, but remains an increasing problem in other regions.

Desertification

Drylands – arid, semi-arid and dry sub-humid – comprise some 41% of global lands (MEA, 2005b). It is thought that at least a quarter of drylands are already degraded and heading toward desertification (Safriel, 1997).

Human activity contributes directly to dryland degradation (and desertification) through changes in the use of the topsoil in vulnerable areas. This leads to the loss of recycled minerals, organic matter, moisture-retention potential and seed bank capacity. In many areas, irrigation causes dryland salinisation: where irrigation water is sufficient to bring up salts under the soil, but not sufficient (partially due to high evaporation) to leach them back down. When such croplands or rangelands are abandoned due to salinisation, the low level of tolerance of the original species to the salty soils makes it impossible to recover the original conditions. Desertification thus becomes irreversible without large scale human intervention.

Climate change is also thought to contribute indirectly to the degradation of drylands, although this is more difficult to quantify rigorously since local climate impacts from GHG emissions are difficult to separate from natural variability.

Figure 9.4. **Change in agricultural activity in arid areas, 2005-2030**

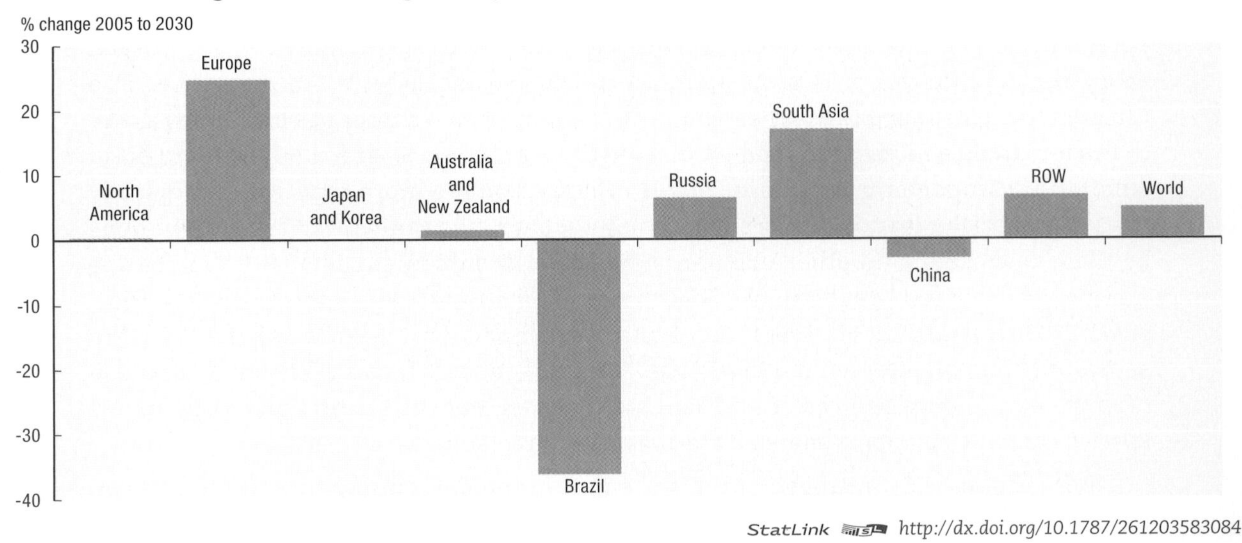

StatLink http://dx.doi.org/10.1787/261203583084

Source: OECD Environmental Outlook Baseline.

In the *Outlook* Baseline, future agricultural activity is expected to change in response to growing demand; this includes a substantial expansion of agricultural lands. Figure 9.4 shows the part of that expansion that is expected to occur in arid areas. Desertification, of course, is not an automatic outcome, but without special care it becomes a distinct possibility. The change shown for Europe is mostly in Turkey, where a significant expansion is projected in the Baseline. In Brazil, the small amount of agriculture that is in arid zones is gradually being phased out in favour of other, more profitable, areas. The results for Russia and South Asia are explained by a general expansion of agriculture, but because South Asia can only expand into arid zones, the impact is greater there.

Policy implications

While most of the policies to protect biodiversity are enacted at the national or sub-national level, the benefits of biological diversity, and some of the pressures on it, extend beyond national boundaries. By 2006, 190 countries had ratified the Convention on Biological Diversity (CBD) with the aim of conserving biodiversity as well as ensuring the sustainable use of its components. A range of other multilateral environmental agreements also help to protect biodiversity, for example the Convention on International Trade in Endangered Species (CITES), the Convention on Wetlands (Ramsar Convention), the World Heritage Convention, and the Convention on the Conservation of European Wildlife and Natural Habitats. These measures attempt to ensure a co-ordinated process for addressing biodiversity loss. Implementation is generally done at a national level through policies that address the sources of impacts on biodiversity. Valuation helps prioritise and set objectives so that policies are set at the right level and directed at the most pressing issues. Underpinning most of the policy discussion in this section, therefore, is an implicit assumption that priorities and objectives are being addressed through means such as valuation (Box 9.3).

Box 9.3. **The need to value biodiversity**

Policies to protect biodiversity aim directly or indirectly to move the cost of biodiversity-affecting activities to levels that reflect social values for biodiversity. With market-based instruments, it is the market price that is being targeted.

For example, taxes impose a cost on users of biodiversity-related resources to reflect the loss faced by others by that use (*i.e.* the social cost). Taxes are "indirect" because they require policy-makers to obtain additional information about the level of this collective loss by some means other than observing the market itself – the level of tax is meant to exactly internalise the non-marketed cost of the activity. To set the tax at the socially optimum level, information is needed about the (incremental) social cost of using the biodiversity-related resource. Economic *valuation* provides a monetary measure of the (monetary and non-monetary) impacts and thus helps set the tax. Other policy instruments, such as regulations, scientific information provision and gathering, also need to be based on some measure of biodiversity value to justify the expenditure of resources toward stated goals.

Regulatory approaches and protected areas

Restrictions or prohibitions on the harvesting or use of wildlife species are common in many countries to protect threatened or endangered species or specific ecosystems of value. Globally, CITES[4] regulates international trade in products of endangered species of wild animals and plants.

The creation of protected areas is another important policy instrument to conserve biodiversity. Figure 9.5 shows that there has been particularly rapid growth in protected areas in the last three to four decades. By 2003, just under 12% of the world's land area was devoted to protected areas (Chape *et al.*, 2003).

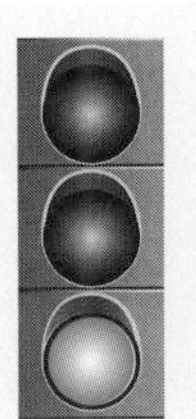

Worldwide, almost 12% of land area is devoted to protected areas.

Of course, the number of locations and the area that is protected are only rough indicators of policy success in conserving and sustainably using biodiversity. Policy optimisation would call for setting the cost of protecting an additional area to its (general) incremental benefit. Such an analysis has not been undertaken as it would require a lot of information, but there is reason to believe that even existing protected areas are under-funded (Balmford *et al.*, 2002). A main reason for this under-funding is the traditional sources of market failure identified by economists: the mismatch between those who benefit from, and those who incur the costs of, maintaining biodiversity (OECD, 2007).

A few biomes are well represented in protected areas, but others less so. Tropical humid forests, subtropical/temperate rainforests and mixed island ecosystems have seen large increases in the area protected, while lake systems and temperate grasslands are poorly covered. One area that is thought to be under-represented is marine ecosystems, for which only a few protected areas exist. Based on a number of studies of marine protected areas, Halpern (2003) shows that in terms of density, biomass, size of organisms and diversity, marine protected areas do deliver benefits.

Some governments are moving towards ecosystem-based fisheries management systems. To appreciate how difficult it will be to fully implement sound management globally, it is worthwhile recalling that the "tragedy of the commons" is often invoked to describe incentives facing fishermen. Unsustainable harvesting in the fisheries industry is thus systemic and changing behaviour to implement good management will be an

Figure 9.5. **Cumulative change in protected areas worldwide, 1872-2003**

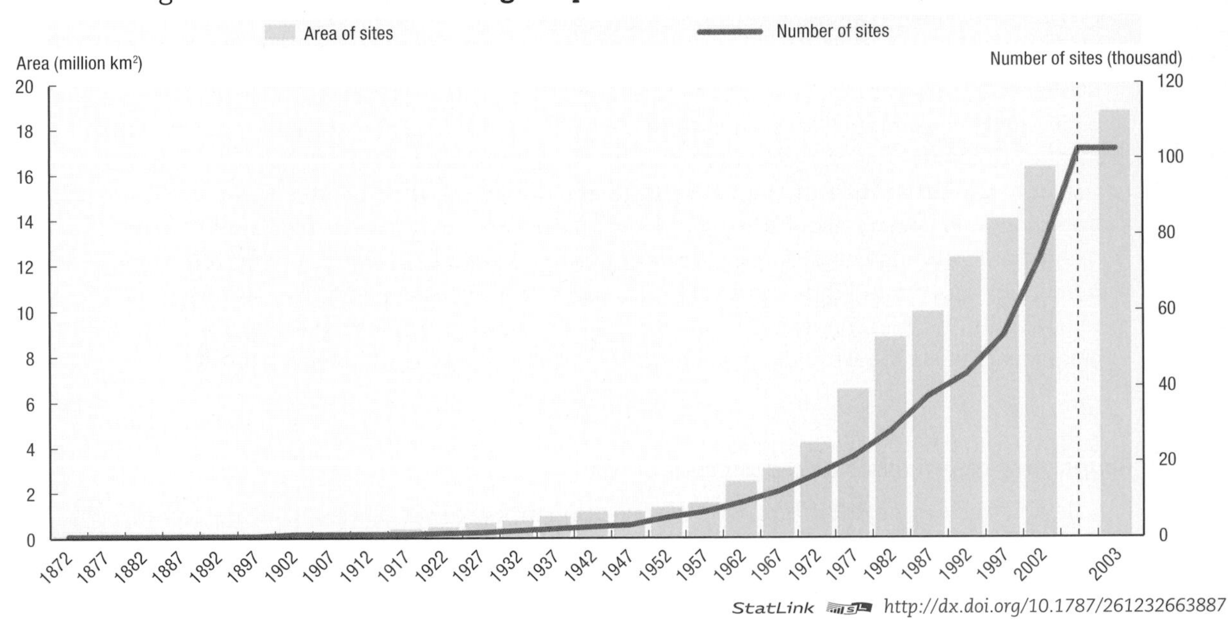

StatLink http://dx.doi.org/10.1787/261232663887

Source: Chape *et al.*, 2003.

undertaking of considerable proportions. Given the rate at which marine ecosystems are being disturbed, immediate action through the development of more marine protected areas is justified from a biodiversity perspective, while continuing to work towards sound long-term management (see also Chapter 15 on fisheries and aquaculture).

Of course, establishing a protected area is only a first step. If protection is not enforced then the biodiversity may still be lost. The World Conservation Union (IUCN) has established seven categories of protected areas, ranging from those where human activity is severely limited, to those where only certain aspects of the natural environment are prohibited from being altered. These categories explicitly recognise that protection and sustainable use are complex objectives that have to be achieved in different ways to serve various social goals. Integrating protected areas into an overall sustainable use agenda is important to ensure long-term viability and compatibility with development goals. Often, however, even the level of protection that an area is intended to receive does not actually happen. Adequate resources for the management of protected areas are just as important as the extent of such areas. Some protected areas have been called "paper parks" because there is nothing to distinguish them from other areas; monitoring and enforcement are essentially non-existent.

Protecting an area from certain types of development is only one of a number of regulatory measures that can be used to achieve biodiversity goals. Though in the past regulatory measures were often the instrument of choice and were over-used in many public policy areas, they nonetheless have a place in the difficult terrain of biodiversity policy-making. Information and transaction costs may sometimes favour regulatory measures since they can minimise the costs of public administration, monitoring and enforcement, as well as the private costs of implementation. Some regulatory measures available to governments for encouraging biodiversity conservation and sustainable use include:

- Non-compliance fees and penalties (*e.g.* for certain types of forestry activities).
- Liability frameworks for harm to certain species.
- Liability fees for the rehabilitation or maintenance of ecologically-sensitive lands.
- Implementation of biodiversity-related labelling schemes.
- Community-based measures that facilitate regional co-operation.
- Providing research and development that facilitate knowledge expansion of biodiversity.
- Providing rigorous monitoring and enforcement.

Economic incentives and market creation

Incentive measures can be used to try to reconcile differences between the market value of biodiversity-related goods and services to individuals and the value of biodiversity to society as a whole. They can increase the cost of activities that damage ecosystems, and reward biodiversity conservation and enhancement/restoration. Since the main policy problem facing biodiversity conservation is the problem of the global commons, economic incentives that close the gap between private and public values of biodiversity are, in principle, all that are needed.

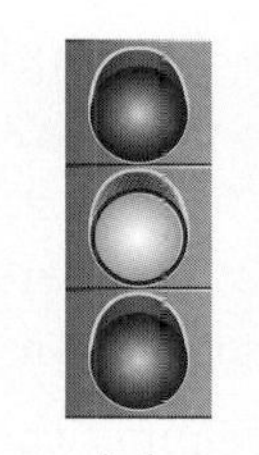

Economic incentives are increasingly used to protect biodiversity, but are clearly insufficient given the scope of continued biodiversity loss.

Markets for biodiversity are created by removing barriers to trade of goods or services derived from biodiversity and creating public knowledge of their special characteristics. Important steps to remove barriers are taken with the establishment and assignment of well-defined and stable property and/or use rights, and the creation of information instruments for the products. Market creation is based on the premise that holders of these rights will maximise the value of their resources over long time horizons, thereby optimising biodiversity use, conservation and restoration.

The range of economic incentives available to governments for encouraging biodiversity conservation and sustainable use includes:

- Financial instruments that optimise the purchase of biodiversity "services", *e.g.* auctions.
- Offset schemes that allow an overall level of biodiversity to be maintained, with local tradeoffs.
- Fishing license fees or taxes.
- Levies for the abstraction of surface water or groundwater.
- Charges for:
  - use of public lands for grazing in agriculture;
  - use of sensitive lands;
  - hunting or fishing of threatened species;
  - tourism in natural parks.
- Market-based support for activities that improve biodiversity quality and quantity.
- Access and benefit sharing regimes which create value for high biodiversity areas.

One of the more important approaches to creating markets and incentives for biodiversity is payments for ecosystem services (PES). The idea is that by requiring people to pay for services they otherwise obtained for free (because they were otherwise unsuitable for markets), overuse of these services would diminish. In recent years the use of PES schemes has been increasing and they are expected to continue to grow in popularity. One good example is watershed services. Many cities derive their water from watersheds in which agriculture puts pressure on water quality. Payments to farmers or other watershed users to modify their activities have helped maintain watersheds and reversed downward trends in water quality. Prominent examples can be found in France, Costa Rica and the United States (OECD, 2004).

Information and other instruments

The creation of specific markets for biodiversity-friendly products is based on the premise that informed consumers will choose products friendly to biodiversity. The growing popularity of organic agriculture, eco-labelled timber, fish certified as being sourced from sustainable fisheries, shade-grown coffee, and eco-tourism opportunities are examples of where consumers have chosen to pay more for a good or service because of a perceived environmental benefit.

In general, good physical and economic data and indicators on biological diversity are scarce, and where they do exist there is little comparable information over time or between countries. This has hampered efforts to design appropriate policies to protect biodiversity. Efforts are underway in many countries and international bodies to improve both the physical understanding of ecosystems and biodiversity, and to measure them. The recent Millennium Ecosystem Assessment (2005a) provides a state-of-the-art assessment of the status of different types of ecosystems worldwide, and the pressures on them.

A number of techniques to value the economic benefits of biodiversity and ecosystems have also been developed, and are gaining in rigour and acceptability in decision-making (OECD, 2002). Once economic values of biodiversity or ecosystem services are established, these can be used to inform policy decisions or in the development of appropriate economic incentives to internalise the full costs of natural resource use.

Costs of inaction

Biodiversity has high economic value. Some of the more obvious sources of value include: bio-prospecting, carbon sequestration, watersheds and tourism. These are direct sources of biodiversity value and do not include indirect aspects such as protection against major pathogens, sources of innovation in agricultural production, the existence value of biodiversity, etc. The pharmacological value of biodiversity may be in the multi-billion dollar range; a successful product can be worth USD 5 to USD 10 billion per year in revenues net of production costs, with a present value over its life of perhaps USD 50 to USD 100 billion. Indeed, finding just a small number of additional blockbuster drugs from the remaining biodiversity would justify significant conservation for bio-prospecting. Biodiversity's carbon storage value may also be in the tens of billions of dollars since it is a significant reservoir of carbon: there are now markets for carbon that allow the implicit pricing of stored carbon. The services provided by biodiversity through watersheds and charismatic megafauna are harder to estimate in total, but again clearly run to billions of dollars. New York City alone saved hundreds of millions of dollars by maintaining its source watershed rather than building a water purification plant (Heal, 2000).

The costs of biodiversity loss through continued policy inaction will thus be significant in both measurable economic loss and difficult-to-measure non-marketed terms. Getting a precise total figure for that loss is not possible, but there is good reason to suspect that it is large.

Notes

1. Mean species abundance (MSA) captures the degree to which biodiversity, at a macrobiotic scale, remains unchanged. If the indicator is 100%, the biodiversity is similar to the natural or largely unaffected state. The MSA is calculated on the basis of estimated impacts of various human activities on "biomes". A reduction in MSA, therefore, is less an exact count of species lost, than an indicator that pressures have increased.
2. In the US, for example, it takes one hectare of maize to produce 3 100 litres of ethanol (IEA, 2004). This is roughly one third of the annual fuel requirement of a small North American car that is driven 18 000 km/year (a rough North American average), so each small car requires three hectares of cropland to support its fuel use. Since the entire US maize crop was 32 million hectares in 2000, this would produce enough fuel to support roughly 10 million small cars – about one tenth of all cars (big and small) in the US.
3. The extinction of a species of mountain-top frog that succumbed to changing precipitation and humidity (Pounds and Savage, 2004) is a good example of this type of study.
4. Convention on International Trade in Endangered Species of Wild Fauna and Flora.

References

Atkinson, I.A.E. and E.K. Cameron (1993), "Human Influence on the Terrestrial Biota and Biotic Communities of New Zealand", *Trends in Ecology and Evolution*, 8: 447-51.

Balmford, A. *et al.* (2002), "Economic Reasons for Conserving Wild Nature", *Science*, Vol. 297, pp. 950-53.

Chape, S. *et al.* (2003), *United Nations List of Protected Areas*, IUCN, Gland, Switzerland and Cambridge, UK and UNEP-WCMC, Cambridge.

Committee on Biological Diversity in Marine Systems (1995), *Understanding Marine Biodiversity*, Commission on Life Sciences, National Research Council, National Academy Press, Washington, DC.

Diaz, R.J., J.A. Nestlerode and M.L. Diaz (2003), "A Global Perspective on the Effects of Eutrophication and Hypoxia on Aquatic Biota", in G.L. Rupp and M.D. White (eds.) *Proceedings of the 7th International Symposium on Fish Physiology, Toxicology and Water Quality*, Tallinn, Estonia, May 12-15.

FAO (UN Food and Agriculture organisation) (2004), *The State of World Fisheries and Aquaculture: 2004*, UN Food and Agriculture Organization, Rome.

FAO (2005), *State of the World's Forests*, 2005, Rome.

Frank, K.T. *et al.* (2005), "Trophic Cascades in a Formerly Cod-Dominated Ecosystem", *Science*, Vol. 308 (5728), 10 June.

Gattuso, J.-P. *et al.* (1998), "Effect of Calcium Carbonate Saturation of Seawater on Coral Calcification", *Global Planetary Change* 18, pp. 37-46.

GISP (Global Invasive Species Programme) (2001), *Global Strategy on Invasive Alien Species*, GISP, IUCN, Gland, Switzerland.

Halpern, B. (2003), "The Impact of Marine Reserves: Do Reserves Work and Does Reserve Size Matter?", *Ecological Applications*, Vol. 13, No. 1, pp. s117-s137.

Heal, G. (2000), *Nature and the Marketplace: Capturing the Value of Ecosystem Services*, Island Press, Washington, DC.

Heilig, G.K., G. Fischer and H. van Velthuizen (2000), "Can China Feed Itself? An Analysis of China's Food Prospects with Special Reference to Water Resources", *The International Journal of Sustainable Development and World Ecology*, Vol. 7, pp. 153-172.

Hole, D.G. *et al.* (2005), "Does Organic Farming Benefit Biodiversity?", *Biological Conservation*, vol. 122, pp. 113-30.

Howarth, R.W. *et al.* (1996), "Regional Nitrogen Budgets and Riverine N and P Fluxes for the Drainages to the North Atlantic Ocean: Natural and Human Influences", *Biogeochemistry*, Vol. 35, pp. 1-65.

Hughes *et al.* (2003), "Climate Change, Human Impacts, and the Resilience of Coral Reefs", *Science*, Vol. 301, No. 5635, pp. 929-933.

IEA (International Energy Agency) (2004), *Biofuels for Transport: An International Perspective*, International Energy Agency, Paris.

IPPC (Intergovernmental Panel on Climate Change) (2007), *Fourth Assessment Report: Working Group 1 Summary for Policymakers*, Intergovernmental Panel on Climate Change, Geneva.

ISSG (Invasive Species Specialist Group) (2000), *Aliens 12*, IUCN, Gland, Switzerland.

Kelleher, G., C. Bleakley and S. Wells (1995), *A Global Representative System of Marine Protected Areas*, Volume 1, World Bank.

Keller, W., J.M. Gunn and N.D. Yan (1999), "Acid Rain – Perspectives on Lake Recovery", *J. Aquat. Ecosys. Stress Recov.* 6: 207-216.

Lenihan, J.M. *et al.* (2003), "Climate Change Effects on Vegetation Distribution, Carbon, and Fire in California", *Ecological Applications*, Vol. 13, pp. 1667-81.

MEA (Millennium Ecosystem Assessment) (2005a), *Ecosystems and Human Well-Being*, Island Press, Washington, DC.

MEA (2005b), *Ecosystems and Human Well-being: Desertification Synthesis*, World Resources Institute, Washington, DC.

Mitchell, R.J. *et al.* (2007), "*England Biodiversity Strategy: Towards Adaptation to Climate Change*", Final Report to Defra for contract CRO327, Department for Environment, Food and Rural Affairs, London.

OECD (2002), *Handbook of Biodiversity Valuation: A Guide for Policymakers*, OECD, Paris.

OECD (2004), *Handbook of Market Creation for Biodiversity: Issues in Implementation*, OECD, Paris.

OECD (2007), *People and Biodiversity Policies: Impacts, Issues, and Strategies for Policy Action*, OECD, Paris, forthcoming.

Parmesan, C. (1996), "Climate and Species' Range", *Nature*, Vol. 382, pp. 765-66.

 OECD ENVIRONMENTAL OUTLOOK TO 2030 – ISBN 978-92-64-04048-9 – © OECD 2008

Parmesan, C. (2005), "Range and Abundance Changes", in Lovejoy, T.E. and L.J. Hannah (eds.) *Climate Change and Biodiversity*, Yale University Press.

Perrings, C., M. Williamson and S. Dalmazzone (2000), *The Economics of Biological Invasions*, Edward Elgar, Cheltenham, UK.

Pimentel, D. *et al.* (1999), *Environmental and Economic Costs Associated with Non-Indigenous Species in the United States*, College of Agriculture and Life Sciences, Cornell University, NY.

Pimm, S. L. *et al.* (1995), "The Future of Biodiversity", *Science*, Vol. 269, pp.347-350.

Pounds, A. and J. Savage (2004), "Bufo periglenes", *IUCN Red List of Threatened Species*, IUCN World Conservation Union, IUCN Species Survival Commission, Cambridge, UK, available at *www.iucnredlist.org/search/details.php/3172/all*. Retrieved on 28 February 2007.

Safriel, U.N. (1997), "Relations Between Biodiversity, Desertification and Climate Change", Report submitted to the Ministry of the Environment: *Israel Environment Bulletin*, Winter 1997-5757, Vol. 20, No. 1.

SCBD (Secretariat of the Convention on Biological Diversity) (2006), *Global Biodiversity Outlook 2*, CBD, Montreal, Available at *www.biodiv.org/GBO2*.

SCBD (2007), "Cross-roads of Life on Earth. Exploring Means to Meet the 2010 Biodiversity Target. Solution-oriented Scenarios for Global Biodiversity Outlook", *CBD Technical Series*, No. 31, Montreal.

Seneca Creek Associates and World Resources International, (2004), *Illegal Logging and Global Wood Markets: The Competitive Impacts on the US Wood Products Industry*, Report prepared for the American Forest and Paper Association, Poolesville, Maryland, USA.

Wilson, E.O. (2002), *The Future of Life*, A. E. Knopf, New York.

World Bank (2006a), *Strengthening Forest Law Enforcement and Governance: Addressing a Systemic Constraint to Sustainable Development*, Report No. 36638-GLB, Washington, DC.

World Bank (2006b), *World Development Indicators 2006*, April, World Bank, Washington DC.

ISBN 978-92-64-04048-9
OECD Environmental Outlook to 2030

Chapter 10

Freshwater

Significant water scarcities already exist in some regions of the OECD and many regions of non-OECD countries. More than 3.9 billion people (47% of the world population) are expected to live in areas with severe water stress by 2030, mostly in non-OECD countries. This chapter examines trends and projections in water stress, public water supply, urban waste water treatment, nitrogen pollution and soil erosion by water. It highlights the good policy principles to address the main water challenges. Much progress remains to be made to integrate water management into sectoral (e.g. agriculture) and land use policies, ensure a more consistent application of the polluter pays and user pays principles through water pricing and reduce subsidies that increase water problems.

KEY MESSAGES

Significant water scarcities already exist in some regions of the OECD and many regions of non-OECD countries. An estimated 3.9 billion people (47% of the world population) are expected to be living in areas with high water stress by 2030, mostly in non-OECD countries (see graph).

People living in areas of water stress, by level of stress (millions of people)

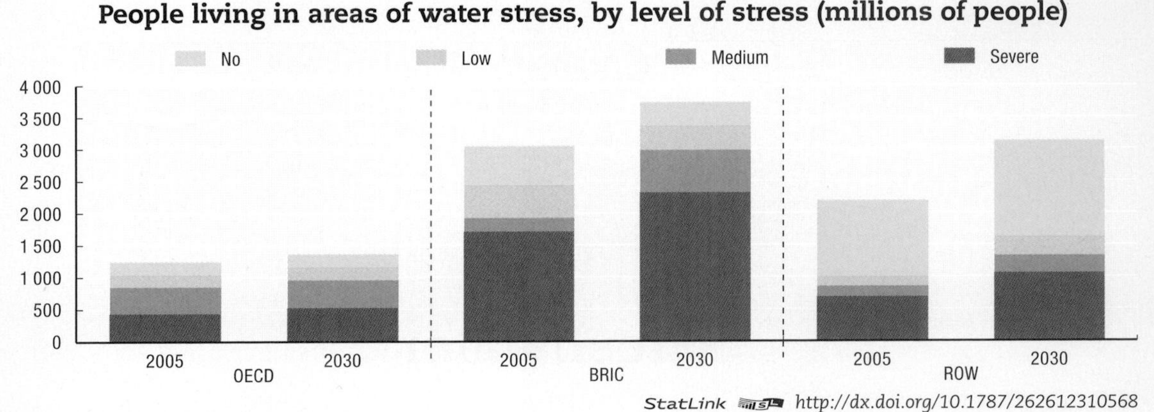

StatLink http://dx.doi.org/10.1787/262612310568

More than 5 billion people (67% of the world population) are expected to be without a connection to public sewerage in 2030 – 1.1 billion more than today.

Nearly 55 million tonnes of nitrogen are projected to reach coastal waters from inland sources by 2030 (an increase of 4% since 2000). Soil erosion by water will increasingly undermine soils' capacity to support food production. Areas with a high level of erosion risk from water are projected to increase by more than a third to some 27 million km^2 in 2030 (21% of the world land area).

Many OECD countries in recent years have successfully reduced water use per capita and in total, indicating that the right policies can lead to more efficient water use and a decoupling of water use from economic growth/population growth, while taking account of social factors.

OECD countries are committed to increasing official development assistance to the water sector, though recent trends are not sufficient to meet the Millennium Development Goal (MDG) of halving the world's population without access to water and sanitation by 2015.

Policy options

- Put in place the necessary policy frameworks to secure the substantial financing required for non-OECD countries to build and operate water supply and waste water treatment infrastructure and for OECD countries to upgrade theirs.
- Address nutrient pollution of water from diffuse sources (agriculture, atmospheric deposition) and point sources (urban sewage), in both OECD and non-OECD countries.
- Develop policy mechanisms to take into account the economic, environmental and social costs and benefits of water used in agriculture, and to ensure that it is sustainable in the long run. Agriculture is by far the major user of water and is responsible for much of its pollution.
- Improve water governance and river basin management, and ensure optimal pricing for water services worldwide.
- Foster international co-operation on shared river basins to avoid major disruptions to countries' water supply and address transboundary water pollution issues.

Consequences of inaction

- Achieving the MDG of halving the population without access to water and sanitation by 2015 is expected to cost about USD 10 billion per year. But this figure could be far outweighed by the costs of inaction if the MDG is not achieved, in terms of impacts on human health and economic productivity.
- Climate change will pose new challenges for water management, including through impacts on water systems and hydrology, and the potential for increased stress for human populations and ecosystems. Government will need to factor adaptation to long-term climate change predictions into national water management strategies.

Introduction

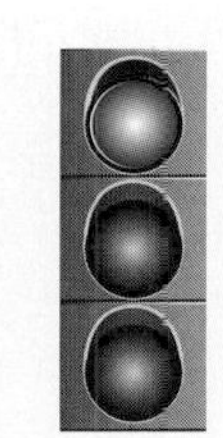

1.1 billion people lack access to clean drinking water, and 2.6 billion lack access to improved sanitation.

Clean water sustains human life and ecosystems. Water scarcity[1] affects human health and contaminated drinking water kills an alarming 1.7 million people a year, mostly children under the age of five in non-OECD countries (see also Chapter 12 on health and environment). Lack of, or inadequate, water policy is an indicator of poverty the 2.6 billion people without access to improved sanitation and the 1.1 billion without access to improved drinking water sources[2] are to be compared, respectively, with the 2.5 billion people who earn less than USD 2/day and the 1.5 billion people with less than USD 1/day.

The world's soaring demand for freshwater and pressures on water quality are also causing increasing environmental stress.[3] Some 24% of mammal and 12% of bird species associated with inland waters are threatened. About a third of known freshwater fish species[4] are also thought to be threatened.

Water policy is attracting growing international attention (Box 10.1) and more and more countries aim to enshrine the right of access to (sufficient, affordable and safe) drinking water in national legislation. But water continues to be used inefficiently in many areas. Key challenges identified by UN-Water as priorities for the decade include coping with water scarcity; access to drinking water, sanitation and hygiene; and disaster risk reduction (UN World Water Assessment Programme, 2006). Overcoming the crisis in access to water and sanitation is one of the greatest human development challenges of the early 21st century (UNDP, 2006). Other future challenges include adaptation to climate change and more severe and frequent weather events, such as flooding and droughts; impacts on food security and increasing risk of human migrations, often adding to water supply problems; as well as contamination threats by chemicals, heavy metals and other toxic contaminants.

Key trends and projections[5]

Water stress

In the OECD area, the greatest demands for water come from irrigation (43%), electrical cooling and industry (42%), and public water supply (15%) (OECD, 2007a). But because of losses through evaporation and plant transpiration, the share of irrigation in total water consumption is much higher. In the developing world, agriculture is by far the main user.[6] According to the OECD *Outlook* Baseline, agricultural production will increase two times faster in developing countries than in OECD countries, further exacerbating water scarcity in those regions (see also Chapter 14 on agriculture). Almost all of the projected 34% population increase to 2030 will occur in developing countries, and growing urbanisation in both OECD and non-OECD countries will also increase demand for public water supply (see also Chapter 5 on urbanisation). Electricity and industrial production will increase

Box 10.1. **The advent of water as an international priority**

The Global Water Partnership (GWP) and the World Water Council (WWC) were created in 1996 in the wake of the Dublin Conference on Water and the Environment and the Rio Earth Summit, both held in 1992. The GWP brings together government agencies, public institutions, private companies, professional organisations, multilateral development agencies and others committed to the Dublin-Rio principles. It is financed by governments. The WWC is an "international multi-stakeholder platform" for over 300 member organisations representing more than 50 countries. Since 1997 the WWC has organised the World Water Forum, held every three years and in which the OECD participates; the latest was in Mexico in 2006 and the next will be in Istanbul in 2009. In 2002, the GWP and WWC joined forces to address the key issue of financing water infrastructure.

In 2003, the "Camdessus Panel" warned that the Millennium Development Goals would not be achieved unless annual investments in water and sanitation services in developing countries were doubled from the 2003 level (Winpenny, 2003). These conclusions were adopted at the G8 meeting in Evian (2003) and the need to implement them in Africa was endorsed at the G8 meeting in Gleneagles (2005). In 2004 the UN Secretary-General's Advisory Board on Water and Sanitation was set up to galvanise action on water and sanitation, and to help mobilise resources to achieve the water and sanitation MDG. Chaired by Angel Gurría, now Secretary-General of the OECD, a "Task Force on Financing Water for All" was established in 2005 to continue the work initiated by the Camdessus Panel. A Task Force report was presented to the 4th World Water Forum in Mexico. It highlights the need to harness local financing and to fund necessary investments in agricultural water management (van Hofwegen, 2006). Some of the key areas for further work identified in this report have since been taken up in the OECD's horizontal work programme on water for 2007-2008.

much faster in non-OECD countries than in OECD countries (see also Chapter 17 on energy). Overall, pressures on water use are thus projected to increase at a much higher pace in developing countries than in OECD countries.

According to the Baseline,[7] 44% of the world population already lives in areas of high water-stress[8] and the situation is projected to worsen, with an additional 1 billion people projected to be living in areas with severe water stress by 2030 (Table 10.1). More than half the population affected by severe water stress is (and will continue to be) found in the BRICs. The main increase in population affected will be in India and, to a lesser extent, in China, Africa and the Middle East. The latter region is projected to experience its fastest population growth in the most arid areas. The relatively limited projected increase in China to 2030 (an additional hundred million people affected) reflects China's one-child policy and lack of land for expanding agriculture. The *OECD Environmental Outlook* Baseline projects a decrease in agricultural water consumption in China to 2050, largely as a result of uptake of improved irrigation technology. But this is expected to be more than offset by a dramatic increase in non-agricultural water uses associated with economic development, most prominently industrial use and, to a lesser extent, urban household demand (Chinese Academy of Sciences, 2000).

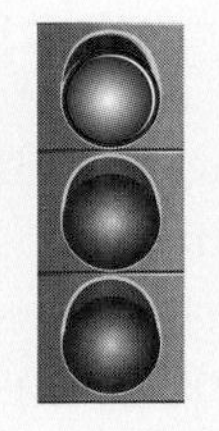

More than 40% of the world's population lives in areas affected by high water-stress, and this will increase to 2030.

Table 10.1. **Population and water stress, 2005 and 2030**
Millions of people

Region	Degree of water stress	2005	% of total in 2005	2030	% of total in 2030	% change 2005-2030
OECD	**Severe**	**438**	35%	**525**	38%	**20%**
	Medium	415	33%	434	32%	5%
	Low	186	15%	198	14%	6%
	No	211	17%	211	15%	0%
	Total	**1 250**	100%	**1 368**	100%	**9%**
BRIC	**Severe**	**1 710**	56%	**2 319**	62%	**36%**
	Medium	216	7%	661	18%	207%
	Low	506	17%	381	10%	–25%
	No	619	20%	378	10%	–39%
	Total	**3 051**	100%	**3 740**	100%	**23%**
ROW	**Severe**	**688**	31%	**1 057**	34%	**54%**
	Medium	164	7%	272	9%	66%
	Low	143	7%	287	9%	101%
	No	1 198	55%	1 512	48%	26%
	Total	**2 193**	100%	**3 128**	100%	**43%**
World	**Severe**	**2 837**	44%	**3 901**	47%	**38%**
	Medium	794	12%	1 368	17%	72%
	Low	35	13%	866	11%	4%
	No	2 028	31%	2 101	26%	4%
	Total	**6 494**	100%	**8 236**	100%	**27%**

StatLink http://dx.doi.org/10.1787/257213423814

Source: *OECD Environmental Outlook* Baseline.

Public water supply and urban waste water treatment

Most OECD countries have been able to ensure adequate access to a safe water supply for human needs and significant efforts have been made in the OECD area to treat organic pollution from urban waste water. Since 2000 sustainable access to improved drinking water sources and sanitation facilities have become key policy objectives for non-OECD countries, as part of the UN Millennium Development Goals (MDG) and as agreed at the 2002 World Summit on Sustainable Development.[9]

According to the OECD *Outlook* Baseline, some of the recent progress in public sewerage connection rates is projected to continue to 2030. Despite this, it is projected that there will be 1.1 billion more people worldwide in 2030 who are not connected to public sewerage compared with 2000 (Figure 10.1). By 2030 the situation will have (further) improved in the OECD area but deteriorated in the BRIC area, with the latter still accounting for half of the world population not connected. The situation is of most concern in the rest of the world where the number of people not connected will dramatically increase (to 2.4 billion), accounting for 80% of the 2030 population in these regions. In many areas of the developing world, waterborne sanitation systems may not be the most sustainable option, and other improved facilities may be more suitable. Even when considering other solutions, in 2004 only around one-third of the sub-Saharan African population had access to improved sanitation facilities (WHO/UNICEF, 2006).[10] According to UN projections based on 1990-2004 trends, the MDG target for sanitation will not be met by 2015.

Figure 10.1. **People not connected to public sewerage systems, 2000 and 2030**

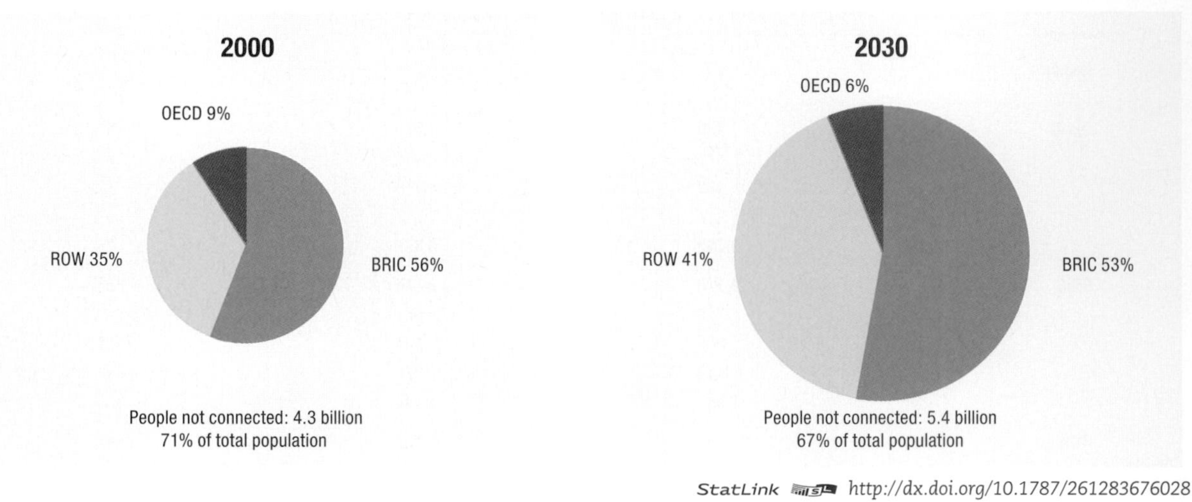

StatLink http://dx.doi.org/10.1787/261283676028

Source: OECD Environmental Outlook Baseline.

Nitrogen pollution[11]

In the OECD a third of main rivers have phosphorus and nitrate concentrations higher than, respectively, 0.2 mg P/litre and 2.5 mg N/litre (OECD, 2007a). In combination, these levels contribute to algal growth in receiving coastal waters. Point source discharges to surface water have been reduced in OECD countries, especially from industrial and urban waste water systems, though the level of treatment (nutrient removal in sensitive areas) could still be improved. In contrast, little progress has been made to tackle pollution arising from agricultural runoff and other non-point sources of pollution. Similar to OECD countries, non-OECD countries are expected to expand the share of houses connected to sewage systems for public health reasons, though waste water treatment may only be considered at a later stage. Many industrial sources are not yet equipped with/connected to waste water treatment plants in non-OECD countries, thereby adding to the projected total nutrient loading.

According to the Baseline, the global release of nitrogen compounds by rivers to coastal marine systems (and associated risks of eutrophication of coastal waters) is projected to increase by 4% to 2030. This masks differences between the OECD area, where some improvement is expected, and other areas (BRIC, rest of the world) where the projected increase is a continuation (though at much lower rate) of the trend observed in past decades (Table 10.2). The highest increase in nitrogen pollution will be in the BRIC countries and, to a lesser extent, in the rest of the world (outside the OECD area). There are, however, large differences between sub-regions and countries. River nitrogen exports will decrease by 5% in North America, 4% in OECD Europe and over 20% in Japan (agricultural land contraction) and Russia (reduced atmospheric deposition). In contrast, river nitrogen exports are projected to increase by 5% in Oceania, 3% in Brazil, 16% in China and over 40% in India.

Nitrogen surplus from agriculture is projected to increase significantly in China and India, while in the United States and OECD Europe it may decrease or stabilise, following a marked increase in US voluntary agri-environmental incentive programmes and the introduction of cross-compliance in EU agricultural policies (see Chapter 14 on agriculture). There is a wide range in nitrogen surplus per hectare, driven by intensity and management practices; highest surpluses occur in Asian regions and OECD Europe.

Table 10.2. **Source of river nitrogen exports to coastal waters, 2000 and 2030**

Million tonnes

Area	2000				2030				% change (total)	
	Nature[a]	AGR[b]	Urban[c]	Total	Nature[a]	AGR[b]	Urban[c]	Total	2000-30	1970-2000
OECD	6.4	4.4	1.8	12.6	5.7	4.3	2.0	12.0	*–5*	*10*
BRIC	11.9	8.6	1.4	21.9	9.0	12.9	2.4	24.3	*11*	*57*
ROW	12.7	5.0	0.9	18.6	10.8	6.5	1.6	18.9	*2*	*26*
Total (world)	31.0	18.0	4.1	53.1	25.5	23.7	6.0	55.2	*4*	*33*

StatLink http://dx.doi.org/10.1787/257313201371

a) N deposition on and biological fixation in non-cultivated areas.

b) N surplus on cultivated areas.

c) N effluents from public sewerage.

Source: OECD Environmental Outlook Baseline.

Only a small share of the world population is (9%) or will be in 2030 (16%) connected to advanced (N-removal) sewage treatment plants according to the *Outlook* projections. Most household sewage is therefore discharged untreated (or treated without N-removal) into rivers.[12] Nitrogen from urban sewage is projected to increase very strongly in India, China and the Middle East, where population and urbanisation are likely to outstrip the construction of public sewerage and waste water treatment plants.

Soil erosion by water

The capacity of soils to support food production can be seriously impaired by surface water runoff. Worldwide, areas with a high level of erosion risk from water are projected to increase from 20 million km^2 in 2000 to nearly 30 million in 2030 (Figure 10.2). The increase will occur in all regions

Figure 10.2. **Land area under high soil erosion risk by surface water runoff, 2000-2030**

Million km^2

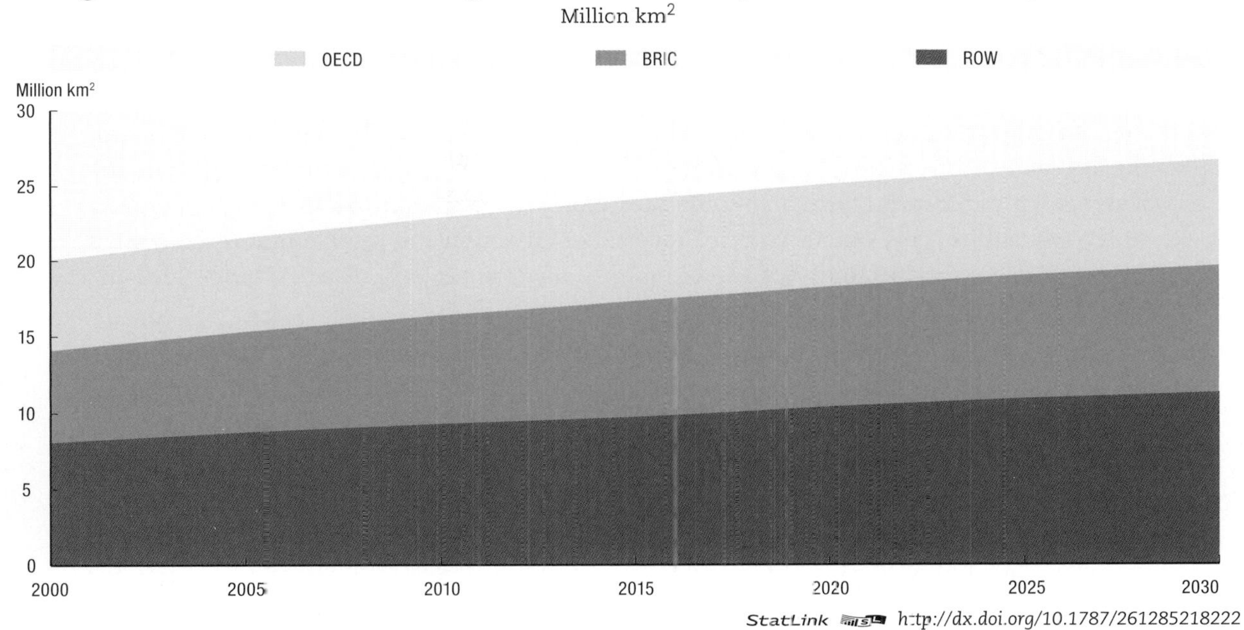

StatLink http://dx.doi.org/10.1787/261285218222

Source: OECD Environmental Outlook Baseline.

Policy implications

Both OECD and non-OECD countries must apply "good policy principles" to address the main water challenges. Some of these good policy principles have been successfully applied in OECD countries, while some have not. For example, OECD countries have made progress towards a whole-basin approach and towards expanding the use of water pricing mechanisms to manage demand (OECD, 2006b). Nevertheless, much progress remains to be made to: *i)* co-ordinate water management policies with sectoral (*e.g.* agriculture) and land use policies; *ii)* ensure a more consistent application of the polluter-pays and user pays principles; and *iii)* reduce subsidies that increase water problems (*e.g.* over-abstraction, pollution). A major remaining challenge is to design and implement water management policies that better reflect ecosystem needs for freshwater, as well as human needs. There is a need to assess the economic efficiency and environmental effectiveness of water pollution abatement measures in different sectors (municipal, industrial, agricultural) in the context of river basin management. Making wider use of markets and improving the coherence of decision-making (water governance), together with technological developments,[13] have been identified as key elements of effective water management (OECD/IWA, 2003). These options are described in detail below.

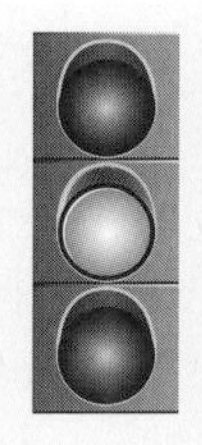

Water use can be decoupled from economic growth with the right policies – about half of all OECD countries have stabilised or decreased their total water use since 1980.

Market-based approaches

Currently, many water supply systems do not include their own investment and running costs in water prices. This reduces the financial sustainability of service provision. In the OECD area, such below-cost pricing is prevalent in publicly-funded irrigation systems, and households' drinking water bills do not fully reflect actual supply costs for many utilities. Moreover, the scarcity value of water resources is rarely reflected in the pricing of public water supply (though there are some cases of seasonal pricing). This can create incentives for consumers to over-use water.[14] The costs of not incorporating scarcity into groundwater pricing (and thus not matching the rate of groundwater extraction with the available resource base in a given region) may include the adjustment costs of changing water supply infrastructure or the need to relocate population to areas with adequate water availability. This vicious circle of government inaction and market failure, combined with the low substitutability of the resource, can trap regions in a "resource lock-in" (OECD, 2008).

As water becomes ever scarcer (due to both drought and degraded water quality), water pricing is increasingly seen as a necessary public policy instrument to encourage more responsible use (Jones, 2003). OECD countries should work towards full cost recovery pricing (where the price of water services should at least cover the capital, operation, maintenance and environmental costs[15]). Fuller cost recovery can provide an incentive to use water more efficiently, while generating revenues to support necessary investment in infrastructure. In Denmark, water consumption decreased from 155 litres per head per day in 1993 to 125 litres per day in 2003, following a 54% increase of the water bill. A similar pattern was observed in the Czech Republic. Both countries now rank among the "low-use" group by OECD standards.

A key issue in water pricing is how to ensure affordable access by the poorest communities to adequate water supply and sanitation services. Available evidence suggests that in half of OECD countries, affordability of water charges for low-income households is either a significant issue now or might become one in the future (OECD, 2003a). There is a wide range of practice in OECD countries to address affordability, including targeted support to low-income groups,[16] which is more efficient and environmentally effective than providing across-the-board subsidies through low water prices.

Only about half of OECD countries charge for abstracting surface water or groundwater or for the direct pollution of water (*i.e.* outside the public water supply and sanitation system). Abstraction charges can create incentives for efficient water use and can reduce water withdrawals. Similarly, water-pollution charges are likely to reduce discharges efficiently (see Box 10.2) – provided they are set at similar rates across sectors (which is often not the case). The charges should be levied according to the quality of receiving waters; quantitative information on the benefits of reducing pollution is a prerequisite to the formulation of efficient water quality objectives.

The use of tradable water rights can help to allocate limited water resources to their most productive uses. For example, Australia has been reforming its water policies since 1994 to introduce a fully market-based system for apportioning the amount of water available. But this potential remains largely unexploited, and the capacity to enhance efficient resource allocation is often hampered (*e.g.* by poor documentation). In Mexico, for example, water trading between irrigators and other users, such as industrial plants, requires government approval. In the arid west of the United States, trade of abstraction rights for surface water is subject to complicated rules. In Spain, the Environment Ministry is in the process of clarifying historical abstraction rights to enhance water trading. OECD-wide, transactions in water rights have remained largely marginal, and there are few examples of trade other than between farmers. As a consequence of recent droughts, however, there is increased trading between farmers and public water supply utilities.

While the long-term objective of optimal pricing – *i.e.* "internalising" the full marginal social costs (including environmental costs) into decisions that affect the use of water and water quality – is also valid in developing countries,[17] achieving it is probably unrealistic for most in the short term. In areas where more than 60% of the population lives on less than USD 2 per day, public budgets and external finance will need to play a role in covering capital costs (OECD, 2005, and see below). In areas where non-payment of water bills is widespread due to the poor quality of water services, cost recovery should correspond with noticeable improvements in service quality in order to gain consumer trust.

Water governance and the whole-basin approach

A number of countries (*e.g.* Australia, France, Spain) aim to manage water resources and pollutant discharges in a common, consistent framework at the river-basin level. An important development in this area is the European Union Water Framework Directive which calls for integrated river basin management planning in all EU member countries by 2009.[18] Because such integrated policies clarify the link between water use and water pollution, they are likely to be more efficient in meeting water management objectives. For example, they can enable a comparison between the costs of cleaning water downstream before it is supplied with the costs of discouraging pollution upstream. Integrated policies also facilitate cost recovery (OECD, 2004). When river-basin authorities have access to the cost of treatment for water supply operators, this provides them with a wealth of

Box 10.2. **Policies for water management by agriculture**

OECD countries are at very different stages in developing water pricing systems in agriculture (OECD, 2006c). However, most of the costs of investment in irrigation fall on the taxpayer and on other water users (through cross-subsidies). And it is mainly national treasuries that have financed dams, reservoirs and delivery networks, as well as a large part of the cost of installing local and farm infrastructure. Governments generally attempt to recover some of these costs through user charges, but revenues are rarely enough to cover even operation and maintenance costs. As a rule, in the absence of water rights farmers have free access to (or are charged only a nominal fee for) water that they pump themselves. And several countries (including Mexico, Turkey and the United States, at least in some federal irrigation districts) continue to offer preferential tariffs for electricity used to pump water for irrigation.

Not enough has been done to address diffuse pollution from agriculture. Even though the switch to low-dose agents has significantly reduce pesticide consumption in the OECD area, most surface water and groundwater samples still contain pesticides, sometimes at levels harmful for human health and the environment. In the few OECD countries where they have been introduced, pesticide taxes have not created enough incentives to reduce treatment frequency. Taxes should apply rates that reflect the products' human and environmental toxicity. Even though the use of fines has helped reduce the use of nitrogen in the few OECD countries where farm fertiliser accounts have been introduced, it would be more cost-effective to replace the complex mix of regulatory and incentive measures used by most countries by a tax based on the nitrogen surplus for the whole agricultural sector, as measured by the soil surface nitrogen balance (OECD, 2007b). A rebate could be granted to farmers based on the nutrient content of their output, thereby applying the polluter-pays principle while leaving flexibility in the choice of crops and farming techniques. Moreover a tax on phosphorous surpluses could be piggy-backed on the administrative set-up for the tax on nitrogen surpluses.

The economic distortions caused by the underpricing of water used in agriculture have been compounded in many instances by other agricultural support policies, particularly those linked to the production of particular commodities. Such linked support draws resources, including water, into the activity being supported, thereby driving up both the price of water to other users and the volume of agricultural subsidies. Moreover, since fertiliser use is highly responsive to the price of commodities, agricultural support misaligns farmer incentives and aggravates pollution of water (OECD, 2006c). See Box 10.3 for information on policy simulations which included reducing agricultural production support.

information on the costs of upstream pollution, which they can use to estimate the rates at which pollutant releases should be charged. River basin management also facilitates water allocation among competing uses within the basin as well as the control of inter-basin transfers. In Spain, river basin authorities are purchasing water rights for over-exploited water bodies.

There is a need to extend water policy to risk management to address the trend of increasing flood/drought damage.[19] With respect to floods, a more proactive land use policy across an entire watershed combined with enforcement of zoning provisions (making "room for rivers") can help. But a lot remains to be done. Measures such as "green corridors" along rivers and streams, reinstatement of flood control plains, or better control of deforestation and preservation of wetlands often are not binding and the issuance

Box 10.3. **Policy package simulations: impacts on water projections**

Chapter 20 on environmental policy packages describes how a mix of policies was simulated to reflect global action to address many of the key environmental challenges identified in this *Outlook*. A number of the policies simulated in the policy mix would affect the water projections to 2030, including the scaling back of agricultural support measures, increasing connections to public sewerage at the same rate as urbanisation, and increased removal of nitrogen from waste water.

of building permits continues to be left to local authorities' discretion. The insurance and re-insurance industry may have an increasing role to play in facilitating the management of natural hazards (OECD, 2003b). More broadly, in the absence of proper (enforcement of) land use planning, and with increasing incidence of extreme weather events due to climate change (see Chapter 7), it may become necessary for potential flood/drought victims to assume a greater share of the risk through higher flood/drought insurance premiums or reduced compensation. There is also a need for early warning systems and observatories to enhance risk management. For example, based on experience gained in actuarial science, the Australian government has developed innovative information technology tools to improve drought risk management in agriculture (Grant *et al.*, 2007).

Parties to the Helsinki Convention on Transboundary Watercourses have recently agreed to implement pilot projects of payments for ecosystem services that would apply to water related ecosystems like forests and wetlands, which are constituent parts of river basins (UN-ECE, 2006). However, policies to enhance forests' role in water management ("ecosystem services") should not imply giving more subsidies to forest owners (to improve forest management) or to farmers (to convert farmland to forest). That would run the risk of repeating in the forestry sector the mistakes that policy reforms are now seeking to address in the agricultural sector. The reform of agricultural policy underway in OECD countries has in itself important implications for farmland conversion to forests: where price support to commodities is reduced, there is less incentive to expand agricultural production on marginal land. Instead of seeking compensation for any foregone revenues (from timber sales or from farming), any forestry payments should reward the provision of well-targeted (climate and/or water-related) environmental services.

Financing investment in infrastructure

Countries will need to mobilise significant financial resources in the next few decades, including in the OECD, to replace ageing water infrastructure to extend services to those currently unserviced (especially in non-OECD countries), and to meet increasingly stringent environmental and health standards.[20] Based on income categories,[21] projected annual (current and investment) expenditure on water and waste water services by 2025 has been estimated at around USD 600 billion for OECD countries (half of which is for Mexico and the United States) and USD 400 billion for BRIC countries (half of which is for China; OECD, 2007c).

Though estimates vary significantly, the investment cost of implementing the Millennium Development Goals (MDG) for drinking water and sanitation would be around USD 10 billion a year over 15 years. This is more than three times the current level of official development assistance[22] (ODA) to water supply and sanitation, which has only slightly increased in recent years after a downward trend in the second half of the 1990s (OECD/DAC, 2006). The WHO and UNICEF estimate that meeting these MDGs would mean doubling the efforts of the past 15 years for the sanitation target and by one-third for the MDG drinking water target (WHO/UNICEF, 2006). However, the potential economic benefits of meeting the MDG for drinking water and sanitation far outweigh the costs (see also Chapter 12 on health and environment). In developing regions, the WHO estimated the economic return on one USD investment to be USD 5 to USD 28 (WHO, 2004). This is mainly due to time savings associated with better access to water supply and sanitation services, although avoided health impacts are also important. The cost of not meeting this MDG (cost of inaction) has been estimated at some USD 130 billion a year (Hutton and Haller, 2004).

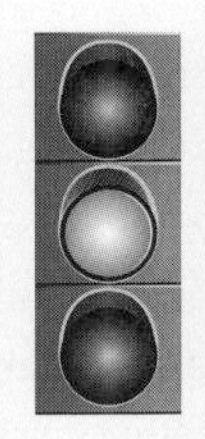

OECD countries are committed to increase ODA to the water sector, though recent efforts are not sufficient to meet the MDG of halving the world's population without access to water and sanitation by 2015.

Key drivers of water infrastructure development include financing, demand management, economies of scale, public involvement and equity, competition and climate change (OECD, 2007c). Services liberalisation can also contribute to achieving universal access to water and sanitation services (OECD/World Bank, 2006). However, both the 2003 Camdessus Panel and the 2006 Gurría Task Force on Financing Water for All (Box 10.1) highlight that problems with the governance of the water sector hamper its ability to mobilise and to attract finance from a range of possible sources, including public spending, international development assistance, private financing and through charging for the use of water services. Over the longer term, a sustainable financing system should rely primarily on water charges, with provisions for affordable access by the poor. Local capital markets and innovative financing mechanisms also have a strong role to play in harnessing sufficient financing for water supply and sanitation infrastructure. The first step for enhancing access to finance for local governments is to increase their capability and creditworthiness to engage in financial actions (van Hofwegen, 2006). In Africa, ensuring adequate financing remains a key challenge for improving the water and sanitation sector, which has been the infrastructure sector least attractive to private investors (OECD/African Development Bank, 2007). An OECD Task Team of officials from development agencies and environment ministries was set up in 2006 to work on developing guidance for sustainable financial planning of water supply and sanitation investments in developing countries, with a particular focus on Africa.

Water management in the context of climate change

As a natural resource, water is obviously influenced by climatic factors. The projected change in climate will significantly affect the hydrological cycle, and in response water management frameworks will need to adapt to the impacts of climate change (see also Chapter 7 on climate change). A warmer climate will be accompanied by shifts in precipitation patterns and increased rates of evapotranspiration which are likely to aggravate water stress, especially in regions where available water is heavily managed and demand for water is growing rapidly. Extreme weather events will be exacerbated by climate change, including an increase in warm spells and heat waves, extreme precipitation, the area affected by drought[23] and coastal and river delta flooding. Warmer weather is likely to translate into increased occurrence and intensity of water quality problems (*e.g.* harmful algal blooms as surface waters warm and salt-water intrusion resulting from storm surge and coastal flooding) (IPCC, 2007).

Water policies have potentially large implications for climate change and *vice versa*. For example, saving water also means saving energy, as extracting, transporting and treating water comes at a high energy cost. For example, the EU Nitrates Directive aims to reduce nitrogen run-off from agriculture to freshwater resources. These measures would also reduce N_2O emissions from the agriculture sector; N_2O is a potent GHG (UNFCCC, 2006). Water policies also affect the vulnerability of water systems to changes in the climate. For example, subsidising agricultural or urban water use leads to inefficient and excessive water use, which in turn aggravates vulnerability to any temporary or long-term changes in physical supply of freshwater due to climate change. Co-benefits of appropriate water pricing or water pollution policies include both sustainable water resources management as well as resilience to climate change.

Climate change policy also has significant spillovers to other policy areas (*e.g.* energy, agriculture, forestry, urban development) that affect water management. For example, measures to preserve forest areas reduce greenhouse gas emissions, increase sink activity from forests, conserve water (*e.g.* by reducing runoff), and regulate water storage and flows. Similarly, restoring wetlands, natural waterways or coastal zone management can re-establish natural habitats for plants and animals, provide flood protection, protect freshwater supplies (*e.g.* from saltwater intrusion) and build resilience to future climate change.

Most OECD government sustainable water management strategies are developed to address current problems in the water sector looking 10 to 20 years ahead and have yet to seriously factor in long-term climate change predictions (Levina and Adams, 2006). However, some attention to these issues is emerging in OECD countries. For example, Germany recently hosted a conference on climate change and the European water dimension to discuss the need for adaptation plans in water-related sectors (Federal Ministry for the Environment, Nature Conservation and Nuclear Safety, 2007). The EU has identified an initial set of policy options to mitigate the impacts of and adapt to water scarcity and drought in a context of climate change (Commission of the European Communities, 2007). Information on the nature of climate change (regional temperature and precipitation predictions under plausible futures) and on the costs and benefits of climate change measures in the water sector could contribute to better water management in the face of climate change. The latter entails looking at the direct benefits and costs of policy as well as at the nearer term[24] co-benefits of adaptation (or mitigation) choices in other policy areas.

Notes

1. Although most of the planet is covered by water, only 2.5% of it is fresh, while the rest is salt. Of the freshwater, two-thirds is locked up in glaciers and permanent snow cover (although this is changing with the decrease in snow and ice extent).
2. 2004 data (WHO/UNICEF, 2006).

3. The flows of about 60% of the world's largest rivers have been interrupted by dams.
4. Only 10% of freshwater fish species have been studied in detail.
5. This section only includes the four themes for which OECD modelling work has been carried out (water stress, public water supply and urban waste water treatment, nitrogen pollution and soil erosion by water).
6. Globally agriculture uses roughly 70% of available water resources (see also Chapter 14 on agriculture).
7. These projections are likely to underestimate water stress in some regions, as the WaterGap model used assumes no impact of climate change on rainfall distribution to 2030. See this chapter's annex for a discussion of the assumptions and uncertainties regarding the projections.
8. Areas with a ratio of withdrawals to available resources that exceeds 0.4 – see annex to this chapter.
9. For both improved drinking water sources and improved sanitation, the MDG goal is to halve the proportion of people who lack access by 2015, from the reference year 1990. Achieving this would require providing services to an additional 1.1 billion people and sanitation to an additional 1.6 billion people between 2004 and 2015 (WHO/UNICEF, 2006).
10. Defined as facilities which are not shared/public and consist of: *i)* flush or pour-flush to piped sewer system, septic tank or pit latrine; *ii)* ventilated improved pit latrine; *iii)* pit latrine with slab; and, *iv)* composting toilet.
11. Even though phosphorus equally contributes to eutrophication, this section focuses on nitrogen because nitrogen compounds are relatively mobile and easy to measure, and the load on the environment easier to model.
12. Even less sewage is treated for phosphorus removal.
13. A recent study commissioned by the European Commission estimates that water efficiency in the EU could be improved by nearly 40% through technological improvements alone. This includes reduction of leakage in water supply networks and more efficient household appliances; conveyance efficiency of irrigation systems; application efficiency of irrigation water; changes in irrigation practices, use of more drought-resistant crops and reuse of treated sewage effluent in agriculture; changes in industrial processes, higher recycling rates or the use of rainwater by industry. *http://ec.europa.eu/environment/water/quantity/scarcity_en.htm.*
14. The widespread failure to charge for irrigation water at rates that reflect the scarcity of the resource has resulted in the over-use of water in agriculture.
15. Pursuant to the OECD Council *Recommendation on Water Resource Management Policies: Integration, Demand Management and Groundwater Protection* [C(89)12/Final] (*http://webdomino1.oecd.org/horizontal/oecdacts.nsf/linkto/C(89)12*).
16. For example, additional direct income support, appropriately designed increasing block water tariffs where those who only use a small amount of water pay very little for it, subsidised connection fees, *etc.*
17. Where buying water from "water sellers" is often more expensive than paying for a public water supply. The other alternatives are also costly in terms of social or opportunity costs, in particular drinking unsafe water or walking long distances to public water pumps as many do in less developed countries.
18. Another key objective of the Water Framework Directive is to achieve good chemical and ecological status of all EU surface water bodies by 2015.
19. The EU directive on the assessment and management of flood risks requires drawing up of flood risk maps and flood management plans.
20. For example, in the European Union the lead concentration limit of 10 μg/l of the 1998 EU Drinking Water Directive is to be met by 2013, which will involve replacing mains in the private part of the water supply system.
21. Based on an assumption that 0.35 to 1.20% of GDP is required to finance water and waste water services in high income countries; 0.54 to 2.60% of GDP in middle income countries, and 0.70 to 6.30% of GDP in low income countries. Other estimates also exist in the literature.
22. Including the 22 DAC countries' bilateral ODA as well as multilateral ODA.
23. In the EU, the number of areas and people affected by droughts went up by almost 20% between 1976 and 2006 (Commission of the European Communities, 2007). One of the most

widespread droughts occurred in 2003 when over 100 million people and a third of the EU territory were affected. The cost of the damage to the European economy was at least EUR 8.7 billion. The total cost of droughts over the past 30 years amounts to EUR 100 billion. The yearly average cost quadrupled over the same period.

24. The benefits of mitigation are long-term. Even if strong action was taken today, there will be no discernible effect (identifiable benefit) on rates of warming (and rainfall distribution) for considerable periods of time (Pearce, 2000).

References

Alcamo, J., *et al.* (2003), "Developing and Testing the WaterGap 2 Model of Water Use and Availability", *Hydrological Sciences*, Vol. 48, pp. 317-337.

Bouwman, A.F. *et al.* (1997), "A Global High-resolution Emission Inventory for Ammonia", *Global Biogeochemical Cycles*, Vol. 11, pp. 561-587.

Bouwman, A.F., L.J.M. Boumans and N.H. Batjes (2002), "Estimation of Global NH3 Volatilization Loss from Synthetic Fertilizers and Animal Manure Applied to Arable Lands and Grasslands", *Global Biogeochemical Cycles*, Vol. 16(2), 1024, doi:10.1029/2000GB001389.

Bouwman, A.F. *et al.* (2005), "Exploring Changes in River Nitrogen Export to the World's Oceans", *Global Biogeochemical Cycles*, Vol. 19, GB1002.

Bruisnma, J.E. (2003), *World Agriculture: Towards 2015/2030 – an FAO Perspective*, Earthscan, London.

Chinese Academy of Sciences (2000), "Analysis of Water Resource Demand and Supply in the First Half of the 21st Century", in *China Water Resources*, US Department of Commerce, 2005, Washington, DC.

Cleveland, C.C. *et al.* (1999), "Global Patterns of Terrestrial Biological Nitrogen (N_2) Fixation in Natural Ecosystems", *Global Biogeochemical Cycles*, Vol. 13, pp. 623-645.

Commission of the European Communities (2007), *Communication from the Commission to the European Parliament and the Council Addressing the Challenge of Water Scarcity and Droughts in the European Union*, COM(2007) 414 final, CEC, Brussels.

Dentener, F. *et al.* (2006), "The Global Atmospheric Environment for the Next Generation", *Environment Science and Technology*, Vol. 40, pp. 3586-3594.

Federal Ministry for the Environment, Nature Conservation and Nuclear Safety (2007), *Time to Adapt – Climate Change and the European Water Dimension, Vulnerability – Impacts – Adaptation*, International Symposium organised by Ecologic in co-operation with the Potsdam Institute for Climate Impact Research, February 12-14, Berlin.

Grant, C. *et al.* (2007), "Farming Profitably in a Changing Climate: a Risk Management Approach", paper presented at the 101st European Association of Agricultural Economists (EAAE) Seminar on *Management of Climate Risks in Agriculture*, July 5-6, Berlin.

Hofwegen, van P. (2006), *Enhancing Access to Finance for Local Governments, Financing Water for Agriculture*, Task Force on Financing Water for All, chaired by Angel Gurría, presented at the 4th World Water Forum, Mexico, World Water Council, Marseille, *www.financingwaterforall.org/fileadmin/Financing_water_for_all/Reports/Financing_FinalText_Cover.pdf*

Hutton, G. and L. Haller (2004), *Evaluation of the Costs and Benefits of Water and Sanitation Improvements at the Global Level*, Water, Sanitation and Health, Protection of the Human Environment, World Health Organization, Geneva.

IPCC (Intergovernmental Panel on Climate Change) (2007), *Climate Change 2007: Impact, Adaptation and Vulnerability*, Contribution of Working Group II to the 4th Assessment Report of the Intergovernmental Panel on Climate Change, Cambridge University Press, New York.

Jones, T. (2003), "Pricing Water", *OECD Observer*, No. 236, OECD, Paris.

Levina, H. and H. Adams (2006), *Domestic Policy Frameworks for Adaptation to Climate Change in the Water Sector: Part I: Annex I Countries*, [*www.oecd.org/env/cc/aixg*], OECD/IEA, May 2006, Paris.

OECD (2003a), *Social Issues in the Provision and Pricing of Water Services*, OECD, Paris.

OECD (2003b), "Environmental Risks and Insurance: a Comparative Analysis of the Role of Insurance in the Management of Environment-related Risks", *Policy Issues in Insurance* No. 6, OECD, Paris.

OECD (2004), *Sustainable Development in OECD Countries: Getting the Policies Right*, OECD, Paris.

OECD (2005), *Financing Strategy for the Urban Water Supply and Sanitation Sector in Georgia*, OECD, Paris.

OECD (2006a), "Keeping Water Safe to Drink", *OECD Observer Policy Brief*, March 2006, OECD, Paris.

OECD (2006b), "OECD Environmental Performance Reviews, Water: the Experience in OECD Countries", paper presented at the *4th World Water Forum* held on March 16-22 in Mexico City, OECD, Paris.

OECD (2006c), *Water and Agriculture, Sustainability, Markets and Policies*, OECD, Paris.

OECD (2007a), *OECD Environmental Data, Compendium 2006*, OECD, Paris.

OECD (2007b), *Instrument Mixes Addressing Non-point Sources of Water Pollution*, [COM/ENV/EPOC/AGR/CA(2004)90/FINAL] [*www.oecd.org/env*], OECD, Paris.

OECD (2007c), *Infrastructure to 2030 Volume 2: Mapping Policy for Electricity, Water and Transport*, OECD, Paris.

OECD (2008), *Cost of Inaction: Technical Report*, OECD, Paris.

OECD/African Development Bank (2007), *African Economic Outlook 2006/2007*, OECD, Paris.

OECD/DAC (2006), "Measuring Aid for Water, Has the Downward Trend in Aid for Water Reversed?", paper presented at the *Meeting of the OECD Development Assistance Committee and the Environment Policy Committee at Ministerial Level*, 4 April 2006, OECD/DAC, Paris, *www.oecd.org/dac/stats/crs/water*.

OECD/IWA (2003), *Improving Water Management, Recent OECD Experience*, OECD, Paris.

OECD/The World Bank (2006), *Liberalisation and Universal Access to Basic Services: Telecommunications, Water and Sanitation, Financial Services, and Electricity*, OECD Trade Policy Studies, Paris.

Palaniappan, M., *et al.* (2006), *Assessing the Long-term Outlook for Current Business Models in the Construction and Provision of Water Infrastructure and Services*, [ENV/EPOC/GF/SD(2006)3], *www.oecd.org*, OECD, Paris.

Pearce, D. (2000), "Policy Frameworks for the Ancillary Benefits of Climate Change Policies", in OECD (2000), *Ancillary Benefits and Costs of Greenhouse Gas Mitigation*, Proceedings of an IPCC Co-Sponsored Workshop held on 27-29 March 2000 in Washington DC, pp. 517-560, OECD, Paris.

UNDP (United Nations Development Programme) (2006), *Human Development Report 2006, Beyond Scarcity: Power, Poverty and the Global Water Crisis*, UNDP, New York.

UN-ECE (UN Economic Commission for Europe) (2006), *Nature for Water, Innovative Financing for the Environment*, Convention on the Protection and Use of Transboundary Watercourses and International Lakes, UN-ECE, Geneva.

UNFCCC (United Nations Framework Convention on Climate Change) (2006), *Synthesis of Reports Demonstrating Progress in Accordance with Article 3, Paragraph 2, of the Kyoto Protocol*, FCCC/SBI/2006/INF.2, 9 May, UNFCCC, Bonn.

UN World Water Assessment Programme (2006), *Water, A Shared Responsibility, The United Nations World Water Development Report 2*, UNESCO and Berghahn Books, Paris and New York. *www.unesco.org/water/wwap/wwdr2/*.

WHO (World Health Organization) (2004), *Evaluation of the Costs and Benefits of Water and Sanitation Improvements at the Global Level*, World Health Organization, Geneva.

WHO/UNICEF (2006), *Meeting the MDG Drinking Water and Sanitation Target, The Urban and Rural Challenge of the Decade*, WHO, Geneva. *www.who.int/water_sanitation_health/monitoring/jmpfinal.pdf*.

Winpenny, J. (2003), *Financing Water For All*, Report of the World Panel on Financing Water Infrastructure, chaired by Michel Camdessus, presented at the 3rd World Water Forum, Kyoto, World Water Council, Marseille. *www.adb.org/Water/Forum/PDF/CamdessusReport.pdf*.

 OECD ENVIRONMENTAL OUTLOOK TO 2030 – ISBN 978-92-64-04048-9 – © OECD 2008

ANNEX 10.A1

Key Assumptions and Uncertainties in the Water Projections

The degree of water stress is assumed to be proportional to the ratio between annual average water abstractions and annual average water availability in a river basin. The WaterGap model (Alcamo *et al.*, 2003) projects water abstractions by households, industry and irrigation as a function of population, GDP and technology. It projects water availability[1] as a function of land cover and climatic conditions (assuming no impact of climate change on rainfall distribution between now and 2030). Data on withdrawals are well established in the OECD area, as well as those on water availability for half of the world area (where there are long-term hydrological gauges). However, while irrigation is the main water user in most river basins, there is strong uncertainty about future development of irrigated areas and volumes. Moreover, the water stress indicator does not take account of seasonal patterns in water supply and demand, a main factor driving irrigation.

Projections of connection to public sewerage are a function of income and World Health Organization (WHO) projections regarding the Millennium Development Goals. Data of the WHO/UNICEF Joint Monitoring Programme (JMP) on improved sanitation facilities were used to estimate the share of the population connected to public sewerage in non-OECD countries. However, the JMP may underestimate the number of people who do not have access to sanitation (OECD, 2006a).

River nitrogen (N) exports to coastal waters are assumed to be 70% of the sum of *i)* runoff and leaching from non-cultivated areas, fed by atmospheric N deposition and biological fixation; *ii)* N surplus from agriculture (diffuse pollution); and, *iii)* N effluents from public sewerage (point sources). It is therefore underestimated as it excludes diffuse urban sources (population not connected to public sewerage) and direct discharges from (large) industry into water bodies. Based on empirical studies in Europe, the remaining 30% of the N load is the assumed share of retention in-stream and from leaching, assuming a half-life of nitrate in groundwater of two to three years.

Atmospheric N deposition from natural origins (in particular lightning) and sectoral emissions (transport, power generation, agriculture) is based on estimates (Dentener *et al.*, 2006) applied to projections for NOx and NH_3 emissions, using the global atmospheric transport model TM3 (see also Chapter 8 on air pollution). Biological fixation is estimated based on coefficients for the various natural ecosystems (Cleveland *et al.*, 1999).

The agricultural soil surface N surplus is estimated as the annual balance between N "inputs" (biological fixation, atmospheric deposition, use of chemical fertilisers and livestock manure) and N "outputs" (removals by crop harvest and forage grazing and ammonia volatilisation) at the country level.[2] Regional changes in crop production are downscaled to the country level on the basis of distribution in the FAO projection to 2030. Projections on fertiliser use are also derived from the FAO (N in crop harvest as a share of N fertiliser inputs, Bruinsma, 2003). Crop N content is obtained from crop-specific data (Bouwman *et al.*, 2005). Biological fixation is estimated for both leguminous crops and free living organisms in farmland. Ammonia (NH3) volatilisation is estimated for animal housing and grazing systems (Bouwman *et al.*, 1997), on the basis of crop type, manure or fertiliser application mode, soil type and climate (Bouwman *et al.*, 2002). The extent to which the N surplus ends up in surface water is uncertain, as the (soil surface) balance does not take account of changes of N in soil organic matter.

Projections of N loads from urban sewage (including industry connected to public sewerage) are a function of GDP per capita. Part of the N load is discharged into sewers, of which part is removed in waste water treatment (WWT) plants. N effluents are estimated as the part that is not removed during treatment plus the amount that is collected via public sewerage but not treated. WWT plants are distinguished according to their N-removal rates (up to 80% for most advanced treatment). It was assumed that the N-removal rates would be doubled by 2030 (up to the current maximum of 80%).

Risks of soil erosion from water runoff are a function of the land erodibility index (based on soil properties and topography), rainfall erosivity index (based on monthly precipitation) and land cover. However, this compound index does not capture cultivation practices, such as tillage (bound to exacerbate the erosion risk) or contour ploughing and terracing (both enhancing soil conservation).

Notes

1. Defined as precipitation net of evapotranspiration (from vegetation and soils) at the grid cell level.
2. The balance is calculated for each grid cell and then aggregated to the country level. It includes areas used for biofuel production.

ISBN 978-92-64-04048-9
OECD Environmental Outlook to 2030

Chapter 11

Waste and Material Flows

This chapter focuses on the material basis of the global economy, and municipal waste generation and management in OECD and non-OECD countries. With continuous growth in the global demand for materials and the amounts of waste generated and disposed of, conventional waste policies alone may not be enough to improve material efficiency and offset the waste-related environmental impacts of materials production and use. New integrated approaches – with stronger emphasis on material efficiency, redesign and reuse of products, waste prevention, recycling of end-of-life materials and products and environmentally sound management of residues – could be used to counterbalance the environmental impacts of waste throughout the entire life-cycle of materials.

KEY MESSAGES

Illegal shipments and unsound management of end-of-life materials and products constitute a considerable risk for human health and the environment.

Management of rapidly increasing municipal waste in non-OECD countries will be an enormous challenge in the coming decades.

Municipal waste generation is still increasing in OECD countries, but at a slower pace since 2000. There has been a relative decoupling of municipal waste generation in OECD countries from economic growth, but waste generation is continuing to increase (see graph).

With continuous growth in the global demand for materials and the amounts of waste generated and disposed of, conventional waste policies alone may not suffice to improve material efficiency and offset the waste-related environmental impacts of materials production and use.

With continuous growth in global demand for materials and amounts of waste generated and subsequently disposed of, conventional waste policies alone may not suffice to improve material efficiency and offset the waste-related environmental impacts of materials production and use.

Current waste policies have been successful in diverting increasing amounts of valuable materials from landfills to further use, remanufacturing and recovery, thereby reducing considerably the associated environmental impacts, including greenhouse gas (GHG) emissions.

OECD country municipal waste generation, 1980-2030

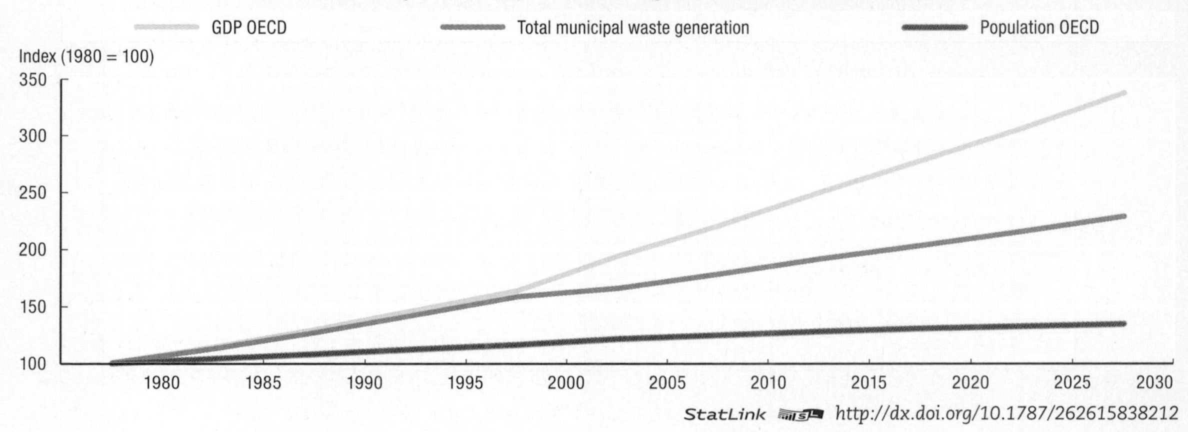

StatLink http://dx.doi.org/10.1787/262615838212

Policy options

- Develop new integrated approaches to address the environmental impacts of waste throughout the entire life-cycle of materials. Place stronger emphasis on material efficiency, redesign and reuse of products, waste prevention (reduction of both amount and hazard), recycling of end-of-life materials and products and environmentally sound management of residues.
- Support these integrated approaches with sound and reliable information on waste, material flows and resource productivity, including improved data quality and availability.
- Increase policy approaches which combine economic, regulatory and information instruments, as well as public-private partnerships, to address the negative environmental impacts of increasing waste volumes, and to encourage waste prevention and economically efficient and environmentally sound recovery of waste.
- Urgently address shipments of problematic end-of-life materials and products, such as electric and electronic appliances, ships and hazardous waste, to ensure they are managed in an environmentally sound manner. Recent incidents also call for intensified enforcement of existing rules and regulations, aimed at eliminating illegal shipments of these materials and products.
- Develop and transfer waste management technologies and know-how from OECD countries to developing countries.

 OECD ENVIRONMENTAL OUTLOOK TO 2030 – ISBN 978-92-64-04048-9 – © OECD 2008

Introduction

Recent decades have seen unprecedented growth in human population and economic well-being for a good portion of the world. This growth has been fed by equally unprecedented resource and material consumption and related environmental impacts, including conversion of large portions of the natural world to human use, prompting concerns about whether the world's natural resource base is capable of sustaining such growth (Huesemann, 2003; Krautkraemer, 2005).

This chapter focuses on two key issues: the material basis of the global economy, and municipal waste generation and management in OECD and non-OECD countries.[1]

Key trends and projections

Material basis of the global economy

Since 1980, global resource extraction (by mass) has increased by 36%, and is expected to grow to 80 billion tonnes in 2020.[2] Growth rates and extraction intensities vary by material categories and among world regions, reflecting different levels of economic development and endowment in natural resources, varying trade patterns and industrial structures, and different socio-demographic patterns. OECD countries as a group figure substantially in both global resource use and raw materials supply, although non-OECD economies, especially the BRIICS countries (Brazil, Russia, India, Indonesia, China and South Africa) are catching up to OECD levels (Figure 11.1).

Anticipated growth in resource extraction is also unevenly distributed among the main material categories. Metal ores exhibit the highest rates, and these are expected to almost double – from 5.8 billion tonnes in 2000 to more than 11 billion tonnes in 2020 (see also Chapter 19 on mining). With projected growth of only 31%, extraction of biomass (agriculture, forestry, fishery, grazing) is expected to expand less than all the non-renewable resource categories combined, indicating a decreasing share of renewable resources in the production and use of materials at the global level (Figure 11.1).

On a per capita basis, resource extraction levels are highest in the OECD area, in particular in North America and the Asia-Pacific region, and are expected to grow further to reach about 22 tonnes per capita in 2020, mainly because of growing demands for coal, metals and construction minerals. Extraction levels in the BRIICS countries are expected to grow much more rapidly over this period, to 9 tonnes per person in 2020, a growth of 50% (Giljum *et al.*, 2007).

On a per unit GDP basis, OECD countries have decreased their extraction intensity in recent decades, reflecting some decoupling of extraction from economic growth. This trend is expected to continue until 2020. The main drivers of this decoupling are structural changes away from the primary and secondary sectors towards the service sector (structural effect), increased applications of more material efficient technologies (technology effect), and increases in material intensive imports (trade effect), due to outsourcing of material-intensive production stages to other world regions.

Figure 11.1. **Global resource extraction, by major groups of resources and regions, 1980, 2002 and 2020**

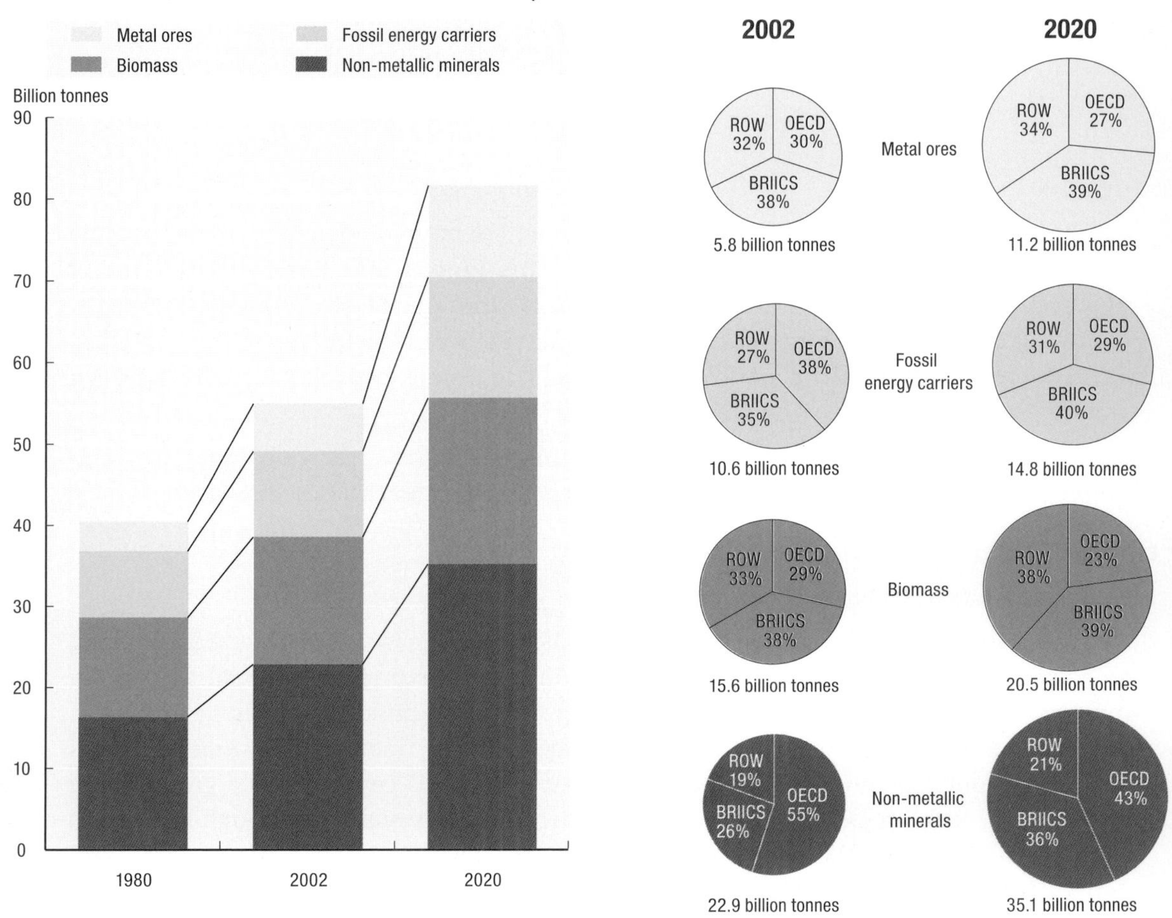

	World				OECD			BRIICS*			ROW*		
		Rate of change				Rate of change			Rate of change			Rate of change	
	2002	1980-2002	2002-2020	1980-2020	2002	1980-2002	2002-2020	2002	1980-2002	2002-2020	2002	1980-2002	2002-2020
Amounts extracted (billion tonnes)													
Total	55.0	36%	48%	102%	22.9	19%	19%	17.7	67%	74%	14.4	35%	64%
Metal ores	5.8	56%	92%	200%	1.8	41%	69%	2.2	110%	100%	1.9	30%	105%
Fossil energy carriers [a]	10.6	30%	39%	81%	4.1	12%	5%	3.7	58%	59%	2.9	31%	60%
Biomass [b]	15.6	28%	31%	68%	4.5	11%	5%	5.9	49%	33%	5.2	25%	51%
Non-metallic minerals [c]	22.9	40%	54%	114%	12.6	21%	21%	5.9	81%	115%	4.4	58%	64%
Per capita (tonne/cap)													
Total	8.8	-4%	22%	17%	20.0	0%	8%	6.0	19%	51%	6.7	-16%	21%
Metal ores	0.9	11%	58%	75%	1.5	19%	54%	0.7	51%	73%	0.9	-19%	51%
Fossil energy carriers [a]	1.7	-8%	14%	5%	3.6	-6%	-4%	1.3	13%	38%	1.3	-18%	18%
Biomass [b]	2.5	-9%	8%	-2%	3.9	-6%	-5%	2.0	7%	15%	2.4	-22%	11%
Non-metallic minerals [c]	3.7	-1%	27%	25%	11.0	2%	10%	2.0	30%	86%	2.0	-2%	21%
Per unit of GDP (tonne/1000 USD [d])													
Total	1.6	-26%	-14%	-36%	0.8	-33%	-24%	4.6	-35%	-32%	4.5	-21%	-26%
Metal ores	0.2	-15%	11%	-5%	0.1	-20%	9%	0.6	-18%	-23%	0.6	-24%	-7%
Fossil energy carriers [a]	0.3	-29%	-19%	-43%	0.1	-37%	-32%	1.0	-38%	-38%	0.9	-24%	-27%
Biomass [b]	0.4	-30%	-24%	-47%	0.2	-37%	-33%	1.5	-42%	-48%	1.6	-27%	-32%
Non-metallic minerals [c]	0.6	-24%	-11%	-32%	0.4	-32%	-22%	1.5	-29%	-17%	1.4	-8%	-26%

StatLink http://dx.doi.org/10.1787/261323245151

a) Crude oil, natural gas and peat.
b) Harvests from agriculture and forestry, marine catches, grazing.
c) Non-metallic industrial and construction minerals.
d) Constant 1995 USD.
* BRIICS = Brazil, Russia, India, Indonesia, China and South Africa; ROW = Rest of world.
Source: MOSUS MFA database, Sustainable Europe Research Institute, Vienna, *http://materialflows.net*; Giljum *et al.*, 2007.

When resources are extracted or harvested, huge amounts of materials are moved, but not all of them are used in the economy (*e.g.* mining overburden, by-catch from fishing, harvest losses). Although not visible in production statistics, these movements of unused materials (or resources) may add to the environmental burden of resource extraction, disrupt habitats or ecosystems, and alter landscapes in the supplying region (see also Chapters 9 on biodiversity, 15 on fisheries and aquaculture, and 19 on mining). The amounts of unused materials are particularly high for energy carriers (some 3.5 tonnes per tonne of fossil fuel extracted) and metals (some 2 tonnes per tonne of metal ore extracted).

Under the *Outlook* Baseline, it is expected that the world population will continue to grow by about one-third to 2030 while the economy will double, placing increasing strains on the global environment. This raises the question of how to sustain economic growth and welfare in the longer term, while keeping negative environmental impacts under control and preserving natural capital – in other words, how to further decouple environmental degradation from economic growth. Against this background, managing the environmental impacts of extracting, processing, using, recovering and disposing of materials will be critical, not only from an environmental perspective but also from an economic and trade perspective. More coherent management policies will be needed, based on a mix of integrated demand and supply-oriented measures. To be successful, such policies will need to be supported by reliable information on waste and material flows, and on resource productivity, and with sound analysis (material flow analysis, input-output analysis, life-cycle analysis, cost-benefit analysis; OECD, 2007a and Box 11.1).

Waste generation and management

In line with continuously growing global demand for raw materials, the amount of waste being generated by economic activity has been rising. Consequently, many valuable material and energy resources are being wasted, and/or disposed of, and thus will be lost to the economy. This has consequences for both the efficiency of material use and for the quality of the environment in terms of land use, water and air pollution, and greenhouse gas (GHG) emissions. Conventional waste policies have been successful in diverting many valuable materials from landfills and in promoting further use, remanufacturing and recovery. They may not, however, be sufficient to improve material efficiency and to offset the waste-related environmental impacts of materials production and use in the longer term. Broader approaches, considering the whole life-cycle of materials, are needed.

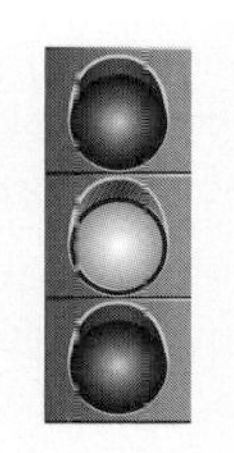

With growing demand for materials and rising amounts of waste, waste policies alone may not suffice to offset the negative environmental impacts of materials use.

Hazardous waste

Although reliable data are difficult to obtain, best available estimates suggest that the amount of hazardous waste generated in OECD countries was some 115 million tonnes in 1997, or 2.5% of total waste (OECD, 2001a). This amount may have increased slightly in the 1997-2001 period. Within this time-frame, 19 OECD countries reported increased generation of hazardous waste, 3 reported decreased generation, 3 reported no change, and 5 provided no data. At the same time, GDP increased by 18% and industrial production grew by 19% (OECD, 2005). In the EU-25, hazardous waste generation increased between 1998 and 2002 by 13%, while gross value-added grew by 10% (Eurostat, 2005).

Box 11.1. A common knowledge base on material flows and resource productivity

Improving resource productivity and putting in place effective and integrated materials management policies within the context of economic development and globalisation are not easy. They require a good understanding of the economic efficiency and environmental effectiveness with which resources and materials are used throughout their life-cycle, and need to be supported by reliable information on material flows.

Existing information is insufficient to give a coherent view of how different materials flow through the economy (from their extraction or import to their final disposal). It does not give many insights into how these flows relate to environmental risks and impacts and to resource productivity, or how globalisation and foreign outsourcing affect international flows of materials and related environmental impacts. Knowledge gaps also remain about waste and recyclable materials.

This is why OECD countries decided to work together, and with other international partners, to establish a common knowledge and information base on material flows and resource productivity. In 2004, OECD governments adopted an OECD Council Recommendation to this effect, following requests from heads of state and government of G8 countries (Evian Summit, 2003 and Sea Island Summit, 2004).

The objective is to enable sound, fact-based material flow analysis (see definition opposite) at the national and international level and to inform related policy debates. The work is proceeding along two tracks:

1. Improving the quantitative knowledge base, by providing guidance to countries on how to construct material flow accounts and indicators in a coherent framework and by compiling material flow information from existing data sources.
2. Improving the analytical knowledge base, by using material flows information in policy analysis and evaluation, including in OECD country environmental performance reviews (EPRs), in work on sustainable materials management (SMM) and in 3R (reduce, reuse, recycle) activities.

Material Flow Analysis (MFA) refers to the monitoring and analysis of physical flows of materials into, through and out of a given system (usually the economy), and is generally based on methodically organised accounts in physical units (OECD, 2007a). It analyses the relationships between material flows, human activities (including economic and trade developments) and environmental changes. It helps identify unnecessary waste of materials, in the economy or in process chains, which go unnoticed in conventional monitoring systems, and analyse opportunities for efficiency gains.

Material flows can be analysed at various scales and with different instruments depending on the issue of concern. The term MFA therefore designates a family of tools encompassing a variety of analytical approaches and measurement tools (economy-wide MFA, material system analysis, life-cycle analysis, input-output analysis, etc.).

Problematic waste

The globalisation of trade has made transboundary movement of waste an attractive and cost-efficient option for the recovery and disposal of problematic end-of-life materials and products, such as electric and electronic appliances and ships (Box 11.2). These end-of-life materials and products are defined differently in different countries: some countries consider them to be "hazardous waste", some see them as "non-hazardous waste"; others consider them to be "used products"; while still others control their movements, but without classifying them as hazardous waste.

OECD ENVIRONMENTAL OUTLOOK TO 2030 – ISBN 978-92-64-04048-9 – © OECD 2008

Box 11.2. **Dealing with ship waste**

Before 1980, most dismantling of ships (vessels or other floating structures) for recycling was taking place in the US and Europe. Since then, this work has been occurring mainly in India, China, Pakistan and Bangladesh. The risks associated with hazardous materials contained by ships destined for scrapping are currently a source of debate, and the issue has risen on the international agenda. Much work is being carried out in many international and regional fora to come up with a sustainable ship dismantling industry that safeguards those employed in it, and protects the environment, while recognising the vital role the dismantling industry plays in the economies of certain countries. The International Maritime Organization is currently developing a new legally-binding instrument on ship recycling (see: *www.basel.int/ships/index.html*).

The development of a legal instrument may, however, take several years, so the Parties to the Basel Convention will soon be exploring possibilities for effective short- and medium-term measures (Basel Convention, Decision VIII/11).

End-of-life electric and electronic appliances ("e-waste") are also creating an increasingly important management challenge in both developed and developing countries. Markets in electronic equipment change rapidly and the useful life of such appliances is constantly shrinking, resulting in an exponential growth in e-waste. Globally, some 20-50 million tonnes of e-waste are estimated to be generated every year.[3] Electric and electronic waste is the fastest growing waste stream in the EU, totalling some 6-7 million tonnes every year.[4]

At the 8th Conference of Parties to the Basel Convention on the Control of Transboundary Movements of Hazardous Wastes and their Disposal (November 2006), ministers and other heads of delegations agreed to a *Nairobi Declaration on the Environmentally Sound Management of Electrical and Electronic Waste*. While acknowledging that all countries benefit from increasing access to modern information and communications technologies, they noted that the rapid expansion of production and use of electric and electronic goods results in an increase in e-waste, and transboundary movements of end-of-life electronic products – even to countries which do not posses the capacity for the environmentally sound management of these materials and waste. This situation requires the urgent attention of the international community in general, and of the OECD countries (which are usually the source of this e-waste), in particular.

Illegal shipments of waste

Unfortunately, illegal shipments of end-of-life materials and products are also rather common. For example, one study found that 51% of inspected transboundary movements of waste within and from the EU area were illegal between 2004 and 2006 (IMPEL, 2006). While some illegal shipments from EU countries stay within Western Europe, many go to developing regions, such as Africa and Asia. The most prominent reasons for these illegal shipments seem to be the lack of enforcement and the high costs of treatment or disposal in the exporting country (IMPEL, 2005).

Although very little is known about the actual volume and number of illegal shipments, their environmental and health impacts may be considerable. In order to be able to reduce these threats, effective compliance and enforcement of existing obligations, as well as increased border controls for shipments of end-of-life materials and products, should be considered.

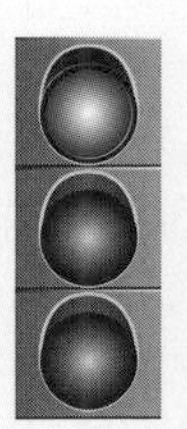

Illegal shipments and unsound management of end-of-life materials and products constitute a considerable risk for human health and the environment.

Non-hazardous industrial waste

The generation of non-hazardous industrial waste has largely stabilised in OECD countries since the late 1990s, as has industrial production (OECD, forthcoming). Reasons for this could include the increased implementation of pollution reduction measures; the economic downturn in early 2000; or the relocation or outsourcing of waste-intensive OECD industry to non-OECD countries, and the subsequent increase in imports of semi-finished or finished products from non-OECD to OECD countries (Bringezu, 2006; Giljum *et al.*, 2007; ETC/RWM, 2007a). Studies of the EU-15 (EEA, 2005) suggest, however, that the volume of non-hazardous wastes from industry will increase by about 60% between now and 2020. There is little or no information available on the management of such waste.

Municipal waste[5] trends and outlook

OECD countries. Table 11.1 provides data and projections from 1980 to 2030 for population, real GDP, and generation of municipal waste for the OECD and its regions. OECD data for municipal waste exist for 1980-2005, and these form the basis of the OECD *Outlook* projections to 2030.

Table 11.1. **Municipal waste generation within the OECD area and its regions, 1980-2030**

	1980	1995	2000	2005	2015	2020	2030	Estimated annual increase 2005-2030
Population (billions) in OECD	1.1	1.2	1.2	1.3	1.3	1.3	1.4	0.4%
(Index)	100	112	116	119	125	127	130	
Real GDP (trillion USD) in OECD	14.4	21.0	23.5	28.0	36.2	40.2	49.0	
(Index)	100	146	163	195	251	279	340	2.3%
Municipal waste generation in OECD								
(million tonnes/year)	395	561	624	653	754	800	900	1.3%
(Index)	100	142	158	165	190	202	228	
(Kg/capita/year)	376	476	512	522	576	600	658	
(index)	100	127	136	137	153	160	175	
OECD Pacific								
(million tonnes/year)	12	15	16	17	19	20	22	1.1%
(Index)	100	124	133	142	154	167	182	
OECD Asia								
(million tonnes/year)	55	68	69	74	84	88	97	1.1%
(Index)	100	124	126	135	153	160	176	
OECD Nafta								
(million tonnes/year)	164	242	272	284	326	347	389	1.3%
(Index)	100	147	166	173	199	212	237	
OECD Europe								
(million tonnes/year)	170	236	267	279	328	352	400	1.5%
(Index)	100	139	157	164	192	207	235	

StatLink http://dx.doi.org/10.1787/257332365178

Source: OECD Environmental Outlook Baseline.

Within the OECD region, the increase in municipal waste generation was about 58% (2.5%/year) from 1980 to 2000, and 4.6% (0.9%/year) between 2000 and 2005 (Table 11.1). During the latter period, the number of OECD households increased by some 4% (0.8%/year) (OECD estimate), population increased by 3.6% (0.7%/year), GDP grew by 11% (2.2%/year), and private final consumption (PFC) rose by 13% (2.6%/year). These data therefore suggest a rather strong relative decoupling of municipal waste generation from economic growth.[6] However, as discussed in Box 11.3, the observed reduction in the growth of municipal waste generation with respect to economic growth between 2000 and 2005 may not really reflect an improving situation.

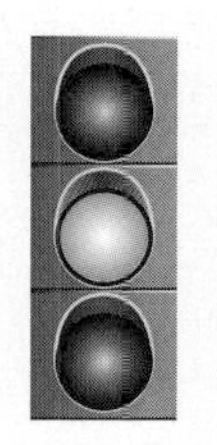

Municipal waste generation is still increasing in OECD countries, but at a slower pace since 2000.

Box 11.3. **Key uncertainties and assumptions**

The GDP and population trends contained in Table 11.1 are from the economic Baseline for this *Outlook* (see Chapters 2 and 3). Historical trends of municipal waste generation in the OECD and its regions have been calculated on the basis of OECD data (OECD, forthcoming). Waste generation projections in Table 11.1 have been extrapolated from observed municipal waste generation between 2000 and 2005. The figures for OECD and its regions in Table 11.2 are partly taken from Table 11.1, and partly calculated on the basis of Table 11.1 figures. The figures for BRIICS countries and the rest of the world (ROW) have mainly been calculated on the basis of municipal waste generation figures found in the literature.

In general, the lack of frequent, consistent and reliable waste data remains a serious problem. For the OECD, only the data on municipal waste allows the establishment of trends, and even these may be questioned. The most recent OECD data (OECD, forthcoming) indicate that the increase in generation of municipal waste has been considerably reduced in 2000-2005, compared to previous years. However, this may not necessarily reflect the real situation, especially given that the conclusion seems inconsistent with recent trends in the economic or social "drivers" of municipal waste generation. It could be that the observed breaks in time series of several countries' data during this time-frame partly cause the lower trends. It is also possible that municipal waste has become "lighter" over the years (with more packaging and related reductions in food waste volumes), but there is no convincing data to support this hypothesis. Another explanation could be that some of the household waste (*e.g.* bulky waste, electric and electronic appliances), as well as commercial waste, are increasingly escaping municipal waste statistics, perhaps because they are returned to retailers or submitted to private industrial waste management systems.

There are also weak indications that the generation of hazardous waste is increasing within the OECD area, but (due to missing time series) this cannot be verified. Concerning the non-OECD countries, the situation is even more unclear, since practically no time-series data exist. Therefore, the values presented in Table 11.2 are "educated guesses" of the current and future status of the non-OECD municipal waste generation and management problem. The order of magnitude is probably broadly correct, but the details remain highly uncertain.

Using these assumptions, and assuming no new policies, the generation of municipal waste is projected to increase from 2005 to 2030 within the OECD region by 38% (1.3%/year). This is less than the projections that were made in 2001, reflecting the recent downturn in

Table 11.2. **Current municipal waste generation in OECD, BRIICS and the rest of the world (ROW)**

	Million tonnes/year	kg/capita	kg/capita/day	Inappropriate collection and/or treatment %	Urbanisation % 2005[h]
OECD (2005)	653	559	1.5		
OECD Pacific	17	702	1.9		
OECD Asia	74	421	1.2		
OECD NAFTA	284	650	1.8		
OECD Europe	279	523	1.4		
BRIICS	~446	151	0.4		
Brazil (2000)[a]	58	339	0.9	60	81
Russia (2004-05)[b]	50	340	0.9	20	73
Indonesia (1995)[c]	56	280	0.8	60	42
India (2001)[d]	108	102	0.3	40	27
China (2004)[e]	154	118	0.3	48	37
South Africa (2005)[f]	20	430	1.2	58	53
ROW (early 2000)[g]	~537	255	0.7		
Total	~1636				

StatLink http://dx.doi.org/10.1787/257332441322

Source:
a) IBGE, 2004.
b) OECD, 1999; Federal Statistical Service of Russia, 2006.
c) World Bank, 1999.
d) Kumar, 2005.
e) OECD, 2007c.
f) Statistics South Africa, 2005; von Blottnitz, 2005.
g) UNEP, forthcoming.
h) PRB, 2005.

municipal waste generation (OECD, 2001a; OECD, forthcoming). In 2001, it had been estimated that there would be some 835 million tonnes of waste being generated annually by 2020; it is now estimated that this figure will be closer to 800 million tonnes. A recent projection by the European Topic Centre on Resource and Waste Management (ETC/RWM, 2007b) seems to support the new estimate, since it projects that (within the EU15) the generation of municipal waste will increase by only 33% to 2030. However, in the new EU member states, municipal waste generation is projected to grow faster than this – by about 66% to 2030. The primary variable explaining the increase in municipal waste generation within the ETC/RWM projections was either the total final private consumption or the sub-categories of final private consumption such as food, beverages and clothing (ETC/RWM, 2007b).

The annual per capita generation of municipal waste within OECD countries seems to be stabilising. It was 556 kg in 2000 and 557 kg in 2005. However, if municipal waste generation increases by 38% (and population by 11%) between now and 2030, as projected here, municipal waste generation per capita will increase to 694 kg in 2030 (up 25% from 2005) (OECD, forthcoming).

Municipal waste management practices vary widely among OECD countries. In the mid-1990s, approximately 64% of municipal waste was destined for landfills, 18% for incineration, and 18% for recycling (including composting) (OECD, 2001a). In 2005, the situation looked rather different, with only 49% of municipal waste being disposed of in landfills, 30% being recycled or composted, and 21% being incinerated or otherwise treated (OECD, forthcoming). Even more remarkable is that not only did the relative share of landfilling decrease considerably within OECD countries during this 10-year period, but the absolute amount of landfilled waste also apparently decreased almost 8% (from 346 to 320 million tonnes per year). Even so, in 2005, seven OECD countries still landfilled more than 80% of their municipal waste, and two did so for almost all of their waste (OECD, forthcoming). On the other hand, six countries landfilled less than 10% of their municipal waste in 2005, and another six countries considerably reduced their landfilling rate between 1995 and 2005.[7]

The OECD (2001a) projected that about 45% of municipal waste within the OECD area would be landfilled in 2020, 25% would be incinerated, and 30% would be recycled or composted. Since most of the current waste management policies, such as diversion of biodegradable waste from landfills within the EU, will be implemented by 2020, it is assumed here that the recycling rate will continue to increase until 2020, but will then gradually slow down in the Baseline situation. In fact, it has been observed in the US that the recycling rate of municipal waste in 2005 was already about 32% – up from 16% in 1995. In EU15, the recycling rate in 2005 was about 41% – up from 22% in 1995. Hence, it is assumed here that recycling will continue increasing within OECD countries, and will reach an average rate of 40% in 2030. However, the recycling rate may increase even more rapidly than this, due to the emerging recognition of the economic and environmental benefits of recycling, compared to other waste management options (Box 11.4).

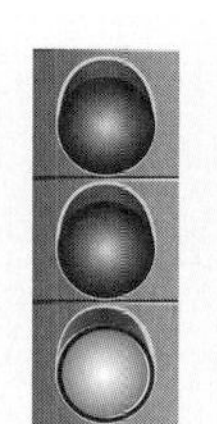

Recovery of municipal waste will continue increasing, while landfilling is expected to considerably decrease by 2030 in OECD countries.

Box 11.4. **Environmental and economic benefits of recycling**

A recent international review of life-cycle analysis (LCA) work on key materials that are collected for recycling clearly demonstrated that recycling usually has more environmental benefits and lower environmental impacts than other waste management options. Whilst the review also highlighted important differences in how the LCAs were constructed, from 188 scenarios that included recycling, the overwhelming majority (83%) favoured recycling over either landfilling or incineration (WRAP, 2006). Recycling can also provide considerable economic and social (*e.g.* increased employment) benefits (*e.g.* US REI, 2001).

Between 1990 and 2004, global emissions of methane increased by more than 10%, with the largest growth coming from Latin America and Asia, whereas emissions from OECD countries as a whole remained almost constant. The latter was mainly due to increased methane recovery from landfills and underground coal mines, and increased diversion of organic waste from landfills to recovery. Within the EU15, the waste sector contributes 2.6% to total EU GHG emissions. Between 1990 and 2004, total waste-related emissions from these countries fell by 33%, mainly due to methane recovery from landfills and wastewater treatment, and diversion of organic waste from landfills to recovery (EEA, 2006). It is estimated that the municipal waste sector in EU15 has a GHG reduction potential of 134 million tonnes of CO_2 equivalent from 2003 to 2020, which represents 11% of the planned total EU15 GHG reductions of CO_2 equivalent. The main contribution (close to 100 million tonnes) to this potential waste-related emissions reduction would come from diversion of organic waste from landfills to recovery (UBA, 2005).

Non-OECD countries. Table 11.2 summarises global municipal waste generation in early 2000. The OECD countries had at that time 18% of the world population, but generated 40% of municipal waste. This situation is changing rapidly – in 2030, the non-OECD area is expected to produce about 70% of the world's municipal waste, mainly due to rising incomes, rapid urbanisation, and technical and economic development (UNEP, forthcoming; World Bank, 2005). It is estimated that in 2030 the mean daily per capita municipal waste generation will be 1.8 kg in the OECD region, about 0.75 kg in the BRIICS countries, and about 0.9 kg in the rest of the world (ROW). Total annual waste generation in 2030 is projected under the Baseline to be about 900 million tonnes for OECD countries, about 1 billion tonnes in the BRIICS countries, and around 1.1 billion tonnes in the rest of the world (ROW).

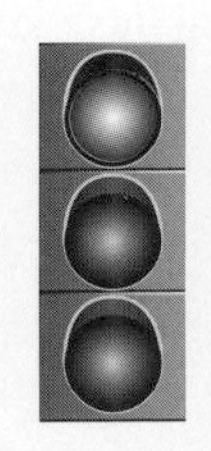

Management of the increasing municipal waste in non-OECD countries will be an enormous challenge in the coming decades.

Some BRIICS countries (Brazil, Russia, Indonesia and South Africa) have already exceeded the estimated mean daily generation of municipal waste (0.75 kg/capita/day) that is projected for 2030 for this grouping of countries, although China and India still have a long way to go in this regard. On the other hand, municipal waste generation in urban China is already some 444 kg/capita/year (1.2 kg/capita/day), while the generation rate in rural areas is largely unknown.[8] However, increasing incomes, rapid urbanisation, population and GDP growth will greatly accelerate municipal waste generation rates in India and China. It is estimated that in 2030 some 60% of the Chinese population will live in urban areas; in India, the urbanisation rate will be about 35%. Thus, in 2030 in China, annual urban municipal waste generation is expected to be at least 485 million tonnes (up 214% from 2004). In India, it will be around 250 million tonnes (up 130% from 2001; World Bank, 2005). This would mean that the daily per capita generation of municipal waste would be 1.5 kg in urban China, and 1.4 kg in urban India.

In Brazil, it is estimated that some 60% of all municipal solid waste is disposed of inappropriately (Leslie and Utter, 2006). In China, 48% of municipal waste is not treated (OECD, 2007c). In India, up to 40% of the municipal waste across urban areas remains uncollected (Joardar, 2000). In Indonesia, Malaysia, Myanmar, Philippines, Singapore, Thailand and Vietnam, 50-80% of municipal waste is simply dumped (UNEP, 2004). Only some 5-30% of municipal waste in these countries is properly landfilled and about the same amount is composted. Informal recycling plays an important role in Latin America and Asia (Nas and Jaffe, 2004; Leslie and Utter, 2006).

Considering the huge increase in municipal waste generation expected in non-OECD countries by 2030, appropriate management of this waste will be an enormous policy challenge. This will likely require that integrated waste management practices be introduced and that the large number of informal waste recyclers be integrated into the official waste management infrastructure (McDougall *et al.*, 2001; World Bank, 2005).

Policy implications

During the late 1990s, it became evident that waste policies which addressed only the end-of-life products and materials were not effective in reducing increasing amounts of waste. This stimulated a new emphasis on integrated waste and materials policies, addressing environmental impacts along the whole life-cycle of products and materials, such as the OECD approach to sustainable materials management (*www.oecd.org/env/waste*). There are also several other examples of "new generation" waste and materials management policies, such as Japan's 3R-approach (Reduce, Reuse, Recycle), China's Circular Economy, the European Union's Thematic Strategy on Sustainable Use of Natural Resources and on Waste Prevention and Recycling (recycling society), and the US's Beyond RCRA: Waste and Materials Management in the Year 2020.

Common elements of these integrated policies are: *i)* targeting primarily the environmental impacts rather than material use *per se*; *ii)* putting wastes into the material balance context of societies; *iii)* taking an integrated life-cycle approach; *iv)* increasing use of economic instruments, such as taxes and tradable permits; and *v)* building partnerships with stakeholders, rather than using command-and-control approaches (OECD, 2001b and c).

These integrated policies normally target the most environmentally harmful products, materials and activities. They place stronger emphasis on material efficiency, redesign and reuse of products, recycling of end-of-life materials and products (*i.e.* considering end-of-life materials and products as resources rather than waste), and environmentally sound management of residues (management standards). These integrated policies also take the "carbon agenda" into account, and therefore put particular emphasis on minimising organic waste in landfills.

Further action should be considered within OECD countries to address the continuous increase in municipal waste generation, as well as to strengthen the implementation of existing waste management policies. This would require examination of the wider use of instrument mixes containing economic, regulatory and information instruments, as well as public-private partnerships, to address the negative environmental impacts of increasing waste amounts and to encourage economically efficient and environmentally sound recovery of waste. Concerning municipal waste, policy instruments such as extended producer responsibility programmes can considerably improve recovery rates and efficiencies, in particular when associated with variable "unit-based waste collection charges" (OECD, 2006).

In the BRIICS countries, strong legislation and policies are broadly in place to support integrated waste management (McDougall *et al.*, 2001; World Bank, 2005). However, their implementation is weak and the waste management infrastructure is still underdeveloped. The result is that approximately 50% of all waste is not collected and/or treated appropriately. A key for the future will therefore be to ensure a higher status for waste and resource-related issues on the political agenda in these countries, as well as increased enforcement of current legislation. OECD countries could make a major contribution by sharing information on the costs and benefits of practices aimed at the environmentally sound management of waste.

In the rest of the world, people living in urban areas (76% in Latin America, but only 30% in South Central Asia) generally have some sort of waste collection. In rural areas, however, there is hardly any organised waste collection (PRB, 2005; Leslie and Utter, 2006). Even if it is collected, the majority of this waste is still not disposed of properly. For example, in Venezuela, some 4.1 million tonnes of municipal waste are generated annually and disposed of at around 200 sites around the country, mostly just in open-air dumps. In Southeast Asia, only an estimated 10-30% of municipal waste is landfilled, the rest is dumped. Informal recycling is flourishing – with severe health impacts (Nas and Jaffe, 2004; Cuadra *et al.*, 2006). For these countries, the priority is therefore to develop strong waste legislation, and to receive access to know-how (and funding) for capacity-building in appropriate waste management infrastructure. OECD countries are already playing an important role in providing funding, developing waste technology and know-how, and transferring them to developing countries (Box 11.5).

Box 11.5. **Technology development and transfer**

Waste prevention, recycling, collection and management technologies are developing rapidly in OECD countries – in part driven by new regulations related to waste disposal changes. Cleaner production, low-waste technologies, automatic garbage bin weighing during collection, automatic sorting of waste, producing oil from plastic, etc. are some examples of new technologies. However, finance and transfer of these innovations to developing countries is still a major challenge. The UNIDO/UNEP National Cleaner Production Centre (NCPC) programme was established in 1994 to build local capacity to implement cleaner production in developing countries and economies in transition. To date, 24 UNIDO/UNEP NCPCs exist around the world (*www.uneptie.org/pc/cp/*). The Basel Convention has also established 14 Regional Centres for training and technology transfer for managing hazardous and other waste, and for minimising their generation in the first place (*www.basel.int/centers/centers.html*).

Notes

1. While it is recognised that the generation and management of other material and waste streams, such as electric and electronic appliances, ships and other problematic material, have gained considerable importance over the past decade and are likely to continue to do so in the years ahead, data for these waste streams are in very short supply, so the focus here is mainly on the key management issues associated with them.
2. Projections of global resource extraction and materials use are only available up to 2020, based on the BASE scenario of the GINFORS model (Giljum *et al.*, 2007).
3. See: *www.unep.org/Documents.Multilingual/Default.asp?DocumentID=496&ArticleID=5447&l=en.*
4. See: *www.basel.int/meetings/cop/cop8/docs/16eREISSUED.pdf.*
5. "Municipal waste is waste collected and treated by or for the municipalities. It covers waste from households, including bulky waste, similar waste from commerce and trade, office buildings, institutions and small businesses, yard and garden waste, street sweepings, the contents of litter containers, and market cleansing waste. The definition excludes waste from municipal sewage networks and treatment, as well as municipal construction and demolition waste" (OECD, 2007b).
6. Decoupling is said to be "relative" when the growth rate of the environmentally relevant variable (*e.g.* waste generation) is positive, but less than the growth rate of the economic variable (OECD, 2002). "Absolute" decoupling occurs when the environmentally relevant variable is stable or decreasing, while the economic variable is growing.
7. Landfilling of secondary waste is not included in these figures (*e.g.* waste from incineration).
8. The figure for China (154 Mt/year) in Table 11.2 may represent only the municipal waste generation in urban areas, rather than the situation in the country as a whole.

References

Blottnitz, von H. (2005), *Solid Waste*, background briefing paper for the National Sustainable Development Strategy, University of Cape Town, South Africa.

Bringezu, S. (2006), "Materializing Policies for Sustainable Use and Economy-wide Management of Resources: Biophysical Perspectives, Socio-economic Options and a Dual Approach for the European Union", *Wuppertal Papers* No. 160, Wuppertal Institute, Wuppertal, Germany.

Cuadra, S.N. *et al.* (2006), "Persistent Organochlorine Pollutants in Children Working at a Waste-disposal Site and in Young Females with High Fish Consumption in Managua, Nicaragua", *AMBIO* 35:109-115.

EEA (European Environment Agency) (2005), *European Environment Outlook,* EEA Report No. 4/2005, Copenhagen, *http://reports.eea.europa.eu/eea_report_2005_4/en.*

EEA (2006), *Annual European Community Greenhouse Gas Inventory 1990-2004 and Inventory Report 2006,* EEA Technical Report No. 6/2006, EEA, Copenhagen, *http://reports.eea.europa.eu/technical_report_2006_6/en.*

ETC/RWM (European Topic Centre on Resource and Waste Management) (2007a), *Environmental Input-Output Analyses based on NAMEA Data: A Comparative European Study on Environmental Pressures Arising from Consumption and Production Patterns,* ETC/RWM Working Paper 2007/2, Copenhagen, *http://waste.eionet.europa.eu/publications.*

ETC/RWM (2007b), *Environmental Outlooks: Municipal Waste,* ETC/RWM Working Paper 2007/1, Copenhagen, *http://waste.eionet.europa.eu/publications.*

Eurostat (2005), *Waste Generated and Treated in Europe: Data 1995-2003,* Eurostat, Luxembourg.

Federal Statistical Service of Russia (2006), *Main Environmental Indicators,* Statistical Bulletin (in Russian), Moscow 2006.

Giljum, S. *et al.* (2007), "Modelling Scenarios Towards a Sustainable Use of Natural Resources in Europe", *Seri Working Papers,* No. 4, January 2007, Sustainable Europe Research Institute, Vienna *www.seri.at/index.php?option=com_docman&task=search_result&search_mode=phrase&search_phrase=PE.SGJ&Itemid=39.*

Huesemann, M.H. (2003), "The Limits of Technological Solutions to Sustainable Development", *Clean Techn. Environ Policy* 5:21-34.

IBGE (Brazilian Institute of Geography and Statistics) (2004), *PNSB 2000* (National Survey of Basic Sanitation in 2000), information received in January 2007 from IBGE.

IMPEL (2005), *IMPEL-TFS Threat Assessment Project: The Illegal Shipment of Waste Among IMPEL Member States,* Project Report, May 2005, *http://ec.europa.eu/environment/impel/tfs_projects.htm.*

IMPEL (2006), *IMPEL-TFS Seaport Project II: International Co-operation in Enforcement Hitting Illegal Waste Shipments,* Project Report September 2004-May 2006, June 2006, *http://ec.europa.eu/environment/impel/tfs_projects.htm.*

Joardar, S.D. (2000), "Urban Residential Solid Waste Management in India: Issues related to Institutional Arrangements", *Public Works Management and Policy,* Vol. 4, No. 4:319-330.

Krautkraemer, J.A. (2005), *Economics of Natural Resource Scarcity: The State of the Debate,* Discussion Paper 05-14, Resources for Future, Washington DC.

Kumar, S. (2005), *Municipal Solid Waste Management in India: Present Practices and Future Challenge,* Asian Development Bank, Manila, *www.adb.org/Documents/Events/2005/Sanitation-Wastewater-Management/paper-kumar.pdf.*

Leslie, K. and L. Utter (eds.) (2006), *Recycling and Solid Waste in Latin America: Trends and Policies 2006,* Raymond Communications, Inc., College Park, Maryland.

McDougall, F. *et al.* (2001), *Integrated Solid Waste Management: A Life Cycle Inventory,* Blackwell Science Ltd, London.

Nas, P.J.M. and R. Jaffe (2004), "Informal Waste Management: Shifting the Focus from Problem to Potential", *Environment, Development and Sustainability* 6:337-353.

OECD (1999), *OECD Environmental Performance Reviews: Russian Federation,* OECD Paris.

OECD (2001a), *OECD Environmental Outlook,* OECD, Paris.

OECD (2001b), *Sustainable Development – Critical Issues,* OECD, Paris.

OECD (2001c), *Policies to Enhance Sustainable Development,* OECD, Paris.

OECD (2002), *Indicators to Measure Decoupling of Environmental Pressure from Economic Growth,* SG/SD(2002)1/FINAL, OECD, Paris, *www.oecd.org/dataoecd/0/52/1933638.pdf.*

OECD (2005), *OECD Environmental Data, Compendium 2004,* OECD, Paris.

OECD (2006), *Impact of Unit-based Waste Collection Charges,* ENV/EPOC/WGWPR(2005)10/FINAL, OECD, Paris, *www.oecd.org/dataoecd/51/28/36707069.pdf.*

OECD (2007a), *Measuring Material Flows and Resource Productivity – An OECD Guide,* OECD, Paris.

OECD (2007b), *OECD Fact Book 2007, Economic, Environmental and Social Statistics,* OECD, Paris.

OECD (2007c), *Environmental Performance Review of China*, OECD, Paris.

OECD (forthcoming), *OECD Environmental Data Compendium,* OECD, Paris.

PRB (Population Reference Bureau) (2005), *2005 World Population Data Sheet*, Population Reference Bureau, Washington, DC, *www.prb.org/*.

Statistics South Africa (2005), *Non-Financial Census of Municipalities for the Year Ended,* Statistical Release P9115, 30 June 2005, *www.statssa.gov.za*.

UBA (German Federal Environmental Agency) (2005), *Waste Sector's Contribution to Climate Protection,* Research Report 205 33 314, UBA-FB III, UBA, Berlin, *www.wte.org/docs/2005AugGermanyclimate.pdf.*

UNEP (United Nations Environment Programme) (2004), *State of Waste Management in South East Asia,* United Nations Environment Programme UNEP/IETC, Paris, *www.unep.or.jp/Ietc/Publications/spc/State_of_waste_Management/index.asp*.

UNEP (forthcoming), *Global Environment Outlook* 4, United Nations Environment Programme, Nairobi.

US REI (2001), *US Recycling Economic Information Study,* A study prepared for The National Recycling Coalition by R. W. Beck, Inc., *www.epa.gov/epaoswer/non-hw/recycle/jtr/econ/rei-rw/pdf/n_report.pdf.*

World Bank (1999), *What a Waste: Solid Waste Management in Asia,* World Bank, Urban Development Sector Unit, East Asia and Pacific Region, *http://web.mit.edu/urbanupgrading/urbanenvironment/resources/references/pdfs/WhatAWasteAsia.pdf.*

World Bank (2005), *Waste Management in China: Issues and Recommendations,* Urban Development Working papers, East Asia Infrastructure Department, Working Paper No. 9, 146 p, World Bank, Washington DC. *http://siteresources.worldbank.org/INTEAPREGTOPURBDEV/Resources/China-Waste-Management1.pdf.*

WRAP (Waste and Resources Action Programme) (2006), *Environmental Benefits of Recycling*, WRAP, Banbury, UK. *www.wrap.org.uk/applications/publications/publication_details.rm?id=698&publication=2838.*

ISBN 978-92-64-04048-9
OECD Environmental Outlook to 2030

Chapter 12

Health and Environment

Without more stringent policies to better address environmental concerns, the adverse health effects of air and water pollution are likely to increase in the future. The economic burden of environmental health is significant in both OECD and non-OECD countries, and recent analysis suggests that health damage associated with air and water pollution represents a significant share of GDP. This chapter explores the health impacts of outdoor air-pollution, unsafe water, sanitation and hygiene, as well as the costs and benefits of policy enhancements in these areas. Improving environmental conditions upstream in order to prevent downstream environment-related health outcomes, is often cost-efficient.

KEY MESSAGES

The *OECD Environmental Outlook* Baseline projects that between 2000 and 2030, premature deaths caused by ground-level ozone will increase by a factor of 4 (see left-hand figure, below) and premature deaths caused by PM_{10} (particulate matter) will increase by more than 2 (see right-hand figure).

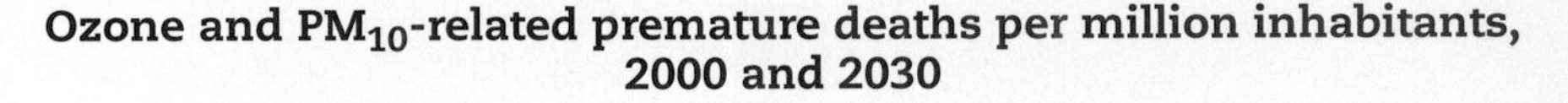

Ozone and PM_{10}-related premature deaths per million inhabitants, 2000 and 2030

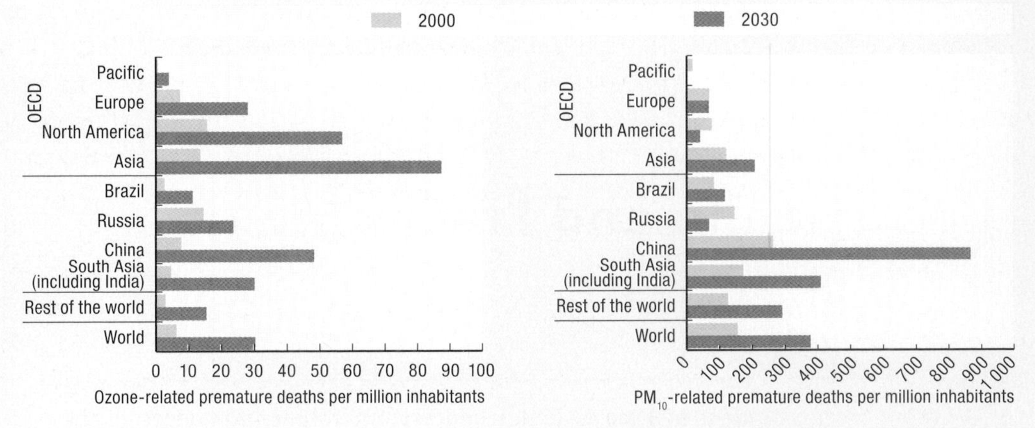

StatLink http://dx.doi.org/10.1787/262616733406

In OECD countries, the health impacts of water-related diseases remain very low except for some countries where water supply and sanitation coverage are still relatively poor.

Improving environmental conditions upstream, in order to prevent downstream environment-related health outcomes, is often cost-efficient.

Policy options

- Continue to support environmental policies as a key way of reducing health problems and healthcare costs caused by environmental degradation.
- Design efficient environmental policies, such as those which target several air pollutants together or which improve water quality and waste treatment at low cost.
- Strengthen OECD countries' air quality policies to further reduce air pollutant emissions. This will strongly decouple emissions from GDP and limit populations' exposure.
- Invest to improve drinking water quality and sewage treatment in OECD countries.
- Increase financing substantially, through both development aid and direct investment, to allow developing countries to achieve the Millennium Development Goal of halving the number of people without access to water and sanitation by 2015.

Consequences of inaction

- Without new (or more stringent) policies to better address environmental health issues, the adverse health effects of the most harmful environmental pollutants (*e.g.* air and water pollution) are likely to increase in the future.

Introduction

Environmental pollution and degradation exert significant pressure on human health. Exposure to air, water and soil pollution, to chemicals in the environment, or to noise, can cause cancers, respiratory, cardiovascular, cerebrovascular and communicable diseases, as well as poisoning and neuropsychiatric disorders. A recent World Health Organization (WHO) study (Prüss-Üstün and Corvalán, 2006) suggests that 24% of the global burden of disease and 23% of all deaths are attributable to environmental factors. The proportion of environment-related diseases in non-OECD countries is higher (24%) than in high income OECD countries (14%).[1]

Although environmental risk factors can affect the health of the whole population, some groups are particularly vulnerable, including children, pregnant women, the elderly and people with pre-existing diseases (Box 12.1). Low-income households are often more exposed to environmental pollution than middle- and high-income households and thus are also more vulnerable (Scapecchi, 2008).

Box 12.1. **Children's health and the environment**

Children are more susceptible to the impacts of environmental pollution than adults (OECD, 2006a). Metabolic activity is higher in children, as their bodies are still developing. Children's bodies respond differently than adults to the same apparent levels of exposure and are less able to metabolise or remove pollutants. In addition, adults and children are exposed to different types of risk, mainly because of their different daily activities. For example, children tend to spend more time outdoors and are more exposed to soil and outdoor air pollution. They are also less aware of the environmental risks surrounding them. As such, children can be exposed to higher levels of pollutants than adults.

Examples of impacts of environmental pollution on children's health include (Tamburlini *et al.*, 2002):

- cancer (*e.g.* skin cancer from excessive exposure to UV radiations or leukaemia resulting from in-utero exposure to pesticides);
- asthma (exacerbated by outdoor air pollution);
- birth defects (from drinking water contaminants);
- neurodevelopmental disorders (resulting from lead poisoning).

Prüss-Üstün and Corvalán (2006) estimate that 33% of diseases among 0 to 14-year-old children can be attributable to environmental factors; this figure increases to 37% for the 0-4 age group.

Despite a large number of actions undertaken in OECD countries to protect children's health from environmental hazards, most existing environmental legislation does not take account of children's specific vulnerability to the various environmental risks.

The economic burden of environmental health is also significant in both OECD and non-OECD countries. Two recent economic studies (Muller and Mendelsohn, 2007; World Bank, 2007) estimated the total health costs of selected environmental risk factors in the US and China, respectively. These analyses suggest that health damage associated with air and water pollution represents a significant share of GDP.

There are cost-efficient environmental interventions available for improving outdoor air and water quality, based on the associated health impacts. Outdoor air pollution and unsafe water, sanitation and hygiene are two key health impacts affecting both OECD and non-OECD countries.

Key trends and projections: outdoor air pollution

Air pollution results from a "cocktail" of several pollutants, such as particulate matter[2] (PM), carbon monoxide (CO), nitrogen dioxide (NO_2), sulphur dioxide (SO_2), ozone (O_3) and volatile organic compounds (VOC), which come from fixed or mobile anthropogenic and natural sources (see Chapter 8 on air pollution). The major man-made sources of air pollution include transport (see Chapter 16 on transport), industries and domestic housing. According to the European Environmental Agency (EEA, 2003), in 2000 the greatest sources of PM_{10} emissions in Europe were energy production (30%), road transport (22%), industry (17%) and agriculture (12%).

Despite significant decreases in emissions of major air pollutant concentrations in recent years (see Chapter 8 on air pollution), many urban areas in OECD countries still suffer from high levels of outdoor air pollution. This is a major concern due to their adverse effects on human health. Concentrations of PM_{10} pollution in most OECD countries, especially in Mexico, Greece and Turkey (World Bank, 2006), still exceed WHO guidelines, which recommend the following maximum levels (WHO, 2006):

- $PM_{2.5}$: 10 $\mu g/m^3$ annual mean
- PM_{10}: 20 $\mu g/m^3$ annual mean
- O_3: 100 $\mu g/m^3$ for daily maximum 8-hour mean
- NO_2: 40 $\mu g/m^3$ annual mean
- SO_2: 20 $\mu g/m^3$ for 24-hour mean.

The health effects of outdoor air pollution can be either acute (*i.e.* resulting from short-term exposure) or chronic (*i.e.* resulting from long-term exposure). They range from minor eye irritations to upper respiratory symptoms, chronic respiratory diseases, cardiovascular diseases and lung cancer, and may result in hospital admission and even death (WHO, 2006). The severity of the effects depends on the pollutant's chemical composition, its concentration in the air, the length of exposure, the synergy with other pollutants in the air, as well as individual susceptibility (Box 12.1).

Projected health impacts from exposure to PM pollution

As part of the *Outlook* Baseline to 2030, the health impacts caused by exposure to PM_{10} were projected, in terms of cardiopulmonary disease (CPD) and lung cancer (LC) in adults, and acute respiratory infections (ARI) in children aged 0-4 years. The relative risks of mortality and the coefficients of the concentration-response functions used in this assessment were taken from Cohen *et al.* (2004). The OECD *Outlook* assessment also assumed that there is no excess risk with PM_{10} below 15 $\mu g/m^3$and no further increase in risk above 150 µg/m^3. National demographic data (such as age groups and disease incidences) have been taken either directly or downscaled (from the regional level to the national level) from the world population prospect (UN, 2005) and the WHO burden of disease project (see Bakkes *et al.*, 2008 for further details).

Mortality-related health damages can be expressed either as "premature deaths" or as "years of life lost". For the year 2000, the *OECD Environmental Outlook* Baseline estimated that exposure to PM_{10} caused approximately 960 000 premature deaths and 9.6 million years of life lost worldwide. The largest contribution to premature deaths came from cardiopulmonary disease in adults (80% to more than 90%, depending on the regional cluster).

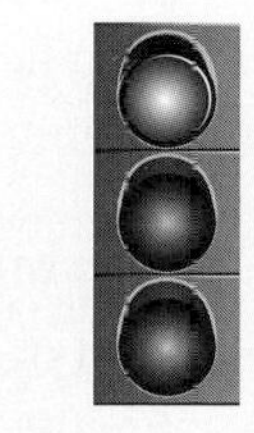

The number of premature deaths caused by PM_{10} and O_3 pollution is projected to significantly increase by 2030.

The *OECD Environmental Outlook* Baseline also projects estimates of the premature deaths associated with PM_{10} pollution to 2030. Figure 12.1 shows premature deaths are likely to increase for most world regions by 2030, even those where PM_{10} levels are anticipated to decrease (for example, the regional clusters OECD Asia, and Brazil – see Chapter 8 on air pollution). For 2030, the worldwide number of premature deaths and years of life lost are estimated to be 3.1 million and 25.4 million, respectively.

Figure 12.1. **Premature deaths from PM_{10} urban air pollution for 2000 and 2030**

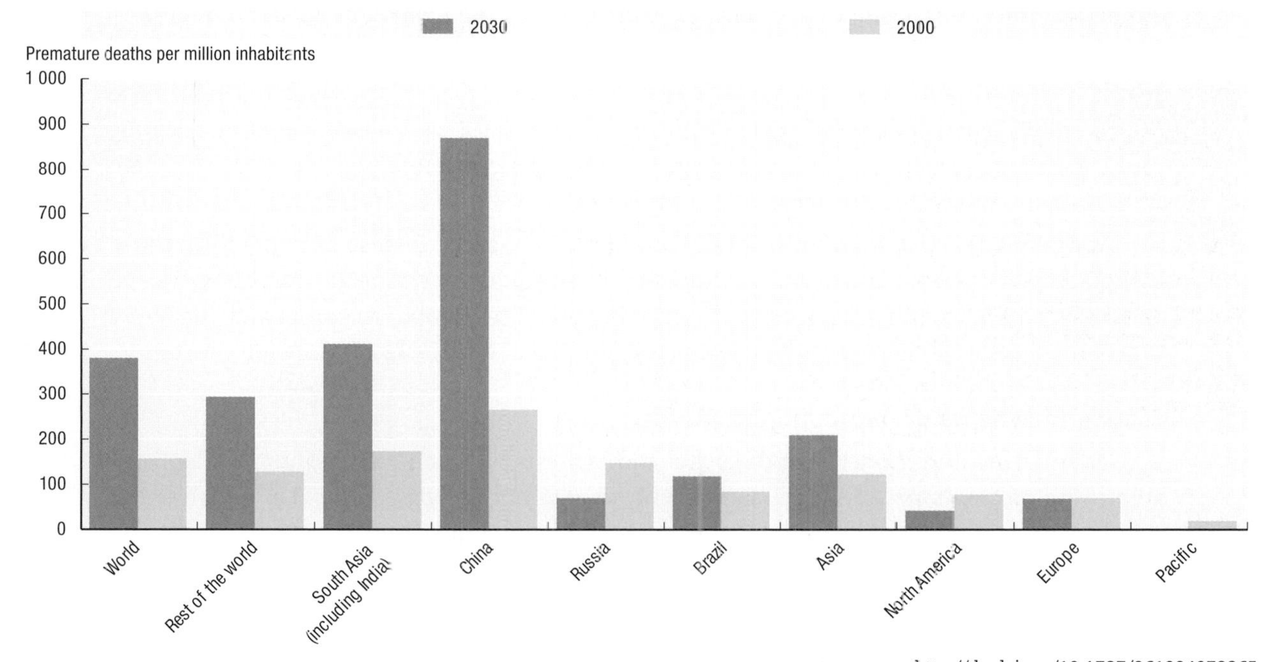

StatLink http://dx.doi.org/10.1787/261324078865

Source: *OECD Environmental Outlook* Baseline.

Factors other than PM_{10} levels and the population's exposure to these levels of pollution are also thought to influence this outcome. Increasing urbanisation, especially in China and South Asia, as well as ageing of the population (since the elderly are generally more susceptible to air pollution) could be potential contributors to this phenomenon.

In 2030, premature deaths from lung cancer are projected to be multiplied by four; premature deaths from acute respiratory infection in children would decrease both in absolute and relative numbers.

The *OECD Environmental Outlook* Baseline also projects large variations between OECD countries, with OECD Asia being relatively more affected than Europe and North America (in the OECD Pacific, no premature deaths from PM_{10} are estimated for 2030 because the concentrations are projected to be below the minimum threshold of 15 $\mu g/m^3$).

Projected health impacts from exposure to ozone

Ground-level ozone (O_3) is also known for its morbidity impacts, such as aggravation of respiratory ailments, but its effect on mortality was only recently clearly identified. Levy *et al.* (2007) reviewed three meta-analyses which found strong evidence of an association between short-term exposure to ozone and mortality. On average, their analysis suggests that a 10 part per billion (ppb) increase in 1-hour ozone maximum level over the year results in a 0.4% increase in short-term mortality, with most risks being concentrated in the warmer months. Other recent multi-city studies also provide strong evidence of the linkages between ground-level ozone and mortality (Gryparis *et al.*, 2004; Health Effects Institute, 2003). However, the estimated effects were of a smaller magnitude than those obtained by the meta-analyses.[3]

These findings suggest that reducing O_3 concentrations would probably help prevent some premature deaths. However, some uncertainties remain, such as the magnitude of risk reduction, the use of ambient data as a proxy for personal exposure and how to separate out other confounding factors (such as particulate matter, see Box 12.2), which inhibit accurate values being placed on the impact of reduced exposure to O_3 on health.

Figure 12.2 shows the *OECD Environmental Outlook* Baseline's estimates of premature deaths attributable to ozone exposure. These estimates suggest a strong increase in these deaths in all regions between 2000 and 2030, although with significant variations among countries. For example, OECD Asia is projected to experience more premature deaths than the European and North American OECD countries.

At first glance, the health impacts of ozone may appear to be less extreme than those associated with exposure to PM_{10}. However, the impact of ozone on health could be underestimated, as the assessment assumed a cut-off of 35 ppb. According to the WHO (2006), it is not possible to identify a lower threshold for the effects of ozone on mortality.

Costs and benefits of improving air quality

Quantifying and monetising the costs and benefits of specific environmental policies help indicate whether these policies would be economically efficient (*i.e.* whether the social benefits outweigh the social costs). As health is one of the most important benefits of environmental regulation (94% according to Muller and Mendelsohn, 2007), its valuation is crucial – albeit controversial (Box 12.2). From an economic perspective, health benefits are usually expressed either as:

- values of avoided cost of illness (COI) which include direct medical costs and indirect costs (*e.g.* productivity losses); or
- willingness-to-pay (WTP) values, which include direct and indirect costs of illness; and intangible aspects, such as pain and suffering, time spent caring for sick people and the impossibility of leisure or domestic activities when sick.

Box 12.2. **Key uncertainties**

Key uncertainties involved in quantifying environmental health impacts include limited epidemiological evidence of the health effects of environmental pollution and degradation (in particular for children), as well as a limited knowledge of the long-term health impacts of air pollution. Differences between outdoor concentrations and personal exposure to air pollutants can also be a source of uncertainty for ground-level ozone (Levy *et al.*, 2007). "Confounding factors" (*e.g.* temperature or other air pollutants) also complicate the linkages between outdoor air pollution and human health. The use of data collected from human bio-monitoring is increasing, and may provide relevant information for policy-makers to develop, adapt and evaluate environmental policies.

Different methodologies can be used to calculate the environment-related burden of disease, resulting in different estimates. The approach presented in this chapter follows the WHO methodology. However, measurement of the burden of disease attributable to specific risk factors can be influenced by many parameters, such as the exposure assessment and associated data sources; the exposure-response relationship; and the method used to extrapolate data to subpopulations (Prüss-Üstün *et al.*, 2003). For outdoor air pollution, the estimates cited for the global level vary by less than two-fold (Cohen *et al.*, 2004). For unsafe water, sanitation and hygiene (WSH), the level of uncertainty may be influenced by the method used, as the relative risk used to determine the fraction attributable to WSH is first estimated at the regional level (14 sub-regions), and then applied to national levels.

However, the lack of robust data and the uncertainties associated with the methodologies should not preclude public intervention. It is important to recognise that this chapter's examples of cost-benefit analyses done on environmental policies generally consider only the health benefits of the policies, thereby potentially under-estimating the total social benefits, which also include benefits to the environment.

Finally, in most health impact assessments, the shape of (and coefficients used in) the concentration response function is mainly based on epidemiological research in North America and Europe. The composition of PM pollution is different in the various regional clusters and, with changing emissions in primary PM and its precursors, its composition will change over time. However, due to lack of information it was assumed here that the relative risk factors do not vary by time or region.

Figure 12.2. **Premature deaths from urban ozone exposure for 2000 and 2030**

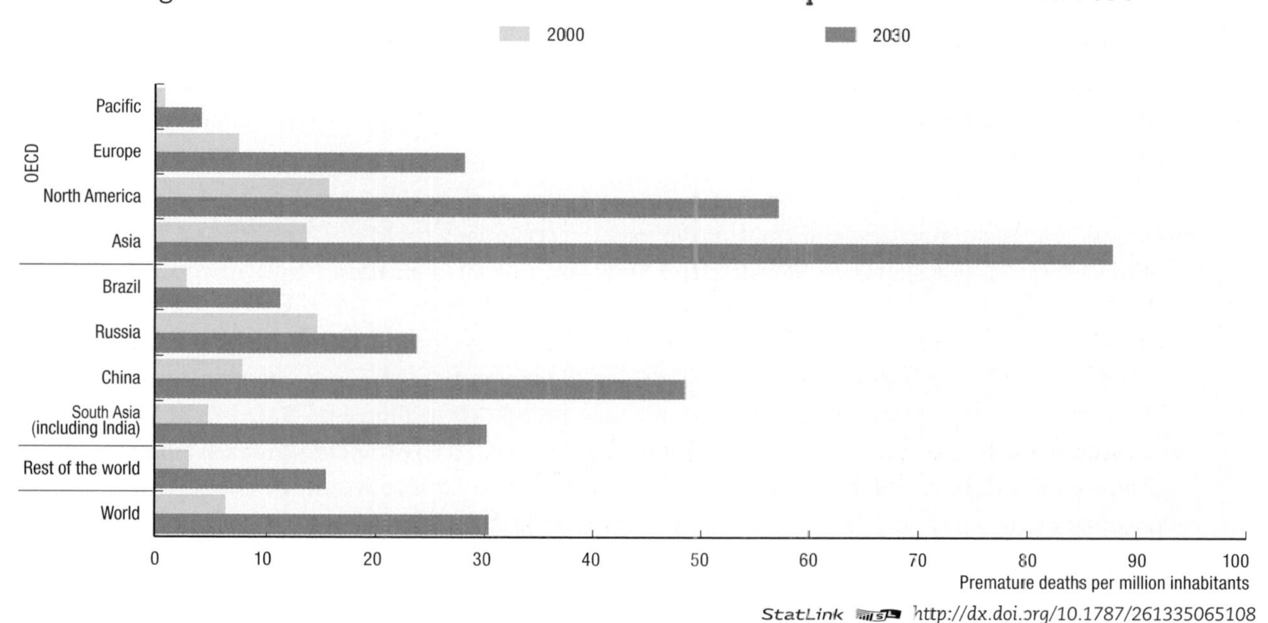

StatLink http://dx.doi.org/10.1787/261335065108

Source: OECD Environmental Outlook Baseline.

The methodologies underlying the estimation of COI and WTP differ widely: the COI is usually estimated *ex post*, whereas the WTP is generally estimated *ex ante* (for a more technical discussion, see OECD, 2006b). Although both measures can be used in policy-making, the use of WTP values is recommended because of its broader coverage. When WTP figures cannot be used, cost-of-illness values should be used instead.

There is an extensive literature on valuing the health benefits of reducing environmental hazards, mainly focusing on adult populations (see OECD, 2006b). For example, Markandya *et al.* (2004) showed that, on average, the French are willing to pay almost USD 600 per year per person to reduce the mortality risks associated with air pollution by 5/1000, while Italians and the British are willing to pay USD 900 and USD 480 respectively for a similar risk reduction.[4] Similarly, Hammitt and Ibarraràn (2002) showed that Mexicans are willing to pay USD 181 per year to reduce mortality risks of air pollution by 1/10 000. These WTP values can be used to evaluate the health benefits of air quality policies in those countries.

More recently, Muller and Mendelsohn (2007) estimated the gross annual damages in the US associated with six different air pollutants: ammonia, nitrogen dioxide, PM_{10}, $PM_{2.5}$, sulphur dioxide and volatile organic compounds. Depending on the approach used for modelling the human health effects, the study estimated the gross annual damages to range between USD 71 billion (0.7% of GDP) and USD 277 billion (2.8% of GDP) per year. A plausible scenario led to an annual total estimate of USD 74.3 billion (0.7% of GDP). Human health impacts made up 94% of the total damages, including premature deaths (USD 53 billion or 71% of global annual damages) and illnesses (USD 17 billion or 23% of global annual damages).

Similarly, the World Bank (2007) used a WTP approach to estimate the health costs of air pollution in China. Total air pollution damages to health represented 3.8% of China's GDP (CNY 519.9 billion – approximately USD 69 billion). The costs associated with mortality were estimated to be CNY 394 billion (approximately USD 52 billion), while the costs associated with morbidity were evaluated at 126 billion YUAN (approximately USD 17 billion). The study also highlights the importance of premature mortality associated with air pollution, which represents 75% of total health costs.

Governments have different policy options available for improving air quality, *e.g.* regulating fuel quality or imposing stringent standards on emissions of specific air pollutants (see Chapter 8 on air pollution). Transport policies (see Chapter 16 on transport) may also be changed in order to better internalise their effects on health and the environment.

A review of the literature on the different policy options for reducing air pollution (Scapecchi, 2008) summarised the anticipated (or observed) costs and benefits of these policy options. For instance, Pandey and Nathwani (2003) estimated that introducing standards for PM_{10}, $PM_{2.5}$ and ozone in Canada would result in net benefits (*i.e.* total benefits minus total costs) of USD 3.6 million per year. As another example, the annual benefits of reducing the sulphur content of fuels in Mexico were shown to be significantly larger than the associated implementation costs (USD 9 700 million and USD 648 million, respectively; Blumberg *et al.*, 2004). Similarly, Stevens *et al.* (2005) projected that introducing filters on vehicles to reduce diesel-related PM pollution in Mexico would be cost efficient with between USD 1 and USD 7 of benefits for every dollar spent (*i.e.* benefit-cost ratios of between 1 and 7). Finally, reducing air pollution in Europe slightly more than is currently done under the *EU Thematic Strategy on Air Pollution* would generate net benefits of between USD 42 billion and USD 168 billion over 20 years (AEA Technology Environment, 2005).

These examples suggest that policies which improve air quality are often cost efficient: the benefits outweigh the costs. Reductions in PM air pollution levels are highly beneficial in health terms, probably due to the relatively strong link between PM exposure and premature mortality. The fact that most of these cost-benefit analyses only consider the health impacts of specific interventions further suggests that total benefits (including benefits to the economy and the environment as well) may be underestimated.

Although there is wide variation between these policy interventions in terms of their benefit-cost ratio (BCR), some lessons can be learned from the Scapecchi (2008) review:

- Less stringent policies can be very effective: the current *EU Thematic Strategy on Air Pollution* has a BCR of 6 to 20.
- Simple policies can sometimes be the most efficient: Mexico's fuel policies that require ultra-low sulphur content have a BCR of 10 to 19.
- There is evidence of a "first mover and laggard" effect: policies introduced recently benefit from the experience of countries which implemented similar policies a few years earlier (*e.g.* BCR of 6-20 for the *EU Thematic Strategy on Air Pollution* in 2005, *versus* the BCR of 3 for Canada-wide standards in 1999).
- Policies targeting several pollutants at the same time are more efficient than single-pollutant policies (*e.g.* BCR of 1.4-20 for the *EU Thematic Strategy on Air Pollution*, which targets PM_{10}, $PM_{2.5}$ and ozone, *versus* a BCR of 1-7 for reducing diesel-related PM emissions in Mexico), suggesting opportunities for economies of scope in abatement policies.
- Total benefits vary across countries, mainly because of GDP differences.
- A comparison of *ex ante* and *ex post* evaluations of environmental policies suggests that *ex ante* costs are often overestimated, while *ex ante* benefits are underestimated due to information failures, partly as a result of strategic behaviour by involved industries (AEA Technology Environment, 2005).

Environmental policies targeting air pollution are generally both cost-efficient and beneficial in health terms. These policies should therefore continue to focus on reducing emissions and concentrations of air (and other environment-related) pollutants that have the most severe impacts on health, such as PM_{10}, $PM_{2.5}$ and O_3. However, although environmental policies themselves do have a significant effect on reducing air pollution, other factors can also contribute, including the role of traffic growth, urban planning, and the behaviour of firms and individuals. Therefore, adopting a more integrated approach by complementing environmental policies with other types of interventions would likely result in improved air quality and health.

Policy scenarios for air pollution: health impacts

The effects that air pollution abatement policies will have on health were simulated in three policy "packages" for the *OECD Environmental Outlook*. Like other simulations for this *Outlook*, the first package (ppOECD) assumed that enhanced air pollution measures are implemented in OECD countries; the second package (ppBRIC + OECD) assumed also that BRIC countries move to a similar air quality target as OECD countries; and the third package (ppglobal)[5] assumed further that other countries eventually adopt comparable air pollution control policies.

The health impacts of these three policy scenarios are summarised in Figure 12.3. Despite anticipated decreases in concentrations in some regional clusters (*e.g.* South Asia), total premature deaths associated with PM_{10} exposure are not projected to be significantly reduced in any of the policy scenarios, although substantial reductions in excess deaths could be anticipated in OECD countries and the Russian region.

Two reasons can explain this result. First, by 2030 the reduction in concentrations would be still limited in some regions (*e.g.* in South Asia and other Asian countries), though further reductions are anticipated by 2050. Second, these health impact assessments truncated PM_{10} concentrations at 150 μg/m^3 (annual mean). Thus, even though concentrations were largely projected to fall, for example from 250 μg/m^3 to 150 μg/m^3, this will not be reflected in the health benefits because of the truncation point. In other words, the difference between "dirty" and "very dirty" air is neglected, although this may underestimate the total number of premature deaths.

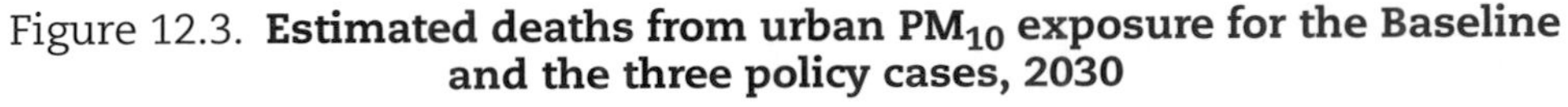

Figure 12.3. **Estimated deaths from urban PM_{10} exposure for the Baseline and the three policy cases, 2030**

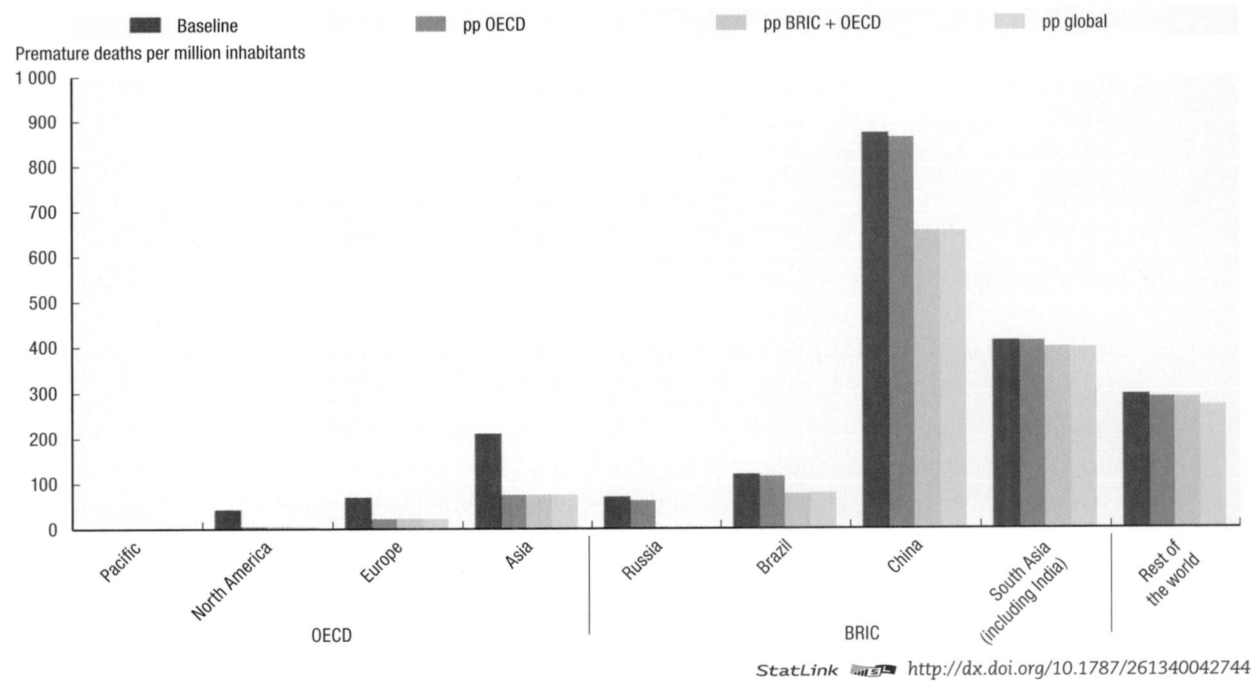

StatLink http://dx.doi.org/10.1787/261340042744

Source: *OECD Environmental Outlook* Baseline and policy simulations.

These simulations suggest that, even if all countries adopted air pollution abatement policies similar (or slightly enhanced) to those currently undertaken in OECD countries, PM_{10} pollution would still cause substantial health damages worldwide by 2030. Efficient environmental policies are therefore needed to reduce these adverse impacts on health.

Key trends and projections: water supply, sanitation and hygiene

Health impacts of unsafe water, sanitation and hygiene

In 2004, 17% of the world's population did not have access to an improved water supply[6] and 41% lacked access to improved sanitation[7] (see also Chapter 10 on freshwater). Although the situation is much better in OECD countries than the rest of the world, unsafe

 OECD ENVIRONMENTAL OUTLOOK TO 2030 – ISBN 978-92-64-04048-9 – © OECD 2008

water, sanitation and hygiene (WSH) remain a major environmental health issue for those parts of the population that are not yet connected to safe water supplies.

The *OECD Environmental Outlook* Baseline projects that worldwide there will be more than 1 billion additional people without access to public sewerage in 2030 than in 2000, when about 4.3 billion people lacked access to public sewerage (see Chapter 10 on freshwater). Sewage treatment is still not universal in many countries, and many OECD countries are facing several new challenges in this field, including a demand for more advanced microbiological purification. Inadequate sewage treatment and poor sanitation mainly result in diarrhoeal diseases caused by bacteria (*e.g.* cholera, E. coli, shigellosis), viruses (*e.g.* norovirus, rotavirus) or protozoan parasites (*e.g.* cryptosporidiosis, giardiasis).

In OECD countries, the health impact of water-related diseases remains very low except for some countries where water supply and sanitation coverage are still poor.

The greatest health risk from pathogenic micro-organisms comes from unsafe drinking water. A number of waterborne disease outbreaks occur mainly in developing countries, but have also occurred in OECD countries in recent years. It is also likely that other waterborne disease outbreaks have occurred without necessarily being recognised as such. The poor effectiveness of current surveillance systems may explain this.

Although chemical contamination (*e.g.* by nitrates and pesticides) is also of concern, the estimates of premature deaths and disability-adjusted life years (DALYs)[8] reported in Figure 12.4 were restricted to water-related diseases caused by pathogenic micro-organisms (due to limited available data).

Figure 12.4. **Percentage of total mortality and burden of disease due to unsafe water, sanitation and hygiene, 2002**

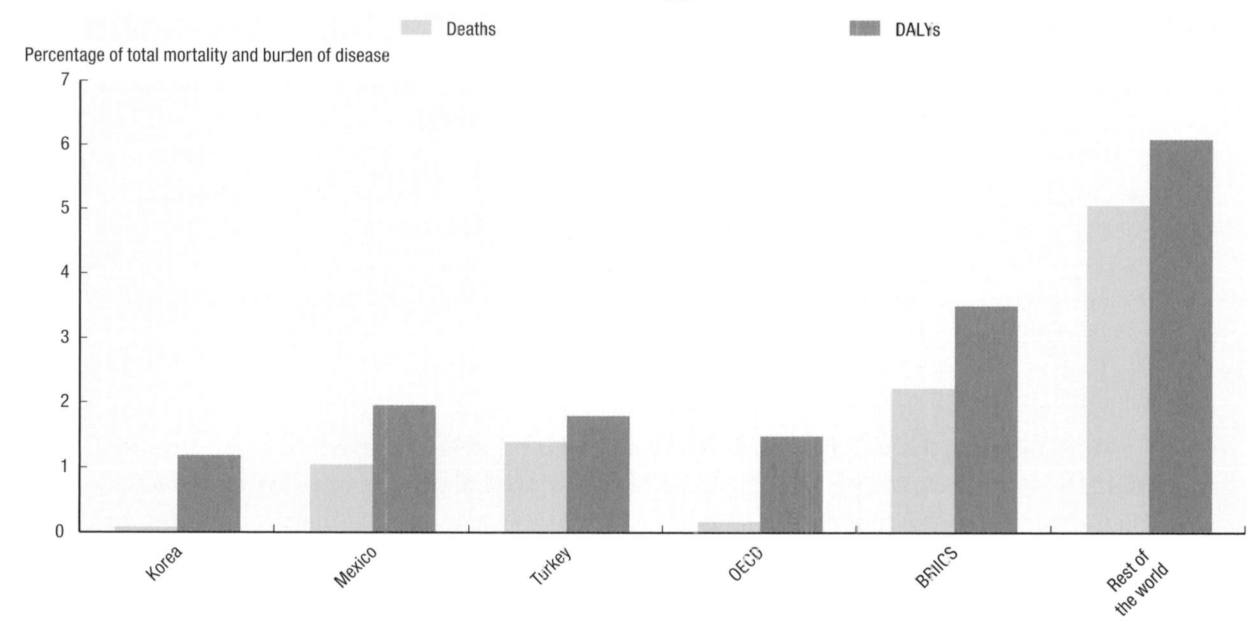

StatLink http://dx.doi.org/10.1787/261343317284

Source: Prüss-Üstün *et al.*, 2004.

In the OECD region, the health impacts of water-related diseases remain very low (around 0.2% of deaths and 0.5% of all DALYs). But some OECD countries – such as Mexico, Turkey and Korea – are affected more than others (Figure 12.4). According to Prüss-Üstün *et al.* (2004), these countries account for 78.6% of the total OECD burden of disease due to unsafe WSH. In the BRIICS[9] countries, unsafe WSH was responsible for 2.2% of all deaths, and accounted for 3.5% of the total burden of disease in these countries in 2002 – of which 87% occurred in India and China alone (Gagnon, 2008).

Unsafe WSH accounts for 3% of all deaths and 4.4% of all DALYs (EEA, 2003; World Bank, 2003). Around 99% of these deaths and DALYs occur in non-OECD countries; 90% of those dying are children. Indeed, unsafe WSH is the world's biggest child killer after malnutrition. When these figures are normalised according to population size, deaths from unsafe WSH in the rest of the world (ROW in Figure 12.4) are 40.5 times higher than in the OECD region, and 2.7 times higher than in the BRIICS region.

The poorest developing countries are significantly affected by poor water supply, inadequate sanitation facilities and insufficient hygiene. The burden of disease due to unsafe WSH could therefore be greatly reduced by implementing appropriate interventions in these countries (Box 12.3).

Box 12.3. **Effectiveness of interventions to reduce the incidence of diarrhoea**

A recent review and meta-analysis assessed the effectiveness of interventions to reduce the occurrence of diarrhoeal diseases in non-outbreak conditions (Fewtrell *et al.*, 2005). The interventions examined included improvements in drinking water, sanitation facilities and hygiene practices in developing countries as well as in several lower income OECD countries (Hungary, Korea, Mexico, Poland and Slovak Republic). The estimated health benefits of these improvements were as follows:

- Water supply interventions, including the provision of new or improved water supply systems or improved distribution (at the public level or household level), can reduce diarrhoea morbidity by up to 25%.
- Water quality interventions, *e.g.* providing water treatment to remove microbial contaminants (at source or within houses) can reduce diarrhoea incidence by between 35% and 39%.
- Sanitation interventions, such as providing some means of excreta disposal (usually public or household latrines), can reduce diarrhoeal morbidity by 32%.
- Hygiene interventions, which include hygiene and health education and the encouragement of hygienic behaviour (like hand washing), can reduce diarrhoeal morbidity by up to 45%.

More recently, the World Bank (2007) produced estimates of the health costs of water pollution in China. Based on a WTP approach, the costs of diarrhoea were estimated to be CNY 14 billion (approximately USD 1.9 billion), while cancer-related costs amounted to CNY 52 billion (approximately USD 6.9 billion). Total costs (CNY 66 billion or USD 8.7 billion) represent 1.9% of GDP, suggesting that the health impacts of water pollution represent a significant economic burden in China.

Costs and benefits of improving water supply, sanitation facilities and hygiene

OECD countries

Economic studies have demonstrated that environmental interventions to improve the quality of drinking water and sanitation facilities can have significant health benefits, reducing mortality and morbidity-related health costs of waterborne diseases. Drinking water quality improvements in the US and recreational water quality improvements through sanitation (*i.e.* sewage treatment) in France, Portugal, the US and the UK show that huge health benefits can result from these interventions (Gagnon, 2008).

Moreover, the health benefits of drinking water quality and sewage treatment often outweigh the costs of policy implementation (Gagnon, 2008). For example, the US Environmental Protection Agency (US EPA, 2006) estimated the annual cost of the *Long Term 2 Enhanced Surface Water Treatment Rule* to improve drinking water quality to be between USD 93 and 113 million, while the annual health benefits range from USD 177 million to 2.8 billion. In this case, even where drinking water quality was already good, water treatment and monitoring interventions still appear to be cost-efficient.

Sewage treatment improvements are also cost-efficient. For example, Georgiou *et al.* (2005) showed that in the UK the total health benefits of the revised *EU Bathing Water Directive* varied between USD 19.3 and 37 billion over a 25-year period, whereas the costs of improvements ranged between USD 3.9 and 8.4 billion. Even though sewage treatment is usually more expensive than drinking water treatment, the former appears to be more cost-efficient than the latter.

Non-OECD countries

Target 10 of the Millennium Development Goals is to halve by 2015 the proportion of people without sustainable access to safe drinking water and basic sanitation.[10] A cost-benefit analysis of this target[11] shows that meeting it would be cost-efficient even though only the health benefits were taken into account. The BCR would be more than 11 if total benefits are considered (such as time saved and productivity gains of being nearer a water supply). The option with the highest BCR in terms of health (3.1) and total benefits (12.5) is minimal water disinfection at the point of use, on top of improved water supply and sanitation facilities (Hutton, personal communication, March 2006). These interventions are still cost-efficient, despite the analysis only considering those benefits which have a market value.

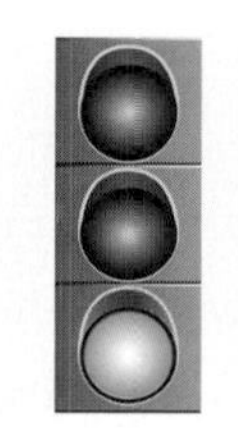

Improving environmental conditions upstream, in order to prevent downstream environment-related health outcomes, is often cost-efficient.

In summary, a large part of the burden of disease due to unsafe WSH can be prevented through cost-efficient environmental policies. Hygiene interventions (*e.g.* awareness campaigns on hand-washing) can also be cost-efficient in developing countries, because these are generally cheaper than water supply and sanitation interventions. Economic studies of water supply and sanitation interventions reviewed in both OECD and non-OECD countries have demonstrated that benefit-cost ratios vary from 1 to 3.1, suggesting significant cost savings for healthcare. Moreover, as these examples focus only on the health benefits, the total social benefits of these interventions (including benefits to the ecosystem) may be underestimated.

Policy implications

The economic evidence shows that there could be significant net benefits from limiting air and water pollution (and more generally environmental degradation), not only for human health, but also for the economy. This is especially significant in low-income OECD and developing countries.

Examples of selected cost-benefit analyses suggest that treating environmental health issues upstream (*i.e.* improving the environmental conditions to prevent environment-related health problems) rather than downstream (*i.e.* treating the health problem) can be cost-efficient. The cost of these interventions is covered (sometimes several times over) by the health benefits they generate. When other benefits are considered (including benefits to the economy and the environment) the benefit-cost ratios of environmental interventions are even larger.

Therefore, environmental policies targeting outdoor air and water pollution could be cost-efficient in the long term. This finding is particularly true for those OECD and non-OECD countries which have significant levels of air pollution and unsafe water, sanitation and hygiene.

OECD countries should therefore:

- Continue to support environmental policies as a key vector for reducing health damages and healthcare costs caused by environmental degradation.
- Strengthen their efforts to further reduce outdoor air pollution emissions to meet the WHO guideline levels (WHO, 2006). Such efforts could include more stringent legislation and implementation of appropriate pollution control policies, cleaner and more efficient energy policies and environmentally sustainable transport policies (see Chapter 16 on transport).
- Commit significant financial resources in the coming decades to upgrading water supply and sanitation infrastructure.
- Improve the effectiveness of surveillance systems for waterborne disease outbreaks.
- Increase international development aid and encourage internal investment towards helping developing countries achieve MDG Target 10.

More specifically, additional efforts will be needed for low-income OECD countries to reach the levels of drinking water quality and sewage treatment currently observed in OECD countries as a whole.

Given current trends in population growth coupled with limitations in connections to water supply, sanitation and sewage, unsafe WSH is expected to continue to have substantial health impacts in developing countries (see WHO/UNICEF, 2006). In addition, the rapid rise in transport and energy use projected for non-OECD countries is likely to increase air pollution levels (see Chapter 8 on air pollution), causing a growing number of health problems in these countries. Emerging environmental challenges, such as climate change and indoor air quality, may result in new, significant impacts on human health in the near future. OECD countries also continue to be concerned about the environment-related health risks of exposure to chemicals (see Chapter 18 on chemicals).

Without sufficient mitigation efforts, the costs of healthcare from environmental pollution are likely to become greater in the years ahead. Appropriate environmental policies should therefore be implemented to address those environmental issues that cause the strongest effects on human health.

Notes

1. The definition of "environment" used in this WHO survey was quite broad, and includes many risk factors not commonly referred to as "environmental", such as injuries (*e.g.* burns, poisoning, falls, etc.), physical inactivity, sexually transmitted diseases, etc.
2. "Particulate matter" refers to fine suspended particulates. These can have a diameter of less than 10 microns (PM_{10}). When their diameter is less than 2.5 microns they are referred to as "fine PM" ($PM_{2.5}$). They are referred to as "ultrafine PM" when their diameter is less than 1 micron (PM_1).
3. For a comparison of the two approaches, see Bell *et al.* (2005).
4. Although the study prepared by Markandya *et al.* (2004) was not specific to outdoor air pollution, the conclusions drawn in the study (and the WTP values estimated therein) can be used to design environmental policies targeting outdoor air pollution.
5. Referred to as the *Environmental Outlook* (or EO) policy package in Chapter 20.
6. "Improved water supply" is defined as having "reasonable access" to protected water resources, such as rainwater collection, protected springs and dug wells, and household connections. It also includes the application of measures to protect the water source from contamination (Hutton and Haller, 2004).
7. "Improved sanitation" involves access to sanitation facilities which allow for safe disposal of excreta.
8. DALYs are defined as the sum of years of life lost and years of life lost due to disability. They provide a measure of the burden of disease associated with a specific health risk (Prüss-Üstün and Corvalán, 2006).
9. Brazil, Russia, India, Indonesia, China and South Africa.
10. The Millennium Development Goal focuses on access to safe drinking water, while the sanitation target was agreed at the World Summit on Sustainable Development in 2002.
11. The findings reported here come from a report commissioned by the WHO which examined the economic benefits of different interventions aimed at improving WSH at the global level (Hutton and Haller, 2004).

References

AEA Technology Environment (2005), *CAFE CBA: Baseline Analysis 2000 to 2020*, Report to the European Commission DG Environment, Brussels.

Bakkes, J. *et al.* (2008), *Background Report to the OECD Environment Outlook: the Modelling Baseline*, Netherlands Environmental Assessment Agency (MNP), Bilthoven, The Netherlands.

Bell, M.L., F. Dominici and J.M. Samet (2005), "A Meta-Analysis of Time-Series Studies of Ozone and Mortality with Comparison to the National Morbidity, Mortality, and Air Pollution Study", *Epidemiology*, Vol. 16, No. 4, pp. 436-445.

Blumberg, K., M.P. Walsh and D. Greenbaum (2004), "The Potential Benefits of Reducing Sulphur in Gasoline and Diesel Fuel and Tightening New Vehicle Standards in Mexico", Presentation made at the *Roundtable on Sulphur in Fuels and Tailpipe Standards*, April 13, 2004, Mexico.

Cohen, A.J. *et al.* (2004), "Urban Air Pollution", in M. Ezzatti, A.D. Lopez, A. Rodgers and C.U.J.L. Murray, (eds.) *Comparative Quantification of Health Risks: Global and Regional Burden of Disease due to Selected Major Risk Factors*, Vol. 2, pp1353-1433, World Health Organization, Geneva.

EEA (European Environmental Agency) (2003), *Europe's Environment: the Third Assessment, Environmental Assessment Report*, No. 10, European Environment Agency, Copenhagen.

Fewtrell, L. *et al.* (2005), "Water, Sanitation, and Hygiene Interventions to Reduce Diarrhoea in Less Developed Countries: a Systematic Review and Meta-analysis", *The Lancet Infectious Diseases*, 5: 42-52.

Gagnon, N. (2008), *Background Report to the OECD Environmental Outlook: Unsafe Water, Sanitation and Hygiene*, OECD, Paris, forthcoming.

Georgiou, S., I.J. Bateman and I.H. Langford (2005), "Cost-benefit Analysis of Improved Bathing Water Quality in the United Kingdom as a Result of a Revision of the European Bathing Water Directive", in R. Brouwer and D. Pearce (eds.), *Cost-benefit Analysis and Water Resources Management*, Edward Elgar, Cheltenham, UK.

Gryparis, A. B. Forsberg, K. Katsouyanni *et al.* (2004), "Acute Effects of Ozone on Mortality from the Air Pollution and Health: A European Approach' Project", *American Journal of Respiratory Critical Care Medicine*, Vol. 170, pp. 1080-1087.

Hammitt, J.K. and M.E. Ibarraràn (2002), "Estimating the Economic Value of Reducing Health Risks by Improving Air Quality in Mexico City", newsletter on *Integrated Program on Air Pollution*, Vol. 2, Fall 2002, Massachusetts Institute of Technology.

Health Effects Institute (2003), *Revised Analyses of Time-Series Studies of Air Pollution and Health: Revised Analyses of the National Morbidity, Mortality, and Air Pollution Study, Part II, Revised Analyses of Selected Time-Series Studies*, Cambridge, MA.

Hutton, G. and L. Haller (2004), *Evaluation of the Costs and Benefits of Water and Sanitation Improvements at the Global Level*, Water, Sanitation and Health, Protection of the Human Environment, World Health Organization, Geneva.

Levy, J., J. Schwartz and J.K. Hammitt (2007), "Mortality Risks from Ozone Exposure", *Risk in Perspective*, Vol. 15, No. 2.

Markandya, A., A. Hunt, R. Ortiz, and A. Alberini (2004), *EC NewExt Research Project: Mortality Risk Valuation – Final Report – UK*, Brussels, European Commission.

Muller, N.Z. and R. Mendelsohn (2007), "Measuring the Damages of Air Pollution in the United States", *Journal of Environmental Economics and Management*, Vol. 54, pp. 1-14.

OECD (2006a), *Economic Valuation of Environmental Health Risks to Children*, OECD, Paris.

OECD (2006b), *Cost-benefit Analysis and the Environment – Recent Developments*, OECD, Paris.

Pandey, M.P., and J.S. Nathwani (2003), "Canada Wide Standard for Particulate Matter and Ozone: Cost-Benefit Analysis Using a Life Quality Index", *Risk Analysis*, Vol. 23, No. 1, pp 55-67.

Prüss-Ustün, A. and C. Corvalán (2006), *Preventing Disease through Healthy Environments – Towards an Estimate of the Environmental Burden of Disease*, World Health Organization, Geneva.

Prüss-Üstün, A. *et al.* (2003), "Introduction and Methods: Assessing the Environmental Burden of Disease at National and Local Levels", *WHO Environmental Burden of Disease Series*, No. 1, World Health Organization, Geneva.

Prüss-Üstün, A., D. Kay, L. Fewtrell and J. Bartram (2004), "Unsafe Water, Sanitation and Hygiene", in M. Ezzati, *et al.* (eds.), *Comparative Quantification of Health Risks, Global and Regional Burden of Disease attributable to Selected Major Risk Factors*, World Health Organization, Geneva.

Scapecchi, P. (2008, forthcoming), *Background Report to the OECD Environmental Outlook: Health Costs of Inaction with Respect to Air Pollution*, OECD, Paris.

Stevens, G., A. Wilson and J.K. Hammitt (2005), "A Benefit-Cost Analysis of Retrofitting Diesel Vehicles with Particulate Filters in the Mexico City Metropolitan Area", *Risk Analysis*, Vol. 25, No. 4, pp. 883-899.

Tamburlini, G., O.S. von Ehrenstein and R. Bertollini (eds.) (2002), "Children's Health and the Environment: A Review of Evidence", *Environmental Issue Report* No. 29, World Health Organization Regional Office for Europe and European Environment Agency, Copenhagen.

UN (2005), *World Population Prospects: The 2004 Revision. CD-ROM Edition – Extended Dataset*, (United Nations publications, Sales No. E.05.XIII.12), United Nations, Department of Economic and Social Affairs, Population Division, New York.

US EPA (2006), "National Primary Drinking Water Regulations: Long Term 2 Enhanced Surface Water Treatment Rule; Final Rule", *Federal Register*, Vol. 71, No. 3, pp. 653-786.

WHO (2006), *Air Quality Guidelines; Global Update 2005: Particulate Matter, Ozone, Nitrogen Dioxide and Sulphur Dioxide*, World Health Organization regional office for Europe, Copenhagen, Denmark.

WHO/UNICEF (2006), *Joint Monitoring Programme for Water Supply and Sanitation*, *www.wssinfo.org/en/welcome.html*, accessed October 2006.

World Bank (2003), "Water, Sanitation and Hygiene", *At a Glance Series*, November 2003, available at: *http://siteresources.worldbank.org/INTPHAAG/Resources/AAGWatSan11-03.pdf.*

World Bank (2006), *World Development Indicators*, World Bank, Washington DC.

World Bank (2007), *Cost of Pollution in China – Economic Estimates of Physical Damages*, World Bank, Washington DC.

ISBN 978-92-64-04048-9
OECD Environmental Outlook to 2030

Chapter 13

Cost of Policy Inaction

This chapter provides information on the "costs of policy inaction", i.e. the costs associated with the negative environmental impacts of the existing policy framework. It highlights three key environmental challenges: health impacts of water and air pollution; fisheries management; and climate change. Estimates of aggregate "costs of inaction" can help to identify important environmental policy problems, but they are not sufficient on their own to determine policy priorities. Non-linear impacts, including the existence of ecological thresholds and irreversible changes, can have significant effects on the total costs of inaction.

KEY MESSAGES

The costs of policy inaction in a number of environmental areas are significant and are already affecting economies in a manner which shows up in market prices and national accounts both directly and indirectly. For example:

- The costs of inaction on *water pollution* are especially high in developing countries, where the health impacts of inadequate water supply and sanitation are particularly acute.
- The costs of inaction associated with *air pollution* are as much as a few percentages of GDP in the US, the EU, and China. Many of these costs are not reflected in market prices or national accounts (*e.g.* "pain and suffering" through poor health).
- While the costs of *unsustainable natural resource management* first affect those who previously exploited the (now-depleted) resources, others may also bear significant costs. For example, large sums of public finance have been used to support unemployed fishers and to facilitate sectoral adjustment as fish stocks have declined.
- The estimated costs of inaction associated with *climate change* vary widely, according to coverage of issues, modelling and valuation approaches. Assuming that emissions remain unmitigated, estimated costs range from less than 1% of global output, to more than 10%. Existing estimates are still partial, however, often excluding, for example, costs associated with increases in extreme weather events due to climate change.

Environmental policy action in the OECD and elsewhere has begun to limit environment costs of inaction, making these costs generally lower than they would otherwise have been.

Key policy and analytical issues

- Assess both the costs of inaction and the costs of the associated interventions to determine policy priorities. Estimates of aggregate "costs of inaction" can help to identify important environmental policy problems, but they are not sufficient on their own.
- Remember that non-linear impacts, including the existence of ecological thresholds and irreversible changes, can have significant effects on the total costs of inaction.
- Consider the use of declining (but not zero) discount rates to address uncertainties about both long-term environmental impacts and economic development.

Introduction

This chapter summarises issues related to estimating the "costs of policy inaction", with particular focus on three key environmental challenges: health impacts of water and air pollution; fisheries management; and climate change.

"Costs of inaction" are understood here to mean the costs associated with the negative environmental impacts that result from the existing policy framework. In general, OECD countries have well-developed policy frameworks in place to address significant environmental impacts. Therefore "inaction" often already incorporates significant levels of policy intervention, as in the areas of air and water pollution. But there are still "residual" impacts from existing policies, and these can generate significant costs. There are also areas in which the policy regime is less well-developed. For instance, in many cases, inaction from past years may have left a significant legacy (*e.g.* contaminated sites, accumulated stock of greenhouse gases, unregulated groundwater extraction). And there are likely to be new challenges emerging in the future.

This chapter does not include a discussion of the costs of implementing the existing policy framework, nor of strengthening this framework. However, it is clear that for all environmental concerns, there is a point at which the economic costs of reducing adverse environmental impacts will exceed the benefits. There are some areas for which this may already be the case, particularly if the policies used to address the environmental concern in question are badly-designed. Efficient environmental policy depends upon carefully balancing the marginal benefits and costs of that policy, as well as choosing the most efficient policy instrument.

Environmental policy frameworks in the OECD and elsewhere have begun to limit environmental costs of inaction, making these costs generally lower than they would otherwise have been.

From the perspective of a policy-maker considering introducing new environmental policies, the most useful approach is to assess the *marginal* social costs and benefits associated with an incremental change in environmental quality, relative to the current policy situation (*i.e.* the Baseline). This approach can provide information that can be directly used in decisions about the allocation of scarce resources. However, estimates of the *total* costs of inaction have significant value in terms of highlighting the economic impacts of not addressing pressing environmental problems. It is these latter (total) costs that are the main focus of this chapter.

The total costs of environmental policy inaction involve several different types of costs (Figure 13.1). These include public finance expenditures (*e.g.* health service costs, restoring contaminated sites); direct financial costs borne by households and firms (*e.g.* increased insurance costs, reduced productivity in resource-based sectors); indirect costs, such as those which arise through markets affected by environmental factors (*e.g.* employment markets, real estate markets); and social welfare costs, which are not reflected in market prices or national accounts at all – including some non-use values of environmental damage (*e.g.* ecosystem degradation).

Several different units (or metrics) can be used to describe the costs of inaction, but the broadest distinction that can be made is between "physical" (ecological, health, etc.) metrics and "monetary" metrics (*e.g.* willingness to pay). However, even this distinction is somewhat artificial, since assessing the former is always a precursor to assessing the latter.

The costs of policy inaction in a number of environmental areas are significant and are already affecting OECD economies in a manner which shows up in market prices and national accounts, both directly and indirectly.

Inaction on a particular environmental concern is likely to lead to a varied set of impacts. For instance, unconstrained climate change will eventually lead to aggregate productivity losses in agriculture and food security problems, water stress, sea-level rise and risks to coastal settlements and summer tourism, loss of human life and sickness due to extreme heat events, loss of biodiversity and other ecosystem services. An impact assessment will give some indication of the nature and size of these impacts, expressed in different units, *e.g.* reduced agricultural yields in m^3; millions of people at risk of food or water shortages; biodiversity loss in terms of number of species threatened; lost tourism days; etc. A key challenge will therefore be to estimate the physical linkages between human actions and environmental change.

Figure 13.1. **Defining the "costs of inaction" of environmental policy**

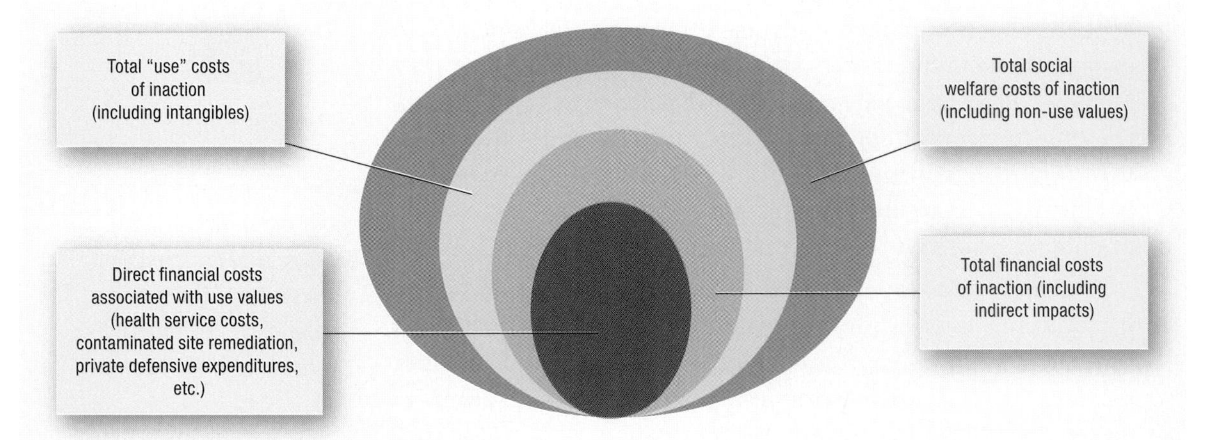

Even if these relationships are known with precision, the non-commensurability of the units involved will mean that they cannot easily be aggregated. Taking the additional step to try to value these impacts in monetary terms therefore allows comparisons across types of environmental impact (*i.e.* loss of biodiversity and human health impacts) using a common metric, and also provides a basis for later comparing the benefits of inaction (*i.e.* avoided investment and other costs) with the costs of inaction.[1]

Actually taking this "valuation step" requires care, since many environmental impacts do not have a readily identifiable market value. Two approaches can be used to place a

 OECD ENVIRONMENTAL OUTLOOK TO 2030 – ISBN 978-92-64-04048-9 – © OECD 2008

value on environmental assets at this stage: revealed preferences and stated preferences. In the case of revealed preferences, efforts are made to derive the value of environmental assets from behaviour in existing markets for "associated" goods and services. For instance, the cost of polluted air may be reflected indirectly in real estate markets. Efforts to value environmental assets through stated preference techniques posit a hypothetical market, for which respondents are requested to value changes in environmental conditions directly (Pearce *et al.*, 2006).

Issues in valuation (key assumptions and uncertainties)

Dealing with the very long run adds an additional level of complexity to estimating the "cost of inaction". Carbon dioxide emitted today has an atmospheric lifetime of over 200 years; air pollutants to which people are exposed today can generate adverse health impacts in 50-60 years; over-exploited fish stocks can take decades to recover (if ever). Costs today also have a higher value than those borne in the future, both because of "pure time preference" (preference for immediate over postponed consumption) and "declining utility of income" (with growing per capita consumption). The further into the future a cost occurs, the lower the weight that will tend to be attached to it. Indeed, the estimated present value of the costs of inaction can vary by orders of magnitude with small changes in the discount rate that is applied.[2] Some people even find the practice of discounting morally unacceptable, because it seems to suggest that future costs are less important than present ones, and is therefore unfair to future generations. Temporal considerations such as these lie at the heart of concerns about climate change, as well as the fisheries management problem, and the choice of a particular discount rate will determine to a great extent the estimated (present) value of the (future) damages.

Environmental pressures can also embody complicated non-linear impacts, including thresholds and irreversible changes. Three issues seem to be especially important in this regard:

- *Cumulative effects:* Some environmental impacts will become significantly greater as a result of cumulative environmental pressures over time. Many health-related impacts exhibit such an effect, such as bio-accumulation of hazardous substances in the food chain.
- *Thresholds:* Impacts may increase sharply once a particular level (threshold) of environmental pressure is exceeded. In the area of climate change, thermohaline circulation is one example; in effect, there may be a "tipping point" after which an inversion might arise, with significant implications for the total costs of inaction.[3]
- *Irreversible changes:* While some environmental impacts are potentially "reversible" (allowing for the restoration of environmental conditions to their prior state), there are many areas in which this is not the case (once degraded, environmental values are lost permanently). Species loss associated with unsustainable fisheries management is one example.

In the presence of such non-linear effects, the costs of preventing environmental degradation in the first place (mitigation) will be less than the costs of addressing the impacts of the environmental problem once it has occurred (restoration). For many types of impacts – and particularly for those involving irreversible changes – it is not possible to restore the environment to its previous state.

Uncertainty can also complicate efforts to value the cost of inaction. In some cases the probabilities of different outcomes may be known. Different weights can then be attached to these outcomes, depending upon the probability of their arising. However, some forms of uncertainty are more fundamental than this, so it may not even be possible to assign credible probabilities to different possible environmental outcomes. For instance, there is considerable uncertainty about the likelihood of certain catastrophes arising as a result of climate change, and the available information is insufficient to posit probabilities for them. In cases where probabilities can *not* be reasonably attached to different outcomes, sensitivity analysis will be appropriate, in which different values are assumed for key parameters.

Another important factor concerns the treatment of the distributional impacts of environmental degradation. Different environmental impacts can affect individual countries (and individuals within individual countries) very differently. In some cases, one group of individuals may benefit, while others will bear costs. There are good ethical and political reasons (*i.e.* social aversion to inequality) to weight impacts relatively more heavily if they particularly affect poorer households. These issues are particularly relevant for climate change, where equity weighting will have a significant impact on estimated costs. However, social concerns may also relate to specific communities above and beyond the distributional implications in terms of income levels. In the area of fisheries management, specific concerns of this kind are also common (*i.e.* employment in fishing communities).

Finally, valuing the costs of environmental policy inaction will depend on how households, firms and farmers, among others, are likely to respond in the face of changing environmental conditions. This "adaptation" can take many forms, and can arise spontaneously (or endogenously). For example, with changing temperatures and precipitation due to climate change, farmers may change their choice of inputs, crops and tilling practices. With rising sea levels and more frequent extreme weather events, there are likely to be investments made in protective infrastructure and changing spatial patterns of development. In the case of local air pollutants or contaminated sites, choices related to residential location will be affected. With groundwater depletion, alternative sources of water (and alternative means of livelihood) will be explored. Assuming that households, firms and farmers are completely "myopic" and do not adjust in any way to changing environmental conditions is, of course, unrealistic, and will likely result in a significant overestimate of the "costs of inaction".

Selected examples of the costs of inaction

Drawing on OECD (2008a and b), this section highlights the costs of inaction in three areas of environmental policy: *i)* health impacts from air and water pollution; *ii)* fisheries management; and *iii)* climate change. A few examples from other areas are noted at the end of the section.

Air pollution, water pollution and the health "costs of inaction"

The costs of inaction in the area of air and water pollution include a wide variety of "use" (*e.g.* the effects of ambient ozone on agricultural productivity) and "non-use" values (*e.g.* the existence value of affected species habitats). These costs can be further distinguished between costs which are generally reflected in existing "market" prices for different goods and services (*e.g.* lost employee productivity, medical costs, increased raw water treatment costs) and those which are not (*e.g.* health costs in terms of "pain and suffering").

Table 13.1. **Selected types of costs related to air and water pollution**

Air pollution	Water pollution
Material damages (including cultural heritage)	Increased drinking water treatment
Reduced agricultural yields	Reduced commercial fish stocks
Polluted freshwater sources	Reduced recreational opportunities
Reduced visibility	Loss of biodiversity
Loss of biodiversity	Adverse health impacts
Adverse health impacts	

Table 13.1 illustrates the diversity of impacts that are involved. While all impacts from policy inaction in the area of water and air pollution are potentially difficult to value, the most difficult are probably those relating to ecosystems (*e.g.* airsheds, water courses) which are not directly related to some downstream economic activity. Valuation of some of the costs of inaction associated with human health (*i.e.* mortality) can also be very controversial.[4]

Pearce *et al.* (2006) suggested that the health-related costs are typically more than 80% of the total costs of air pollution (and sometimes much more). They also found that reduced health impacts were at least one-third (possibly extending to nearly 100%) of the total social benefits of pollution control. However, only a sub-set of the non-health costs are usually included in studies in which the health costs exceed 90%. For instance, in a study by Dziegielewska and Mendelsohn (2005), it was found that estimated ecosystem and cultural heritage costs comprise more than 13% of total damage; these costs were not even included in many of the other studies that were reviewed.

Many intangible health costs of environmental degradation are difficult to value, and may not be reflected in any market. For instance, the "personal pain and suffering" associated with being ill will not be reflected in financial expenditures.[5] Where intangible costs are significant – and the empirical evidence suggests that they frequently are – it is particularly important to rely on stated preference techniques (OECD, 2008a and b).

In a study of acute cardio-respiratory cases in Canada, Stieb *et al.* (2002) estimated that, for some impacts (*e.g.* emergency department visits, asthma symptom days, etc.), "pain and suffering" represented 40% or more of the total health costs of particulate matter. In a French study, Rabl (2004) found that, for other types of impacts attributable in part to pollution levels (*e.g.* cancer), the proportion of costs represented by "pain and suffering" may even exceed 90%.

Health impacts of water pollution[6]

Table 13.2 summarises the main health effects of selected water pollutants. The principal sources include municipal wastewater collection and treatment systems, runoff from agricultural practices, and effluent from manufacturing facilities (see Chapter 10 on freshwater). Particular industrial sectors in which the potential contribution to water pollution is significant include the chemicals sector, the food and beverage sector, and the pulp and paper sector. Mining and the mineral processing sectors can also have significant implications for water quality, as can direct household discharge of hazardous substances into drains.

The policy framework for regulating industrial point sources of water pollution is well-developed in most OECD countries, although some pollutants such as heavy metals and chlorinated solvents remain a concern. Increasing attention is being paid to "non-point"

Table 13.2. **Health effects associated with selected water pollutants**

	Disease/pollutant	Health impacts
Bacterial	Amoebic dysentery	Abdominal pain, diarrhoea, dysentery
	Campbylobacteriosis	Acute diarrhoea
	Cholera	Sudden diarrhoea, vomiting. Can be fatal if untreated
	Cryptosporidiosis	Stomach cramps, nausea, dehydration, headaches. Can be fatal for vulnerable populations
Chemical	Lead	Impairs development of nervous system in children; adverse effects on gestational age and foetal weight; blood pressure
	Arsenic	Carcinogenic (skin and internal cancers)
	Nitrates and nitrites	Methaemoglobinaemia (blue baby syndrome)
	Mercury	Mercury and cyclodienes are known to induce higher incidences of kidney damage, some irreversible
	Persistent organic pollutants	These chemicals can accumulate in fish and cause serious damage to human health. Where pesticides are used on a large scale, groundwater gets contaminated and this leads to the chemical contamination of drinking water.

Source: EEA/WHO-Europe, 2002.

sources such as agricultural runoff, which are more difficult to regulate. In addition to efforts to reduce run-off of organic pollutants from fertilisers and manure, organophosphates and carbonates from pesticides are a concern.

The percentage of the population connected to sewerage systems has increased in OECD countries in recent decades (see Chapter 10 on freshwater). However, there are still deficiencies in collection and treatment systems in some countries. Total investment in the water sector for the 30 OECD countries – which already exceeds USD 150 billion per year (over 0.5% of GDP) – is likely to increase further in the years ahead (OECD, 2001).

The studies reviewed in OECD (2008a&b) suggest that national measures to reduce agricultural runoff and storm water management – including introducing targeted measures to reduce a variety of different pollutants (*i.e.* arsenic, nitrates, etc.) – could yield health benefits in excess of USD100 million in large OECD economies. Many of these estimates are *lower-bound* estimates, since they are obtained from cost-of-illness studies that do not account for "pain and suffering". In some cases, the non-financial opportunity costs for caregivers (and others) are not included either.

In a study of the Chesapeake Bay, Poor *et al.* (2007) found that a one mg/litre increase (approximately 8%) in total suspended solids resulted in a fall in coastal property prices of USD 1 086 (approximately 0.5%). For dissolved inorganic nitrogen, a one mg/litre change (300%) resulted in a USD 17 642 fall (approximately 9%). Gibbs *et al.* (2002) found that a one metre decrease in underwater visibility in New England led to a decrease in property values of 6%.

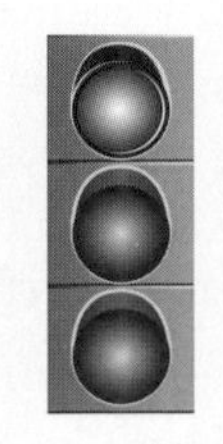

The costs of inaction with respect to water pollution are especially high in developing countries.

In non-OECD countries, the costs of inaction with respect to unsafe water supply and sanitation are particularly acute. At the global level, about 1.1 billion people still do not have access to a safe water supply; 2.6 billion people do not have access to adequate sanitation facilities (WHO/UNICEF, 2006). The associated health impacts are alarming: 1.7 million deaths per year, of which 90% are children under 5 years of age (see also Chapter 12 on health and environment). Indeed, unsafe WSH is the world's biggest child killer after malnutrition (Prüss-Üstün *et al.* 2004). In addition to the direct health impacts, the resources (time and money) devoted to obtaining safe drinking water can have appreciable negative impacts on employment opportunities and schooling.

Health impacts of air pollution[7]

The main health effects associated with selected air pollutants are outlined in Table 13.3. Although the epidemiological evidence related to air pollution is uncertain, particulate matter (PM) appears to be the most health-damaging air pollutant – with well-recognised effects in terms of both morbidity and mortality (see also Chapter 8 on air pollution and 12 on health and environment).

Table 13.3. **Health effects associated with selected air pollutants**

Pollutant	Short-term effects	Long-term effects
PM	• Increase in mortality • Increase in hospital admissions • Exacerbation of symptoms and increased use of therapy in asthma • Cardiovascular effects • Lung inflammatory reactions	• Increase in lower respiratory symptoms • Reduction in lung function in children and adults • Increase in chronic obstructive pulmonary disease • Increase in cardiopulmonary mortality and lung cancer • Diabetes effects • Increased risk for myocardial infarction • Endothelial and vascular dysfunction • Development of atherosclerosis
O_3	• Increase in mortality • Increase in hospital admissions • Effects on pulmonary function • Lung inflammatory reactions • Respiratory symptoms • Cardiovascular system effects	• Reduced lung function • Development of atherosclerosis • Development of asthma • Reduction in life expectancy
NO_2	• Effects on pulmonary structure and function (asthmatics) • Increase in allergic inflammatory reactions • Increase in hospital admissions • Increase in mortality	• Reduction in lung function • Increased probability of respiratory symptoms • Reproductive effects

Source: Adapted from WHO, 2004; 2006.

At the aggregate level, the health costs associated with air pollution can be considerable. Muller and Mendelsohn (2007) have estimated that the total damages associated with emissions of air pollution from 10 000 major sources in the US are between USD 71 billion and 277 billion (0.7-2.8% of GDP). For China, which has a much less well-developed environmental policy regime, the relative costs are correspondingly higher. The World Bank (2007) estimated that the health impacts associated with air pollution in China were about 3.8% of GDP, with much of the impact occurring in urban areas (water pollution costs may also represent between 0.3 and 1.9% of rural GDP, depending on the estimated value of statistical life that is applied).

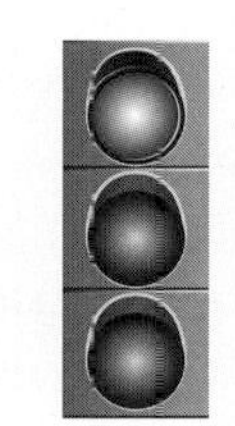

The costs of inaction associated with air pollution may be as high as a few percentage points of GDP. Most of these costs are not reflected in market prices or national accounts (e.g. "pain and suffering").

AEA Technology Environment (2005) has estimated that 3.7 million life years are lost annually in the EU25 countries due to PM. This is equivalent to 348 000 estimated premature deaths; 21 000 deaths were also estimated to occur earlier than normal due to ozone episodes (O_3). The total health damages associated with prevailing EU legislation for O_3 and PM in 2000 for these same countries was estimated to be between EUR 276 and EUR 790 billion, with the mortality impacts of PM being responsible for over two-thirds of these costs. This is equivalent to 3% to 10% of GDP for the EU25 region. According to the OECD Baseline, the worldwide number of premature deaths from PM is projected to be over 3 million in 2030 (see Chapter 12 on health and environment). Samakovlis *et al.* (2004) have estimated that an increase of 1 $\mu g/m^3$ in NO_2 emissions in Sweden resulted in a 3.2% increase in respiratory-related restricted activity days – approximately 685 637 additional restricted activity days. Hansen and Selte (2000) found that the effect of reducing PM_{10} concentrations in Oslo from 24.5 $\mu g/m^3$ to 12.3 $\mu g/m^3$ would reduce the sick-leave ratio by 7%.

Several studies report on the negative effects of O_3 pollution on agricultural yields. In Europe, for example, it has been estimated that the costs of not having introduced the Gothenburg Protocol[8] in terms of lost agricultural output would have been EUR 462 million/year (Holland *et al.* 2002).

Incidence of health costs

Given that health costs can be a significant proportion of the total costs of inaction on air and water pollution, environmental policy in this area can be understood as a form of "upstream prevention". The costs of inaction associated with not undertaking *ex ante* prevention are reflected in the health costs that are borne *ex post*. However, the *incidence* of the costs associated with these health impacts varies (Table 13.4).

Table 13.4. **Types and incidence of health costs from air and water pollution**

Cost	Examples	Incidence
Pain and suffering	Direct welfare loss	Individual sufferer
Restricted activity	Inability to undertake certain physical activities	Individual sufferer, dependents
Lost productivity	Sick leave, less efficiency	Individual sufferer, employer, insurance (public and/or private)
Preventive behaviour	Residential location, bottled water, lead-free paint	Individual sufferer
Caregiver resources	Compassionate leave, time and effort	Family/friends, employer
Medical service costs	Admission costs, operating costs	Individual sufferer, health insurance, public health service costs
Medicines	Prescription costs	Individual sufferer, health insurance, public health service costs

While the costs of "pain and suffering" are borne directly by exposed individuals, the financial costs may be diffused more widely. Indeed, one study of the costs of respiratory problems associated with air pollution (Chestnut *et al.*, 2005) found that only a small proportion of the financial and opportunity costs are borne directly by the individual sufferer.

While this example provides a general indication of the breakdown of the "costs of illness" by type of cost and bearer, it is clear that institutional factors are also important. For example, a study by the (Canadian) Ontario Medical Association (2005) estimated that the healthcare costs associated with $PM_{2.5}$ and ozone in Ontario were CAD 507 million per annum. However, the incidence of these costs will depend upon how public health services are financed. Other institutional factors (*e.g.* labour market policy) can also affect the incidence of health-related costs of inaction.

Fisheries

The fisheries sector is an important source of employment (see Chapter 15 on fisheries and aquaculture) – about 40 million fishers and fish farmers depend on fisheries worldwide (FAO, 2005). An overwhelming majority of these people (about 95%) are located in developing countries (FAO, 1999). In many of these countries, fish is an essential part of the diet, providing 22% and 19% of animal proteins consumed in Asia and Africa, respectively (FAO, 2005). The recreational opportunities associated with fishery resources also contribute to the livelihoods of coastal or island communities. The impacts of fisheries on aquatic ecosystems are also being increasingly recognised. For all of these reasons, it is important that fishery resources be managed sustainably.

FAO (2007) has reported that exploitation of the world marine fishery resources intensified rapidly during the 1970s and 1980s. The proportion of over-exploited and depleted stocks rose from 10% in 1974 to 25% in 2005, although this trend has moderated in the last 10-15 years, even if increased rates of exploitation have been reported for some fish stocks and specific areas. Excess fishing pressure exerted on these stocks in the past leaves no possibilities in the short- or medium-term for further expansion, with an increased risk of further declines or even commercial extinction (Figure 13.2).

Figure 13.2. **Status of world fish stocks (2005)**

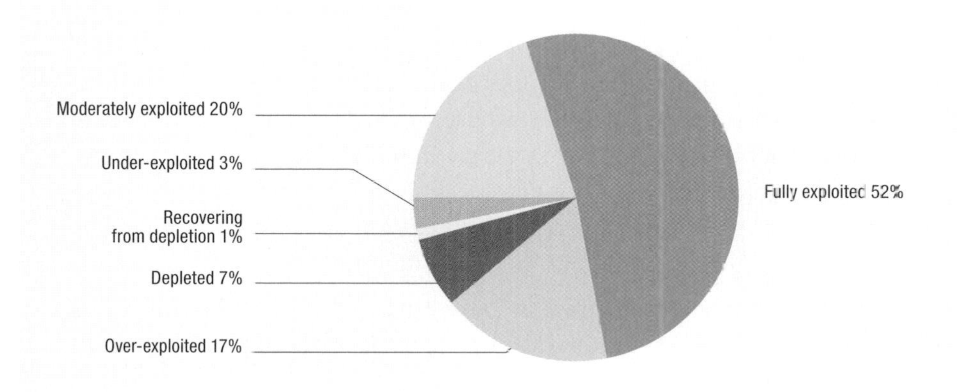

StatLink http://dx.doi.org/10.1787/261346154127

Source: Data from FAO, 2007.

Policy "inaction" in the context of fisheries management can be best described as unsustainable resource management (*i.e.* where the stock is being exploited at a rate which is greater than that which can be supported). In practice, few (if any) fisheries are currently unregulated. Regulation of fisheries typically involves some constraints via gear restrictions; spatial and/or temporal restrictions on fishing; and volume restrictions on fish harvest and fishing effort. If the combination of regulatory measures in place is not sufficient to ensure sustainable resource management, the economic consequences can be considerable.

Fisheries management takes place against a backdrop of imperfect information and imperfect control. The size of the stock, its growth rates, and its relationship with other stocks are not known with precision. Even if this was not the case, regulation of the sector would still be imperfect, particularly in areas not controlled by any one government (*e.g.* high-seas fisheries). In the face of imperfect information and control, precaution should be exercised – if thresholds are breached, a given stock can be fished into commercial extinction, with the permanent loss of all of the benefits indicated earlier. Therefore, the fisheries sector is also an example of where environmental pressures can have potentially "irreversible" consequences.

There are many different types of costs arising from unsustainable fisheries management. These include direct economic consequences, such as lost receipts for fishers and vessel owners from falling catches. There are also indirect consequences, such as lost earnings for workers and foregone profits of fish-processing and related industries. Then there is the additional loss of "use values", including those costs which can be difficult to value due to their non-market characteristics, such as reduced recreational opportunities. And finally, there are costs associated with damage to marine ecosystems.

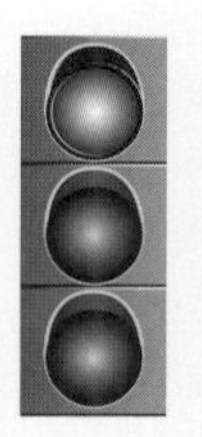

While the costs of unsustainable natural resource management fall mainly on those who previously exploited the resource, others may also bear significant costs.

The costs of unsustainable fisheries management can be considerable:

- Bjorndal and Brasao (2005) have estimated that net present value (NPV) associated with retaining the existing ineffective fishery management regime (*i.e.* total allowable catch and restrictions on gear selection) for East Atlantic Bluefin tuna is only one-third of what would be achieved from an optimal regime. This is estimated to result in a total loss of USD 2 billion.
- Based on a study of 13 "overfished" fish stocks in US waters, Sumaila and Suatoni (2006) compared the lost direct use values (commercial fishery yields and recreational fishing) associated with continued excessive fishing with a case in which the stock "rebuilding" plans developed by Regional Fishery Management Councils were adopted. They found that the lost NPV of continuing the existing excessive fishing management regime was USD 373 million (USD 193.7 million, instead of USD 566.7 million).

The incidence of costs is also an important policy consideration for fisheries managers. Those who exploit a resource are often those who bear the highest cost from unsustainable management. However, others may also bear some of the costs, including taxpayers. In response to the collapse of the cod stock in Canada, for example, substantial public funds were spent on income support (including fishers' unemployment benefits) and government assistance programmes (expenditures towards restructuring, sectoral adjustment, and regional economic development). An estimated CAD 3.5 billion was spent on these programmes (OECD, 2006b).

Climate change

The total economic damage costs associated with climate change are likely to be significant. The projected consequences include:

i) The *market impacts* associated with the agriculture, forestry and energy sectors.

ii) The market and *non-market impacts* associated with human health, *e.g.* diarrhoea and heat stress, and on marine and terrestrial *ecosystems*.

iii) The impacts associated with extreme weather events (rather than with mean climate change), such as more frequent flooding and more intense hurricanes.

Climate change might also lead to a variety of social impacts, such as political instability or migration of people from one location to another. Finally, in the very long-run, climate change might lead to non-linear or catastrophic events, such as the shutting down of the thermohaline circulation in the North Atlantic, sudden and rapid release of methane emissions, or deglaciation of Antarctic or Greenland ice sheets.

 OECD ENVIRONMENTAL OUTLOOK TO 2030 – ISBN 978-92-64-04048-9 – © OECD 2008

The *OECD Environmental Outlook* Baseline estimates that the expected "costs of inaction" or of delayed action for climate change are significant, at least in terms of physical environmental change. Delayed policy action is accompanied by a significantly faster rate of warming in 2030, of more than 0.22°C/decade, compared to 0.16°C/decade under the 450 ppm stabilisation policy (Figure 13.3; and Chapter 7 on climate change). By 2050, the difference between the 450 ppm stabilisation scenario compared to the Baseline ("no additional policy") projection is about 0.6°C.[9] Extrapolating *Outlook* projections to the end of the century suggests that the difference in the increase in the global mean temperature by 2080-2090 under the two scenarios is likely to be roughly 1-3°C (see Chapter 7 on climate change). The latest Intergovernmental Panel on Climate Change (IPCC) report (2007: WG1 and WG2) suggests greater risks than previously for temperature increases between 1 and 3°C (above pre-industrial levels).

Figure 13.3. **Global mean temperature change under the Baseline, an aggressive mitigation scenario, and delayed action, 1970-2050**

Global mean temperature change (2050 compared to preindustrial)

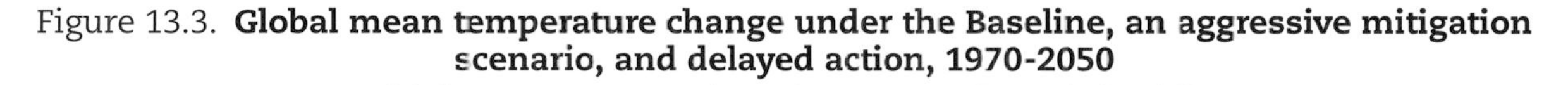

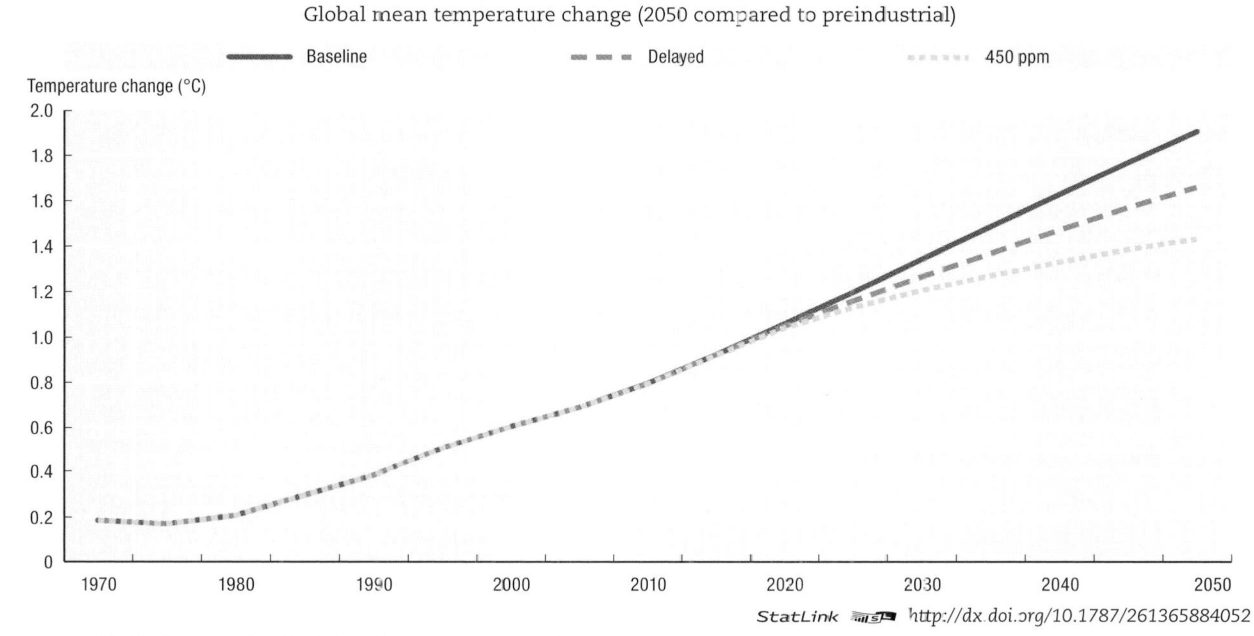

StatLink http://dx.doi.org/10.1787/261365884052

Source: OECD Environmental Outlook Baseline and policy simulations.

Estimates of the total costs of inaction on climate change are relatively few, due to the significant modelling requirements associated with generating these estimates. In recent work using the PAGE2002 Model, Stern (2007) estimated the costs of inaction in terms of reductions in "per capita consumption equivalents".[10] Taking into account all potential impacts (market, non-market, extreme weather events, and catastrophic events), the discounted value of the costs of inaction with respect to climate change were estimated by Stern (2007) to be 14.4% in terms of per capita consumption equivalents, relative to the "no additional policy" baseline scenario.

Kemfert and Schumacher (2005) estimated damage costs associated with a reference scenario in which no new climate policies are introduced. The total damage costs in 2100 represented 23% of global world output. The damages associated with "delayed action" were also assessed. In this latter case, no measures are undertaken until 2030, at which point measures are introduced to ensure that the increase in temperature is no larger than 2°C. In this case, the damages in 2100 are equal to approximately 15% of world GDP.

Since the early 1990s, Nordhaus has produced a series of estimates based on the Dynamic Integrated Model of the Climate and Economy (DICE), the most recent of which are contained in Nordhaus (2007). His baseline scenario is one in which "no policies are taken to slow or reverse greenhouse warming", consistent with the definition of "inaction" that is applied here. The discounted present value of damages for selected runs from the DICE model are USD 22.65 trillion. As a percentage of the discounted value of total future income, this is less than 1%. With a 50-year delay assumed in the implementation of "optimal" policies, the damages estimated by Nordhaus fall by approximately 20% relative to the "no policy" scenario.

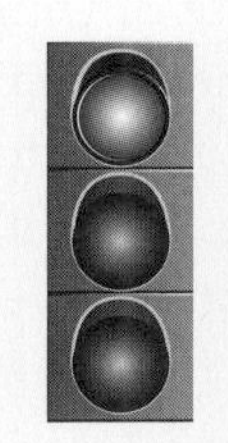

Assuming that emissions remain unmitigated over time, estimated costs associated with climate change range from less than 1% to more than 10% of global output.

Climate change may affect aggregate levels of investment and savings, which affect the entire economy. Fankhauser and Tol (2005) undertook simulations which took into account the prospect of future damages on capital accumulation and savings rates. They found that these "indirect" costs can even exceed the "direct" costs of climate change, with the difference becoming greater over time. In the face of rigidities in capital and labour markets, these costs are likely to be greater still, particularly if the change in environmental quality is sudden. Using a model which allows for market rigidities in the adjustment to an extreme weather event "shock", Hallegatte *et al.* (2006) found that the overall impacts were much greater than if a smooth adjustment was assumed (as is the case in many models). Ultimately, with sufficient frequency and intensity of extreme weather events, an economy may therefore find itself in "perpetual reconstruction", with the economic impacts again being amplified over time.

Generally, estimates of aggregate damages and of social costs of carbon are thought to be underestimated, and to be growing over time (IPCC WG2, 2007). This recognises that studies in the literature generally omit extreme events and non-market impacts – as well as potentially high-consequence, low-probability events such as the deglaciation of Greenland or West Antarctic ice sheet, which could raise sea levels by over several meters in the long-term (IPCC WG2, 2007; Tol, 2005). On the other hand, many studies fail to consider potentially positive amenity *benefits* of climate change (*i.e.* warmer climates in northern Europe) or the offsetting impacts of higher levels of economic development over time, both of which are expected to increase adaptation capacity to climate change (Tol, 2005). However, the additional negative effects are expected to outweigh the positive ones, leading to the conclusion that the current literature is biased downward.

Because ecosystem impacts are often excluded from economic estimates of the costs of climate change inaction, it is useful to consider these explicitly (using both physical and economic metrics). Even with very low temperature increases (in the order of those already being experienced), there is evidence of coral bleaching and shifts in species habitat. In addition, there is high confidence that the extent and diversity of polar and tundra ecosystems is already in decline, and pests and disease have been spreading to higher latitudes and altitudes. Arnell *et al.* (2002) have reported that the "vegetation dieback" under the IPCC "unmitigated" scenario (IS92a) could amount to 1.5-2.7 million km^2 in 2050, rising to 6.2-8.0 million km^2 in 2080.

Figure 13.4 provides an overview of some of the areas for which there is reasonable confidence of impacts associated with different temperature increases. Globally, it is estimated that "net ecosystem productivity" would peak with a warming of 2°C. Beyond that point, terrestrial vegetation is "likely to become a net source of carbon".[11] It has been estimated that up to 43% of species in 25 biodiversity "hotspots" would be at risk from a warming of about 3-4°C (IPCC WG2, 2007).

Figure 13.4. **Temperature increases and likely impacts on marine and terrestrial ecosystems**

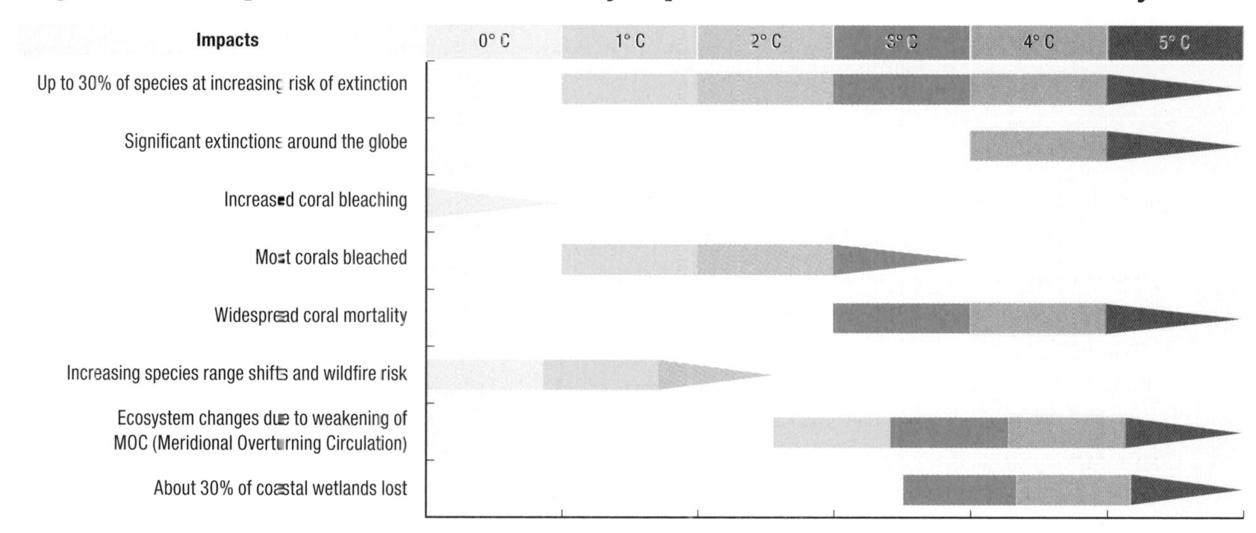

Source: Derived from IPCC WG2 (2007).

The damages to different ecosystems will depend to a great extent on their capacity to adapt to changing climatic conditions, and on the rate at which the climate is changing. For instance, grasslands and deserts can adapt quickly, while forests will adapt more slowly (particularly at higher latitudes) – not more rapidly than 0.05°C per decade (Arnell, 2006). Assuming a temperature increase of 2°C, Leemans and Eickhout (2004) estimated that over 15% of the total area of ecosystems would be affected,[12] with 40% of this area being able to adapt. However, almost 20% of nature reserves will be affected, with less than 40% being able to adapt. Warming, of course, is not the only climate change-related determinant of changes in ecosystems. Changing precipitation will also have important implications for ecosystem health and biodiversity, especially in Central Asia, the Mediterranean, Africa and Oceania. A small change in precipitation in desert ecosystems can also have devastating implications for local species.

Based on several previous studies which have valued willingness-to-pay (WTP) for species, ecosystem and landscape preservation, Tol (2002) estimated the costs of ecosystem damages of a 1°C increase in temperature in different regions, indicating wide variation (from USD 17 billion in OECD North America to roughly USD 100 million in Africa, or South and South East Asia). However, these estimates are extremely crude – due to uncertainties about impacts and their valuation. For example, Hitz and Smith (2004) point out that the evidence is not clear whether particular ecosystem impacts of increased warming will be linear or exponential. Similarly, "wllingness to pay" values are frequently transferred across regions using methods which are (at best) approximate.

Cost incidence is also an important dimension of climate change damages. The extent to which households are compensated for losses depends in part upon the "insurance density" of the response, and this varies widely across and within countries. For example, the data suggest that the ratio of insured losses to overall losses for natural disasters was about 38% in the US, *versus* about 27% in Europe between 1980 and 2005 (OECD, 2006a). However, these figures vary by incident. While the "insurance density" in the US is thought to be about 25-50% (OECD, 2006a), in the case of Hurricane Andrew, the relevant figure was approximately 65%. For Katrina, it was 27-33% (OECD, 2006a). The extent of insurance coverage can affect the rate at which reconstruction is undertaken, and thus the adjustment costs.

Partly because markets and capacities to adapt to climate change vary widely across countries, climate change damage costs are projected to be unevenly distributed across world regions, with the highest costs likely to be occurring in developing country regions. For example, Tol (2002) reviewed estimates of the effects on agriculture of a 2.5°C increase in global mean temperature above 1990 levels. For many of the studies reviewed, European and North American OECD regions are seen to experience net aggregate benefits, but in the developing regions of Africa, and South and South-East Asia, negative impacts are almost always found. In addition to ecological factors, higher losses occur in developing countries because they are more vulnerable to climate change due to lack of institutional and economic resources to deal with these impacts. This raises questions about equity in the assessment of global response strategies – an issue which remains prominent in international policy discussions.

Other issues

This section briefly outlines a few of the costs of inaction likely to be associated with other environmental problems than those discussed above.

Groundwater depletion (or pollution) can have significant impacts on agricultural yields – due to reduced irrigation possibilities (see Chapter 10 on freshwater). Indeed, in some cases, groundwater depletion may even render existing agricultural land unviable. It has been estimated, for example, that between 1982 and 1997, 1.435 million acres of irrigated cropland in Texas were brought out of cultivation, due to ground water depletion (USDA, 2007). Costs associated with the depletion of groundwater are also likely to be reflected in the availability and costs of drinking water. According to one estimate almost half of the world's population relies on groundwater for drinking water (Shah *et al.*, 2007). In many large cities falling water tables are resulting in sharply increased costs of drinking water, even if these are not always passed on to consumers.

In the case of *natural disasters*, some of the most visible costs of inaction relate to the need to reconstruct damaged physical infrastructure. Although data on reconstruction costs are not readily available, figures from Swiss Re and the Insurance Information Institute suggest that during the 1970s and 1980s, annual insured losses from natural disasters were in the USD 3-4 billion range (Kunreuther and Michel-Kerjan, 2007).[13] Since the 1980s, the scale of insured losses from major natural disasters has exhibited a steep upward trend. The World Bank (2006) has estimated that, for the poorest countries, the cost of natural disasters represents more than 13% of GDP. While only some of this cost can be

 OECD ENVIRONMENTAL OUTLOOK TO 2030 – ISBN 978-92-64-04048-9 – © OECD 2008

attributed to environmental factors, which can in turn be influenced directly by public policy (*e.g.* flood control, GHG mitigation, etc.), "inaction" concerning natural disasters is clearly resulting in significant costs. The World Bank and the US Geological Survey have estimated that the worldwide economic losses from natural disasters in the 1990s could have been reduced by USD 280 billion, if USD 40 billion had been invested in disaster preparedness, mitigation and prevention strategies (World Bank, 2004).

The link between environmental policy inaction and industrial hazards is more apparent and better understood. The "costs of inaction" in this area can take a variety of forms. Even the "first-order" restoration and clean-up costs associated with *industrial hazards and accidents* can be significant. Restoration costs associated with oil spills are revealing. In the case of *Erika*, these direct costs were estimated to be EUR 100 million (Bonnieux and Rainelli, 2003); for the *Prestige*, they were valued at over EUR 500 million (Loureiro *et al.*, 2006; Garza-Gil *et al.*, 2006). For the *Exxon Valdez*, clean-up costs alone were over USD 2 billion (Carson *et al.*, 1992). Of course, this ignores all of the other impacts of oil spills, such as effects on ecosystems, the fisheries sector, and tourism – which are likely to be considerable. In a similar vein, the costs associated with the remediation of contaminated sites can also be high, representing a significant negative "legacy" of past inaction.

Concluding remarks

There are several issues that complicate the valuation of the "costs of inaction", including:

- Incomplete information and significant uncertainty associated with the likelihood and magnitude of different environmental impacts.
- The existence of ecological thresholds and irreversibilities which can lead to sudden and significant environmental impacts.
- The long-run nature of many impacts arising out of environmental degradation and resource depletion.
- The degree of substitutability which exists between environmental resources and other factors of production, and the implications this has for economic sustainability.
- The importance of the distribution of environmental impacts, and thus, the links between environmental impacts and social concerns for equity.
- The nature of responses of households, firms and governments to changing environmental conditions.

Given the uncertainties involved, and the fundamentally tendentious nature of the problem of estimating the costs of inaction, it would be foolhardy to attempt to "cost" environmental policy inaction in some aggregate sense. However, it is clear that there are many environmental problems for which the costs of not taking policy action are significant, and are already directly affecting OECD economies in a variety of ways. Some of these costs are reflected, for example, in public sector budgets, – *e.g.* public expenditures on health services, unemployment benefits and adjustment programmes for out-of-work fishers, remediation costs for contaminated sites, etc.

However, other elements of the costs of inaction are less apparent (and more difficult to quantify), such as the costs associated with the loss of marine and terrestrial biodiversity; and the "pain and suffering" associated with ill-health. Some components of the costs of inaction may also be reflected in existing markets, even though they are not readily perceived as costs of environmental policy inaction *per se*. Examples include the effects of contaminated sites on adjacent property prices, or the effects of air pollution on agricultural yields.

Focusing on the costs of inaction without taking into account key non-market and intangible issues (such as the "existence value" of biodiversity) can result in a gross underestimate of reality. Nonetheless, in some cases, the assessment of the more tangible market impacts alone may be sufficient to warrant additional policy interventions (*i.e.* above and beyond those policies already in place). Since these "more direct" costs are often easier to estimate with confidence, this is important to bear in mind.

OECD countries have made significant strides in addressing many of the environmental concerns discussed in this chapter. The term "inaction" must therefore be interpreted in this context. Even if the full costs of inaction are deemed to be significant, identifying those areas in which new environmental policies should be undertaken would still require a careful balancing of the marginal costs of inaction with the marginal costs of further reducing the associated impacts. Although an assessment of some of the elements of one side of this equation is instructive, this important additional step would also need to be taken before arriving at sound policy decisions.

The point of incidence of the costs of environmental policy inaction has direct implications for incentives to avoid future negative environmental legacies, and thus for the design of policy. Inaction is a reflection of the non-internalisation of environmental externalities. It is important that price and regulatory signals which reflect the costs of inaction be transmitted to those who are in a position to reduce such impacts, since *ex ante* prevention is often much less costly than *ex post* remediation or adaptation. In many cases (climate change, high-seas fisheries, etc.), this will imply the need for significant international co-ordination.

Notes

1. This chapter does *not* review estimates of the costs of action (*i.e.* the costs of environmental policy interventions). See Morgenstern *et al.* (2001) for one particularly useful example from this vast literature.
2. In the face of uncertainty about both future interest rates and future economic conditions, the appropriate discount rate to apply will vary over the life of the impact. Where such uncertainty exists, a declining rate should be applied through time (see Weitzman, 2001). See Hepburn (2007) for a discussion of the implications for estimating the costs of policy inaction.
3. According to the International Panel on Climate Change (IPCC) the probability of this happening this century is "very unlikely", but slowing of the current is "very likely" (IPCC WG2, 2007).
4. See Pearce *et al.* (2006) for a review of approaches to valuing human morbidity and mortality, and a summary of estimates from the literature of key assumptions (*e.g.* value of a statistical life).
5. Except, perhaps in terms of *ex ante* "defensive" expenditures. One example is time and energy spent collecting drinking water from an uncontaminated source, rather than from a more polluted source nearby. The purchase of bottled water to avoid lead contamination would be an example of private defensive expenditures. However, some "defensive" expenditures may actually overestimate the health costs of inaction, if other non-health benefits are also obtained (*e.g.* better taste).
6. This section draws heavily on Gagnon (2007a and b).
7. This section draws heavily on Scapecchi (2007).
8. The 1999 Gothenburg Protocol to Abate Acidification, Eutrophication and Ground-level Ozone. The protocol sets emission ceilings to 2010 for four pollutants: sulphur, NOx, VOCs and ammonia.

Under the negotiated solution, parties whose emissions have a more severe impact and whose emissions are relatively less costly to reduce have an obligation to achieve the biggest reductions. Compared to 1990 levels, Europe's sulphur emissions should be cut by at least 63%, its NO_x emissions by 41%, its VOC emissions by 40% and its ammonia emissions by 17%.

9. These uncertainty ranges are presented for key *Baseline* estimates and are based on a scaling approach, using the MAGICC model and IMAGE result – see Chapter 7, Climate change, Table 7.4c.
10. The "metric" used in Stern (2007) has caused some confusion, but is an elegant way to express a complex issue. Assuming future growth rates in the absence of any economic impacts from climate change, Stern first calculated the consumption path associated with that growth rate. Next, he considered climate change impacts, which in his model translated into lower future growth rates, and thus a correspondingly lower future consumption path. The "cost of inaction" was thus the difference between these two consumption trajectories (see Sterner and Persson (2007) for additional clarification).
11. IPCC WG2 (2007: Chapter 19).
12. This is based on an indicator derived from the net change in extent of a particular ecosystem, expansion into other areas, and disappearance from existing areas.
13. The extent to which this reflects market losses depends in part upon the "insurance density", and this varies widely across (and within) countries.

References

AEA Technology Environment (2005), *CAFE CBA: Baseline Analysis 2000 to 2020*, Final Report to the European Commission DG Environment, April 2005, Oxford.

Arnell, N.W. (2006), "Global Impacts of Abrupt Climate Change: An Initial Assessment", *Tyndall Centre for Climate Change Research Working Paper* 99, Norwich.

Arnell, N.W. *et al.* (2002), "The Consequences of CO_2 Stabilisation for the Impacts of Climate Change", *Climate Change*, Vol. 53, pp. 413-446.

Bjorndal, T. and A. Brasao (2005), "The East Atlantic Bluefin Tuna Fisheries: Stock Collapse or Recovery", *Working Paper SNF* No. 34/05, Institute for Research in Economics and Business Administration, Bergen.

Bonnieux, F. and P. Rainelli (2003), "Lost Recreation and Amenities: The Erika Spill Perspectives", in *International Scientific Seminar: Economic, Social, and Environmental Effects of the Prestige Spill*, Santiago de Compostela, 7-8 March 2003.

Carson, R.T., *et al.* (1992), *A Contingent Valuation Study of Lost Passive Use Values Resulting from the Exxon Valdez Oil Spill* Attorney General of the State of Alaska, Anchorage.

Chestnut, L.G., *et al.* (2005), "The Economic Value of Preventing Respiratory and Cardiovascular Hospitalizations", *Contemporary Economic Policy*, Vol. 24, No. 1, pp. 127-143.

Dziegielewska, D.A.P. and R. Mendelsohn (2005), "Valuing Air Quality in Poland", *Environmental and Resource Economics*, Vol. 30, pp. 131-163.

EEA and WHO/EURO (2002), "Water and Health in Europe: A Joint Report from the European Environment Agency and the WHO Regional Office for Europe", WHO Regional Publications, *European Series*, No. 93.

Fankhauser, S. and R.S.J. Tol (2005), "On Climate Change and Economic Growth", *Resource and Energy Economics*, Vol 27, pp. 1-17.

FAO (1999), *The State of World Fisheries and Aquaculture 1998*, FAO, Rome.

FAO (2005), *Increasing the Contribution of Small-scale Fisheries to Poverty Alleviation and Food Security*, FAO Technical Guidelines for Responsible Fisheries, No. 10, FAO, Rome.

FAO (2007), *The State of World Fisheries and Aquaculture 2006*, FAO, Rome. (Accessed on 11 May 2007 at *www.fao.org/scf/sofia.*)

Gagnon, N. (2007a), *Health Costs of Inaction with Respect to Water Pollution in OECD Countries*, background document prepared for OECD projects on Costs of Inaction and Environmental Outlook, OECD, Paris.

Gagnon, N. (2007b), *Unsafe Water, Sanitation and Hygiene: Associated Health Impacts and the Costs and Benefits of Policy Interventions at the Global Level*, background document prepared for OECD projects on Costs of Inaction and Environmental Outlook, OECD, Paris.

Garza-Gil, M.D., A. Prada-Blanco and M.X. Vázquez-Rodríguez (2006), "Estimating the Short-term Economic Damages from the Prestige Oil Spill in the Galician Fisheries and Tourism", *Ecological Economics*, Vol. 58, pp. 842-849.

Gibbs, J.P. *et al.* (2002), "A Hedonic Analysis of the Effects of Lake Water Clarity on New Hampshire Lakefront Properties", *Agriculture and Resource Economics*, Vol. 31, No. 1, pp. 39-46.

Hallegatte, S., J.-C. Hourcade and P. Dumas (2006), "Why Economic Dynamics Matter in Assessing Climate Change Damages: Illustration on Extreme Events", *Ecological Economics*, Vol. 62, No. 2, 20, pp. 330-340

Hansen, A.C. and H.K. Selte (2000), "Air Pollution and Sick-Leaves", *Environmental and Resource Economics*, Vol. 16, pp. 31-50.

Hepburn, C. (2007), *Use of Discount Rates in the Estimation of the Costs of Inaction with Respect to Selected Environmental Concerns*, [ENV/EPOC/WPNEP(2006)13/FINAL], OECD, Paris.

Hitz, S. and J. Smith (2004), "Estimating Global Impacts from Climate Change", in OECD (2004), *The Benefits of Climate Change Policies,* OECD, Paris.

Holland, M. *et al.* (2002), *Economic Assessment of Crop Yield Losses From Ozone Exposure*, paper prepared for the UNECE International Cooperative Programme on Vegetation. *www.airquality.co.uk/archive/reports/cat10/final_ozone_econ_report_ver2.pdf*

IPCC (Intergovernmental Panel on Climate Change) (2007), 4th Assessment (WG1), *The Physical Science Basis*, IPCC, Geneva.

IPCC (2007), 4th Assessment (WG2), *Impacts, Adaptation and Vulnerability*, IPCC, Geneva.

Kemfert, C. and K. Schumacher (2005), *Costs of Inaction and Costs of Action in Climate Protection*, DIW Berlin Final Report of Project FKZ 904 41 362 for the German Federal Ministry of the Environment.

Kunreuther, H.C. and E.O. Michel-Kerjan (2007), "Climate Change, Insurability of Large-Scale Disasters and the Emerging Liability Challenge", *NBER Working Paper* 12821, National Bureau of Economic Research, Inc., Cambridge, Massachusetts.

Leemans, R. and B. Eickhout (2004), "Another Reason for Concern: Regional and Global Impacts on Ecosystems for Different Levels of Climate Change", *Global Environmental Change*, Vol. 14, pp. 219-228.

Loureiro, M.L., Ribas, A., E. López and E. Ojea (2006), "Estimated Costs and Admissible Claims Linked to the Prestige Oil Spill", *Ecological Economics*, Vol. 59, pp. 48-63.

Morgenstern, R.D., W. A. Pizer and J.-S. Shih (2001), "The Cost of Environmental Protection", *Review of Economics and Statistics*, Vol. 83, No. 4, pp. 732-738.

Muller, N.Z. and R. Mendelsohn (2007), "Measuring the Damages of Air Pollution in the United States", *Journal of Environmental Economics and Management*, Vol. 54, pp. 1-14.

Nordhaus, W.D. (2007), "The Challenge of Global Warming: Economic Models and Environmental Policy", Yale University, *Department of Economics Discussion Paper* (*http://nordhaus.econ.yale.edu/dice_mss_072407_all.pdf*).

OECD (2001), *Establishing Links between Drinking Water and Infectious Diseases*, [DSTI/STP/BIO(2001)2/FINAL], OECD, Paris.

OECD (2006a), *Reducing the Impact of Natural Disasters: The Insurance and Mitigation Challenge*, [DAF/AS/WD(2006)29], OECD, Paris.

OECD (2006b), *Subsidy Reform and Sustainable Development: Economic, Environmental and Social Aspects*, OECD, Paris.

OECD (2008a) (forthcoming), *Cost of Inaction: Technical Report*, OECD, Paris.

OECD (2008b) (forthcoming), *Costs of Environmental Policy Inaction: Summary for Policy-makers*, OECD, Paris.

Ontario Medical Association (2005), *The Illness Costs of Air Pollution,* OMA, Toronto.

Pearce, D. *et al.* (2006), *Cost-Benefit Analysis and the Environment*, OECD, Paris.

Poor, R.J., K.L. Pessagno and R.W. Paul (2007), "Exploring the Hedonic Value of Ambient Water Quality: A Local Watershed-Based Study", *Ecological Economics*, Vol. 60, pp. 797-806.

Prüss-üstün, A., D. Kay, L. Fewtrell and J. Bartram (2004), "Unsafe Water, Sanitation and Hygiene", in M. Ezzati *et al.* (eds.), *Comparative Quantification of Health Risks, Global and Regional Burden of Disease Attributable to Selected Major Risk Factors*, World Health Organization, Geneva.

Rabl, A. (2004), *Valuation of Health End Points for Children and for Adults*, Working Paper, École des Mines, Paris.

Samakovlis, E., Huhtala, A., T. Bellander and M. Svartengren (2004), "Air Quality and Morbidity: Concentration-Response Relationships for Sweden", The National Institute of Economic Research, *Working Paper* No. 87, January 2004, Stockholm.

Scapecchi, P. (2007), *Health Costs of Inaction with Respect to Air Pollution*, OECD Environment Directorate, [ENV/EPOC/WPNEP(2006)17/FINAL], OECD, Paris.

Shah, T. *et al.* (2007) "Groundwater: A Global Assessment of Scale and Significance", in International Water Management Institute (ed.) *Water for Food, Water for Life: A Comprehensive Assessment of Water Management*, Earthscan, London.

Stieb, D., *et al.* (2002), "Economic Evaluation of the Benefits of Reducing Acute Cardiorespiratory Morbidity associated with Air Pollution", *Environmental Health: A Global Access Science Source* 2002, Vol. 1, p. 7.

Stern, N. (2007) *The Economics of Climate Change: The Stern Review*, CUP, Cambridge.

Sterner, T. and U.M. Persson (2007), "An Even Sterner Review: Introducing Relative Prices into the Discounting Debate", *RFF Discussion Paper* 07-37, Resources for the Future, Washington, DC.

Sumaila, U.R. and L. Suatoni (2006), "Economic Benefits of Rebuilding US Ocean Fish Populations", *Fisheries Centre Working Paper No.* 2006-04, The University of British Columbia, Vancouver.

Tol, R.S.J. (2002), "Estimates of the Damage Costs of Climate Change: Part 1 Benchmark Estimates", *Environmental and Resource Economics*, Vol. 21, pp. 47-73.

Tol, R.S.J. (2005), "The Marginal Damage Costs of Carbon Dioxide Emissions", *Energy Policy*, Vol. 33, pp. 2064-2084.

USDA (US Department of Agriculture) (2007), *Long Range Planning For Drought Management: The Groundwater Component*, USDA Natural Resource Conservation Service, Washington DC. (*http://wmc.ar.nrcs.usda.gov/technical/GW/Drought.html*, accessed on June 27, 2007).

Weitzman, M.L. (2001), "Gamma Discounting", *American Economic Review*, Vol. 91, pp. 260-271.

WHO (World Health Organization) (2004), *Health Aspects of Air Pollution*, results from the WHO Project Systematic Review of Health Aspects of Air Pollution in Europe, WHO, Geneva.

WHO (2006), *WHO Air Quality Guidelines: Global Update 2005*, WHO, Geneva.

WHO/UNICEF (2006), *Joint Monitoring Programme for Water Supply and Sanitation*, (*www.wssinfo.org/en/welcome.html*, accessed October 2006).

World Bank (2004), *Natural Disasters: Counting the Cost*, feature story on March 2, 2004, World Bank, Washington DC.

World Bank (2006), *Hazards of Nature, Risks to Development*, World Bank, Washington DC.

World Bank (2007) *Cost of Pollution in China: Economic Estimates of Physical Damages*, World Bank, Washington DC.

POLICY RESPONSES

III. SECTORAL DEVELOPMENTS AND POLICIES

14. Agriculture
15. Fisheries and Aquaculture
16. Transport
17. Energy
18. Chemicals
19. Selected Industries
    - Steel and cement
    - Pulp and paper
    - Tourism
    - Mining

ISBN 978-92-64-04048-9
OECD Environmental Outlook to 2030

Chapter 14

Agriculture

This chapter examines agriculture's impact on the environment. It outlines key trends and projected developments in the agricultural sector and the environmental impacts of these developments, and assesses policy options that could reduce negative environmental pressures from the sector. Agriculture is responsible for about 40% of land and freshwater use in OECD countries, and 70% of freshwater withdrawals worldwide. Currently, environmental pressures in OECD countries from agriculture are broadly stabilising, but they are increasing elsewhere, especially in those economies where population and economic growth will be largest to 2030. Measures that could help reduce agriculture's harmful impact on the environment include policies to encourage more efficient use of water resources for agriculture (e.g. through moving towards full cost recovery water pricing) and continuing to de-couple support to farmers from production and environmentally harmful input use.

KEY MESSAGES

Agriculture is responsible for about 40% of land use and freshwater withdrawals in OECD countries. Worldwide, agriculture is responsible for 70% of freshwater withdrawals. It also significantly affects soils, water quality, greenhouse gas emissions and absorption, ecosystems and cultural landscapes.

Large increases in agricultural production are expected to 2030, particularly in non-OECD regions. This is a result of an increasing demand for food products, and in particular a shift towards meat-based diets driven by growing economies and populations, and changing consumption patterns. This will lead to 10% more land used for agriculture worldwide (see table below), and increase pressure on the environment and biodiversity.

The rapidly increasing demand for biofuel is leading to competition for agricultural crops and land conversion, and has some negative impacts on the environment.

Harmful environmental impacts of farming in most OECD countries are expected to diminish over time (per unit of product) as policies reorienting farm payments away from production- and input-based support and towards agri-environmental measures increasingly take effect. However, for the world as a whole, without new targeted policies the adverse impacts of farming on the environment and biodiversity are expected to intensify.

In many regions, there is considerable potential for policies and market mechanisms to improve agriculture's efficiency of water use, making it environmentally sustainable.

Policy options

- Implement water pricing policies that encourage more economically and environmentally efficient use of water resources (*e.g.* through moving towards full-cost recovery water pricing structures).
- Continue to de-couple support to farmers from production and environmentally-harmful input use.
- Put in place safeguards in OECD and non-OECD countries to ensure that reductions in production-linked payments benefit the global environment.
- Ensure that the development of biofuels is guided by market signals and takes account of their impact on food prices and the environment.

	Baseline agricultural land use in 2030 (2005 = 100)
OECD	104
BRIC	109
ROW	113
World	**110**

The *Environmental Outlook* Baseline shows a global agricultural land use increase of 10% to 2030. Simulations against that Baseline show that land use changes in some countries would be significant if production-linked payments were reduced by 50% — though the aggregate global change would be small. Trade policy reforms should thus be combined with enhanced environmental protection (*e.g.* for biodiversity) in countries likely to increase land used for agriculture. Those countries are mainly in developing regions where high levels of biodiversity exist. An approach of enhancing protection for the environment while reducing payments linked to production or inputs could help to ensure both economic gains and reduced environmental impacts.

StatLink http://dx.doi.org/10.1787/262650242113

Consequences of inaction

- In some regions, increasing water scarcity, pollution, and a changing climate threaten to reduce the sustainability and productivity of agricultural activities over the coming decades. Some impacts of climate change on agriculture will occur in the coming decades even if emissions of greenhouse gases ceased today.
- To 2030 the main environmental impacts of agricultural policy inaction (*i.e.* business as usual) are likely to come from nutrient loading, such as by nitrogen, and from the expansion of agriculture in developing countries.

 OECD ENVIRONMENTAL OUTLOOK TO 2030 – ISBN 978-92-64-04048-9 – © OECD 2008

Introduction

To produce the food needed to feed the roughly 1 billion people in OECD countries and provide exports to also feed numerous others, OECD country agricultural practices use around 40% of available land and water withdrawals. Globally, agriculture uses roughly 70% of water withdrawals. As a result, the sector significantly affects ecosystems and landscapes.

To 2030, it is projected that global agriculture will need to increase output by more than 50% in order to feed a population more than 27% larger and roughly 83% wealthier than today's (Figure 14.1). The environmental pressure from that change will be very high in the absence of appropriate policies. Currently, environmental pressures in OECD countries from agriculture are broadly stabilising, but the trend elsewhere is towards further increases, especially in rapidly emerging economies where population and economic growth will be largest over the coming decades.

As the nature of production and demand in the global agri-food sector changes, the environmental impacts will also undergo major changes. For example, the intensification of agricultural production has been driving greater energy and water intensity, greater use of agrochemicals, and accelerated land conversion in many areas. High levels of unconstrained production-linked agricultural support and trade policies continue to distort the relative prices of agricultural inputs and outputs. On the other hand regulations and pricing policies (*e.g.* for water), together with better targeted agri-environmental policy

Figure 14.1. **Expected growth of world population, GDP per capita, agricultural production and agricultural land use percentage change from 2005-2030**

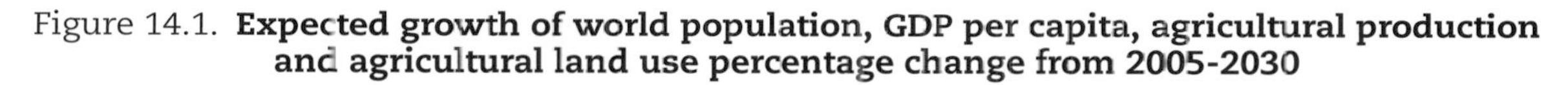

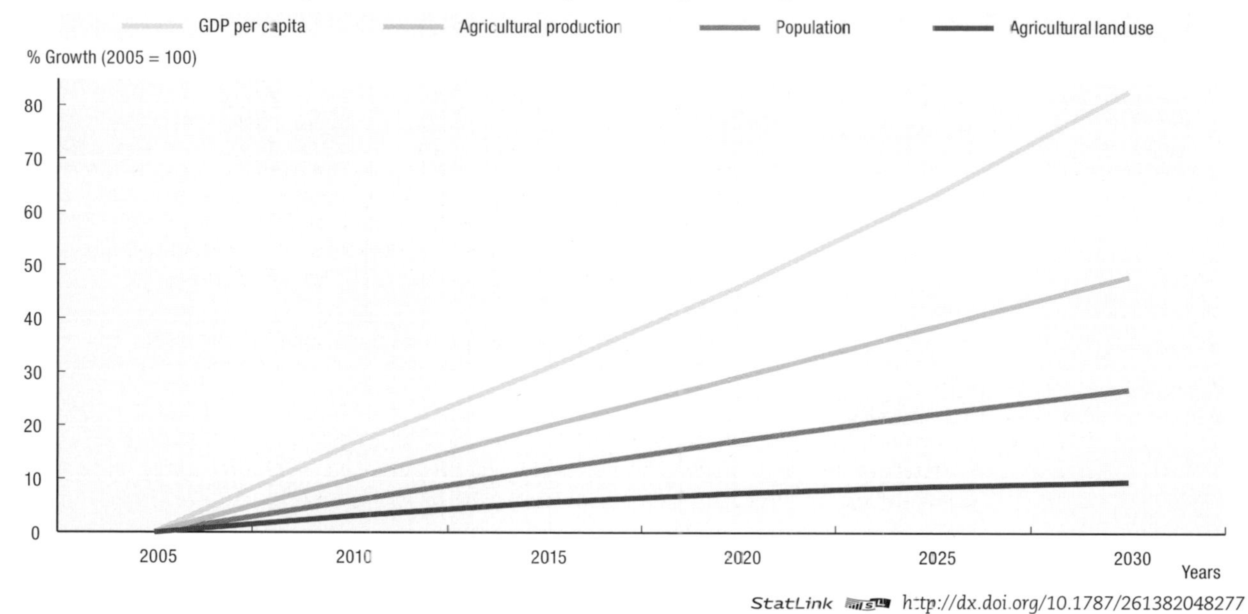

StatLink http://dx.doi.org/10.1787/261382048277

Source: *OECD Environmental Outlook* Baseline.

measures in some OECD countries, as well as constraints on environmentally-harmful input use, have led more recently to a more efficient use of water, fertilisers and pesticides per unit of production, as well as encouraging the adoption of agricultural measures to protect soil, habitat and landscape values.

Key trends and projections

Projected developments in the agricultural sector

Figures 14.2 and 14.3 show some key trends in agricultural production to 2030 under the *OECD Environmental Outlook* Baseline. These projections describe a world where global agricultural production is rising to meet the demands of growing populations whose material wealth is increasing rapidly. Of course there are many uncertainties which may change these general trends (see Box 14.1 and Chapter 6 on key variations to the standard expectation to 2030). These trends assume no new policies that might influence agriculture's impact on the environment, but leave in place existing policies that provide some environmental protection. Environmental impacts discussed here are therefore occurring in areas and countries where existing protection may need strengthening. Given the scale of changes, there will clearly be a need for agricultural and environmental policies that ensure growth is environmentally benign.

Figure 14.2. **Production of food crops, 2005-2030**

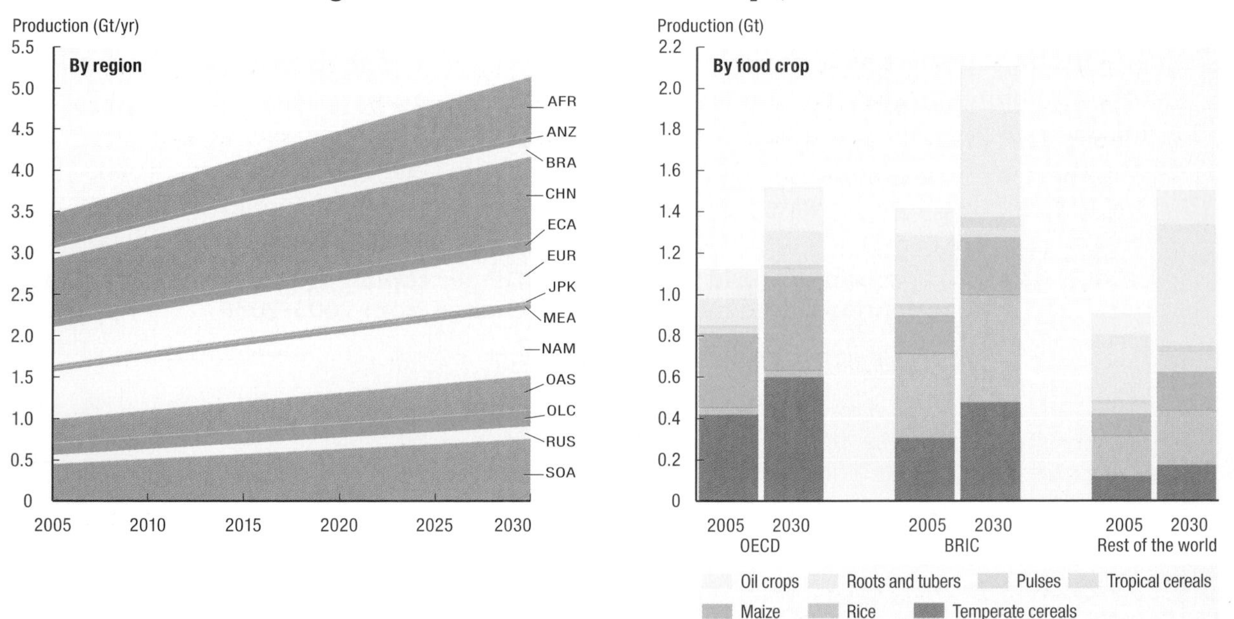

StatLink http://dx.doi.org/10.1787/261460222157

Note: Regions defined as: NAM: North America, MEA: Middle East, EUR: OECD Europe, OAS: Other Asia, JPK: OECD Asia, ECA: Eastern Europe and Central Asia, ANZ: OECD Pacific, OLC: Other Latin America and Caribbean, BRA: Brazil, AFR: Africa, RUS: Russia and Caucasus, SOA: South Asia, CHN: China Region.

Source: OECD Environmental Outlook Baseline.

Some key findings from the Baseline for the development of the agricultural sector include:

- Total land used for agriculture (crops and pasture, and feedstock crops for biofuels) is projected to increase to 2030. For the environment and biodiversity, this implies additional pressure that policy will have to anticipate – especially since other pressures on land use from infrastructure building, etc., will also be increasing.

Figure 14.3. **Production of animal products, 2005-2030**

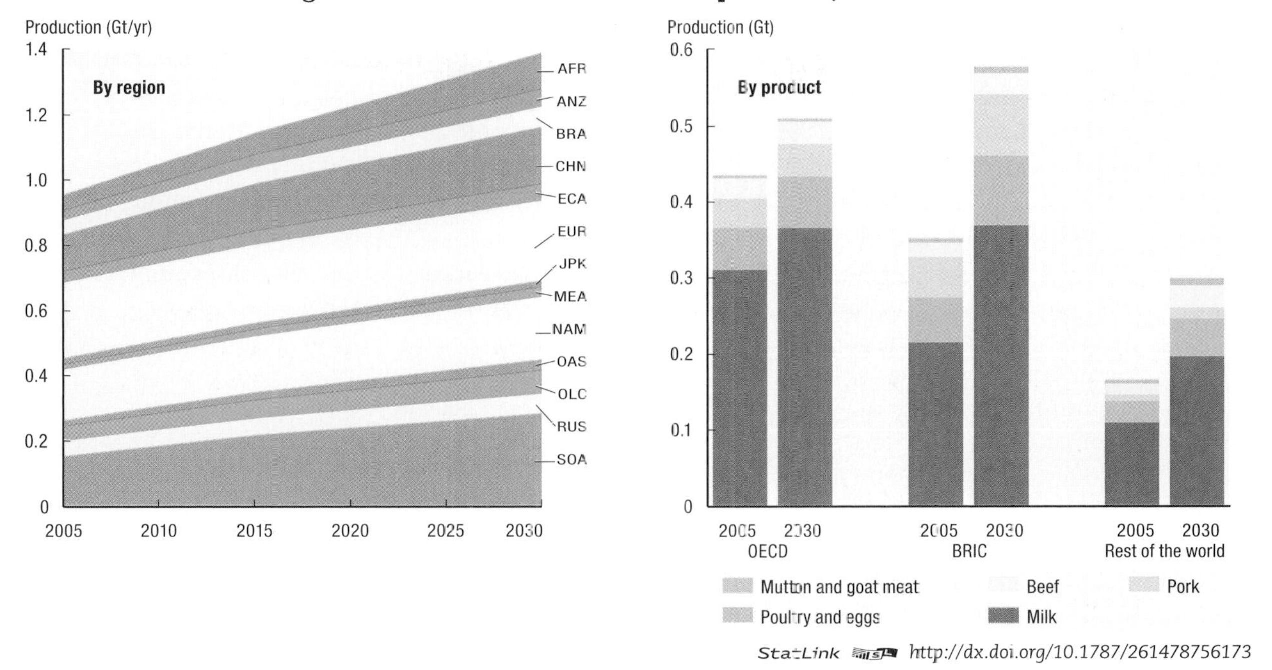

StatLink http://dx.doi.org/10.1787/261478756173

Source: OECD Environmental Outlook Baseline.

- Macroeconomic, population and technological developments and the availability of land in developing countries mean that agricultural production in those countries is growing four times faster than in OECD countries. Per capita consumption of agricultural products is almost stable in OECD countries, while it is projected to grow by 70% in developing countries to 2030 – although almost all of that growth is occurring in the BRIC countries. These pressures are stronger than they have been in the past and thus land used for agriculture will increase despite productivity gains.
- Domestic demand is outstripping domestic production, resulting in increasing agricultural imports by developing and transition countries. Consequently, OECD countries have the highest rates of export growth, exceeding their production and consumption growth rates (clearly these totals mask important differences among countries).
- Oilseed production is projected to grow about 50% faster than overall average agricultural production to 2030. This growth is boosted not only by growing demand for vegetable oils for human consumption, but also for oilseed meal for feeding animals and for biodiesel production. The growth in biofuels production is projected to cause price increases for maize and sugar, but without high oil prices or additional policies, they remain well under 1% of total agricultural land use (see Box 14.2).
- Oilseed trade is also projected to outstrip the trade in grain. The most important importer of oilseed is expected to remain China, which will double its imports from 2001 to 2030. The leading exporters are the United States and Brazil, with the United States almost tripling its oilseed exports by 2030.

Box 14.1. **Key drivers and some uncertainties**

Chapter 3 on economic development outlines some key economic drivers of environmental change, such as productivity and population growth. For agriculture, of critical importance will be the continued increase in yield per hectare. The trends for yields used in developing the Baseline were taken from the UN Food and Agriculture Organization study *Agriculture Towards 2030* (FAO, 2003), which combines macroeconomic prospects with local expert knowledge.

The Baseline projections in this *OECD Environmental Outlook* assume that the strong economic growth in countries such as China, India and Brazil will persist, in turn spurring broader growth in Asia and South America. All three countries have a growing presence in agricultural markets, although India is less of a trader than the other two. If these countries begin to specialise in particular agricultural markets where they may have some comparative advantages, the impact on global trade and local environments could be significant, even if prices are only marginally affected.

The effects of recent drought in Australia illustrate uncertainties and vulnerabilities in agriculture: wheat and coarse grain production fell by more than half in 2006, and, in a context of global cereal production shortfalls, contributed to rising prices for traded commodities. The impacts of numerous such unknowns are thus likely to be significant over the period to 2030.

Technological change is also likely to influence agricultural trends. The growth of modern industrialised production of farm output during the past 25 years was made possible by advances in technology that were incremental, yet which transformed an entire economic sector. Future technological changes may also have important impacts on agricultural development and its impact on the environment (see Box 14.3). For example, Heilig *et al.* (2000) use data from the FAO and International Institute for Applied Systems Analysis (IIASA) to show that by applying existing technologies already in use elsewhere, China could feed itself in 2025 using less land than it did at the turn of the century.

Demand for biofuels is perhaps the greatest uncertainty in the period to 2030 (see Box 14.2). Other uncertainties include the progress and timing of the Doha Development Agenda of multinational trade negotiations and bilateral or regional trade liberalising agreements, as well as the impacts of climate change.

Box 14.2. **Biofuels: the economic and environmental implications**

The recent rapid growth in biofuel production from existing food crops in OECD countries can largely be explained by high levels of government support. Governments support biofuels for a variety of reasons, but the goals most frequently cited are: reduced greenhouse gas (GHG) emissions; enhanced energy security through oil substitution; and support for farm incomes.

However, the benefits achieved in these areas appear so far to be smaller than many anticipated. Life-cycle analyses of the production of biofuels indicate that the GHG emission reductions compared to fossil fuels may not be significant once the upstream inputs and transport costs are taken into account (see also Box 17.2 in Chapter 17 on energy). The extended use of biofuels also has important drawbacks, for example increased demand for water, fertiliser, and pesticides, as well as effects on biodiversity through land use change. Furthermore, the use of food crops to produce biofuels has already caused competition for land and increases in agricultural prices, and is likely to do so even more in the future (OECD and FAO, 2007).

The rapidly increasing demand for biofuels will compete with food production; the environmental impacts of food crop biofuel production are important.

A number of recent studies (OECD, 2006; Doornbosch and Steenblik, 2007; and Tyner and Taheripour, 2007) highlight that without heavy public support for biofuels production, it is not cost-competitive with fossil fuels in most countries. The efficiency of production depends on the type of crop used, with sugar cane a more efficient feedstock than maize or soy. Production is also more efficient in tropical climates than in temperate ones. As such, Brazilian sugar ethanol is one of the few economically competitive biofuels. However, biofuels might become economically competitive without subsidies even in temperate zones if crude oil prices remain high, although increased production would be expected to drive up agricultural prices, including for the biofuel feedstocks themselves.

So called "second generation" or ligno-cellulosic biofuels – which use non-food feedstocks – will probably be more economically competitive and are likely to have environmental benefits. Ligno-cellulosic biomass, such as trees and grasses, can be grown on poorer-quality land than crops, thus avoiding competition for agricultural land. Research and development efforts on second generation biofuel feedstocks are promising, but they are not yet commercially available.

Biofuels policy simulations

To examine some of these issues, the ENV-Linkages model compared four hypothetical biofuel scenarios* with the Baseline (see Annex 14.A1 for more details). The scenarios are: 1) demand for biofuels growing in line with the IEA (2006) scenario; 2) a demand scenario (DS) whereby growth in biofuel demand for transport is entirely driven by exogenous changes, keeping total fuel for transport demand close to the Baseline; 3) a scenario in which crude oil prices remain high (OilS), to determine the profitability of biofuels in the face of increasing costs of producing traditional fossil-based fuels; 4) a subsidy scenario (SubS) in which producer prices of biofuel are subsidised by 50%. This scenario helps to check if price support is enough incentive to drive up biofuel demands endogenously.

The Baseline projects that biofuel's share of total transport fuel increases slightly from the 2006 level, from 2% to 4% in 2030 (see also Box 16.3 in Chapter 16 on transport). In the DS scenario, biofuels displace petroleum more rapidly in OECD countries and Brazil than in the rest of the world. For some developing countries the DS is optimistic. For example, in China and India the biofuel share would grow from less than 1% in 2006 to 23% and 11%

Box 14.2. **Biofuels: the economic and environmental implications** *(cont.)*

respectively in 2030. Globally the demand scenario suggests a 16% displacement of other transport fuels by biofuels by 2030. In both the OilS and SubS scenarios the demand for biofuel is indirectly stimulated by its production becoming more competitive. The more the international price of crude oil increases, or the more biofuels are subsidised, the more financially profitable it becomes to produce biofuel instead of refined oil. The high subsidy policies seem to be as effective in displacing as much conventional fuel as the demand scenario (almost 15% in 2030).

But the continuing subsidy scenario is a costly policy, since the total cost of the subsidies was already USD 14.3 billion (2001-USD) in 2006 for OECD countries and is projected to reach USD 82.5 billion in 2030. Globally the total cost of biofuel support in this scenario would equal 0.45% of world GDP in 2030. The continuing subsidisation of biofuels is already driving investments in biofuels production and infrastructure, locking in the existing less efficient and environmentally beneficial technologies. The higher financial profitability of biofuels resulting from a crude oil price increase appears to be less incentive than a direct subsidy to biofuel production. This is because there is an aggregate demand impact from high prices and various refined petroleum products are inputs into biofuel production. A number of countries have set specific targets for biofuels use in the fuel mix. These simulations indicate that it is likely that both production and demand incentives would need to be applied to attain a 10% or more biofuel use globally.

* Only first-generation biofuels are considered in the model. Our database distinguished three kinds of biofuel: biodiesel from oilseeds and vegetable oils, crops-based ethanol (mainly from corn and wheat) and sugar-based ethanol. Moreover, biofuel trade between regions remains very limited. In ENV-Linkages the trade in biofuel is conditioned by corresponding trade balances in 2001, so countries like China or India are assumed implicitly to only consume their domestic production.

Agriculture and the environment

Land use (potential impact on landscapes)

As a major user of natural resources, agriculture dominates and shapes the rural landscape. In the past, increasing food demand, together with policies in many OECD countries that encouraged production, resulted in intensification of agriculture (more output per unit of land), as well as expansion of agricultural land use onto environmentally sensitive areas. Both these processes can lead to environmental damage, either through higher use of chemicals, or through loss of habitats and landscape features. At the same time, agriculture provides, amongst other things, environmental benefits through on-farm biodiversity, and by providing recreational services and aesthetic values in the landscapes it creates.

Changes in land used for agriculture can thus have important impacts on landscapes. The actual net impact depends on the incentives that farmers face, which can be heavily influenced by policy. Table 14.1 presents the change in land used for agriculture between 2005 and 2030 projected in the Baseline. This includes crops and biofuels, as well as grass and fodder. Changes of this magnitude show a clear potential for agriculture to have a strong impact on landscapes. In many cases, it can not be taken for granted that the impact is environmentally benign and policies may therefore need to be strengthened.

Total land used for agriculture is projected to increase under the Baseline to 2030 in all regions except Japan and Korea. In South Asia, there could be additional loss of remaining forest areas (both tropical and temperate), savannah and scrubland. In Europe, much of the additional land for agriculture is expected to come from its eastern regions – a reversal of the trend during the past 15 years whereby land has been taken out of agriculture in these

Table 14.1. **Change in total land used for agriculture in 2030 (2005 = 100)**

North America	Europe	Japan Korea	Australia New Zealand	Brazil	Russia	South Asia	China	Middle East	South East Asia	Caucasus and Other Central Asia	Other Latin America	Africa	World
104	105	83	104	108	115	124	101	100	127	104	109	118	110

StatLink http://dx.doi.org/10.1787/257346175877

Source: *OECD Environmental Outlook* Baseline.

regions. In the United States, while land used for agriculture has been roughly constant as a proportion of the total area, Lubowski *et al.*, (2006) report that a large compositional change has occurred: land was taken out of agriculture in some parts of the United States, while new land was brought into use in other parts. Table 14.1 suggests that even if the compositional change continues, there will be net additions of land due to increasing demand.

Nitrogen surplus

Intensive agriculture affects the environment in a number of ways. One consequence is an increase in the intensity of surface nitrogen, which has a range of environmental impacts. These include a rise in the nitrogen levels of drinking water supplies, making purification systems necessary. Elevated levels of nitrogen and phosphorous (eutrophication) contribute to algal blooms in freshwater habitats and coastal areas, depriving other species of oxygen (see Chapter 10 on freshwater and Chapter 15 on fisheries and aquaculture), and also change ecosystems by reducing plant diversity. These impacts of nitrogen loading have been observed in particular in parts of North America, Europe and China, where nitrogen and phosphorous use is particularly intensive.

Surface nitrogen[1] surplus is expected to decrease in intensity during the next 25 years in most regions (Figure 14.4) but will increase in a few. For the world as a whole, this intensity is projected to be 0.8% higher by 2030, with most of the increase occurring in non-OECD economies.

Figure 14.4. **Surface agricultural nitrogen losses (2000 and change to 2030)**

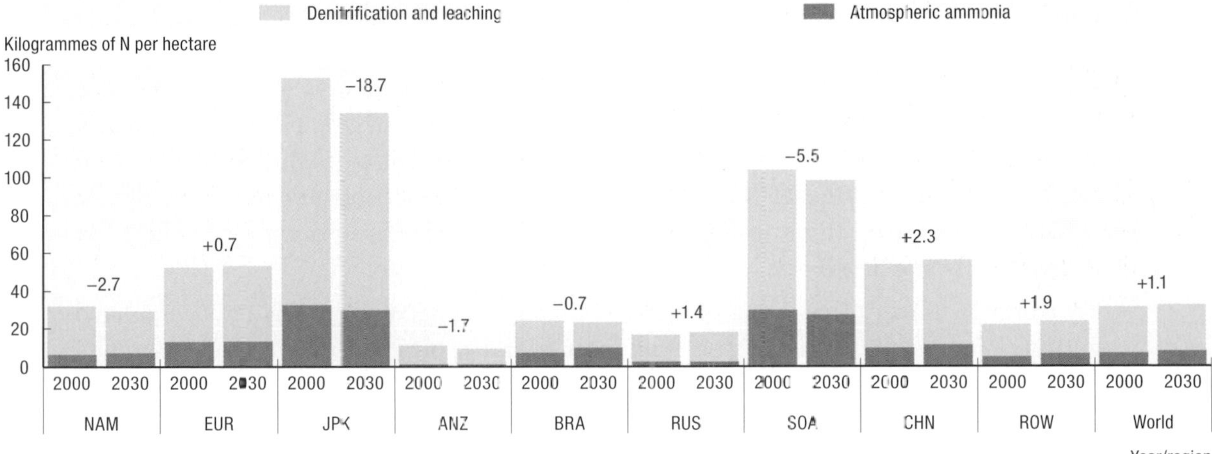

StatLink http://dx.doi.org/10.1787/261486381630

Note: Atmospheric ammonia represents nitrogen from livestock housing, and spreading of manure and inorganic fertiliser. All units are kilogramme per hectare.

Source: *OECD Environmental Outlook* Baseline.

Nitrogen surplus is associated with rising animal stocking densities and changes in the livestock industry towards large confined operations, especially for pigs, poultry and to a lesser extent dairy cattle (OECD, 2003b; 2004). In addition, rising fertiliser demand and growth in nitrogen surpluses is, in part, explained in some countries by the expansion in crop production together with a shift to crops requiring higher fertiliser inputs per unit of output (*e.g.* from wheat to maize, see OECD, 2005a).

Water scarcity

Water scarcity is projected to become worse in some regions by 2030 (Figure 14.5), exacerbated in part by increasing water use by agriculture, in particular irrigation. Changes in precipitation will also be a strong driver of increasing scarcity.[2] Globally, agriculture is responsible for about 70% of water withdrawals, but this varies widely between countries (see Chapter 10 on freshwater). Figure 14.6 shows agriculture's relative share of water use in various parts of the world, as well as regional percentages of the agricultural area that is irrigated. Agriculture is the biggest user of water in most regions, and this is particularly true in developing countries.

Faurès *et al.* (2002) note that in most countries the available water resources will be sufficient to meet demand in 2030 – though not always where and when they are most needed. What is lacking, however, are policies or market signals to provide the incentives for efficient water use; in other words, in most regions there is not a physical scarcity of water, but an economic scarcity.

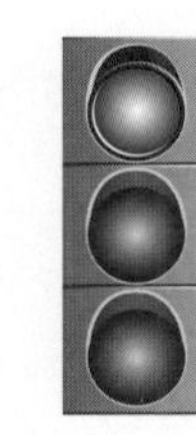

Agriculture remains the largest user and a major polluter of water resources worldwide. Agricultural water use is often inefficient.

Water-conserving techniques such as drip irrigation, for example, could be better used in areas where water is scarce with the right price signals. Drip irrigation systems lose very little water to runoff, deep percolation or evaporation. They also decrease water contact with crop leaves, stems and fruit, thereby helping to prevent disease. Agricultural chemicals, particularly fertilisers, can be used more efficiently with drip irrigation; as only the crop root zone is irrigated, nitrogen already in the soil is less subject to leaching losses. Less insecticide may be required if applied through drip irrigation. However, because agricultural water use remains heavily subsidised in most countries, conversion to drip irrigation will increase production costs, so thus far it is predominantly used for higher-value crops. Drip irrigation systems cost anywhere from USD 1 200 to USD 5 000 per hectare (depending on the type of crop being irrigated) so a strong price signal needs to be given to water users to encourage uptake (Petkov and Kireva, 2003; Shock, 2006). Israel, for example, has instituted fuller-cost pricing of agricultural water which encouraged drip irrigation systems – the resulting efficiency improvement made it an exporter of agricultural products, even though its climate is dry.

A common driver of over use of water is under-pricing. While many countries have water that is priced for domestic and even industrial use, agricultural water use remains heavily subsidised or even free in some OECD countries (OECD, 1999).

Climate change

Agriculture is both a contributor to, and a victim of, climate change. And climate change is likely to have both positive and negative impacts on agriculture. The Intergovernmental Panel on Climate Changes's (IPCC) Fourth Assessment Report

Figure 14.5. **Water stress, 2005 and 2030**

2005

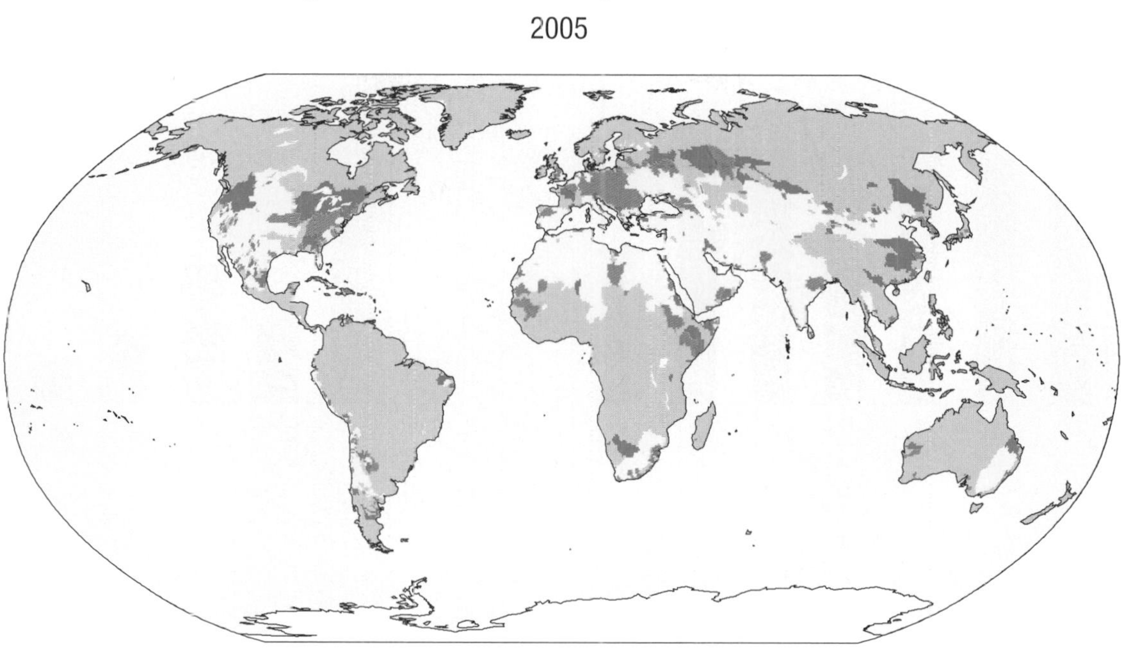

2030

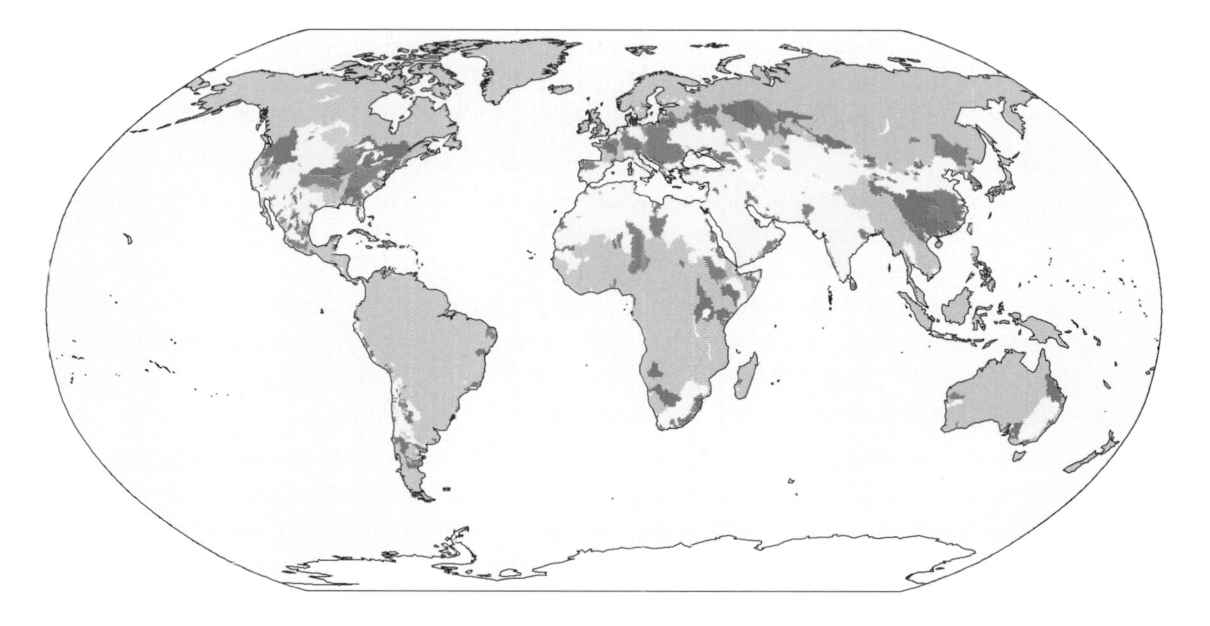

No water stress

Low water stress

Medium water stress

Severe water stress

StatLink http://dx.doi.org/10.1787/

Source: OECD Environmental Outlook Baseline.

Figure 14.6. **Water withdrawals and irrigation**

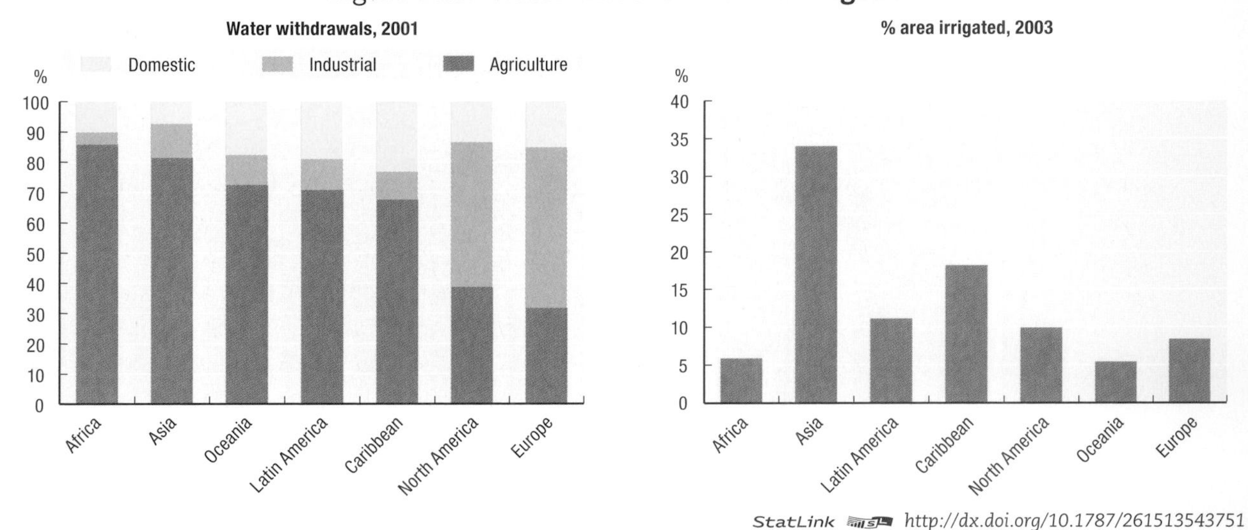

StatLink http://dx.doi.org/10.1787/261513543751

Source: AQUASTAT, 2007.

documents an observed global warming trend and predicts a wide range of climate changes that will affect agricultural production and food security worldwide (IPCC, 2007b and see Chapter 7, this report). With higher temperatures, the hydrological cycle will be intensified as more water evaporates and on the whole more precipitation results. This effect of both temperature and precipitation will be very unevenly distributed and in many areas climate may even become drier.

In the Baseline for this *Outlook*, the results for GHG emissions from land use changes[3] between 2005 and 2030 show only small changes for the world as a whole but large changes in individual regions (Table 14.2). These changes in emissions are dominated by CO_2.

In Brazil and China, deforestation is slowly declining, and therefore deforestation emissions are decreasing as well. In Russia and Central Europe, the decrease in agricultural land in the 1990s led to re-growth in vegetation, which in turn led to carbon uptake that is counted as anthropogenic. From 2005 onwards, this re-growth is expected to stabilise or reverse – partly due to renewed economic growth. The region thus ceases to sequester carbon, and instead starts to emit carbon from land use. In North America, a small increase in agricultural land, combined with an increase in timber demand, is projected to lead to increases in CO_2 emissions by deforestation and timber harvesting.

Looking more broadly at climate change and agriculture, the IPCC reports that for the SRES A1B scenario,[4] 90% of models agree that parts of the temperate Northern and Southern hemispheres will experience lower precipitation.[5] In already water-stressed areas such as Southern Europe, Northern Africa and parts of the Americas, the negative impact (on agriculture and human settlements) could be substantial. Other areas, such as Southern and

Table 14.2. **Percentage differences in GHG emissions from land use changes, 2005 to 2030**

North America	Europe	Japan Korea	Australia New Zealand	Brazil	Russia	South Asia	China	ROW	World
21%	41%	19%	12%	–23%	158%	8%	–18%	–1%	2%

StatLink http://dx.doi.org/10.1787/257355804421

Source: OECD *Environmental Outlook* Baseline.

Eastern Asia and Northern Europe, may experience increases in precipitation. In other words, areas that currently produce a large share of global agricultural output may be significantly affected, potentially causing a shift in patterns of global production (IPCC, 2007a).

Yields are likely to increase with increased temperatures, precipitation and crop fertilisation (from CO_2) in mid and high latitudes. However, these gains are likely to be countered by reductions in the lower latitudes, particularly in Africa and the Indian sub-continent, caused by heat and water stress as well as changing growing seasons. Some crops also seem to be negatively affected by higher night time temperatures (Peng *et al.*, 2004). Crop productivity can increase with elevated CO_2 concentrations in the absence of other climate changes. However, as the severity of climate change increases, these positive effects are likely to be overtaken by the negative factors. Very long-term modelling results show that unmitigated climate change over the next century will lead to higher global and local mean temperatures, altered patterns of precipitation and increased water demand as well as increases in extreme events, such as drought or flooding. Unmitigated climate change will thus lead to progressively depressed yields and increased production risks in most regions over time – not just developing regions (Smith *et al.*, 2007; Rosenzweig and Tubiello, 2007).

The agriculture sector currently contributes 10-12% of global greenhouse gas emissions. Agriculture contributes roughly 50% of global methane (CH_4) emissions and 60% of global nitrous oxide (N_2O) emissions (see Chapter 7 on climate change). Main sources of CH_4 are rice paddies and livestock operations (manure and enteric fermentation), while agricultural soils are the source of N_2O. Agricultural soils can be a source or a sink of CO_2 depending upon cultivation practices; currently the net flux for soils is estimated to be in balance (IPCC, 2007c). Thus agriculture has potential for mitigating some of the emissions of GHGs (Table 14.3; and see IPCC, 2007c), and some of these practices also have indirect effects on ecosystems elsewhere. For example, increased productivity in existing croplands could avoid deforestation and its attendant emissions.

Table 14.3. **Sources of greenhouse gas emission/mitigation potential in agriculture**

Measure	Examples	GHG
Cropland management	Agronomy; nutrient management; tillage/residue management; water management (irrigation, drainage); rice management; agro-forestry; set-aside, land use change	CO_2, CH_4, N_2O
Grazing land management/ pasture improvement	Grazing intensity; increased productivity (*e.g.*, fertilisation); nutrient management; fire management; species introduction (including legumes)	CO_2, CH_4, N_2O
Management of organic soils	Avoid drainage of wetlands	CO_2, CH_4, N_2O
Restoration of degraded lands	Erosion control, organic amendments, nutrient amendments	CO_2, N_2O
Livestock management	Improved feeding practices; specific agents and dietary additives; longer term structural and management; changes and animal breeding	CH_4, N_2O
Manure/biosolid management	Improved storage and handling; anaerobic digestion; more efficient use as nutrient source	CO_2, CH_4, N_2O
Bio-energy	Energy crops, solid, liquid, biogas, residues	CO_2, CH_4, N_2O

Source: IPCC, 2007b.

The global technical mitigation potential of agriculture (excluding fossil fuel offsets from biomass) by 2030 has been estimated to be roughly 9% of all emissions. The economic potential for mitigation, however, is lower. For example, if a price of USD 50 per tonne of CO_2 were levied on GHG emissions (*e.g.* through a tax or tradable permit scheme), the potential for reduced emissions from agriculture is still significant but falls to roughly 4%

Box 14.3. **Agricultural technologies and the environment**

Most modern technological advances are beneficial to larger farms, driving changes in farming structure. Larger farms are able to adopt technologies which can reduce environmental impact and increase productivity. These include computer-aided chemical application, drip irrigation, and computerised feeding, milking and waste management systems. Global Positioning Systems (GPS) and computers can increase efficiency and reduce waste and pollution from farming activities. Computers help producers monitor and respond to weather variability on a day-to-day basis. Solar-powered weather stations in the field can be hooked up to a farmer's computer to relay information about current air and soil temperature, precipitation, relative humidity, leaf wetness, soil moisture, day length, wind speed and solar radiation. When combined with GPS and highly flexible tractors, the application of tillage, seeds, fertiliser, etc., can be adjusted on an ongoing basis for within-field variation. Farms that make extensive use of computer controlled systems can also treat livestock individually for health, feed, etc. – potentially making a very large farm as responsive to individual animals as small family farms are.

These kinds of changes could make it possible to feed the world's population in 2030 and 2050, but minimising the impact on biodiversity and the environment will still have to be ensured through appropriate policy. Farmers will invest in and implement technologies and farm practices if they expect the investment to be profitable, if they have the right education, information, motivation and financial resources, and if government policies set clear goals. Farmers will individually take action to make their livelihoods sustainable over the long run, but ensuring a wider range of environmental benefits requires government policy to create the right incentives.

Although many technological developments favour larger agricultural enterprises, the possibility of co-operative and leasing arrangements means that smaller enterprises can also benefit. In the past, small farms have mainly benefited from improvements in crop strains and animal care, but these have been slower to develop recently – so much of the improvement in productivity has come from re-organising farm activity. In the future, change in farm organisation is likely to continue to be the main source of productivity improvements.

As technological changes continue to remake the agricultural sector, policies will need to adapt. From an environmental perspective, current and future agricultural policies need to focus on the wider objectives of sustainability and minimising environmental damage, rather than the narrower aim of farm income support.

Source: Cooper and Sigalla (1996); OECD (2001).

(Smith *et al.*, 2007). Soil carbon sequestration (enhanced sinks) is the mechanism responsible for most of the mitigation potential, with an estimated 89% contribution. Mitigation of CH_4 and N_2O emissions from soils accounts for 9% and 2%, respectively, of the total mitigation potential. Of course, strategies to mitigate GHG emissions in agriculture change across the range of prices for carbon. At low prices, dominant strategies are those consistent with existing production such as changes in tillage, fertiliser application, livestock diet formulation and manure management.

Policy implications

Some of the environmental impacts of agriculture are taken into account by farmers (because they affect the natural resource base on which agricultural activities depend), but many are not and create externalities and public goods for which markets do not function, or only function poorly. Both production enhancing policies and failure to deal with externalities and public goods lead to prices for agricultural outputs that do not ensure maximum societal benefits from agriculture. Policies need to be put in place to narrow the gap

There is a range of policy approaches for limiting the harmful environmental effects of agriculture. These include economic, non-market and information-based instruments.

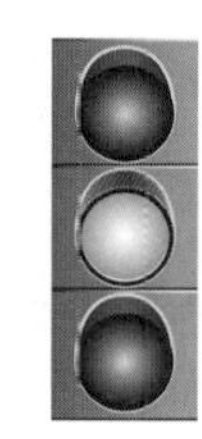

While support to agricultural production remains high in OECD countries, it is shifting away from the more environmentally harmful types of support measures.

Some countries, for example, impose taxes on farm chemicals to limit their use, while others regulate farm practices. OECD countries are increasingly placing greater emphasis on environmental sustainability in agricultural production: *i.e.* water protection, limits to air pollution, reduction of pollution from nutrient sources and chemicals (OECD, 2007b), prevention of soil erosion, and conservation of biodiversity and cultural landscapes (OECD, 2007a). New Zealand, for example, is developing a framework to facilitate implementation of economically and environmentally sound growth in the agriculture sector; that is a Programme of Action – co-led by the Ministry of Agriculture and Forestry and Ministry for the Environment – for sustainable land management, and for water quality and allocation. It is also developing technologies and systems for improving the economic and environmental performance of agriculture via a research consortium. Other countries are also focusing greater effort on research into the effects of climate change on agriculture. The *National Agriculture and Climate Change Action Plan* in Australia identifies four key areas to manage multiple climate change risks to agriculture: adaptation, mitigation, research and development, and awareness and communication. The *Action Plan 2000* in Canada finances programmes to address agricultural sources of emissions of GHGs. In other countries, supplementary payments are paid to producers whose environmental farm practices exceed those required by regulations. Switzerland, Korea, Norway and the EU provide payments for environmentally friendly farming, environmentally friendly livestock practices and improvement of animal welfare.

Economic instruments

A key cause of environmental damage by farming within OECD countries has been unconstrained support payments to farmers which are linked to production; these can encourage overly intensive farming techniques (Box 14.4). Increasingly, however, governments are attaching environmental conditions to these support payments. For example, in some countries in order to be eligible to receive or continue to receive existing agricultural support payments, farmers must adopt certain practices or conform to existing environmental regulations to reduce environmental harm. These are known as environmental cross-compliance requirements. A key limitation is that such payments are not necessarily received by those farming the most environmentally sensitive land, and only apply where support payments exist and environmental problems need to be addressed. Phasing out environmentally harmful production support (see the policy scenario below) and stricter implementations of already existing environmental policies would limit the need for such requirements, although not the need for targeted environmental measures to ensure farmers respect environmental regulations.

Box 14.4. **Progress toward de-coupling farm payments in the OECD**

The most environmentally damaging support transfers are those that encourage overproduction: namely those linked to unconstrained output and variable inputs. However, the share of support based on commodity output and inputs declined from 90% of producer support in 1986-88 to 70% in 2004-06 in OECD countries (OECD, 2007a). The share of support with output constraints fell from 20% to 13% over the same period, while the share with input constraints increased from 4% to 26%. A decrease in production-linked support (including market price support) is also shown by a significant reduction in the gap between domestic producer and border prices (as measured by the producer nominal protection coefficient, NPC). In 1986-88, average OECD producer prices were 51% higher than border prices; by 2004-06 the gap had halved to 25%. Compliance conditions, especially environmental, are increasingly being attached to payments.

The implementation of the 2003 Common Agricultural Policy reform is continuing in the European Union with the introduction of the Single Payment Scheme starting in 2005, and the reform of the sugar regime in 2006. The change in the support regime in the EU is particularly strong, as shown in the components of the Producer Support Equivalent (PSE), with support based on commodity output declining from 91% of the PSE in 1986-88 to 46% of PSE in 2006 (OECD, 2007a).

In Japan, a new basic plan for food, agriculture and rural areas has been established – one of its features is the introduction, from 2007, of new direct payments giving producers more flexibility over production decisions. Korea has begun to revise its rice policy, including abolishing government purchasing and introducing direct payments. Switzerland is gradually phasing out its milk quota production system and the country's new agricultural policy proposals for 2007-11 imply a further move away from measures supporting commodity prices. Iceland is gradually replacing milk payments based on output with headage payments (2005-12).

The United States is in the final year of the 2002-06 Farm Bill and proposals for the new Farm Bill for 2007-11 are underway. The 2002 Farm Bill expanded funding for conservation programmes by 80% and shifted the emphasis of new funding from land retirement to conservation on working lands. Some proposals in the 2007 Farm Bill would increase conservation funding while simplifying and consolidating conservation programmes. In the EU a new Rural Development Regulation was adopted for the 2007-13 period, with EU countries developing their implementation programmes.

Payments to farmers have also been used to reduce pollution (such as installing facilities to deal with animal waste), encourage ecosystem service provision (through field, meadow and wetland management), or support production practices that some governments deem to be favourable to the environment (such as organic systems or production of biomass for energy or materials). Although increasing, on average only about 4% of OECD support to farmers is targeted to agri-environment payments. Overall, these policies that generate benefits associated with the provision of environmental services should be weighed against other policies that contribute to increasing environmental damage.

Other policies that incorporate economic incentives for improved environmental performance in agriculture include (based on OECD, 2003a):

- *Payments based on farm fixed assets*: monetary transfer (including implicit transfers such as tax and credit concessions) to offset the investment cost to farmers of adjusting farm structure or equipment to adopt more environmentally friendly farming practices, or for purchasing conservation easements (not including land retirement).

- *Payments based on land retirement:* monetary transfers (including implicit transfers such as tax and credit concessions) to farmers for retiring eligible environmentally fragile land from commodity production for a given contract duration.
- *Payments based on farming practices:* annual monetary output or input based transfers (including implicit transfers such as tax and credit concessions) to farmers to support site-specific production practices that have the greatest net environmental benefit.
- *Tradable rights/quotas:* environmental quotas, permits, restrictions and bans, maximum rights or minimum obligations assigned to economic agents which are transferable or tradable.
- *Environmental taxes/charges:* taxes or charges relating to pollution or environmental degradation, including taxes and charges on farm inputs that are a potential source of environmental damage. Unfortunately, when it comes to environmental objectives there is a relative absence of environmental taxes and charges. In fact, the dominance of agri-environmental payments (to remunerate farmers for ecosystem service provision or to compensate them for their costs of reducing pollution) in OECD countries, suggests that farmers in some countries may have retained broad implicit rights in the use of natural resources.

One of the biggest challenges facing policy-makers in designing agricultural policies for environmental concerns is to clearly define boundaries – "property rights" – to indicate where farmers should be held liable at their own cost for environmental damage, and where they could be remunerated for providing environmental services that go beyond usual "good farming practices" and for which markets are absent or poorly developed. This is closely linked to the application of the polluter-pays principle in agriculture, which, given the significance of non-point source pollution in agriculture and the historic rights given to farmers, is often only weakly applied in the sector (see Chapter 13, Cost of policy inaction, for a wider-ranging discussion).

Non-market instruments

Non-market instruments include regulatory requirements. Their use is fairly widespread in OECD countries to tackle the environmental impacts of agricultural production, although they are not necessarily the most economically efficient measures to do so. These, for example, can state that animal waste should be dispersed over a wide enough area that nitrogen and phosphorus loading does not exceed absorptive capacity, or that pesticide/herbicide use does not exceed health and safety standards. Fines and penalties for breaching environmental legislation are the usual means of enforcement.

Other non-market instruments include community-based measures. These grant support to public agencies or community-based associations (*e.g.* landcare groups, conservation clubs, environmental co-operatives) to implement collective projects to improve the environment.

Information instruments

Information can aid farmers and consumers to make environmentally sustainable choices:

- *Labelling standards/certification:* voluntary participation measures defining specific eco-labelling standards that have to be met by farm products for certification. The most widely used environmental labelling schemes are for organic foods, which certify that the foods were produced without the use of certain chemical inputs.

- *Technical assistance/extension:* on-farm services providing farmers with information and technical assistance to plan and implement environmentally friendly farming practices.
- *Research:* measures granting support to institutional services to improve the environmental performance of agriculture through research, including in such areas as ecology, engineering, farm management practices, farmer behaviour and economics.
- *Inspection/control:* measures granting support to institutional services controlling the environment associated with agriculture, including monitoring and enforcement of policy measures addressing the environmental effects of agriculture (administration costs).

Information approaches can complement (though not replace) other policies – both economic instruments and regulatory approaches – that aim to reduce the impacts of agriculture on the environment.

Policy scenario: environmental impact of reducing production support and protection

As we have seen above, agricultural producers in the OECD are heavily supported through direct payments and price support. Prices for their products are also protected through the use of tariffs. Subsidies to agriculture allow production to occur in areas where it might not otherwise be viable. This, in turn, causes the agricultural sector to be larger in some places than it would be otherwise and subsidies typically cause overproduction (Box 14.5). The impact of agriculture on the environment and biodiversity can thus be

Box 14.5. **Intensive *versus* extensive agriculture**

Policies that affect market prices for agricultural inputs or outputs influence substantially the methods used for agricultural production and trade. Input and production linked payments, as well as protection from external competition, cause agricultural products to be grown in areas where they might not otherwise be viable or in a manner that can cause significant damage to the environment (OECD, 2005b). That is, the agricultural sector may be larger than it otherwise would be in a particular region, affecting environmentally sensitive lands (Lubowski *et al.*, 2006). Such policies may also encourage farming to be more intensive and environmentally damaging than it otherwise would be, as a result of greater use of chemical inputs. However, neither of these adverse impacts is inevitable. Recent moves to reform agricultural policy have reduced its environmental impact in a number of OECD countries. For example, the Common Agricultural Policy reform in the EU is decoupling payments traditionally given to farmers from both inputs and outputs. In other cases, policies that explicitly target environmental impacts, such as nitrogen loading of groundwater or ammonia evaporation, could potentially mitigate the environmental consequences without having overly strong impact on agricultural intensity by encouraging more efficient application of nutrients. Explicit protection of environmentally sensitive land can also mitigate the effects of expanding agriculture.

The debate concerning intensive *versus* extensive agriculture may thus be somewhat misplaced. The environmental impact of intensive agriculture, and the biodiversity loss from expansion of agricultural lands, occurs because policies for safeguarding the environment were not part of the original policy design. It is, however, possible to make agriculture more intensive (to save on land), or make it more extensive (to save on chemical application), without undue environmental impacts.

magnified by subsidies that prevent production from being rationalised to its most productive areas. Gottshalk *et al.* (2007) show that even the type of subsidy that is applied can have different consequences for biodiversity, with income support having less impact than input/output linked payments. However, in cases where environmental and biodiversity protection is inadequate (or not sufficiently enforced), lowering subsidies in some countries may cause a net loss of biodiversity by having agricultural production migrate and cause even more environmental damage elsewhere. The Secretariat of the Convention on Biological Diversity (SCBD, 2006) finds an increase in pressure on biodiversity from agricultural liberalisation that may cause a net loss of global biodiversity.

Overall in the OECD, while the nominal value of total support to farmers has increased in recent years, there has been a long-term downward trend in support (as a percentage of the value of agricultural revenues; OECD, 2007a). Reducing support would be expected to cause significant changes in the agricultural sector of OECD countries, as well as in non-OECD countries, leading to changes in environmental impact. To explore the results of such changes, a simulation was run in the ENV-Linkages model to lower by 50% the support in OECD countries that directly distorts markets for agricultural products.

The ENV-Linkages model used in this simulation is based on the global trade, assistance, and production (GTAP) database (Dimaranan, 2006), which uses 2001 as the base year. The quantitative results shown below thus reflect policy changes relative to 2001. So, for example, the changes in agricultural input/output-linked payments (OECD, 2007a) made in recent years – in particular the more recent broad move towards decoupling support from production – are not explicitly taken into account in the simulations. Nonetheless, the simulations undertaken on the basis of 2001 input/output-linked payment data (the GTAP database) are qualitatively consistent with actual policy implementation in some OECD countries; they are thus not purely hypothetical analyses, even if the magnitudes differ. Table 14.4 illustrates the magnitudes represented in the GTAP database (using terminology defined in the database).

Table 14.4. **Agricultural input/output-linked payments in selected countries (2001, millions USD)**

	USA	EU15	Canada	Mexico	OECD total
Value of output	198 772	234 150	24 096	79 939	717 013
Subsidies	32 746	36 001	2 347	7 729	87 880
Output	9 841	3 586	265	1 411	17 586
Intermediate input	6 760	–1 344	123	1 290	7 563
Factor	16 145	33 759	1 959	5 028	62 731

StatLink http://dx.doi.org/10.1787/257385057403

Source: GTAP 6 database.

To explore agricultural issues, the ENV-Linkages model has been specially adapted to represent agricultural sectors in more detail. Instead of the three sectors (livestock, crops and rice) used in other chapters of this *Outlook*, ten primary agriculture sectors were distinguished along with eight sectors for products such as dairy products and meats (which are normally classified as non-durable manufacturing). In the analyses below, the primary agricultural sectors are subjected to support reduction.

Production-linked support reduction

For this analysis, support payments were removed only in OECD countries. Some payments to farmers in OECD countries are specifically oriented to non-commodity products such as environmental service provision. Since these payments are currently small, the analysis assumes that these payments are not reduced.

Reducing production-linked agricultural payments by 50%[6] had substantial and, in some cases, surprising results (Table 14.5). In both Brazil and Canada, an increase in output was projected, coming mainly from oilseeds. For the USA, the biggest percentage losses are in oilseeds and rice, but all agricultural sectors suffer some loss.

Table 14.5. **Impact of policy simulation on agriculture and land use types, relative to Baseline in 2030**

Country/region	Change in area for livestock	Change in crop area	Comment
Canada	0.9%	2.3%	Some loss of prairie and forests
Central America	0.1%	2.6%	Additional pressure on forests and rainforests
Brazil	0.1%	2.3%	Loss of some cerrado, pressure on rainforest
Mexico	2.9%	–1.0%	Loss of natural pastureland, less pressure on forests
Latin America	0.0%	1.7%	Additional pressure on forests and rainforests
East Asia	0.2%	1.2%	Additional pressure on forests and rainforests
..	..	..	..
Italy	–4.2%	3.4%	Shifting composition with some use of marginal lands
UK	0.3%	–1.2%	Crops shifting to pastureland, possible loss of landscapes
Iceland, Norway and Switz.	–1.1%	–1.1%	Gain in forested areas with some loss of alpine pastures
Spain	15.3%	–17.6%	Shift of crops to pasturelands
Rest of EU15	1.2%	–3.8%	Increasing forest areas and pastureland
USA	0.0%	–5.2%	Low quality cropland to pastures and natural pastureland

StatLink http://dx.doi.org/10.1787/257451265631

Source: *OECD Environmental Outlook* Baseline and policy simulations. Model used: OECD Env-Linkages.

The results illustrate that the regions with the largest support payments will experience the greatest impact. The land use implications of these payments are significant, given that the results imply changes measured in thousands of square kilometres. The simulation shows little net difference in total land used for agriculture worldwide. OECD (2005a) undertakes a broader analysis of trade liberalisation (reducing support payments, tariffs, as well as other measures) and focuses on the impacts on arable crops. It is thus a useful source for more detailed information.

The aggregate impact on environment and biodiversity of these changes is not clear-cut given the heterogeneity of these regions.Table 14.5 outlines in a general sense the types of agricultural land use changes that would be associated with reducing production-linked support. The increases in agricultural land shown in the top part of the table are strong in tropical countries, while the decreases are strong in temperate zones; so it appears that there is a shift in production from temperate to tropical areas. These changes are relative to the Baseline, which in most cases is different from today's agricultural land use (Table 14.1). The net environmental result of this reform would be strongly dependent on the safeguards that are negotiated along with the payment reductions, and the policies put in place to limit the environmental impacts of agriculture in regions where it would be expected to expand.

Table 14.5 also shows that there is a tendency to increase land used for livestock as a response to the policy, but little can be said of the general trend for crops. The change in distribution and composition of crops suggests that policy needs to be carefully studied for its impact on some key ecosystems in a few countries. Of relevance to these results is some evidence showing that production intensity in countries with historically high levels of fertiliser and pesticide application falls with reductions in production-linked support, lowering environmental stress in these areas (OECD, 2005b).[7] At the same time, in other countries, increasing production by using more agro-chemicals has increased environmental pressure in these areas – except in parts of Africa where agro-chemicals were in such low use that increased use could lead to gains in production with little environmental impact.

Costs of inaction

The costs of policy inaction on the environment in agriculture will be borne through a number of channels and impacts. Nitrogen loading – mostly in non-OECD regions – will imply that surplus nitrogen enters into groundwater, surface water bodies, and the atmosphere. The nitrogen loading from non-point sources in agriculture is exacerbated by industrial and urban point sources, including discharges from sewage water into surface water.

For climate change, Arnell *et al.* (2002) provide forecasts of the effects of climate change on cereal production. Their unmitigated scenario (IS92a) has lower greenhouse gas emissions than the Baseline developed for this *Outlook*, but suggests a reduction of global cereal production of about 30 million tonnes in the 2020s. On the other hand, Fischer *et al.* (2002) actually show an increase in global cereal production on all lands (not just currently cultivated lands). The land use impacts of emissions that underlie their results have been modelled by the UK Hadley Centre's global circulation model. Since the impact from their model is asymmetric (developing countries lose land), the result will be an increase in the number of people going hungry. Other work also supports the projection that by 2030 a small impact will be felt in agriculture from climate change, but its distribution is highly uncertain given the lack of consensus of climate models on where temperature and precipitation changes will occur (IPCC, 2007a).

Nonetheless, the IPCC has summarised studies on the impacts on agriculture of a one to two degree Celsius temperature change (Table 14.6). Such increases in temperature could be seen by mid-century under the Baseline developed for the *Outlook*, with higher temperature increases expected by the end of the century if the trends in the Baseline were to continue (see Chapter 7 on climate change). These impacts do not include the more speculative, though plausible, impacts of spreading tropical pests and diseases to current temperate areas.

These impacts will occur with only marginal changes if mitigation is begun immediately. This is because stopping all emissions today would still lead to roughly 0.2° Celsius[8] of additional warming by 2030, and 0.5 °C of additional warming is likely to result with all but the most aggressive mitigation measures (IPCC, 2007a). When such changes are added to the warming that has already occurred, the impacts shown in Table 14.6 seem inevitable – the only question is where they will occur. Inaction on agricultural policy related to climate change can be expected to lead mostly to impacts on developing countries and future generations.

Table 14.6. **Impact of a 1 to 2 degree Celsius temperature change**

Sub-sector	Region	Finding
Food crops	Temperate	• Cold limitation alleviated for all crops • Adaptation of maize and wheat increases yield 10-15% • Rice yield no change
Pastures and livestock	Temperate	• Cold limitation alleviated for pastures • Seasonal increased frequency of heat stress for livestock
Food crops	Tropical	• Wheat and maize yields reduced below Baseline levels • Rice is unchanged • Adaptation of maize, wheat, rice maintains yield at current levels
Pastures and livestock	Semi-arid	• No increase in net primary productivity • Seasonal increased frequency of heat stress for livestock
Prices	Global	• Agricultural prices: –10% to –30%

Source: Easterling *et al.*, 2007.

Notes

1. Annual surface balance includes nitrogen inputs for agricultural systems: biological fixation, atmospheric deposition, application of fertilisers, application of animal manure and animal manure excreted during grazing. It also includes nitrogen from natural ecosystems such as atmospheric deposition, and biological fixation. The surplus reported in Figure 14.4 represents that which enters the groundwater or atmosphere.
2. However, given the high variability in regional precipitation and temperature changes between models (IPCC, 2007a), these results are useful in demonstrating the potential impacts by 2030, rather than being predictions of future outcomes.
3. Changes in carbon emissions from land use are related to changes in land cover. If agricultural area increases over a period of time, then there will be an increase in the carbon entering the atmosphere – from both decaying and burned trees, as well as from subsequent changes in the soil. Regrowth forest starts absorbing large quantities of carbon some time after land abandonment, initially at low rates, then followed by higher rates before slowing down again when reaching maturity.
4. See IPCC (2000) for more detail on the scenarios.
5. The spatial distribution, however, of changes in temperature and precipitation at any given global mean are subject to large uncertainties; state-of-the-art climate models yield very different patterns.
6. That is, payments reported in the GTAP database (Dimaranan, 2006) to farmers that are linked to variable inputs and outputs from the level that existed in 2001 – the GTAP base year.
7. Moreover, a detailed study by Lubowski *et al.* (2006) suggests that policies that increase incentives for crop cultivation stimulate production on land that has a greater environmental impact.
8. These temperature changes are relative to the 1990s' average. Add 0.5 degrees Celsius to compare to pre-industrial levels.

References

Arnell, N. W. *et al.* (2002), "The Consequences of CO_2 Stabilisation for the Impacts of Climate Change", *Climatic Change*, Vol. 53, pp. 413-446.

Cooper, J.B, and F. Sigalla (1996), *Agriculture, Technology, and the Economy*, Federal Reserve Bank of Dallas, Dallas, Texas.

Dimaranan, V.B., (Ed.) (2006), *Global Trade, Assistance, and Production: The GTAP 6 Data Base*, Center for Global Trade Analysis, Purdue University, Indiana.

Doornbosch, R. and R. Steenblik (2007), *Biofuels: Is the Cure Worse than the Disease*, OECD Report SG/SD/RT(2007)3, OECD, Paris, available at: *www.oecd.org/dataoecd/9/3/39411732.pdf*.

Easterling, W.E., *et al.*, (2007), "Food, Fibre and Forest Products", in M.L. Parry, *et al.*, (Eds.), *Climate Change 2007: Impacts, Adaptation and Vulnerability. Contribution of Working Group II to the Fourth Assessment Report of the Intergovernmental Panel on Climate Change*, Cambridge University Press, Cambridge, UK.

FAO (Food and Agriculture Organization) (2003), *World Agriculture: Towards 2015/2030 – An FAO Perspective*, J. Bruinsma, editor, FAO, Rome.

Faurès, J.M., Hoogeveen, J. and J. Bruinsma (2002), *The FAO Irrigated Area Forecast for 2030*, FAO, Rome, available at: *ftp://ftp.fao org/agl/aglw/docs/fauresetalagadir.pdf*.

Fischer, G., Shah, M., and H. van Velthuizen (2002), *Climate Change and Agricultural Vulnerability*, IIASA, Laxemburg, Austria.

Gottschalk, T.K., *et al.* (2007), "Impact of Agricultural Subsidies on Biodiversity at the Landscape Level", *Landscape Ecology*, Vol. 25, No. 5, pp. 643-56.

Heilig, G. K., Fischer, G., and H. van Velthuizen (2000), "Can China Feed Itself? An Analysis of China's Food Prospects with Special Reference to Water Resources", *The International Journal of Sustainable Development and World Ecology*, Vol. 7, pp. 153-172.

IEA (International Energy Agency), (2006), *World Energy Outlook*, International Energy Agency, Paris.

IPCC (Intergovernmental Panel on Climate Change) (2000), *Special Report on Emissions Scenarios*, Nakicenovic, N., *et al.*, (eds.), Cambridge University Press, Cambridge, UK, available at: *www.grida.no/climate/ipcc/emission/index.htm*.

IPCC (2007a), *Climate Change 2007: The Physical Science Basis*, Intergovernmental Panel on Climate Change, Geneva.

IPCC (2007b), *Climate Change 2007: Climate Change Impacts, Adaptation and Vulnerability*, Intergovernmental Panel on Climate Change, Geneva.

IPCC (2007c), *Climate Change 2007: Mitigation of Climate Change*, Intergovernmental Panel on Climate Change, Geneva.

Lubowski, R.N., *et al.* (2006), "Environmental Effects of Agricultural Land-Use Change: The Role of Economics and Policy", *Economic Research Report No.* (ERR-25), pp. 88, Economic Research Service, United States Department of Agriculture, Washington DC.

OECD (1999), *The Price of Water: Trends in OECD Countries*, OECD, Paris.

OECD (2001), *Adoption of Technologies for Sustainable Farming Systems: Wageningen Workshop Proceedings*, OECD, Paris.

OECD (2003a), *Inventory of Policies Addressing Environmental Issues in Agriculture: Developments of the Website and Its Regular Updating*, Joint Working Party on Agriculture and the Environment, OECD, Paris, 24-26 November, 2003.

OECD (2003b), *Agriculture, Trade and the Environment: The Pig Sector*, OECD, Paris.

OECD (2004), *Agriculture, Trade and the Environment: The Dairy Sector*, OECD, Paris.

OECD (2005a), *Agriculture, Trade and the Environment: Arable Crops Sector*, OECD, Paris.

OECD (2005b), *Environmentally Harmful Subsidies: Challenges for Reform*, OECD, Paris.

OECD (2006), *Agricultural Market Impacts of Future Growth in the Production of Biofuels*, AGR/CA/APM(2005)24/FINAL, OECD, Paris, available at: *www.oecd.org/dataoecd/58/62/36074135.pdf*.

OECD (2007a), *Agricultural Policies in OECD Countries: Monitoring and Evaluation 2007*, OECD, Paris.

OECD (2007b), *Instrument Mixes Addressing Non-Point Sources of Water Pollution*, OECD, Paris, available at: *www.olis.oecd.org/olis/2004doc.nsf/linkto/com-env-epoc-agr-ca(2004)90-final*.

OECD/FAO (2007), *OECD-FAO Agricultural Outlook 2007-2016*, OECD, Paris.

Peng, S., *et al.* (2004), "Rice Yields Decline with Higher Night Temperature from Global Warming", *Proceedings of the National Academy of Science*, United States, Vol. 101, pp. 9971-9975.

Petkov, P. and Kireva, P. (2003), "Use of Drip Irrigation in Bulgaria – Present State and Future Perspectives", paper presented at *Workshop on Improved Irrigation Technologies and Methods: R&D and Testing*, Montpellier, 18 September 2003.

Rosenzweig, C., and F. Tubiello (2007 forthcoming), *Metrics for Assessing the Economic Benefits of Climate Change Policies in Agriculture, ENV/EPOC/GSP(2006)12*, OECD, Paris.

SCBD (Secretariat of the Convention on Biological Diversity) (2006), *Global Biodiversity Outlook 2*, SCBD, Montreal.

Shock, C.C. (2006), "Drip irrigation: An introduction", *Sustainable Agricultural Techniques*: EM 8782, October, Oregon State University, Oregon.

Smith, P. *et al.* (2007), "Agriculture", in B. Metz, *et al.* (eds.) *Climate Change 2007: Mitigation, Contribution of Working Group III to the Fourth Assessment Report of the Intergovernmental Panel on Climate Change*, Cambridge University Press, Cambridge, United Kingdom and New York, USA.

Tyner W.E. and F. Taheripour (2007), "Future Biofuels Policy Alternatives", paper presented at the *Biofuels, Food, and Feed Tradeoffs Conference* organised by the Farm Foundation and USDA, April 12-13, St. Louis, Missouri.

ANNEX 14.A1

Biofuels Simulation Results

The ENV-Linkages model was used to compare four hypothetical biofuel scenarios with the Baseline: *1)* demand for biofuels growing in line with the IEA (2006) scenario; *2)* a demand scenario (DS) whereby growth in biofuel demand for transport is entirely driven by exogenous changes, keeping total demand for fuel for transport close to the Baseline; *3)* a high crude oil price scenario (OilS) to determine the profitability of biofuel in the face of increasing costs of producing traditional fossil-based fuels (Table 14.A1.1); *4)* a subsidy scenario (SubS) in which producer prices of biofuel are subsidised by 50%. This latter scenario helps to check if price support is enough of an incentive to drive up biofuel demands endogenously.

Only first-generation biofuels are considered in the model.* Our database distinguished three kinds: biodiesel from oilseeds and vegetable oils, crops-based ethanol (mainly from corn and wheat) and sugar-based ethanol. Biofuel trade between regions remains very limited. In ENV-Linkages the trade in biofuel is conditioned by corresponding trade balances in 2001, so countries like China or India are assumed implicitly to mainly consume their domestic production.

Table 14.A1.1. **International price of crude oil (2001 USD)**

Scenario	2005	2010	2015	2020	2025	2030
Baseline	46.8	48.2	49.1	49.9	50.8	51.6
OilS	46.8	55.7	60.0	65.4	68.8	68.6

StatLink http://dx.doi.org/10.1787/257468650776

Source: OECD Environmental Outlook Baseline and policy simulations.

Biofuel trends under alternative scenarios

Table 14.A1.2 indicates the evolution of biofuel as a share of transport fuel under the four scenarios. The Baseline projects that to 2030, biofuel's share of total transport fuel will increase slightly from its 2006 level, from 2% to 4%. In the demand scenario biofuels will displace petroleum more rapidly in OECD countries and Brazil than in the rest of the world. Between 2006 and 2015 these trends are consistent with OECD/FAO (2007) forecasts. After 2015, the rhythm of displacement is assumed to be less pronounced. The increase in biofuel use in the DS may appear rather exaggerated, but for some countries and regions,

* Given the uncertainty with second generation biofuels noted in Doornbosch and Steenblik (2007), these simulations have not included them in the analysis.

Table 14.A1.2. **Share of biofuel as a percentage of all transport fuel (volume in gasoline energy equivalent)**

		Baseline			DS			SubS			OilS		
		2006	2015	2030	2006	2015	2030	2006	2015	2030	2006	2015	2030
OECD	Crops ethanol	1.5%	2.0%	3.5%	1.6%	4.5%	10.0%	5.0%	7.8%	13.2%	1.5%	3.4%	10.0%
	Sugar ethanol	0.1%	0.1%	0.7%	0.1%	0.4%	2.8%	0.2%	0.6%	3.4%	0.1%	0.2%	2.3%
	Biodiesel	0.3%	0.4%	0.7%	0.3%	1.3%	3.0%	1.2%	1.8%	3.9%	0.3%	0.6%	2.3%
BRICs	Crops ethanol	0.0%	0.1%	1.4%	0.0%	1.0%	9.4%	0.1%	0.6%	6.5%	0.0%	0.2%	2.6%
	Sugar ethanol	4.9%	3.9%	3.5%	5.0%	5.2%	8.9%	11.4%	9.3%	7.9%	5.0%	5.1%	6.8%
	Biodiesel	0.0%	0.0%	0.0%	0.0%	0.4%	3.9%	0.0%	0.0%	0.1%	0.0%	0.0%	0.0%
ROW	Crops ethanol	0.0%	0.1%	0.5%	0.0%	0.5%	3.8%	0.0%	0.4%	3.0%	0.0%	0.1%	1.6%
	Sugar ethanol	0.1%	0.1%	0.4%	0.1%	0.8%	3.2%	0.3%	0.7%	2.5%	0.1%	0.2%	1.2%
	Biodiesel	0.0%	0.0%	0.4%	0.0%	0.3%	2.7%	0.0%	0.2%	2.0%	0.0%	0.1%	1.3%
World	Crops ethanol	0.9%	1.1%	2.1%	1.0%	2.7%	8.3%	3.0%	4.3%	8.6%	0.9%	1.9%	5.9%
	Sugar ethanol	1.0%	1.0%	1.5%	1.0%	1.7%	5.0%	2.4%	2.7%	4.6%	1.0%	1.4%	3.1%
	Biodiesel	0.2%	0.2%	0.4%	0.2%	0.8%	3.2%	0.7%	1.0%	2.2%	0.2%	0.4%	1.5%

StatLink http://dx.doi.org/10.1787/257475433434

Source: *OECD Environmental Outlook* Baseline and policy simulations. Model used: OECD Env-Linkages.

such as the USA, EU and Brazil, it would correspond to some politically defined targets. For some developing countries the DS is optimistic too. For example, in China and India the biofuel share in transport fuels would grow from less than 1% in 2006 to 23% and 11% respectively in 2030. Globally the demand scenario suggests a 16% displacement by biofuels of other transport fuels by 2030.

In both the OilS and SubS scenarios the demand for biofuel is indirectly stimulated by its production becoming more competitive. The more the international price of crude oil increases, or the more biofuel is subsidised, the more financially profitable it becomes to produce biofuel instead of refined oil. The high subsidy policies seem to be effective in displacing as much conventional fuel as the DS (almost 15% in 2030). But it is a costly policy since the total cost of the subsidies was USD 14.3 billion (2001-USD) in 2006 for OECD countries and is projected to reach USD 82.5 billion in 2030. Globally the total cost would equal 0.45% of world GDP in 2030. The higher financial profitability of biofuels resulting from a crude oil price increase appears to be less incentive than the direct subsidy to biofuel production, so both production and demand incentives would be necessary to attain a higher than 10% target for the entire world.

Impacts of biofuel production on prices

Using more maize (corn), oilseeds or sugar as energy inputs will increase their price (Table 14.A1.3), a trend that has already been emerging in the last couple of years (OECD/FAO, 2007). But it will also indirectly increase the prices of other agricultural products, from the increase of rate of return to land resulting from competition between land surfaces. For livestock production another effect could be added – an increase in the price of livestock (given the increase in price of inputs to livestock production – namely, cereals and other crops).

World price changes are rather sensitive to assumptions about land availability and the possibility of switching easily between crops. Note that this table reflects changes in international prices – in some countries the price changes would be expected to be considerably higher.

Table 14.A1.3. **World prices[a] of agricultural products (% differences from the Baseline)**

	DS			SubS			OilS		
	2010	2020	2030	2010	2020	2030	2010	2020	2030
Other crops (wheat, rice)	0.1%	0.2%	0.3%	0.3%	0.5%	0.6%	0.1%	0.2%	0.2%
Livestock	0.1%	0.3%	0.4%	0.5%	0.8%	0.8%	0.0%	0.1%	0.2%
Oil seeds	0.2%	1.1%	2.1%	1.3%	2.1%	2.5%	0.2%	0.7%	1.3%
Sugar	0.1%	7.2%	25.3%	0.3%	1.4%	3.4%	0.1%	0.6%	1.4%
Cereals (corn)	1.4%	4.8%	8.0%	6.2%	12.0%	15.3%	0.6%	3.7%	7.9%

StatLink http://dx.doi.org/10.1787/257488278825

a) "World prices" here imply a weighted average of import prices.

Source: *OECD Environmental Outlook* Baseline and policy simulations.

ISBN 978-92-64-04048-9
OECD Environmental Outlook to 2030

Chapter 15

Fisheries and Aquaculture

Without better fisheries management, overfishing and ecosystem damage is likely to lead to significantly reduced incomes or even the collapse of a number of fisheries in the coming decades. There will be severe consequences for local populations dependent on these resources for food and economic development. This chapter reviews the environmental pressures both from and on fisheries and aquaculture and projects the global trends in production and consumption. Looking to 2030, it will be important for governments to address gaps in the institutional and legislative framework for managing the environmental impacts of fisheries and aquaculture, and to strengthen implementation of the existing agreements. At the same time, environmental degradation driven by activities in other sectors is also affecting the economic viability of fisheries. Policies are needed to tackle pollution from land based sources and shipping, to reduce or halt the introduction of invasive alien species, and to help fishing communities adjust to the impacts of global climate change.

KEY MESSAGES

Overfishing remains a major challenge. An estimated 25% of world fish stocks are over-exploited or depleted, and 52% of stocks are producing catches near maximum sustainable limits. Marine and freshwater ecosystems also experience a range of other pressures from capture fishing if it is not conducted responsibly, including destruction of habitats and incidental kill of non-target species. Aquaculture increases pressure on species used for fishmeal and fish oil and can contribute to habitat destruction and pollution.

The economic sustainability of both capture fisheries and aquaculture is itself at risk from environmental pressures – including pollution from land-based sources and ships, the spread of invasive alien species and the impacts of global warming. Climate change is likely to affect the number and distribution of fish stocks, the acidity of marine waters, and the resilience of some aquatic ecosystems.

The rapid expansion of aquaculture is expected to continue to 2030, compensating for declining or stagnant wild fish harvests, but its environmental consequences deserve attention.

Policy options

- Reduce the environmental impacts of capture fishing by limiting total catch levels, in particular through setting total allowable catch (TAC) levels and the use of market-based instruments such as individually transferable quotas (ITQs), fishing seasons and zones; regulating fishing methods and gear use; eliminating environmentally harmful subsidies; reducing fishing effort and existing over-capacity; improving the environmental performance of fishing vessels; and ensuring that consumer prices incorporate environmental costs of production. Set total allowable fish catch levels based on scientific advice.
- Reduce the environmental impacts of aquaculture by: developing national aquaculture plans; regulating the location and operation of aquaculture farms to minimise negative environmental impacts (*e.g.* release of nutrients or antibiotics, escape of organisms, destruction of habitat); and developing alternative feeds that reduce the reliance on capture fisheries.
- Increase the resilience of fisheries communities through strengthening policies and increasing enforcement of existing measures to address the impacts of environmental degradation on the fisheries sector, and to help fisheries activities adapt to climate change.
- Continue to pursue international co-operation to strengthen the management of straddling, highly migratory and high seas stocks. Use regional fisheries management organisations (RFMOs) to help co-ordinate management of regional fisheries. OECD countries have a role to play to ensure policy coherence for development and in helping developing countries to build capacity for sustainable fisheries management.
- Implement policies and surveillance systems to prevent illegal, unreported and unregulated fishing.

Consequences of inaction

- Without better fisheries management, overfishing and ecosystem damage are likely to lead to significantly reduced incomes or even the collapse of a number of fisheries in the coming decades, with severe consequences for local populations dependent on these resources for food and economic development.
- Pollution can decrease the value of fish products and can destabilise aquatic ecosystems that provide essential services for the fisheries sector. Consumers are increasingly concerned about the possible impacts on human health, for example from eating fish with high mercury levels.

Global fisheries and aquaculture production increased by 2.6% annually between 1988 and 2004, but limitations in supply are projected to slow this to an average of 2.1% annually between 2005 and 2030. The 2.1% total fisheries growth rate assumes robust growth in aquaculture, but no growth in capture fisheries. This implies an average growth rate in aquaculture of 3.9% annually to 2030.

Projected composition of world fisheries to 2030: capture and aquaculture

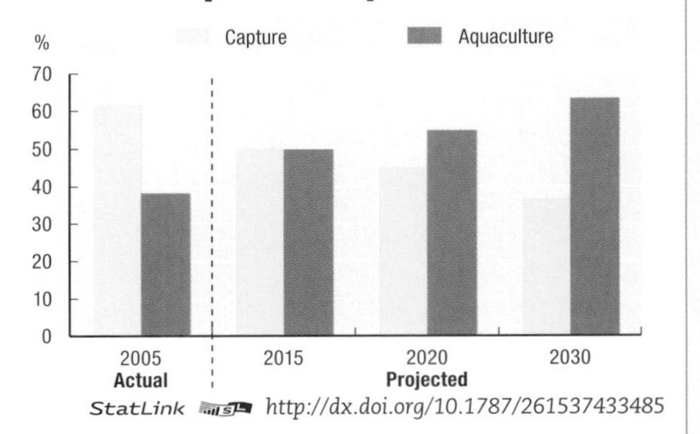

StatLink http://dx.doi.org/10.1787/261537433485

Introduction

Almost one-quarter of the world's capture fisheries are now classified as over-exploited, depleted or recovering (FAO, 2006). In addition to overfishing, fishing can also damage ecosystems through incidental catch of non-target species (by-catch), overfishing of young stocks, pollution and habitat destruction (*e.g.* through bottom-trawling).

Ecosystem change or damage can, in turn, affect the economic viability of fisheries. Some fisheries have faced economic collapse due to decimated fish stocks. Changes to marine ecosystems from climate change or pollution can upset ecosystems and lead to a decline or geographical shift in key fish stocks. The associated economic losses and disruption to fisheries dependent communities may be significant. The first section of this chapter outlines the inter-relationship between fishing and the environment. The second examines the recent trends and projections for the sector, and the third outlines the key policy options.

Environmental pressures from fisheries and aquaculture operations

Overfishing has traditionally been seen as the major environmental pressure from fisheries activities. But the incidental catch of non-target species, physical damage to habitats caused by destructive fishing practices and construction of aquaculture installations also may have significant impacts on aquatic stocks and ecosystems. In short, how many fish are harvested and how fishing is carried out in a given fishery, in addition to the state of the marine environment, are important.

Overfishing and by-catch

While we lack sufficient information for many species, the overall status of marine fish stocks exploited by commercial capture fisheries is of concern. Since the FAO first started monitoring the global state of fish stocks in 1974, there has been a consistent downward trend in the proportions of under-exploited and moderately exploited stocks (FAO, 2006). About 25% of stocks are classified as over-exploited or depleted, 52% as fully exploited, and only about 23% of commercially exploited marine stocks are considered to have some potential for further development (Figure 15.1; FAO, 2006). Similar global data are not available for inland fish stocks, but regional data suggest that the majority are heavily over-fished.

Depletion of fish stocks can disrupt ecosystems by distorting food webs and changing population dynamics. In over-fished regions, as stocks with high commercial value become depleted the size composition of the entire community is likely to change. This increases fishing pressure on smaller fish of the exploited species, as well as on other species of lower commercial value. Over-exploitation of fish stocks can also have severe impacts on income and employment in fishing communities. For example, it has been estimated that the closure of Atlantic cod fishing in 1992 led to income losses in Canada of CAD 250 million in the short-term, and potential long-term losses of CAD 1 billion per year (OECD, 2008 forthcoming).

Figure 15.1. **Global trends in the state of world marine stocks, 1974-2006**

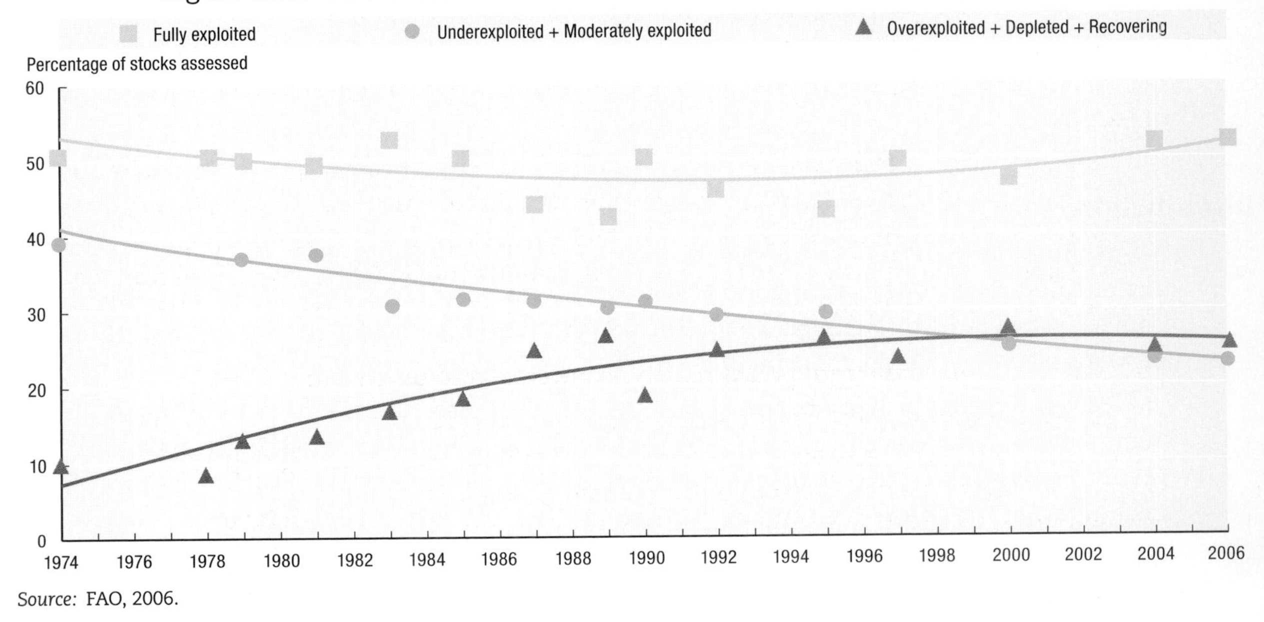

Source: FAO, 2006.

While in some cases heavily depleted stocks have recovered when fishing pressure was reduced, in others important fish stocks have failed to rebound even years after their fishing was reduced. When a species is functionally absent[1] from an ecosystem for an extended time, shifts in predator-prey interactions and in food web structures can lead to alternative states that effectively reduce the likelihood of re-establishment of the depleted species.

Capture fishing can also deplete populations of non-target organisms, including birds, sea mammals, crustaceans and finfish which are inadvertently killed by fishing nets or lines. Such "by-catch" is typically discarded overboard if the organisms have low commercial value, are below minimum size, or do not fit one of the boat's quotas. Trawlers targeting shrimp and flatfish are estimated to discard up to 50% of their catches (FAO, 2004b), although technical measures are available that could reduce these rates substantially. The FAO estimates that global discards declined from 27 million tonnes (Mt) in 1994, to 20 Mt in 1998, and to 7.3 Mt in 2004, although data on by-catch is limited. Such a downward trend in global discards can in part be explained by changes in estimation methods; however, the by-catch intensity of certain fisheries is believed to have diminished in recent years due to the wider use of selective fishing gear and "best practice" fishing techniques.

Habitat destruction and pollution

Fishing and aquaculture can also contribute to the physical degradation of aquatic habitats, sometimes so badly that the local fishing industry may be threatened. Some fishing gear and methods may damage various features of marine communities and habitats. Mobile bottom-contacting fishing gear (*e.g.* bottom trawls, dredges) can be so by-catch-intensive or damaging to ecosystem components such as sea beds that the damage can be irreversible. While for some communities infrequent disturbance (including trawling) may increase biodiversity, very frequent trawling of an area is correlated with a loss of biodiversity. As a result, some countries have restricted or prohibited the use of such gear as a complement to other management measures such as area or time closures.

OECD ENVIRONMENTAL OUTLOOK TO 2030 – ISBN 978-92-64-04048-9 – © OECD 2008

Without the right policies, the development of aquaculture farms can contribute to the destruction of habitats in coastal and inland areas. In a number of marine areas, littoral or estuarine waters that are most commonly developed for aquaculture are also of high ecological importance, having a key role in development and/or recruitment of young organisms.

Discharges from fishing vessels and aquaculture units can contribute to pollution of marine and inland waters. Fishing vessels generate air and water pollution and waste products, and older fishing vessels generally lack modern pollution control equipment. Water pollution from aquaculture farms comes from uneaten food fish, excreta, chemicals and antibiotics used to control diseases.

The impact of environmental pressures on fisheries and aquaculture

The economic viability of fisheries and aquaculture depends on functioning aquatic ecosystems which deliver essential ecosystem services. Long-term climate change, *El Niño* events (Box 15.1) and other environmental changes threaten the sustainability of fisheries and aquaculture. In addition, pollution can degrade the health of aquatic ecosystems, and thus destabilise the resource base supporting fisheries. Contamination of fisheries products by pollutants lowers their economic value.

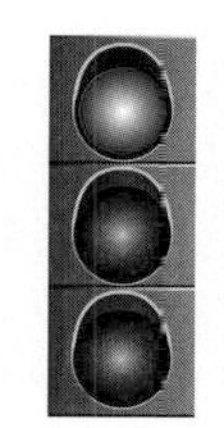

Climate change is likely to affect the number and distribution of fish stocks, the acidity of marine waters, and the resilience of some aquatic ecosystems.

Environmental perturbations

Anthropogenic climate change is expected to increase the mean temperature of sea surface waters and to cause the mean sea level to rise by 2100 (IPCC, 2007). Based on current model simulations, it is very likely that there will be a slowdown of the oceans' thermo-haline circulation by 2100, with severe consequences for fisheries and aquatic ecosystems. As ocean circulation drives larval transport, the recruitment patterns and population dynamics of marine organisms will be altered worldwide.

The pH of ocean surface waters is projected to fall by 0.14 to 0.35 pH units by 2100, due to the uptake of rising levels of atmospheric CO_2 (IPCC, 2007). The consequent acidification of surface waters will change the saturation horizons of aragonite, calcite and other minerals which are essential for calcifying organisms (Feely *et al.*, 2004). While many

Box 15.1. **El Niño Southern Oscillation**

The term "*El Niño*" refers to periods of strong and prolonged warm weather in the Eastern Pacific, accompanied by surface waters that are 0.5 to 3°C warmer than usual. Because changes in air pressure, called "the Southern Oscillation", typically accompany these periods, the whole phenomenon is referred to as the "*El Niño* Southern Oscillation" (ENSO). During an ENSO event, the upwelling of cold, nutrient-rich waters declines significantly, and primary productivity plunges in the eastern Pacific, resulting in a decrease in fish production. At the same time, the phenomenon disrupts weather patterns worldwide, leading to unusually high precipitation along the eastern coasts of the north and south Pacific. The frequency of occurrence of ENSOs is projected to increase with global warming.

aquatic organisms are adapted to thermal fluctuations, the expected changes in pH are higher than any pH changes inferred from the fossil record over the past 200 to 300 million years (Caldeira and Wickett, 2005).

The frequency and severity of a number of extreme weather events, such as tropical cyclones, are expected to increase as a result of global warming in the 21st century (see also Chapter 7). The consequent damage to equipment and infrastructure may compromise the productivity of fishing and aquaculture activities, as did the 2005 tsunami in the Indian Ocean which destroyed fishing boats, aquaculture installations and equipment. Developing countries suffer disproportionately from extreme weather events, as they often have weak response capacities.

Environmental pollution

Elevated levels of nutrients (eutrophication) contribute to algal blooms which cause hypoxic zones (areas deficient in oxygen, often referred to as "dead zones") in marine coastal areas and inland water bodies. The number and extent of such zones have increased since the 1970s, with some 200 persistent dead zones identified in 2006 (UNEP, 2006). Although estuaries and bays are most affected, eutrophication is also apparent in many semi-enclosed seas. For example, eutrophication affects almost all areas of the Baltic Sea, with the frequency and the spatial extent of toxic blooms both increasing since the mid-1990s, reducing the reproductive success of cod and other fish species (EEA, 2002).

Exposure to inorganic pollutants can compromise the breeding success, immunity and health of aquatic organisms. As they have the tendency to bioaccumulate in the body fat of fish, such pollutants can also pose health risks to humans eating them. Since the late 1990s, Baltic Sea countries have faced restricted market access for herring due to dioxin contamination. Inorganic pollutants are often found in fish products from near-shore areas, estuaries and rivers, as well as from regional seas that have relatively little exchange with the open ocean (*e.g.* Baltic, Mediterranean). Such contamination of fish products can lower their market value or block market access altogether (*e.g.* arsenic-contaminated mussels, mercury-contaminated fish). For example, after a 2004 study found that dioxin concentrations were higher in farmed salmon than in wild salmon, consumer concern led to a 25% drop in retail orders (FAO, 2004a).

It is estimated that about 80% of all marine pollution comes from land-based sources (UNEP, 2006). In most OECD countries, considerable progress has been made to reduce land-based discharges to the sea, particularly from municipal wastewater outfalls and industrial effluents (see Chapter 10). Diffuse pollution from agriculture and urban areas remains a big challenge, however, with nitrogen loading to some marine waters degrading ecosystems and damaging coastal fisheries. The Baseline for this *Outlook* projects a 4% increase in the global flux of nitrogen compounds from rivers to coastal marine systems to 2030, with the associated risk of coastal water eutrophication (see Chapter 10 on freshwater). The sources include increasing fertiliser run-off from agriculture and nutrient-loading from untreated urban wastewater. The most notable increases are expected from China and OECD countries, with more moderate increases projected for coastal zones around Africa due to lower fertiliser use in agriculture. Developing countries face a particular challenge in putting in place the necessary regulations and infrastructure to reduce land-based pollution of coastal zones and inland waterways.

The global shipping fleet[2] generates air and water pollution, including considerable operational and accidental discharges of oil. For example, European ships emitted an

OECD ENVIRONMENTAL OUTLOOK TO 2030 – ISBN 978-92-64-04048-9 – © OECD 2008

estimated 2.6 Mt of SO_2 and 3.6 Mt of NO_x to the air in 2000 (Richartz and Corcoran, 2004). Ships are also major sources of solid waste. An estimated 70 000 m^3 of litter enters the North Sea every year, 95% of it non-biodegradable plastics (Richartz and Corcoran, 2004). Exposure to tributyltin (TBT), an anti-fouling compound used worldwide on ship hulls, has been linked to reproductive anomalies in molluscs and other marine life.

Oil and gas production platforms, present on most continental shelves, also contribute to marine pollution levels through operational and accidental discharges of oil and chemicals. In the North Sea, operational discharges from the 475 offshore installations amount to 16 000 to 17 000 tonnes of oil per year (EEA, 2002). Elevated levels of hydrocarbons can be found in the sediment up to 8 km from offshore platforms, and levels of cadmium, mercury and copper are also high in some locations (Richartz and Corcoran, 2004). A number of the chemicals discharged in the "produced water" from platforms have been implicated as endocrine disruptors which reduce the breeding success of certain fish stocks.

Coastal development, aggregate extraction and dredging also destroy or damage key near-shore habitats for juvenile marine organisms.

Introduction of invasive alien species

Invasive species, spread worldwide by "hitch-hiking" in ship ballast waters or on hulls, have in some cases accelerated the collapse of fish stocks (*e.g.* the comb jellyfish and the Black Sea anchovy). Strengthened legislation and better implementation of existing provisions is needed to control the introduction of invasive aquatic species (see also Chapter 9 on biodiversity). Ballast water is essential for the safe and efficient operation of ships, by providing balance and stability, but its transfer worldwide can have serious ecological, economic and health implications.

The spread of invasive species and pathogens can also be facilitated by fisheries and aquaculture. Fishmeal and seed stock used in aquaculture farms are traded internationally, and can spread pathogens and parasites from one marine region to another. Organisms that escape from aquaculture farms often survive in the wild where they compete with native species for habitat and food, and may spread diseases and parasites (*e.g.* sea lice spread by escaped sea trout). In some cases, they can also interbreed with native species, leading to "genetic pollution".

Key trends and projections

Global trends in production and consumption

Average consumption of fish per person has nearly doubled since 1960 worldwide, reaching 16.2 kg per year in 2002. Actual consumption varies widely among regions, with per capita demand highest in OECD countries and in China, and lower in Africa and South America. It is projected that per capita demand for fish will continue to rise by a further 18% to 2015, driven by economic growth and increased awareness of the benefits of consuming fish (FAO, 2004a). Improved access to international markets will further increase pressures on aquatic ecosystems, particularly in developing countries.

Global fisheries production, including both capture fisheries and aquaculture, has risen sharply during the past three decades, reaching 140.5 million tonnes in 2004 (Figure 15.2). Since 1988, total world fisheries production has grown by 2.6% annually.[3] Most of the increase in fisheries has come from new aquaculture development, primarily in non-OECD regions. A very large share of this has come from China.

Figure 15.2. **World fisheries production, 1970-2004**

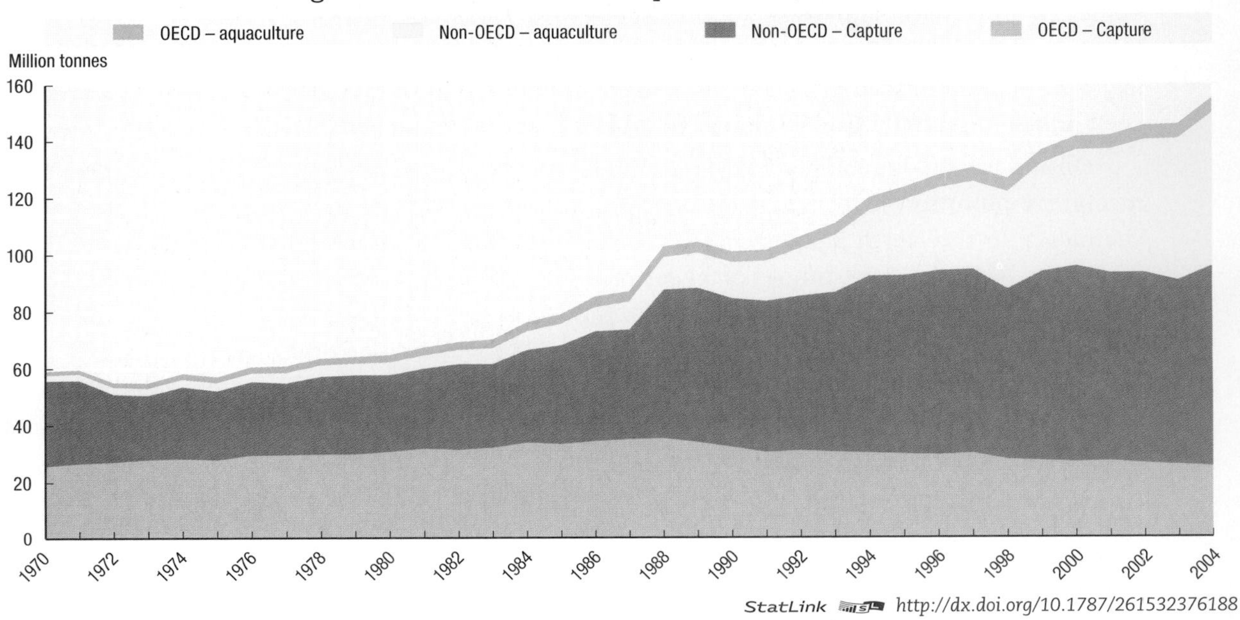

StatLink http://dx.doi.org/10.1787/261532376188

Source: Based on FAO, 2007.

Global capture fisheries production has levelled off at between 90 and 95 Mt since the late 1990s, with marine capture fisheries contributing about 85 Mt, and inland freshwater fisheries the remainder (in Figure 15.2 this is seen by combining OECD and non-OECD capture fisheries). This levelling off reflects the fact that an estimated 52% of the world's fisheries are now fished at their maximum limit, and 24% are overfished, depleted or recovering (FAO, 2006). OECD regions have been reducing their catches in recent years, by 40% between 1988 and 2004. Non-OECD regions increased their capture fisheries production by 35% over the same period.

During the 1970s and 1980s, the rate of growth of the total catch slowed to about 2% per year, before approaching zero in the 1990s, and declining slightly since 2002 (FAO, 2004a). OECD countries landed 27% of the world capture fisheries catch in 2002, with the United States (4.9 Mt), Japan (4.4 Mt) and Norway (2.7 Mt) among the world's top ten producing countries. China (16.6 Mt) and Peru (8.8 Mt) led the list, together landing 27% of world catch (Box 15.2).

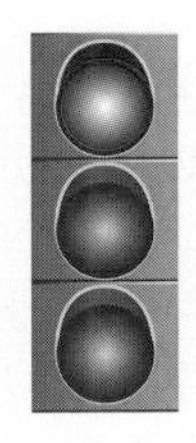

Overfishing is likely to lead to economic collapse of some fisheries and disruption of marine ecosystems.

Inland fisheries

Global landings from inland capture fishing have reportedly been stable at about 8.6 Mt since 2000.[4] The bulk of this inland capture is landed in Asia (66% in 2002) and Africa (24%), with South America (4%), Europe (4%), North and Central America (2%) and Oceania (0.2%) having minor shares. China is the world's top producer, landing about 26% of global inland catch, while other developing countries together produce an additional 68%. In 2002, no OECD countries ranked among the top ten world producers of inland capture fisheries.

Box 15.2. **China: the world's largest producer and consumer of fish products**

China is the world's largest producer of fish, and its per capita fish consumption (27.7 kg per annum) is about twice the world average. Its reported total fisheries production was 44.3 Mt in 2002, roughly one-third of global production.* Two-thirds of the output comes from aquaculture, a sector which is rapidly expanding. From 1970 to 2000, China's inland aquaculture production increased at an average annual rate of 11%, compared with 7% for the rest of the world. Similarly, the country's aquaculture production in marine areas increased at an average annual rate of 11%, compared with 6% for the rest of the world (FAO, 2004a).

China is home to about one-third of the world's fishers and aquaculture workers. In 2002, 8.4 million Chinese worked in capture fisheries, and 3.9 million in aquaculture. But looking to 2030, China's employment in the fisheries primary production sector is expected to decline, as fleet-size reduction programmes are implemented in response to overfishing. Indeed, such programmes implemented from 2000 to 2006 are projected to already shift 4% of Chinese capture fishers to other jobs by 2007 (FAO, 2004a). Policy tools used to accomplish these shifts included scrapping some fishing vessels and training redundant fishers in aquaculture.

* The FAO has issued caveats about the accuracy of statistics on China's capture fisheries and aquaculture production, suggesting that they are likely too high. Thus, these figures should be seen as indicative rather than authoritative.

Aquaculture

In 2004, global production from aquaculture totalled 59 Mt of fish, crustacean and mollusc products, and aquatic plant products,[5] and constituted 38% of global fisheries production by weight. According to FAO simulations, aquaculture will contribute about 43% of global fish production by 2020 (FAO, 2004b). Worldwide, aquaculture has grown at 8.9% per annum since 1970 (compared with 1.2% for capture fisheries and 2.8% for terrestrial farmed meat-production systems). Freshwater aquaculture systems are the main contributors to overall aquaculture output (58% by weight), followed by marine (36%) and brackish aquaculture systems (6%). Since 1990, the growth has been even faster. To a large extent, the rapid increases in aquaculture have been a response to the increasing demand for fish products combined with the biological limits reached in capture fisheries.

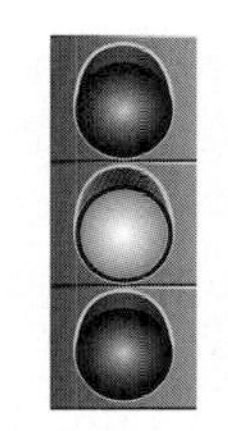

Aquaculture can help to alleviate pressures for fish production from capture fisheries, but its environmental impacts need to be addressed.

Developing countries produce about 90% of aquaculture food fish output, cultivating mainly freshwater species that are herbivorous, omnivorous, or filter feeding. China and India are the world's top two producers of aquaculture, with annual outputs of 27.8 Mt and 2.2 Mt respectively. Three OECD countries (Japan, Norway, United States) rank among the world's top ten aquaculture producers, but OECD countries altogether account for less than 10% of world aquaculture production by weight (20% by value) (OECD, 2004). However, OECD countries may be heavy investors in aquaculture development in developing countries (as well as developing country investors themselves).

In OECD countries, capture-based aquaculture (CBA) has expanded considerably, particularly for high-value fish such as bluefin tuna. CBA involves capturing young organisms, or "seed", from the wild and raising them in captivity to marketable size. For example, high-value species such as bluefin tuna are being caught as juveniles and then raised or fattened in offshore "sea pens". CBA already makes up 20% of food fish production from aquaculture by weight (FAO, 2004a).

The outlook to 2030

The Baseline developed for this *Outlook* determines growth in the demand for fisheries products through increases in population and economic productivity. The Baseline for the *OECD Environmental Outlook* is an analytical tool for projecting developments into the future assuming that no new policies will be introduced. It is thus not a forecast of what is most likely to happen. Under these conditions, and based on recent historical developments, the OECD Baseline does not project as strong a fall in production as forecast by the FAO.[6] The *Outlook* Baseline projects that the supply of fisheries products, particularly from aquaculture, increases as a result of price increases that provide strong incentives for the sector's expansion. The population and wealth increases to 2030 that underlie the Baseline would require much stronger increases in prices to suppress demand sufficiently to lower fisheries growth to the FAO's projected 1.6% (recall that global GDP growth in the Baseline is over 2.5% per year to 2030, and no new policies are assumed in the Baseline that would affect fisheries demand).

Assumed limitations in the supply from capture fisheries lead to a combination of aquaculture output increases and price increases – the price increases implicitly help overcome barriers to the continued expansion of aquaculture. Global fisheries production increased by 2.6% annually between 1988 and 2004, but limitations in supply are projected to slow to an average of 2.1% annually between 2005 and 2030 – a combination of higher growth in the initial years, followed by lower growth in the later years.

Since 2000, catches have decreased or stagnated in marine areas adjacent to most OECD countries, as the majority of fish stocks in their Exclusive Economic Zones (EEZs – see below) are already being exploited fully or beyond maximum sustainable levels. Catches have only been increasing in the tropical Pacific and Indian Oceans, and in high seas areas. Even these areas, however, are not expected to be able to significantly increase output further. Future growth is thus assumed to come from aquaculture. In the *Outlook* projection, therefore, the 2.1% total fisheries growth rate assumes robust growth in aquaculture, but no growth in capture fisheries. This implies an average growth rate in aquaculture of 3.9% annually to 2030 (compared to 8.1% annual growth between 1992 and 2005). This is induced endogenously in the Baseline by a roughly 67% increase in the real price of fish by 2030 (relative to 2001). To understand how strongly this motivates aquaculture development, it is worth pointing out that the real price of almost all fish consumed fell sharply between 1970 and 2000 (Sumaila *et al.*, 2005). Figure 15.3 illustrates the projected evolution of relative shares of capture fisheries and aquaculture production to 2030. Capture fisheries remain roughly constant in landed quantity, but decline as a share of total fisheries production.

Of course, aquaculture is in part dependent on capture fisheries for fishmeal feed. Recent expansion of aquaculture has augmented demand for fishmeal, with 2 to 12 kg of fishmeal feed required to produce one kg of farmed fish or prawns, depending on the species. However, as the price of fish products increases, it is projected that substitutes for

Figure 15.3. **Projected composition of world fisheries to 2030: capture and aquaculture**

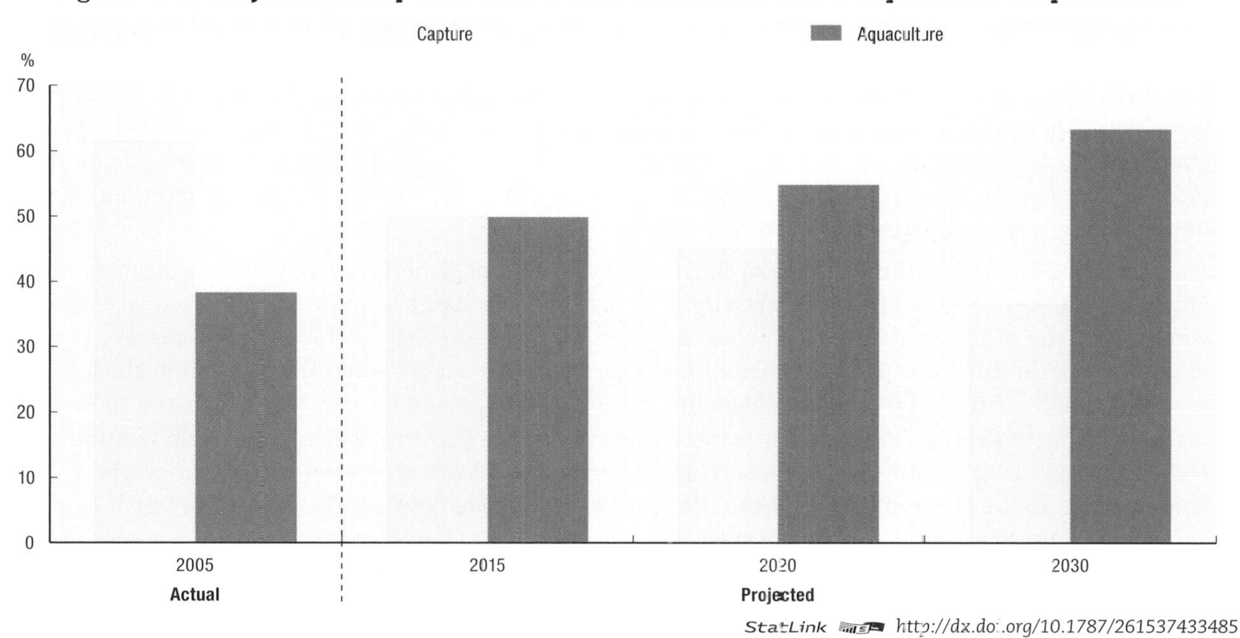

StatLink http://dx.doi.org/10.1787/261537433485

Source: *OECD Environmental Outlook* Baseline.

fish feed for aquaculture, such as soya-based feed, will become more economically viable for those species that can be fed a vegetarian diet. Thus, the FAO projects that the portion of fisheries output used to make fishmeal and oil will decline from 35 million tonnes in 2000 to about 26 million tonnes in 2030 (FAO, 2004a). Other factors influencing demand for fishmeal and oil include trends in the broiler chicken and pork industries, and changes in the price ratio between fishmeal and its close substitutes.

Policy implications

Looking to 2030, it will be important for governments to address gaps in the institutional and legislative framework for managing the environmental impacts of fisheries and aquaculture, and to strengthen implementation of the existing agreements.

At the same time, environmental degradation driven by activities in other sectors is also affecting the economic viability of fisheries. Policies are needed to tackle pollution from land-based sources and shipping, to reduce or halt the introduction of invasive alien species, and to help fishing communities adjust to the impacts of global climate change. The consequences and costs to the fisheries sector of environmental policy inaction in these other sectors should be made explicit in policy decisions (Box 15.3).

Greater understanding is also needed of the potential impacts of climate change and other weather phenomena (*e.g. El Niño*) on fisheries and aquaculture activities. Developing countries may need help to develop appropriate measures to adapt to climate change, and more broadly to support sustainable fisheries management.

International governance

Global governance of fisheries is managed through international bodies such as the UN, the FAO's Committee on Fisheries, and regional fishery management organisations (see below) through which countries agree legally-binding instruments and frameworks for

Box 15.3. **The evolving nature of fisheries management objectives**

Attention has recently been drawn to management objectives in fisheries, especially the need to balance the different objectives of maximising profits, maintaining or increasing employment, ensuring sustainable fish harvests over time, and maintaining a given level of ecosystem integrity. A number of governments have adopted an ecosystem-based approach to fisheries management, recognising the intrinsic link between a sound ecological system and sustainable fisheries over the long-term, although implementing such an approach is often challenging.

Hilborn (2007), and Hilborn *et al.* (2006) illustrate how different fisheries management objectives will engender different levels of fishing activity. Figure 15.4 illustrates a simplified hypothetical fishery similar to many of the world's actual fisheries. The solid line represents yields at various levels of the original stock. In this example, the maximum sustainable yield occurs when the remaining stock is at roughly 20-30% of the original stock. This line roughly corresponds to employment in the fisheries sector – that is, maximum employment generally occurs at the point that maximum yield is attained. Hilborn argues that a number of the world's fisheries have been managed to maximise yields or employment, so the stock of fish has been brought down to these "low" levels. This is borne out by the FAO's reporting that 52% of species are near their maximum yield, while another 25% are either over-exploited, depleted or recovering. The dashed line in the figure represents total economic profits from fisheries. It shows that to maximise profits from fishing, the harvesting of fish should be reduced below the maximum sustainable yield (thereby inducing a higher price for fish caught with less effort). Overfishing leads to economic losses.

Figure 15.4. **Alternative fisheries management profiles**

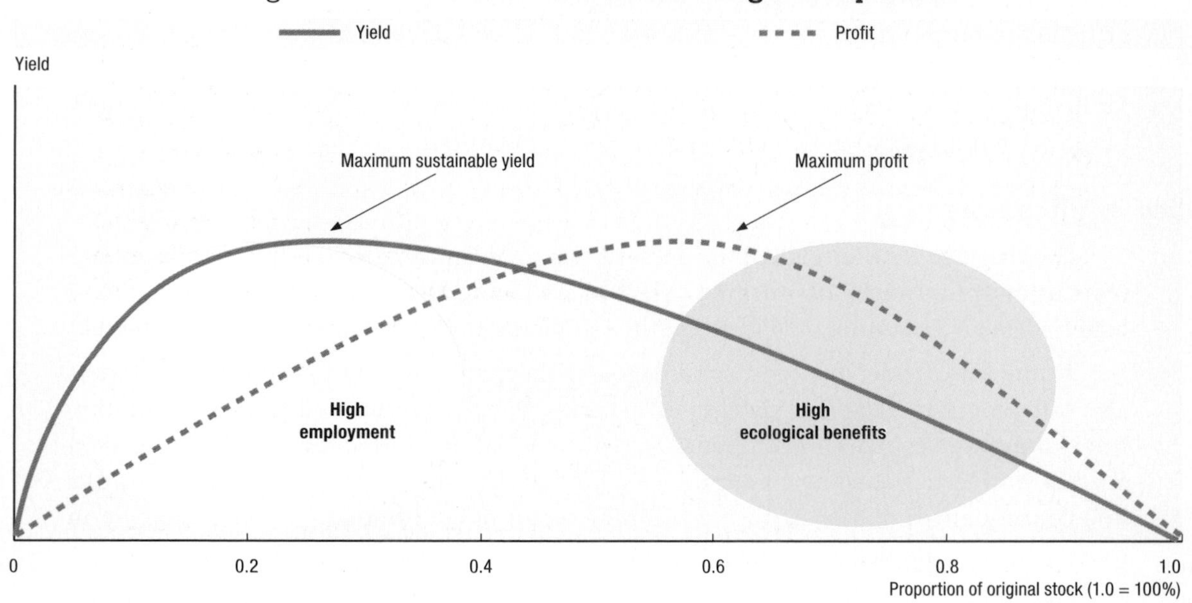

Managing with a biodiversity or ecological perspective would entail a level of fishing that is closer to the profit maximising level, rather than the yield maximising level. A substantial reduction in current fishing levels for many fisheries could allow good management from a biodiversity perspective, as well as maximise profits for the industry. Box 15.4 contains a policy simulation of such an approach.

Box 15.4. **Policy simulation: economic effects of limiting global fisheries catch**

Figure 15.4 above illustrated that managing for maximum profit implied a lower catch level than managing for maximum sustainable yield (MSY). Managing for a high level of biodiversity or ecosystem benefits from fisheries would imply a catch that is closer to, though still lower than, managing for maximum profit. But in unmanaged fisheries, the move first to MSY, then to maximum profit, and finally to high ecosystem benefits is a significant challenge that must work against the "global commons" problem of open access fisheries. Indeed, just to get fisheries to MSY, governments need to implement management schemes and impose catch restrictions to overcome easy access to the harvest areas. Within a managed fishery, movement to the high ecological-value region of Figure 15.4 still requires substantial government commitment to act for the benefit of non-fishers and the environment. This may potentially lead to reduced income to the fishing community in the short-term as catch levels are reduced, although it may lead to a more sustainable economic basis for the fishing industry over the longer-term.

A policy simulation was undertaken using the ENV-Linkages model to examine the impacts of reducing fish catch, as an illustrative example of a policy aiming to manage fisheries in a way that might maximise fisheries profit or even ecological values. The simulation modelled an idealised implementation of internationally tradable quotas, set to bring about a 25% reduction in global fish catch.[1] To actually implement such a reduction, agreement would have to be reached that reduced global fisheries by the right mix (because not all species are at the same point in Figure 15.4). Safeguards of minimum stock levels of specific species would likely have to be put in place to ensure that valuable species were not over-fished. Such safeguards would also have to account for illegal, unregulated, and unreported fishing.

Quotas were applied in the ENV-Linkages model that limited capture fishing to 75% of 2005 levels. The analysis assumed that countries would individually manage fisheries, within their overall quota, so that no individual species was aggressively over-fished. The simulation examined the economic impacts of applying internationally tradable quotas within six geographical areas: trading was permitted *within* those regions, but not *between* those regions. The simulation illustrates – in aggregate – the economic impact of reduced capture fishing and its geographical distribution. Given the projections in the OECD *Environmental Outlook* Baseline for the growth of aquaculture and limited capacity for additional capture fishing (as discussed above), it was found that the 25% reduction in capture fishing would, under this simulation, lead to only a 14% reduction in total fisheries catch[2] (capture plus aquaculture) in 2010 compared with the Baseline. This would fall to a 11% reduction in the value of fisheries catch by 2020 compared to the Baseline, and 9% by 2030.

The policy simulation showed considerable trade in quotas, and thus heterogeneity across countries within a given trading region in terms of the impacts on the fisheries sector. The simulation also showed that these impacts would be expected to evolve over time. Since such trade is always indicative of economic gains relative to the initial allocation of quota, the implication is that any non-quota based international scheme to tackle overfishing would need to have considerable flexibility (mimicking the flexibility inherent in a tradable quota scheme). This flexibility needs to be implemented within a strong framework for co-operative decision-making.

1. There is no clear agreement on how much overfishing is occurring at a globally aggregated level, and thus what level of fish catch reduction would be appropriate to achieve a high ecological outcome. Based on FAO (2004a) estimates that 24% of fisheries are currently over-fished, depleted or recovering, the policy simulation for this *Outlook* was run with a 25% reduction in fish catch as a purely illustrative example of the economic effects such a reduction in catch might have.
2. That is, in terms of the constant-dollar *value* of fish – which approximates fishing tonnage if the composition of fish caught does not change substantially.

managing common fish stocks. The UN's *Convention on the Law of the Sea* (1982) codified the practice of state jurisdiction over marine fisheries resources within 200 nautical miles of their coastlines, referred to as Exclusive Economic Zones (EEZs). It is estimated that EEZs cover about 90% of the world's marine capture fisheries. The creation of EEZs aimed to assign national ownership and management responsibility for fisheries within these zones. International governance remains important for setting the right international legal frameworks for fisheries management, and not least for addressing the management of high seas fisheries (outside the EEZs) and straddling fisheries.

At the 2002 World Summit on Sustainable Development in Johannesburg, governments jointly declared the objectives of restoring global fish stocks to sustainable levels by 2015 and of significantly reducing the rate of biodiversity loss by 2010. In 2006, the UN General Assembly adopted a Sustainable Fisheries Resolution, calling on all nations to apply an ecosystem-based approach to the management of fish stocks, and to protect vulnerable marine ecosystems from destructive fishing practices. A number of specific international fisheries arrangements adopted since the 1992 Earth Summit have helped to strengthen international approaches to fisheries management and global oceans governance, such as the 1995 UN Fish Stocks Agreement, the 1993 FAO Compliance Agreement, the 1995 FAO Code of Conduct for Responsible Fisheries, and the "London Convention" and the UNEP Global Programme of Action for the Protection of the Marine Environment from Land Based Activities. In 2007/2008, the FAO Committee on Fisheries will work to develop an international legally-binding instrument on minimum standards on port state measures, which will be an additional tool in the suite of international fisheries governance measures.

The role of regional fishery management organisations (RFMOs) in managing wild stocks of marine fish has developed considerably in recent years. Whereas in the 1980s the mandates of many RFMOs were limited to research and advisory functions, since the Earth Summit many of these mandates have strengthened and expanded in order to implement modern approaches to fisheries management, including an ecosystem approach, and greater co-operation with developing countries. However, the success of RFMOs largely depends on the ability of their member states to agree co-ordinated approaches to fisheries management, and to delegate sufficient monitoring and enforcement powers to RFMOs to implement their mandates. Non-members can undermine RFMO conservation and management measures. Lack of political will and capacity to implement internationally or regionally agreed fisheries management policies remains a challenge. Efforts are needed to build capacity in developing countries to manage fishery resources in a sustainable manner.

Economic instruments

There is an increasing recognition that market-based instruments can improve the efficiency of fisheries resource allocation and use, and help to align fishers' economic incentives with societal objectives (OECD, 2006a). They do this by limiting fishing pressure (*e.g.* through tradable quotas, access charges), providing fishers with incentives to reduce fishing effort (*e.g.* through vessel buyback schemes), or encouraging compliance with regulations (*e.g.* fees and fines). Limiting access to wild stocks through the allocation of catch permits is a widely used approach to reducing fishing pressure. Some of these measures have proven more effective than others. For example, vessel and license buy-back schemes have often proven ineffective at reducing capacity unless they are accompanied by changes in fisheries management regimes that effectively limit the amount of fishing effort in a fishery (OECD, 2006b).

 OECD ENVIRONMENTAL OUTLOOK TO 2030 – ISBN 978-92-64-04048-9 – © OECD 2008

Historically, some subsidies for shipbuilding and fleet enhancement have contributed to excess fishing capacity. Government financial transfers to the fishery sector in OECD countries amounted to USD 6.4 billion in 2003, or about 21% of the landed value of the catch (OECD, 2006b). Increasingly, government support to fisheries is shifting towards more sustainable fisheries management, rather than increasing fisheries production. Thus, for OECD countries, 38% of government financial transfers now supports research, management and enforcement; 35% goes to infrastructure and the remainder to cost-reducing or income-enhancing measures. Negotiations are underway in the World Trade Organisation (WTO) to clarify disciplines on fisheries subsidies.

Regulatory approaches

Regulatory approaches are being used, for example, to limit fishing effort and gear types, and to optimise the location and operation of aquaculture farms (*e.g.* total catch limits; spatial planning and zoning; and effluent discharge permits). Sensitive habitats or important breeding or feeding grounds for at-risk species could be set aside as conservation areas. Marine Protected Areas declared for fisheries purposes (such as areas closed to specific gear types, or set up to protect key habitats) can also support biodiversity conservation goals, as well as improve the productivity of capture fisheries (Ward and Hegerl, 2003).

Regulatory standards for fishing gear can also be effective means of reducing impacts on habitats and non-target species (*e.g.* requiring turtle excluder devices, seabird-scaring streamers or acoustic deterrent devices for sea mammals and seabirds). But adoption of these "gear fixes" in the global fisheries has been slow, and where regulations for their use do exist, monitoring and enforcement may be poor. Many of these measures are required in southern ocean fisheries, but are not yet mandatory in northern hemisphere fisheries, even though they also have a high bird by-catch intensity (*e.g.* off the coast of Scotland).

Illegal, unregulated, and unreported (IUU) fishing contributes to overfishing by making it more difficult to ensure that fishing limits are respected and by making it harder to develop the robust stock assessments necessary for biologically sound management decisions. IUU fishing has been increasing in recent years, driven by the rising value of certain scarce species and facilitated by new technological developments. IUU fishing is difficult to control due to the open access nature of most open seas fish stocks outside of EEZs and areas controlled by regional fisheries management organisations (RFMOs), and due to the expense and technical challenge of monitoring vast marine zones. The recent introduction of trade and catch certification schemes, coupled with rapidly evolving information technologies, is helping in a number of regions. However, a number of challenges remain for tackling IUU fishing, including ensuring adequate capacity for monitoring and enforcement, and addressing the use of flags of convenience.

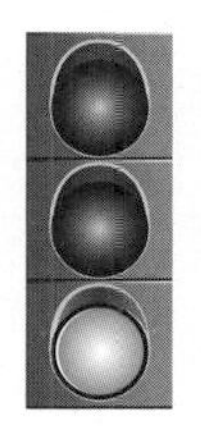

Wider use of remote sensing and GPS technologies can help surveillance and monitoring of illegal fishing activities.

Reducing fishing effort, including through the use of catch limits and fishing capacity, is an important regulatory measure to rebuild stocks of depleted species. Other measures include policies aimed at reducing by-catch; reducing or eliminating environmental degradation; and enhancing factors of growth, for example through stock enhancement and habitat rehabilitation. Species that are particularly vulnerable to fishing pressure, such as those that are long-lived and only start breeding after a relatively long period of immaturity,[7] may especially require long-term management recovery plans.

Regulation of aquaculture has progressed considerably since the 1990s, with most OECD countries now requiring operators to acquire permits or licenses to establish a farm. Environmental impact assessments are generally required for new facilities, and licenses typically specify some operating conditions designed to limit environmental impacts.

Information-based approaches

Voluntary and trade-related approaches are used to encourage the spread of best practices among fishers and fish farmers (*e.g.* codes of practice, eco-labels, and catch certificates). As major consumers and importers of fish products, OECD countries have an interest to promote measures that will ensure the long-term sustainability of capture fisheries in developing countries. The link between pollution and food safety in fish production, including pollution sources from outside the sector, will receive more attention worldwide in the future. Trade related measures can be used to raise the accountability of producer countries (*e.g.* catch certificates, trade certificates).

Eco-labelling of fisheries products began only in the late 1990s. The Marine Stewardship Council (MSC) eco-label is perhaps one of the earliest and best known voluntary schemes, and is used to indicate products sourced from sustainable fisheries, defined as those that "ensure that the catch of marine resources are at levels compatible with long-term sustainable yield, while maintaining the marine environment's biodiversity, productivity and ecological processes". More recently a plethora of different eco-labelling schemes for fish products has emerged, including ones that indicate the origin of fish products, the sustainability of their harvest, whether aquaculture products were produced organically, etc. However, the number of competing schemes, the range of issues that they address and the lack of rigour or clarity about the independent monitoring of some of them have led to some consumer confusion and distrust of eco-labelling. The FAO is working to establish an international set of guidelines for eco-labelling to support more rigorous and reliable eco-labels in fisheries and aquaculture, while the European Commission is looking to develop guidance on eco-labels for use in European Union countries.

Notes

1. A species can be considered to be functionally absent from an ecosystem if the number of individuals is so low that it cannot fill its usual niche in the ecosystem.
2. The global shipping fleet includes some 60 000 vessels with tonnage over 250 gross registered tonnes.
3. The large increase in capture fish production seen in 1988 in Figure 15.2 represents the year when information became available on Russian and Eastern European catches.
4. The FAO warns that global inland catch data are only indicative due to gaps in reporting on catch quantities and species composition.
5. This figure includes aquatic plant production of roughly 13 Mt. Unless otherwise noted, most figures below do not include aquatic plants.
6. The FAO projects an increase in total fisheries production of 43 Mt from 2000 to 2015, the bulk (73%) of it coming from aquaculture. Even so, the FAO expects average annual growth in world fish production to trail off, going from the 2.7% seen in the 1990s to 2.1% per year from 2000 to 2010, before dipping to 1.6% per year from 2010 to 2015 (FAO, 2004b).
7. Sharks, rays and skates, and many species of fish in deep water fall into this category.

References

Caldeira, K. and M. Wickett (2005), "Ocean Model Predictions of Chemistry Changes from Carbon Dioxide Emissions to the Atmosphere and Ocean", *Journal of Geophysical Research*, Vol. 110, C09204.

EEA (European Environment Agency) (2002), *Europe's Biodiversity Bio-Geographical Regions and Seas: Seas Around Europe*, EEA, Copenhagen.

FAO (UN Food and Agriculture Organization) (2004a), *The State of World Fisheries and Aquaculture: 2004*, FAO, Rome.

FAO (2004b), *Future Prospects for Fish and Fishery Products: Medium-Term Projections to the Year 2010 and 2015*, FAO Fisheries Circular FIDI/972-1, FAO, Rome.

FAO (2006), *The State of World Fisheries and Aquaculture: 2006*, FAO, Rome.

FAO (2007), *Fishery Statistics Programme*, Fisheries and Aquaculture Department, FAO, Rome (*www.fao.org/fi*).

Feely, R. *et al.* (2004), "Impact of Anthropogenic CO_2 on the $CaCO_3$ System in the Oceans", *Science*, 305: 362-366.

Hilborn, R. (2007), "Defining Success in Fisheries and Conflicts in Objectives", *Marine Policy*, Vol. 31, pp. 153-158.

Hilborn, R., J. Annala, and D.S. Holland (2006), "The Cost of Overfishing and Management Strategies for New Fisheries on Slow-Growing Fish: Orange Roughy (*Hoplostethus atlanticus*) in New Zealand", *Canadian Journal of Fisheries and Aquatic Sciences*, Vol. 63, pp. 2149-2153.

IPCC (Intergovernmental Panel on Climate Change) (2007), *Fourth Assessment Report, Working Group 1: The Physical Science Basis, Summary for Policymakers*, IPCC, Geneva.

OECD (2004), *Review of Fisheries in OECD Countries: General Survey 2004*, OECD, Paris.

OECD (2006a), *Using Market Mechanisms to Manage Fisheries: Smoothing the Path*, OECD, Paris.

OECD (2006b), *Financial Support to Fisheries: Implications for Sustainable Development*, OECD, Paris.

OECD (2008), *Costs of Inaction: Draft Technical Report* [ENV/EPOC(2007)6/REV2], OECD, Paris, forthcoming.

Richartz, S. and E. Corcoran (2004), *The State of Europe's Regional Seas – Are we Meeting Conservation Targets?* A briefing paper written in preparation for the conference "*Sustainable EU Fisheries: Facing the Environmental Challenges*", European Commission (DG Fish), the Esmée Fairbairn Foundation, FISH and English Nature.

Sumaila, U.R. *et al.* (2005), *Global Ex-Vessel Fish Price Database: Contruction, Spatial and Temporal Applications*, University of British Columbia Fisheries Centre Working Paper Series No. 2005-01, Vancouver, BC.

UNEP (United Nations Environment Programme) (2006), *The State of the Marine Environment: Trends and Processes*, United Nations Environment Programme, Global Programme of Action for the Protection of the Marine Environment from Land-Based Activities, The Hague.

Ward, T. and E. Hegerl (2003), *Marine Protected Areas in Ecosystem-Based Management of Fisheries*, Natural Heritage Trust, Canberra.

ISBN 978-92-64-04048-9
OECD Environmental Outlook to 2030

Chapter 16

Transport

The transport sector is the second largest (and second fastest growing) source of global greenhouse gas (GHG) emissions. If developing countries follow the same path of private car dependence in the future as OECD nations have in the past, technological advances are unlikely to be able to offset the large increase in vehicle related emissions. Maritime shipping is another increasingly important source of environmental concern. Governments should prioritise policy action to reduce the energy intensity of transport. Policy options include applying carbon and fuel taxes, reforming vehicle taxation and regulating vehicle standards. Additional measures, such as implementing road pricing and investing in public transport infrastructure and spatial planning policies, can also help to improve the environmental performance of the transport sector.

KEY MESSAGES

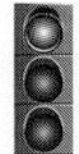

The transport sector is the second largest (and second fastest growing) source of global greenhouse gas (GHG) emissions.

Total CO_2 emissions from transport are still increasing. Emission reductions from technological improvements are being eclipsed by the continuing growth of transport volumes (especially passenger vehicle and air transport).

If developing countries follow the same path of private car dependence in the future as OECD nations have in the past, technological advances are unlikely to be able to offset such a large increase in vehicle related emissions.

Maritime shipping is an increasingly important source of environmental concern.

Emissions of some air pollutants from transport are decreasing; others continue to rise.

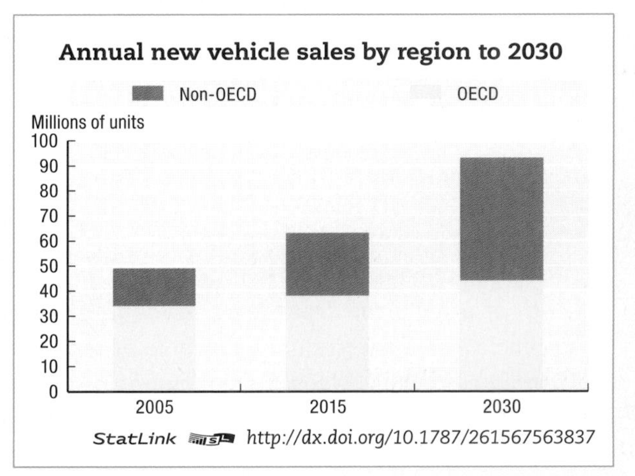

Policy options

- Prioritise policy action to reduce the energy-intensity of transport, which appears to have the greatest potential for reducing CO_2 emissions cost-effectively. Policy options include applying carbon and fuel taxes, reforming vehicle taxation and regulating vehicle standards.
- Ensure that public incentives for biofuels reflect a full life-cycle of their effect on both greenhouse gas emissions and the economy.
- Implement road pricing, and invest in infrastructure and spatial planning policies, all of which can help to improve the environmental performance of the transport sector.

Consequences of inaction

- Poor urban air quality (much of which originates from transport) continues to have negative impacts on human health and the economy, in terms of lost productivity and medical expenses. The human health impacts of transport-related pollution are likely to increase in the next two decades, particularly in rapidly growing developing countries.

 OECD ENVIRONMENTAL OUTLOOK TO 2030 – ISBN 978-92-64-04048-9 – © OECD 2008

Introduction

In recent years, increased trade and investment activities (both of which are closely associated with a more globalised economy) have led to substantial increases in both the volume of goods being shipped and the distance these goods have to travel. Increasing levels of disposable income have also led to significant increases in recreational travel. The result is that total transportation activity in OECD countries has increased much faster in the past 30 years than either population or GDP.

Recent technological developments, partly triggered by the implementation of environmental policies, have helped to improve the environmental performance of the transport sector in a number of areas, in particular reducing vehicle emissions of a number of air pollutants that can damage health and the environment. Despite these developments, transport continues to create significant environmental problems.

The transport sector is the second largest (and second fastest growing) source of global greenhouse gas (GHG) emissions after energy industries. Transport accounted for about 24% of global CO_2 emissions from combustion in 2003. Of that total, road transport contributed 18%, aviation 3%, navigation 2%, and other sources 1% (ECMT, 2007a).

In OECD countries, road transport is responsible for most of the transport sector's impacts on the environment, accounting for over 80% of all transport-related energy consumption, and for most air pollutant emissions, noise and habitat degradation (OECD, 2006a). In Europe,* total external costs of transport (excluding congestion costs and externalities related to maritime transport) have been estimated at EUR 650 billion for 2000, or about 7.3% of total GDP (INFRAS, 2004). Climate change was the most important category, contributing 30% of total costs (Figure 16.1). Air pollution and accidents were the next most significant. In terms of transport mode, road transport has the biggest impact, generating 83% of the total estimated external costs. This is followed by aviation (14%), railways (2%), and inland waterways (0.4%). Road transport accounted for over 89% of the costs in all categories, except for climate change, in which road transport accounted for only 57% of estimated costs. Almost all the remaining costs associated with climate change came from aviation (41%). Two-thirds of all transport-related external costs are caused by passenger transport and one-third by freight transport (INFRAS, 2004).

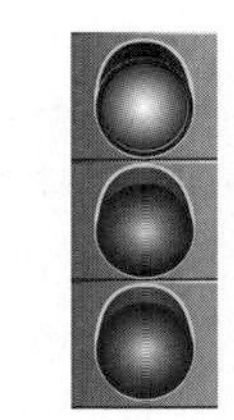

The transport sector is the second largest (and fastest growing) source of global greenhouse gas emissions.

* This includes EU15, Norway and Switzerland.

Figure 16.1. **Transport externalities in Europe in 2004 (by impacts)**

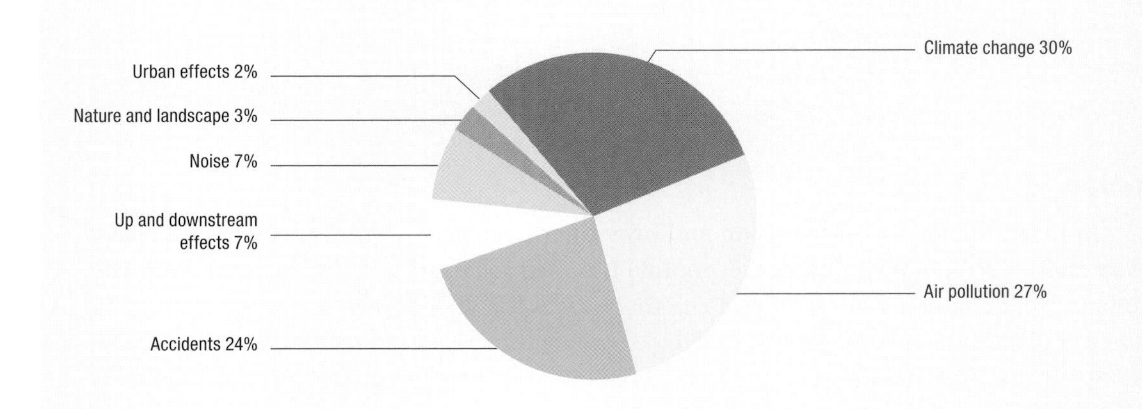

Source: INFRAS, 2004.

StatLink http://dx.doi.org/10.1787/261541473346

Maritime transport, although generally associated with lower environmental impacts, continues to raise concerns, mainly due to oil pollution from major accidents, as well as (accidental or deliberate) discharges of waste products. The maritime shipping sector is also an important contributor to NO_x and SO_2 emissions, as well as to ozone pollution. There is also growing concern over the environmental impacts of air traffic, which continues to increase rapidly mainly due to increased tourism (see also Chapter 19, section on tourism). The rail sector is generally the most environmentally benign form of transport, but is also the least used.

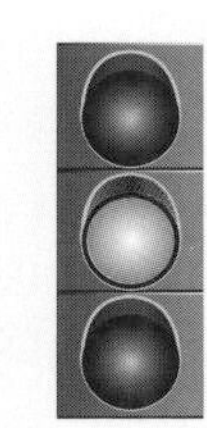

Maritime shipping is an increasingly important source of environmental concern.

Trends and projections

The rapid increase in transportation activity seen in recent decades is expected to continue to 2030 (although see Box 16.1). Between 1970 and 2003, for example, air passenger travel in the US increased by 328% – nearly twice the rate of GDP growth during the same period. Air passenger travel in the EU increased even faster, growing by over

Box 16.1. **Key uncertainties, choices and assumptions**

There are fundamental uncertainties in projecting transport demand and simulating future transport systems. Uncertainties in demographic, economic, technological and institutional factors will affect the actual level of future transport demand, the mix of energy supplies consumed, and the associated rates of (for example) CO_2 emissions. Knowledge is limited of the complex interactions of technological, cultural and political forces that determine the development of national transport schemes. It is therefore not certain that today's relationships will persist unchanged for the next 25 years. For non-OECD countries, it is also difficult to find reliable and consistent data on which to base future projections.

Figure 16.2. **Global air transportation volumes and GDP (1990 = 100)**

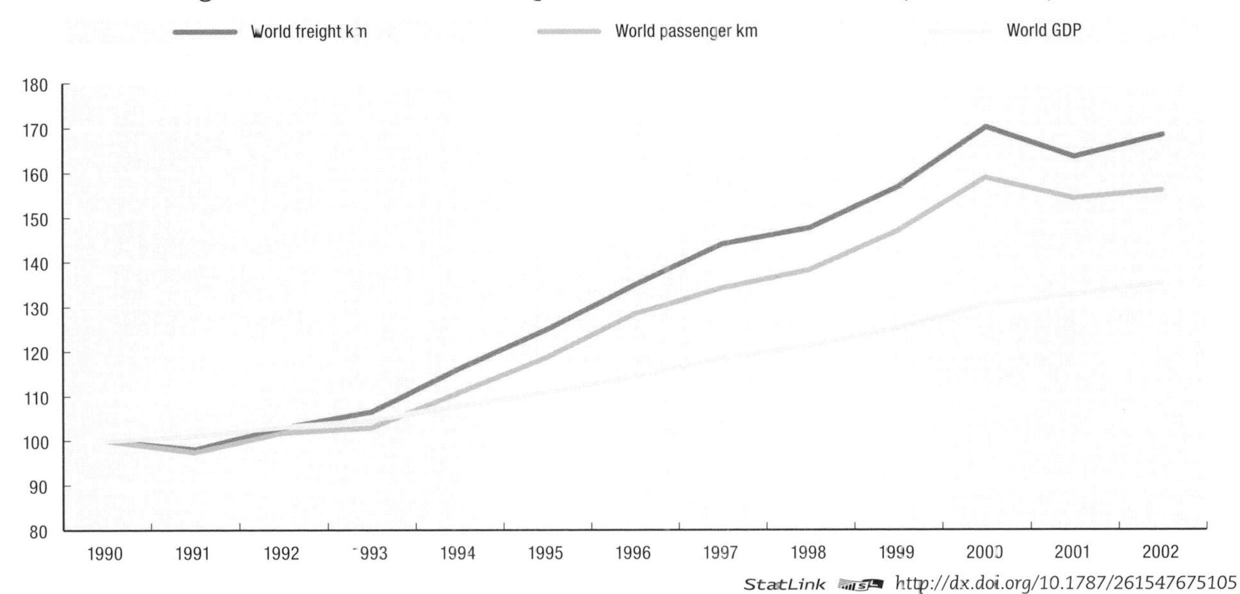

StatLink http://dx.doi.org/10.1787/261547675105

Source: Based on data from the UN Common Database, 2007.

1 200% between 1970 and 2003 (Figure 16.2). Although air travel has been the fastest growing transport mode in recent decades, other modes have increased as well. Road transport, in particular, has grown faster than GDP in both the EU and North America.

Transport growth is not only being driven by people/goods travelling further and more often, but also by an increase in the availability and use of motorised transport. In OECD countries, the private car has been the norm for decades, so only moderate increases in car ownership are predicted over the next 20 years. In non-OECD countries, on the other hand, rapidly rising incomes are expected to lead to large increases in vehicle ownership (Figure 16.3). In some cases, the increase in motorised transport is occurring at the expense of existing modes – some of which are less environmentally damaging than road transport. For example, bicycle use in China has fallen recently as automobile use has expanded.

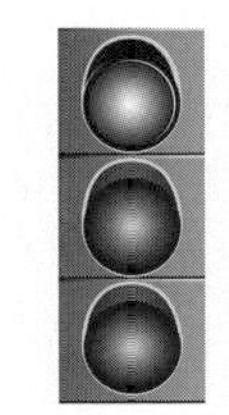

If developing countries follow the same path of car dependence as OECD nations, technological advances are unlikely to offset the large increase in vehicle related emissions.

Air pollution

The transport sector is a major source of air pollution at the local, regional and global levels. It is the dominant source of air pollution in urban areas. In 2002, transportation was responsible for 58% of total US carbon monoxide emissions and 45% of nitrogen oxide emissions. Between 1992 and 2002, however, most transport-based air pollutant emissions in the US actually declined (BTS, 2006). Within the US transportation sector, road transport was the main source of air pollutants over the 10 years ending in 2002. It accounted for 82% of NO_x emissions, 76% of volatile organic compounds (VOCs), and virtually all transport-based carbon monoxide (CO) emissions. Marine vessels and railroad locomotives contributed 11% and 9% of transportation's NO_x emissions respectively, and made minor contributions to other emissions (BTS, 2007).

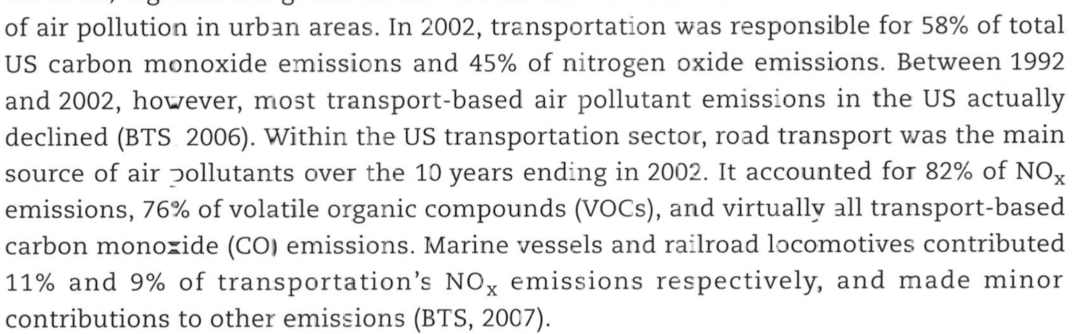

Figure 16.3. **Annual new vehicle sales by region to 2030**

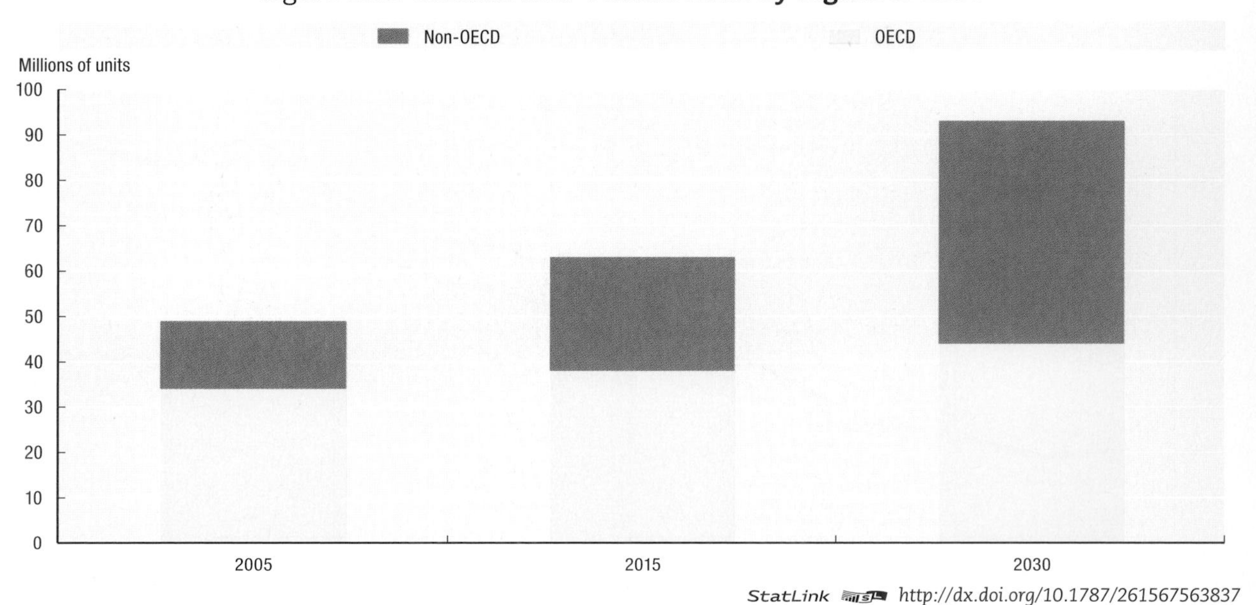

StatLink http://dx.doi.org/10.1787/261567563837

Source: IEA, 2006.

In the EU15, emissions of acidifying substances, particulate matter, and ozone precursors from transport (excluding international aviation and maritime transport) fell by 30-40% from 1990 to 2003 (EEA, 2006). Maritime sources in the EU contributed about 20% of total NO_X and 77% of SO_X emissions from the transport sector (EEA, 2006).

Global use of leaded fuel has declined significantly in recent decades. This trend is being seen in all regions. For example, nearly all countries in Africa have now shifted to unleaded petrol, after adoption of the 2001 Dakar Declaration. The widening use of unleaded petrol has reduced lead-related health problems. For example, mean blood-lead levels in children have fallen by 50% since the phase-out began in India (Singh and Singh, 2006).

Some air pollution from transport is decreasing; other forms of transport-based air pollution continue to rise.

Global sulphur emissions from transport also declined by 18% between 1995 and 2005, mainly through the desulphurisation of fuels. A key barrier to the further penetration of low-sulphur fuels is the high investment costs involved for refineries, particularly in developing countries.

Transport-based emissions of nitrous oxides have decreased by 3% globally since 1995 (23% in OECD countries). This has been achieved mainly through the wider use of new engine technologies and catalytic converters.

Current trends toward reducing sulphur and nitrogen emissions are projected to continue to 2030 globally under the *OECD Environmental Outlook* Baseline, reflecting the positive impact of existing policies. However, without any new policies, some of the less developed regions, such as Africa and parts of Asia, are expected to experience increases in these pollutants in the coming two decades (see Chapter 8 on air pollution).

Exposure to air pollution (from the transport sector or elsewhere) can cause adverse health effects – most acutely in children, asthmatics, and the elderly – and can damage ecosystems and infrastructure (see also Chapter 12 on health and environment; and WHO, 1999). The health effects can range from mild irritation of eyes and lungs, to aggravation of asthma, cancer and premature death. Ground-level ozone can damage vegetation, and acid rain can damage vegetation, buildings and aquatic ecosystems.

The health costs associated with air pollution can be considerable (see Chapter 13 on cost of policy inaction), and much of this pollution is still transport-based. High concentrations of transport-related air pollutants in urban areas continue to present an important challenge (*e.g.* particulate matter, ozone) and are not showing any downward trends despite policy measures to tackle these pollutants.

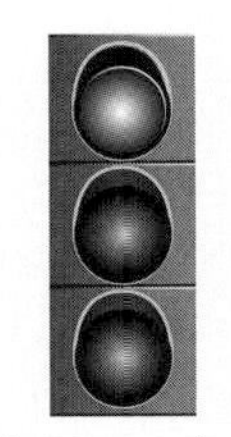

Total CO_2 emissions from transport are still increasing.

Climate change

Transport currently contributes more than one-fifth of global CO_2 emissions. Among the major emitting sectors, transport has the second highest total of CO_2 emissions (after energy industries) (see Chapter 7 on climate change). Transport emissions have had the second highest growth rate over the last 15 years, and are expected to repeat this trend in the near future. If current trends continue, the *OECD Environmental Outlook* Baseline projects that energy-related carbon dioxide (CO_2) emissions from the transport sector will increase by 58% between 2005 and 2030 globally, with emissions more than doubling in China (172% increase), Africa (172%) and South Asia (131%). Increases of this magnitude are inconsistent with the goal of stabilising global atmospheric concentrations of greenhouse gases.

The share of transport in global GHG emissions is expected to remain stable at about 20% over the next 25 years. Among OECD countries, however, transport is expected to account for an increasing proportion of these emissions. In 1995, this share was 20%; by 2020, it is expected to be 30% (OECD, 2006a).

Road transport is by far the largest user of transport fuels in the US and Canada (Figure 16.4). Aviation accounts for a significant proportion, and rail uses a small (but still meaningful) amount.

The share of aviation in total CO_2 emissions from the transport sector has been growing for many years (OECD, 2006a). Emissions of NO_x at high altitudes are also believed to have a significantly larger global warming effect than surface emissions. The Intergovernmental Panel on Climate Change has estimated that the total climate impact of aviation is two to four times greater than the impact of aviation's CO_2 emissions alone (IPCC, 1999). Overall, aviation contributed about 3% of global anthropogenic radiative forcing in 2005 (IPCC, 2007).

Navigation activities (including maritime transport) presently account for about 2% of global GHG emissions. Some projections foresee growth in maritime shipping of 35-45% in absolute levels between 2001 and 2020, based on expectations of continued growth in world trade (Eyring *et al.*, 2005).

In the US, the transportation sector was responsible for 27% of total GHG emissions in 2003. Transport GHG emissions have been growing considerably faster than total US emissions. From 1990 to 2003, these emissions increased by a larger amount than any other economic sector, by 24%. GHGs from all other sectors increased by a total of 9.5% over the same time-frame. Within the transport sector, heavy-duty truck emissions have been

Figure 16.4. **Transport fuel consumption in the US and Canada by mode, 1971-2030**

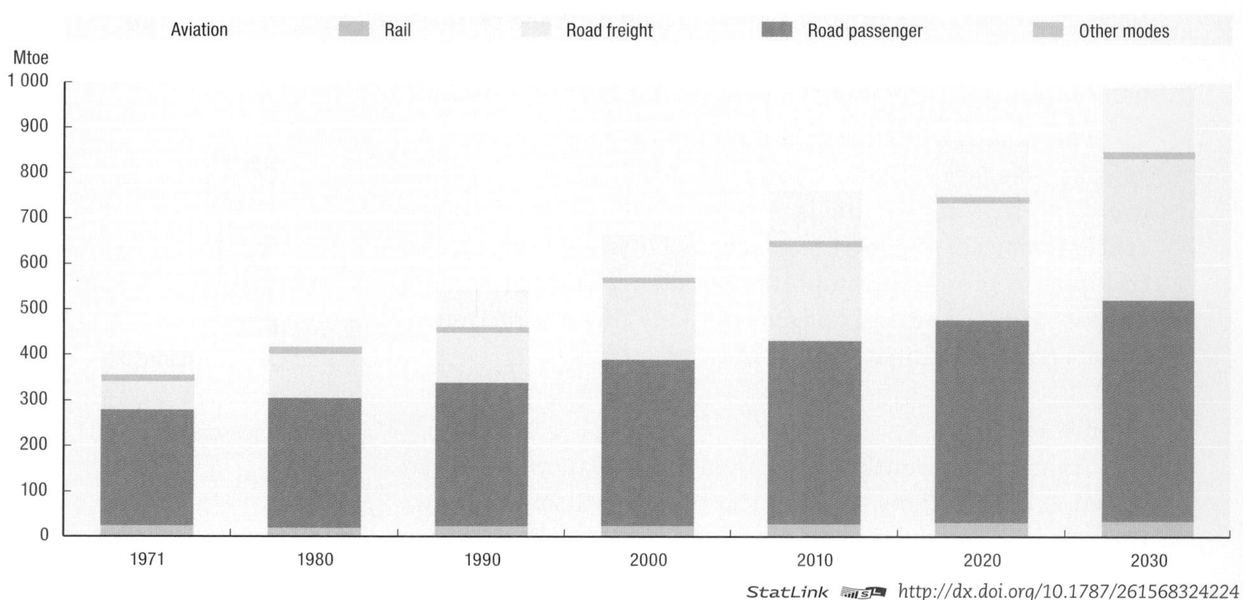

StatLink http://dx.doi.org/10.1787/261568324224

Source: IEA, 2002.

the fastest growing source of GHG emissions, growing by over 50% during the same period. Of all the major US transport modes, air travel experienced the largest reductions in per-passenger-km GHG emissions between 1990 and 2003 (EPA, 2006).

Although OECD countries currently account for the majority of global GHG emissions (both transport-related and other), growth in emissions from the transport sector over the coming years is expected to be driven to a large degree by non-OECD countries. GHG emissions are very closely related to total energy consumption in the transport sector, and this is projected to grow much faster in non-OECD countries than in the OECD (see Chapter 17 on energy).

Energy demand for transportation in the OECD economies is projected to grow at an average annual rate of 1.2% over the next 25 years according to the *Outlook* Baseline. By contrast, energy consumption in non-OECD countries is expected to grow more than three times as fast (3.1% per year). OECD countries currently account for 66% of global energy consumption for transport; by 2030, this is expected to decline to 54%. Figure 16.5 illustrates these trends.

Transport-related GHG emissions are particularly important among the BRIC countries (Brazil, Russia, India and China), who account for more than 60% of all CO_2 emissions from non-OECD countries. China alone accounts for 18% of global emissions. Since 1990, Chinese CO_2 emissions from the transport sector have increased by 156% (IEA, 2006).

Climate change itself will influence the services available within the transport sector (*e.g.* the effects of sea level rise on shipping; the effects of increased weather extremes on aviation). Policies aimed at improving the efficiency of transportation will also increasingly have to address the realities imposed by a changing climate. For example, policies aimed at shifting transport volumes from road to ships could be compromised by the lower water levels in inland waterways that are expected to follow from a warmer climate.

Figure 16.5. **Energy consumption in the transport sector to 2030**

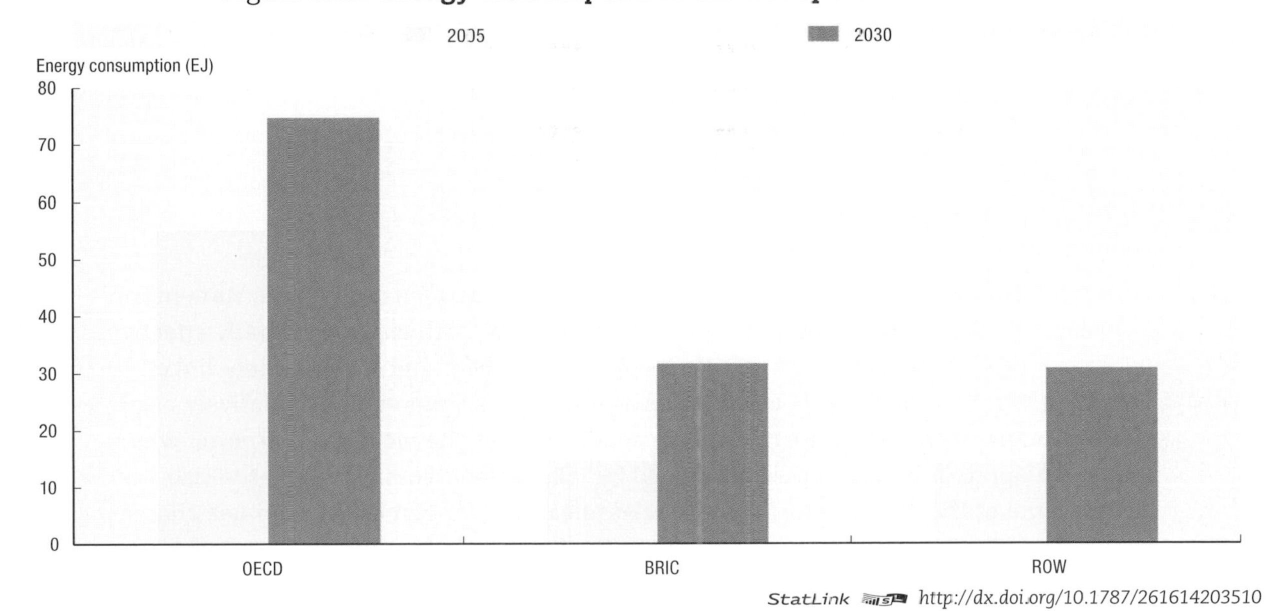

StatLink http://dx.doi.org/10.1787/261614203510

Source: OECD Environmental Outlook Baseline.

Nature, landscape and urban effects

A large proportion of land in built-up regions in OECD countries is already used for transportation infrastructure, mostly roads. The development and extension of transport infrastructure has numerous impacts on soils and water bodies (*e.g.* compaction, soil sealing, diffuse water pollution) and landscape separation effects, leading to habitat fragmentation and destruction, with negative effects on biodiversity. Once this process has started, land fragmentation is extremely difficult to reverse (see also Chapter 9 on biodiversity).

Congestion

In many areas, transport activity has increased much faster than infrastructure capacity has grown, creating severe congestion problems. The largest social costs associated with congestion are the time delays suffered by transport users. Congestion also imposes significant costs on the rest of society, mainly from higher emission levels. A car or truck stuck in congested traffic consumes more fuel for the same distance of travel, and therefore produces more GHG and air pollution emissions per trip. These emissions also tend to be generated in precisely those areas where human exposure levels are the highest. These issues are especially important in the context of urban air pollution, since most traffic congestion occurs in urban areas. The annual external costs of road traffic congestion in the 17 countries that include the EU15, Norway, and Switzerland were estimated at EUR 63 billion in 2004, which corresponds to about 0.7% of the combined GDP of these countries (INFRAS, 2004).

Noise

Transportation is the leading cause of urban noise. Air traffic is the major cause of nuisance noise near airports, while road traffic is the most significant source of noise elsewhere. Although less significant overall than the other externalities discussed above, transport-related noise pollution still imposes many social costs that reduce the quality of life – costs that are reflected, for example, in lower property values around airports or major roads. In OECD Europe, about 30% of the population are exposed to road traffic noise levels above 55 dB(A), and 13% to above 65 dB(A) (EEA, 2001). Persistent exposure to noise levels above 70 dB can result in long-term hearing loss, but even lower levels of exposure can cause irritation, interfere with sleep, and generally detract from the quality of life.

Policy implications

Economic instruments

When the environmental problem being targeted by a particular policy instrument can be closely linked to a taxable item, taxes or charges can be both environmentally effective and economically efficient (Box 16.2). Emissions of CO_2, SO_2, and lead – closely linked to the carbon, sulphur and lead content in various fuels – can therefore be relatively easily priced through taxes. Another example is aircraft noise in the vicinity of airports, which (with some approximation) can be addressed by take-off and landing charges which vary with the time of the day, or with the noise classification of the aircraft. Road use charges can also be linked to distances driven, the time of day (of relevance for congestion and noise impacts), and (roughly) to certain emission characteristics of the vehicle.

Box 16.2. **Efficient prices for transport**

Efficient pricing requires not only that prices reflect all the environmental costs associated with transport, but also that these prices provide incentives to conserve existing transport capacity and to develop future environmentally-sustainable transport options. The European Conference of Ministers of Transport (ECMT, 2003) has estimated that efficient pricing for all modes of inland transport in the three largest EU economies would yield net welfare benefits of more than EUR 30 billion per year.

Market-based approaches can help to ensure that, whatever the environmental objective, it will be achieved at a minimum cost. For example, in the case of fuel taxes, the people who reduce their fuel consumption the most will be those who derive the least benefit from fuel consumption. Flexible mechanisms allow producers and consumers to make the choices that are best for them, and to meet environmental objectives in the way that is least costly for them.

However, taxation is not always practical. For example, finding a suitable tax base for NO_x emissions is difficult. Whereas SO_2 emissions from (road) vehicles are closely related to the sulphur content of the fuel used (end-of-pipe cleaning of these emissions would be very costly), NO_x emissions depend much more on the combustion process being applied, the way the vehicle is driven, as well as on the existence (and maintenance) of end-of-pipe cleaning devices, such as catalytic converters. Some of these aspects can be addressed through taxes on purchases of motor vehicles, or through annual motor vehicle use taxes. For example, the tax could vary according to the type of catalytic converter that is installed. However, additional instruments will probably still be needed to cover these situations.

Fuel taxes are already widespread in OECD countries, but there is considerable variation in the rates being applied (Figure 16.6). From 2002 to 2007, several OECD countries significantly increased their fuel tax rates; nevertheless, most countries still have lower tax rates for diesel fuels than for petrol.

Figure 16.6. **Tax rates on petrol and diesel in OECD countries, 2002 and 2007**

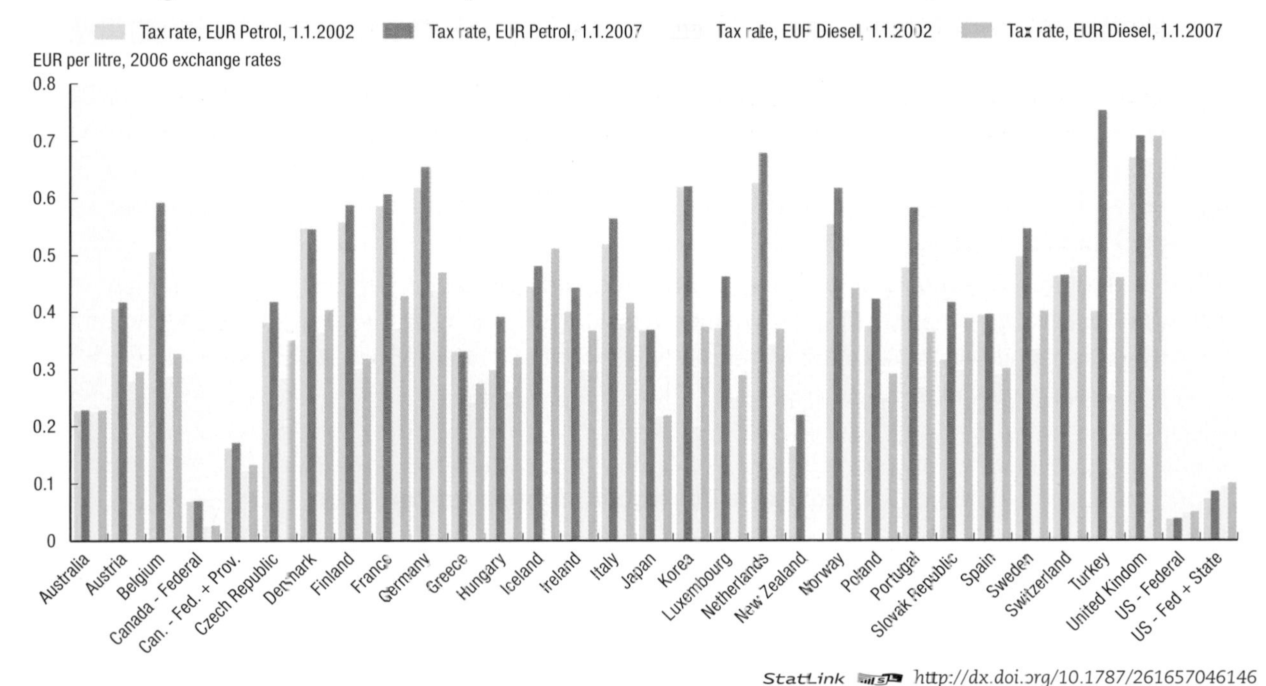

StatLink http://dx.doi.org/10.1787/261657046146

Note: The changes shown here (expressed in EUR per litre) reflect *both* changes in tax rates in national currencies *and* changes in exchange rates (see OECD, 2006b for additional detail).

Source: OECD/EEA database on instruments used for environmental policy, *www.oecd.org/env/policies/database*.

In a recent review of the use of fuel taxes in OECD countries, OECD (2006b) summarised some of the key consequences of applying these taxes as follows:

- Countries with low taxes on petrol and diesel (*e.g.* Canada and the US) tend to have much higher use of these products per unit of GDP produced than countries with higher taxes. On the other hand, countries with high fuel taxes generally have higher fuel efficiencies. Japan is a slight exception – with high fuel efficiency, despite relatively low fuel taxes.
- The recent increases in world market crude oil prices have contributed to improvements in fuel efficiency, even in countries with low fuel taxes.
- OECD countries that have increased their fuel taxes in recent years (*e.g.* Turkey and Germany) have seen very significant improvements in fuel efficiency.
- There has been a general shift from petrol to diesel use in countries that apply lower taxes on diesel than on petrol. Where the tax preference given to diesel is smaller (*e.g.* Canada and the US), the use of diesel is much lower. From the point of view of local air pollution, this is a clear advantage.
- In general, taxes on transport services will tend to affect lower-income households most; however, there are various strategies available for reducing these impacts.

At present, actions to reduce the energy-intensity of transport appear to have the greatest potential for reducing CO_2 emissions cost-effectively. Policies that promote less energy intensive transport modes (*e.g.* more public transport, bicycle use) appear to offer only a very limited potential for controlling greenhouse gas emissions (ECMT, 2006b).

Carbon and fuel taxes are ideal measures for addressing transport-based CO_2 emissions because of their effects on energy intensities. These taxes send clear signals to users, and they are likely to distort the economy less than other measures with the same goals. Significant CO_2 abatement opportunities may also include improving the fuel efficiency of new vehicles; improving the energy efficiency of individual components of vehicles; and improving on-road vehicle performance. In turn, the most cost-effective options within these approaches are promotion of fuel-efficient driving; incentives for car buyers to choose lower emissions vehicles; and regulations for some (currently unregulated) vehicle components. The reform of vehicle taxation is likely to be the highest priority in this context (ECMT, 2006b).

Biofuels also offer potentially significant CO_2 abatement opportunities, but at a high cost (except for ethanol from sugar cane; Box 16.3). Public incentives for biofuels should therefore explicitly reflect a full life-cycle ("well-to-wheel") view of both CO_2 emissions and economic consequences.

Box 16.3. **Prospects for liquid biofuels for transport**

Several countries have recently adopted targets for biofuels use in the transport fuel mix, partly on the grounds that these fuels can significantly reduce emissions of greenhouse gases and partly for energy security reasons. Large investments are therefore now being made in the biofuels option, and subsidies amounting to an estimated USD 14.3 billion were given to support biofuels in OECD countries in 2006 (see Box 14.2 in Chapter 14 on agriculture).

The environmental benefits of biofuels use are uncertain. The emissions savings from replacing conventional transport fuels with biofuels depend upon the amount of energy used in the conversion process and in transporting the raw materials to bio-refineries (see Box 17.3 in Chapter 17 on energy). The production of biofuels can have other negative impacts on the environment, and can compete for land with agricultural food crops. The recent surge in oil prices has made biofuels more cost competitive with conventional oil-based fuels. However, production costs are still above the level of international oil prices in most cases. Without subsidies or other policies to support biofuels use – such as targets for minimum use of biofuels in transport fuel – they are unlikely to be economically competitive with fossil fuels.

Overall, encouraging a switch from fossil-based transport to biofuels seems likely to be a costly way (both environmentally and economically) of addressing the problem of greenhouse gas emissions. Greater policy attention could be paid to the development and introduction of second generation biofuels, which are likely to have greater environmental benefits and lower negative environmental impacts, as well as to ensuring the performance of current generation biofuels.

Sources: ECMT (2007b); OECD and FAO (2007); Doornbosch and Steenblik (2007).

Road pricing systems require motorists to pay directly for driving on a particular road. Raising the cost of using roads discourages some drivers from using them, resulting in less congestion and fewer environmental impacts. Congestion charges have now been applied in a number of urban areas worldwide, and have reduced traffic and traffic-related externalities such as congestion, air pollution and accidents (see also Chapter 5 on urbanisation). Some European road use modelling has concluded that road pricing has a definite impact on traffic volumes, and may also help to modify the pattern of driving behaviour (trip choice, modal choice, etc.) (CANTIQUE, 2001).

Road pricing can also improve the efficiency of road systems. For example, the Swiss distance-related fee for heavy goods vehicles depends on three factors: the distance driven on the Swiss road network (all roads); the maximum permitted weight of vehicle and trailer; and the emissions of the vehicle (there are three emission classes). The effects of the fee, used in combination with weight limits, have been the renovation of the truck fleet, more concentration in the hauler industry, and fewer heavy goods vehicles on the road (OECD, 2005).

Regulatory approaches

In cases where market-based mechanisms are not practical, direct regulation can be an important tool for reducing the environmental impacts of transport. However, regulations aimed at promoting specific transport technologies may be less efficient and effective than regulations on transport-related emissions themselves. For example, regulations that promote hybrid vehicles (on the grounds that they are more fuel efficient) may result only in the production of hybrid vehicles with more powerful engines, without producing any overall emission reductions. Regulations should therefore focus on the desired (environmental) outcome. If this desired outcome is beyond the influence of the proposed regulation, the regulation should still target the particular elements of the problem which most directly influence the desired outcome.

In the US, the EU and Japan, the standards for emissions of air pollutants from vehicles have been regularly tightened since the 1980s. Emissions standards are now in place for carbon monoxide, hydrocarbons, nitrogen oxides, smoke and particulate matter. These standards require the application of end-of-pipe devices like catalytic converters. In some Asian countries, motorised two- and three-wheeled vehicles (with two-cycle, rather than four-cycle engines) contribute disproportionally to transport emissions of particulates, hydrocarbon and carbon monoxide (Faiz and Gautam, 2004).

The availability of sufficiently high fuel quality is crucial for the proper functioning of end-of-pipe devices, and this remains a problem in many parts of Asia and Africa. Inspection and engine maintenance programmes are also important, but while such programmes are common in most OECD countries, they are less in evidence elsewhere.

Many OECD countries also have restrictions on the total number of hours per day (or on the times of day) that large trucks can be on the road. Some cities (*e.g.* Mexico) have also adopted restrictions on the days of the week on which cars can be driven (*e.g.* odd/even-numbered plates on successive days). These regulations help to reduce emissions from heavy trucks (*e.g.* by discouraging the use of trucks when fuel efficiency would be lowest, such as during rush hours). In addition to reducing fuel consumption of road transportation, these kinds of restrictions also have the added benefit of making transportation by rail a more appealing alternative.

Other policies

Infrastructure investments can also have a significant influence on both the efficiency of transport activity, and on modal shifts within the transport sector. For example:

- Improvements in roadways and traffic management can reduce congestion and associated environmental problems. However, this strategy may also lead to increased traffic, rather than to environmental improvements, unless it is properly designed and implemented.

- Investments in the speed and comfort of public transport make that option more attractive to commuters. A shift from personal vehicles to public buses or subways would produce a double environmental benefit – reduced GHG and air pollutant emissions, as well as reduced congestion problems.
- Investments in rail infrastructure, improvements in rail-road connections, and better integration of international rail networks could all make rail a more attractive option for moving freight and people.

Spatial policies (*e.g.* land use planning) can often influence transport decisions much more than transport policy itself. Integrating land use policy with environmental objectives in the transport sector can therefore generate significant environmental benefits. There may be particular advantages to be derived from institutional reforms introduced at the local (municipal) level – especially initiatives aimed at congestion problems. Changes in the regulation of land use will likely be needed to create mixed-use areas (with high density) over time.

Other supporting policies, such as better information management and the promotion of teleworking, can also support more environmentally-friendly transportation. Public outreach campaigns can make consumers aware of the environmental impacts of their actions, and encourage them to make more environmentally-friendly transport decisions. Better communication between government, firms and individuals can help policy-makers develop approaches that work best for citizens. Better communication between different regional governments, different layers of government, and different government departments, can also ensure that environmental and transport related policies in one area support policies in other areas.

The benefits associated with most of the policies discussed above could be increased if the various instruments were used in combination with each other. For example, improvements in public transport are much more likely to increase use if they are accompanied by increased road pricing. Improvement in rail infrastructure would draw more freight to the rail sector if fuel prices were also increased to make trucking less attractive. The positive effects of spatial policies can be strengthened by additional measures to increase the attractiveness of urban areas, such as measures to decrease noise pollution, or to improve cycling and pedestrian infrastructure.

References

BTS (Bureau of Transport Statistics) (2006), *Transport Statistics: Annual Report (2006)*, US Dept of Transportation, Washington DC.

BTS (2007), *National Transportation Statistics*, US Dept of Transportation, Washington DC.

CANTIQUE (Concerted Action on Non-Technical Measures and Their Impacts on Air Quality and Emissions) (2001), *Final Report. www.isis-it.com/doc/progetto.asp?id=16&tipo=Transport.*

Doornbosch, R. and R. Steenblik (2007), *Biofuels: Is the Cure Worse than the Disease?* OECD Roundtable on Sustainable Development, Document SG/SD/RT(2007)3, OECD, Paris.

ECMT (European Conference of Ministers of Transport) (2003), *Reforming Transport Taxes*, ECMT, Paris.

ECMT (2006a), *Reducing NO_x Emissions on the Road*, ECMT, Paris.

ECMT (2006b), *Review of CO_2 Abatement Policies for the Transport Sector*, ECMT, Paris.

ECMT (2007a), *Cutting Transport CO_2 Emissions*. ECMT, Paris.

ECMT (2007b), *Biofuels – Linking Support to Performance*, ECMT, Paris.

 OECD ENVIRONMENTAL OUTLOOK TO 2030 – ISBN 978-92-64-04048-9 – © OECD 2008

EEA (European Environment Agency) (2001), *Traffic Noise: Exposure and Annoyance*, EEA, Copenhagen.

EEA (2006), *Transport and Environment: Facing a Dilemma*, EEA, Copenhagen.

EPA (Environmental Protection Agency) (2006), *Greenhouse Gas Emissions from the US Transportation Sector (1990-2003)*, EPA, Washington DC.

Eyring, V., H.K. Köhler, A. Lauer and B. Lemper (2005), "Emissions from International Shipping: 2. Impact of Future Technologies on Scenarios Until 2050", *Journal of Geophysical Research*, Vol. 110, D17306. doi:10.1029/2004JD005620.

Faiz A. and S. Gautam (2004), "Technical and Policy Options for Reducing Emissions from 2-stroke Engine Vehicles in Asia", *International Journal of Vehicle Design*, 34, 1-11.

IEA (International Energy Agency) (2002), *World Energy Outlook (2002)*, IEA, Paris.

IEA (2006), *World Energy Outlook (2006)*, IEA, Paris.

INFRAS (2004), *External Costs of Transport (Update Study)*, INFRAS, Bern.

IPCC (Intergovernmental Panel on Climate Change) (1999), *Aviation and the Global Atmosphere*, *www.grida.no/climate/ipcc/aviation/008.htm*, IPCC, Geneva.

IPCC (2007), "Transport and its Infrastructure", *Working Group III Fourth Assessment Report* (Chapter 5, page 11), see *www.mnp.nl/ipcc/pages_media/AR4-chapters.html*, IPCC, Geneva.

OECD (2005), *The Window of Opportunity: How the Obstacles to the Introduction of the Swiss Heavy Goods Vehicle Fee Have Been Overcome*, *www.oecd.org/dataoecd/19/36/34351788.pdf*, OECD, Paris.

OECD (2006a), *Decoupling the Environmental Impacts of Transport from Economic Growth*, OECD, Paris.

OECD (2006b), *The Political Economy of Environmentally Related Taxes*, OECD, Paris.

OECD and FAO (2007), *OECD-FAO Agricultural Outlook 2007-2016*, OECD, Paris.

Singh A.K. and M. Singh (2006), "Lead Decline in the Indian Environment Resulting from the Petrol-lead Phase-out Programme", *Science of the Total Environment* 368, 686-694.

UN (United Nations) (2007), *United Nations Common Database*, *http://unstats.un.org/unsd/cdb/cdb_advanced_data_extract.asp*.

WHO (World Health Organization) (1999), *Health Costs due to Road Traffic-related Air Pollution*. An Impact Assessment Project of Austria, France and Switzerland. Synthesis Report, Federal Department of Environment, Transport, Energy and Communications (Bureau for Transport Studies), Bern.

ISBN 978-92-64-04048-9
OECD Environmental Outlook to 2030

Chapter 17

Energy

This chapter examines the recent trends and future projections for energy demand and supply in different regions around the world to 2030. Despite continuing improvements in energy efficiency, world primary energy use is projected to grow by 54% between 2005 and 2030 under the Outlook Baseline. Fossil fuels are expected to continue to dominate the energy mix. Increasing energy production and use will affect the stability of ecosystems, global climate and the health of current and future generations. The chapter also outlines some of the key government policies that are needed to promote a lasting technology shift towards a more sustainable energy path, and examines some of the costs and environmental benefits of specific policy options.

KEY MESSAGES

World primary energy use is projected to grow by 54% between 2005 and 2030 in the Baseline – an average annual rate of 1.8% per year.

Fossil fuels are likely to continue to dominate primary energy use, accounting for most of the increase in energy between 2005 and 2030 (84%). Oil looks to remain the largest single global energy source in 2030, though its share of total energy use is projected to fall from 36% to 33%. Power generation and transport account for most of the increase in energy consumption. Electricity is the fastest growing final form of energy.

For as long as fossil fuels dominate the world energy system, rising energy production and use threaten the stability of ecosystems, global climate and the health of current and future generations. Fossil fuel combustion is the main contributor to air pollution and greenhouse gas emissions, especially carbon dioxide.

Energy intensity – the amount of energy needed to produce one unit of gross domestic product – is projected to continue to decline, thanks to improved energy efficiency and a structural economic shift in all regions towards less energy-intensive activities.

The net environmental effect of switching to renewable energy sources is expected to be positive, despite some adverse environmental effects which need to be addressed through policy.

Policy options

Government policies will be critical to promote a lasting technology shift which steers the world onto a more sustainable energy path. To keep the costs of mitigation low while also stimulating innovation, policies will need to:

- Emphasise market-based instruments in the policy mix to establish a clear price on carbon and other greenhouse gas emissions and encourage mitigation where it is least-cost.
- Reverse growth in energy-related greenhouse gas emissions.
- Encourage more efficient energy use and promote the supply of renewable and low-carbon energy sources.
- Commercialise carbon capture and storage technologies to permit the environmentally acceptable use of coal and other fossil fuels.
- Alter radically the way energy is produced and consumed. Ultimately, the world will need to move away from carbon-intensive fossil fuels towards renewables and/or nuclear power. No one technology or fuel choice will dominate; a mix will be required.

Greater deployment of cleaner technologies in this sector will also deliver a wide range of other benefits, from energy security to environmental benefits (*e.g.* healthier people, cleaner cities, clearer skies).

This figure shows the mix of technologies and mitigation options likely to be important to achieve very low emission levels, *i.e.*, to stabilise atmospheric concentrations of greenhouse gases at 450 ppm CO_2 equivalent. Key approaches in the short term will be low-cost measures that reduce non-CO_2 greenhouse gases, combined with expanding sinks and avoiding emissions from land use and forestry as well as energy efficiency measures. Also essential by 2020 to achieve this objective will be the use of second generation biofuels and carbon capture and storage (CCS) technologies on a worldwide basis, along with increased renewables.

450 ppm CO_2eq emission pathway compared to Baseline: technology "wedges" of emission reduction

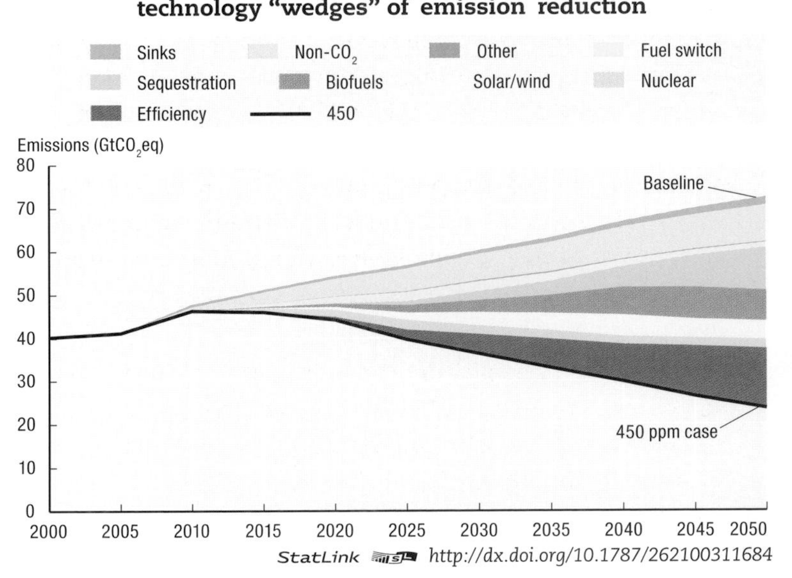

StatLink http://dx.doi.org/10.1787/262100311684

Introduction

The relationships between energy supply and use, economic activity, human development and the environment are extremely complex. Increased energy use is both a cause and an effect of economic growth and development. Energy is essential to most economic activities. Industrialised economies rely on commercial energy to transport goods and people, to heat homes and offices, to power engines and appliances, and to run shops and factories. The prosperity generated by economic development stimulates, in turn, demand for more and better-quality energy services, especially in the early stages of economic development. But the production, transportation and use of energy can have major adverse effects on the environment and on the health and well-being of current and future generations.

Today, energy use is the largest source of air pollution and of the greenhouse gases (GHGs) that threaten to change global climate (see Table 17.1 and Chapter 7). These environmental problems arise principally from the combustion of fossil fuels, which provides the bulk of the world's energy needs. Air pollution occurs through the noxious gases and pollutants – including sulphur dioxide (SO_2), nitrogen oxides (NOx), particulates, methane (CH_4) and volatile organic compounds (VOCs) – emitted either through fuel combustion or in leakages from delivery systems (see Chapter 8, Air pollution). The use of fossil fuels is the leading cause of urban smog, particulate matter air pollution and acid rain. Local and regional air pollution is a major human health problem, especially in the developing world, and also affects the health of natural systems and biodiversity worldwide. Indoor air pollution, caused largely by inefficient and poorly ventilated stoves burning traditional fuels or coal, is a leading cause of health problems in many developing countries. Producing and transporting oil can pollute the sea, freshwater supplies and the soil through accidental leaks or poor management. Combustion of fossil fuels is also the predominant source of greenhouse gases, most notably carbon dioxide (CO_2), while coal mining and natural gas distribution are an important source of methane.

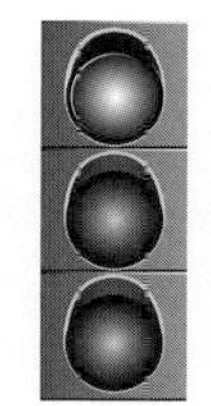

Energy is a leading source of pollution and GHG emissions. Global primary energy consumption is projected to grow rapidly to 2030 in the Baseline.

Alternatives to fossil energy use include renewable and nuclear energy; however, these energy forms are not problem-free either. Renewable energy sources, such as hydroelectric and wind energy, are cleaner, but can also carry limited environmental risks of their own. For example, large-scale hydroelectric dams can be a significant source of CH_4 emissions when they cause deforestation and alter natural river flow, with a range of cascading ecological impacts. Wind energy causes noise pollution and alters the landscape. Nuclear power production gives rise to radioactive waste and waste management problems, raises the risk of accidental contamination as well as a range of national (and international)

Table 17.1. **Environmental impact of the energy sector, 1980 to 2030**

Climate change

		1980 (%)		2005 (%)		2030 (%)		Total % change	
								1980-2005	2005-2030
GHG emissions ($GtCO_2eq$)		32.9	100%	46.9	100%	64.1	100%	43%	37%
CO_2 emissions from energy ($GtCO_2$)	Industry and other[a]	7.6	39%	9.0	32%	12.5	29%	19%	39%
	Power generation	6.2	32%	11.0	39%	18.0	42%	78%	65%
	Residential	2.0	11%	2.3	8%	2.8	7%	14%	22%
	Transport	3.5	18%	6.1	21%	9.6	22%	73%	58%
	Total	19.3	100%	28.4	100%	43.0	100%	47%	52%
CO_2 emissions from energy (t CO_2/per capita)		4.3		4.4		5.2		1%	19%
CO_2 concentration (ppm)		339		383		465		13%	21%
Global mean temperature increase (°C) (above pre-industrial levels)		0.21		0.69		1.34			

Air pollution

	1980	2005	2030	Total % change	
				1980-2005	2005-2030
Nitrogen oxides emission (Mt)	30.5	29.6	29.4	–3%	–1%
Sulphur oxides emission (Mt)[b]	80.5	64.4	67.3	–20%	5%

	2000	2030	Total % change
Loss of health (per million inhabitants)[c]	1 632	3 507	115%
Mortality (deaths/per million inhabitants)[d]	164	412	150%

StatLink http://dx.doi.org/10.1787/257488858217

Note: Totals may not add up due to number rounding.

a) The term "other" includes energy-related emissions of CO_2 from: services, bunkers, energy transformation, losses and leakages, etc.

b) The total sulphur dioxide emission considers both industry related and energy related emissions.

c) The figures for loss of health were obtained by adding loss of health attributable to outdoor exposure to ozone, to loss of health attributable to particulate matter per million inhabitants.

d) Mortality was defined as the sum of deaths related to outdoor exposure to ozone and deaths attributable to particulate matter per million inhabitants.

Source: OECD Environmental Outlook Baseline.

security issues. Beyond economic and technical questions, switching to non-fossil energy sources thus involves trade-offs and consideration of a range of environmental and security consequences, issues that can only be resolved when taking local contexts and preferences into account.

Access to electricity is particularly important for human development. Electricity is needed for activities such as lighting, refrigeration and the running of household appliances. Yet an estimated 1.6 billion people in developing countries, equal to just over a quarter of the world's population, have no access to electricity in their homes (IEA, 2006a). Indeed, 2.5 billion people rely almost exclusively on traditional biomass fuels – such as wood, charcoal, crop waste and dung – for cooking and heating (IEA, 2006a). As incomes rise, households usually switch to modern energy services for cooking, heating, lighting and electric appliances. Rising incomes also boost demand for personal mobility and, therefore, for transport fuels. The shift to modern energy services initially leads to an increase in the energy intensity of the economy – the amount of energy needed for each unit of GDP – through industrialisation, improved comfort levels and increased personal and freight mobility. As industrialisation proceeds, energy intensity eventually peaks and then begins to decline due to structural changes, including a shift to less energy-intensive service activities. Most OECD countries have already reached this stage. Economic development ultimately leads to saturation in demand for bulk industrial goods and increased demand for smaller, less energy-hungry miniaturised products. Technological advances also raise the average energy efficiency of equipment and appliances and reinforce the long-term decline in intensity.

Key trends and projections

Primary energy consumption

Barring a radical change in government policies, major technological breakthroughs, an unexpected change in oil prices or disruption to global economic expansion, the world's energy needs are set to continue to grow steadily over the coming decades. Global primary energy consumption[1] in the *OECD Environmental Outlook* Baseline is projected to increase from 460 exajoules in 2005 to 710 EJ in 2030 and 865 EJ in 2050, which represents an average annual increase of 1.8% in 2005-2030 and 1% in 2030-2050 (Table 17.2; and see Box 17.1 for methodological details). Energy use has grown by 1.7% per year since 1980. Fossil fuels continue to dominate the primary fuel mix. Oil, gas and coal account for 86% of the projected increase in total energy use between 2005 and 2030. The combined share of fossil fuels in total primary energy use remains essentially constant from 2005 to 2030, hovering at about 85%, and then drops to 80% in 2050.

Table 17.2. **World primary energy consumption in the Baseline (EJ), 1980-2050**

					Compound annual growth rate (%)		
	1980	2005	2030	2050	1980-2005	2005-2030	2030-2050
Coal	75.5	129.0	198.1	224.2	2.2	1.7	0.6
Oil	132.4	168.1	239.0	287.8	1.0	1.4	0.9
Natural gas	55.3	98.1	174.9	221.4	2.3	2.3	1.2
Modern biofuels	0.5	2.2	16.4	39.1	6.1	8.4	4.5
Traditional biofuels	33.5	44.4	52.8	50.7	1.1	0.7	–0.2
Nuclear[a]	2.5	9.3	12.9	12.1	5.4	1.3	–0.4
Solar/wind	0.1	0.6	4.9	12.6	7.9	9.1	4.9
Hydro	6.0	10.5	15.1	17.6	2.3	1.5	0.7
Total	305.8	462.3	714.2	865.4	1.7	1.8	1.0

StatLink http://dx.doi.org/10.1787/257501561404

a) These numbers differ by approximately a factor of three from those reported by the International Energy Agency (IEA). This is because the IEA defines the amount of primary nuclear energy consumption as three times the amount of energy produced in the form of electricity – this is the "fossil fuel replacement method" to report primary energy supply associated with this energy carrier. We define all direct electricity options using a simpler method, on the basis of electricity output.

Source: OECD Environmental Outlook Baseline.

Oil remains the single largest fuel in the global primary energy mix throughout the projection period, with consumption growing by 42% between 2005 and 2030. Its share nonetheless stays flat at 33% (Figure 17.1). The bulk of the increase in oil use is projected to come from the transport sector. Natural gas sees the biggest increase in primary consumption in volume terms in 2005-2030, ahead of coal and oil. The share of natural gas in primary energy is projected to grow to 24% by 2030. Nevertheless there is a large projected increase in the volume of coal use over the coming decades (nearly 70 EJ), which drives up GHG emissions. Demand for coal is driven mainly by the power generation sector, especially in China and India (see Box 17.2). Coal's share of world primary energy consumption remains stable at 28% in 2005 and 2030. Nuclear power is projected to grow

Box 17.1. **Key uncertainties and assumptions**

The energy projections presented in this chapter are subject to a wide range of uncertainties, including the rate of economic and population growth, energy prices, the availability and cost of developing energy resources, technological progress, investment trends and government policies on energy and the environment. The Baseline assumes no change in government policies.

The Baseline projections presented here have been calibrated to the Reference Scenario projections of the 2006 edition of the *World Energy Outlook*, which also assumes no new government policies (IEA, 2006a). This was done by running the IMAGE energy model so as to reproduce as closely as possible the *WEO-2006* energy projections based on the population and economic assumptions used in that study. This involved adjusting income elasticities of energy demand and preferences for different fuels. Assumptions about electric power plant efficiency and primary energy prices are comparable. The IMAGE model was then re-run using the Baseline economic growth and population assumptions set out in Chapters 2 and 3 of this *Outlook*, together with the IEA's most recent energy price assumptions. The results of the IMAGE and *WEO-2006* projections differ slightly, mainly because the macroeconomic assumptions differ.

much more slowly than in the past, based on current policies, such that its share in primary consumption falls. The combined share of hydropower and traditional biomass is projected to increase slightly. In aggregate, modern renewables – a category that includes geothermal, wind, solar, wave and tidal energy and biofuels – are expected to grow faster than any other energy source, their contribution to global primary energy rising from nearly 1% in 2005 to 3% in 2030. Modern biofuels (liquid transport fuels derived from biomass) account for most of this increase (see Box 17.3).

Figure 17.1. **World primary energy consumption in the Baseline, to 2050**

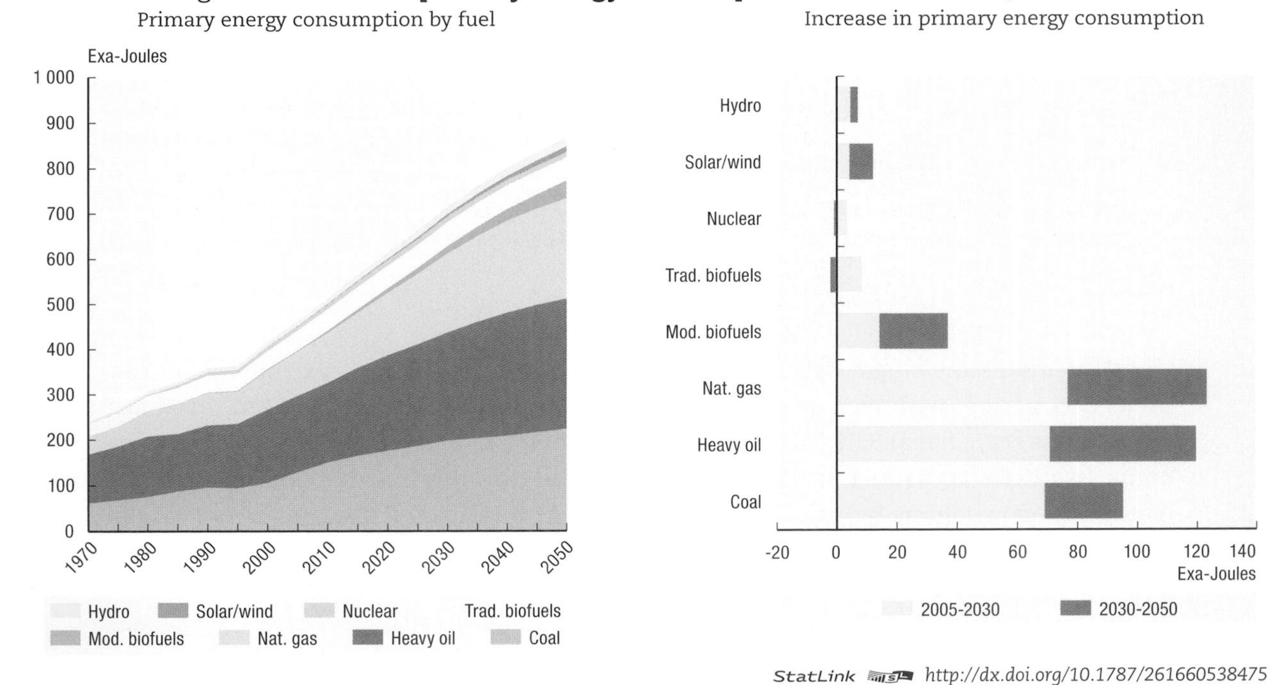

StatLink http://dx.doi.org/10.1787/261660538475

Source: OECD *Environmental Outlook* Baseline.

Box 17.2. **Power generation in China**

Power generation in China has been growing extremely rapidly over the last decades, almost quadrupling from its 1990 level of 650 Terawatt hours (TWh) to reach 2 544 TWh in 2005 (an annual average of growth rate of 9.6%, more than three times the global average for this period). In 2005, Chinese electricity production accounted for 14% of global electricity production. Although Chinese electricity production is large in absolute terms, per capita consumption is still relatively low: only two-thirds of the global average of 2.8 MWh (megawatt hours) per capita in 2005.

The Reference Scenario of the IEA's *World Energy Outlook* (2007a) expects Chinese electricity production to continue to grow at a fast pace after 2005, to reach 8 472 TWh in 2030. This projection is 11% higher than that made a year previously (on which this *Outlook* is based). This increase reflects the extremely rapid growth in recent years: electricity generation in China grew 30% from 2003 to 2005, and 105 gigawatts (GW) of electricity generating capacity (more than the entire capacity of the UK and Netherlands combined) was added in 2006 alone.

Chinese electricity generation is projected to continue to grow, and to account for an increasing proportion of global energy-related CO_2 emissions. Projected growth to 2030 would lead to total electricity generation in 2030 being more than three times the level of electricity generation in 2005 (2 544 TWh), despite growing at a slower rate during this period than in the recent past (4.9% p.a., which is nevertheless almost double the world average for the same time period).* Under such a scenario, Chinese electricity production would account for almost a quarter (24%) of global electricity production, and almost 15% of global energy-related CO_2 emissions, by 2030.

More than three-quarters (78.5%) of Chinese electricity is currently generated from coal. Under the IEA Reference Scenario, this proportion is hardly expected to change by 2030, despite large growth rates for Chinese electricity generation from nuclear power, as well as hydro and other renewables. However, the Reference Scenario projects the CO_2-intensity of electricity production to fall by 18% between 2005 and 2030.** This drop is due almost entirely to the increased share of electricity generation from "clean coal" technologies, such as super-critical and ultra-supercritical pulverised fuel technologies (see below). Further reductions are feasible, for example via an accelerated phase-out of inefficient power generating technologies, more rapid deployment of clean coal technologies, and greater uptake of both nuclear power and renewable electricity. Under such assumptions (the Alternative Policy Scenario of the IEA's 2007 *World Energy Outlook*), emissions from China's electricity sector could be 4.5 billion tonnes of CO_2 in 2030 (1.5 billion tonnes less than in the Reference Scenario).

* The WEO2007 data here are different from data in the *Outlook*, which suggests a 4% annual growth rate for electricity between 2005-2030 (the WEO calculates 2.7%) (IEA, 2007a).

** Personal communication, Maria Argiri, IEA, 19.11.07.

Source: IEA statistics 2007; IEA *World Energy Outlook*, 2007a.

Box 17.3. **Biofuels in the energy mix**

Interest in biofuels is growing in many countries (see also Box 14.2 in Chapter 14 on agriculture and Box 16.3 in Chapter 16 on transport). Many OECD countries are subsidising biofuels production for energy security and climate change reasons. Indigenously produced biofuels can replace imported oil, diversifying the sources of energy and bringing energy-security benefits to importing economies. Biofuels can lead to marginally lower greenhouse gas emissions compared with fossil fuels; however, they may also harm the environment if the biomass raw materials are not produced in an environmentally sustainable manner, and may also raise the cost of food production. Policies to support the production and use of biofuels need to reflect the full life-cycle of their effect on greenhouse gas emissions and the economy.

Today, the overwhelming bulk of biofuels produced around the world are ethanol and esters (known as biodiesel). Ethanol is usually produced from starchy crops, including cereals and sugar, while biodiesel is produced mainly from oil-seed crops, such as rapeseed. Ethanol is usually blended with gasoline (either pure or in a derivative form, known as ETBE), while biodiesel is normally blended with diesel. Global production of biofuels in 2005 amounted to over 640 thousand barrels per day (kb/d), almost 80% of which was in Brazil and the United States, which produce almost exclusively ethanol. Production of biofuels in Europe, the bulk of which is biodiesel, is increasing rapidly thanks to strong fiscal incentives in several countries, notably Germany. Current investment plans point to a continued rapid expansion of biofuels capacity in these regions in the coming years.

In the longer term, the prospects for biofuels in these and other parts of the world hinge on government policies, technological advances and reductions in production costs. New technologies being developed today, including enzymatic hydrolysis and gasification of woody ligno-cellulosic feedstock to make ethanol – the so-called "second generation" biofuels – could be more economically competitive than existing technologies and lead to more certain environmental benefits.

Over three-quarters of the increase in world primary energy consumption through to 2030 is projected to come from non-OECD countries (Figure 17.2), where the economy and population will be expanding faster. As a result, the share of OECD countries in total primary energy consumption looks likely to drop, from 50% in 2005 to 42% in 2030, and to 37% by 2050. Developing Asia[2] sees the fastest rates of growth in energy consumption, increasing by almost 94% between 2005 and 2030. Energy intensity, measured as total primary energy use per dollar of gross domestic product, is projected to decline in all regions. On average, it is projected to fall by 1% per year worldwide between 2005 and 2030, quickening to 1.1% between 2030 and 2050. Intensity falls most quickly in the BRIC developing regions, as energy prices are reformed, more energy-efficient technologies are introduced and wasteful energy practices are discouraged. The shift to service sector economic activity is more advanced in the OECD countries, so there is less scope for them to reduce energy intensity. Per capita primary energy consumption is also projected to continue diverging from income growth in all regions. Globally, per capita energy

Per capita energy consumption is likely to grow more slowly than per capita incomes across the world to 2030.

Figure 17.2. **Primary energy consumption and intensity by region in the Baseline, to 2050**

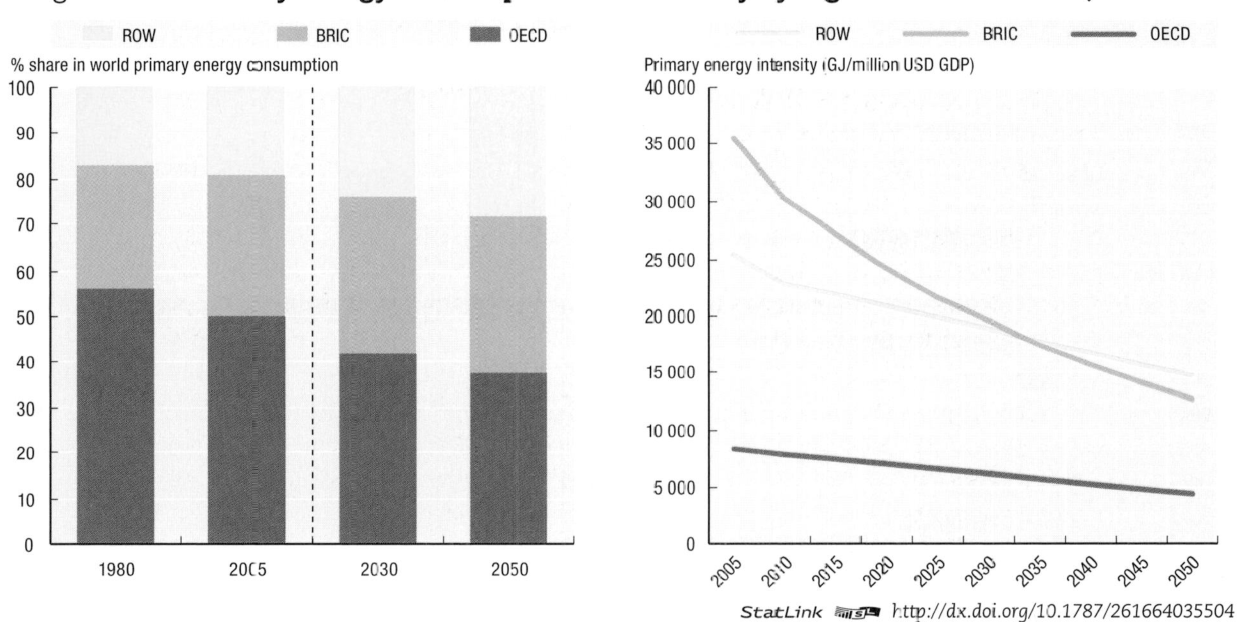

StatLink http://dx.doi.org/10.1787/261664035504

Source: OECD Environmental Outlook Baseline.

consumption looks likely to grow by 0.8% per year on average to 2030 and, as global economies become more fully industrialised, to increase by 0.5% per year between 2030 and 2050 – well below the 1.8% per year rate of per capita GDP growth.

Power generation and other energy uses

Use of primary energy to generate power is projected to continue to grow steadily in every region in the Baseline, driven by strong final demand for electricity.[3] Globally, electricity consumption is projected to grow by 4% per year between 2005 and 2030, down from 5.1% between 1980 and 2005. Non-OECD countries account for 64% of the increase. There is considerable variation across regions in the fuel mix (Figure 17.3). Worldwide, coal accounts for well over half of the total increase in fuel inputs to generation, its share in total generation remains 55% in 2005 and in 2030. Coal-fired power stations are the most competitive generation option for large-scale power generation in the majority of regions, especially developing Asia. In fact, power generation accounts for the bulk of the projected increase in overall coal demand in both the developing world and the OECD countries.

The share of coal in the fuel mix in power generation is expected to rise from 46% to 55% between 2005 and 2030, driving GHG emissions up.

The share of oil, nuclear and hydro-energy in the primary energy mix for power generation is likely to decrease between 2005 and 2030. The share of natural gas is expected to increase from 21% in 2005 to 27% in 2030 and that of coal from 46% to 55%. The share of modern biofuels looks likely to increase from 1% to 4%. As a result of higher prices, the use of oil in power stations is projected to decline in every region, its share of generation worldwide plunging from 7% in 2005 to 1% in 2030. The share of nuclear power drops from 6% in 2005 to 5% in 2030. The decline is expected to accelerate over the projection period, on the assumption that few new reactors are built and several existing

Figure 17.3. **Increase in primary energy use in power generation by fuel and region in the Baseline, 2005-2030**

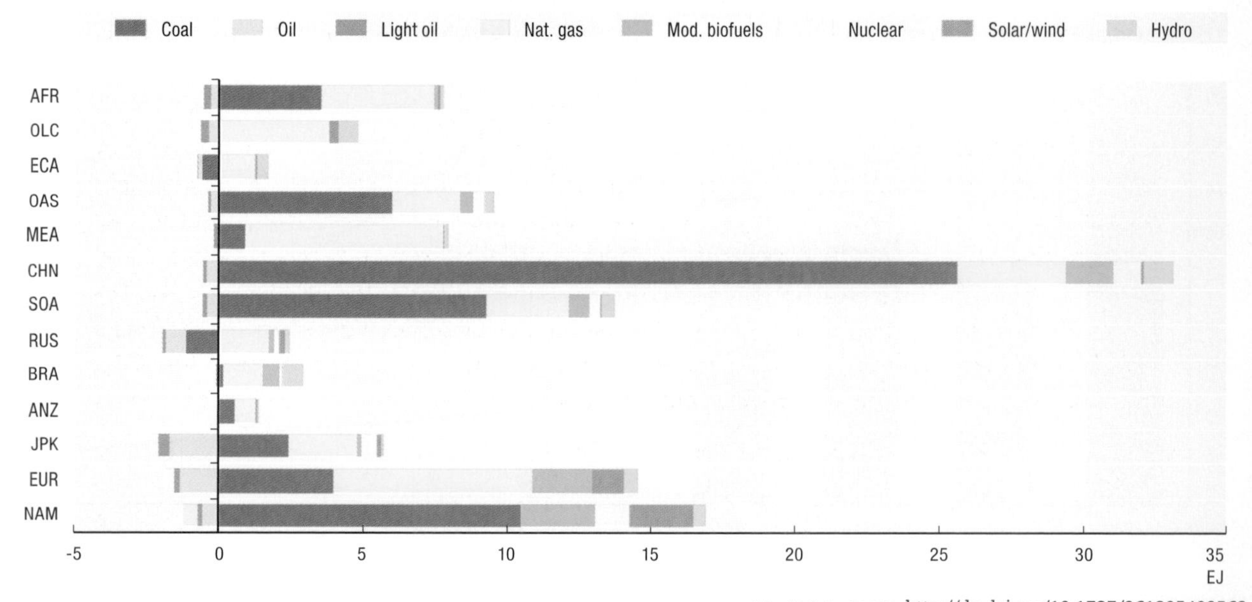

StatLink http://dx.doi.org/10.1787/261805403568

Note: Regional country groupings are as follows: NAM: North America (United States, Canada and Mexico); EUR (western and central Europe and Turkey); JPK: Japan and Korea region; ANZ: Oceania (New Zealand and Australia); BRA: Brazil; RUS: Russian and Caucasus; SOA: South Asia; CHN: China region; MEA: Middle East; OAS: Indonesia and the rest of South Asia; ECA: eastern Europe and central Asia; OLC: other Latin America; AFR: Africa.

Source: OECD Environmental Outlook Baseline.

reactors are retired. However, nuclear power production could turn out to be a lot higher if governments change their policies to facilitate investment in nuclear plants and extend the lifetimes of existing plants.

The relative importance of hydropower is set to diminish. Much of the industrialised countries' low-cost hydropower resources have already been exploited and growing environmental concerns in developing countries will discourage further large-scale projects there. World hydropower production looks likely to grow slowly to 2030, but its share in global electricity generation will drop, from 7% to 6%. Power generation using modern renewable technologies is currently limited, but is projected to grow rapidly in the Baseline. According to the model, the share of such renewables in total generation jumps from 1% in 2005 to 6% in 2030 (including modern biofuels). In absolute terms, the increase is much bigger in the OECD countries, because many of them have adopted strong policy incentives.

The long lifespans of most power plants and the high capital-intensity of power generation mean that changes in the fuel mix occur gradually. Most fossil-fired plants last more than 50 years. Therefore, much of the capacity that will be used to meet electricity demand in 2030 has already been built or is under construction, especially in the industrialised countries. A significant amount of new capacity will nonetheless be needed. Cumulative investment in power stations alone over 2005-2030 is expected to cost USD 5.2 trillion in 2005 prices – just over half of this is likely to be in developing countries (IEA, 2006a).

Primary energy inputs to other transformation activities, including oil refining and district-heat production, will rise broadly in line with final energy demand. A small but growing share of primary natural gas demand will come from gas-to-liquids plants, which convert natural gas and coal into high-value oil middle distillates and other oil products, and from fuel cells for the production of hydrogen. Coal-to-liquids production, which is already under development in China and some other countries, is also expected to increase.

Under Baseline conditions, an important environmental factor is the conversion efficiency of power generation in fossil-fired facilities. This efficiency can vary widely within and between technology types, and will determine the level of local pollutants, as well as the carbon-intensity, of power production. Demonstrated and emerging clean coal technologies offer significant improvements over conventional coal technologies (CIAB, 2006). For example, super-critical or ultra-supercritical pulverised fuel technologies are more efficient than conventional (sub-critical) units and produce significantly less CO_2, SO_2 and NO_x per unit of power generated. Coal gasification technologies promise even greater efficiencies in the future.

Final energy consumption[4]

Global energy consumption in final-use sectors – industry, transport, residential, services, agriculture and non-energy uses – is projected to increase from 308 exajoules (EJ) in 2005 to 472 EJ in 2030 in the Baseline, an average annual growth rate of 1.7%. This means that final consumption grows at roughly the same rate as primary consumption. Transport sees the most rapid growth, 2% per annum, and is projected to overtake industry by 2050 as the largest final-use sector. Transport demand increases fastest in the developing countries, where car ownership rates are still very low (see Chapter 16, Transport). In OECD countries, transport demand slows because of saturation of vehicles for a given population; however, it still remains the fastest growing major energy end-use by 2050.

The use of electricity is projected to grow faster than that of any other final form of energy worldwide, by 2.8% per year from 2005 to 2030. Electricity consumption more than doubles over that period, while its share in total final energy consumption rises from 18% to 23%. Electricity use expands most rapidly in non-OECD countries as the number of people with access to electricity and per capita consumption increase. Natural gas consumption in end uses continues to increase steadily, mainly driven by industrial demand in developing countries and by the residential sector in OECD countries. The share of gas in total final consumption is projected to slightly increase from 17% in 2005 to 18% in 2030. Although conventional oil-based fuels are expected to remain the dominant source of energy for road, sea and air transportation, biofuels are also expected to make a growing contribution to transport energy needs over the projection period (Box 17.3). The final use of coal is projected to rise slowly, its share of total final consumption falling from 9% in 2005 to 8% in 2030.

In per capita terms, final energy consumption is projected to rise in all regions (Figure 17.4). Between 2005 and 2030, per capita consumption is projected to increase 20% in the OECD, 32% in BRIC and 36% in the rest of the world. It increases by more than 50% in Asia. Per capita demand rises less in OECD countries because of saturation effects and slower economic growth, yet in absolute terms is still projected to be much higher than in the rest of the world in 2030. Final energy use remains below 60 gigajoules (GJ) per capita in 2030 in the medium-income regions, such as Brazil, China region, and Latin America, and below 30 GJ/capita in the poorest regions – South Asia and Africa. In the OECD, it reaches 148 GJ/capita by 2030.

Figure 17.4. **Final energy consumption in the Baseline, 1970-2050**

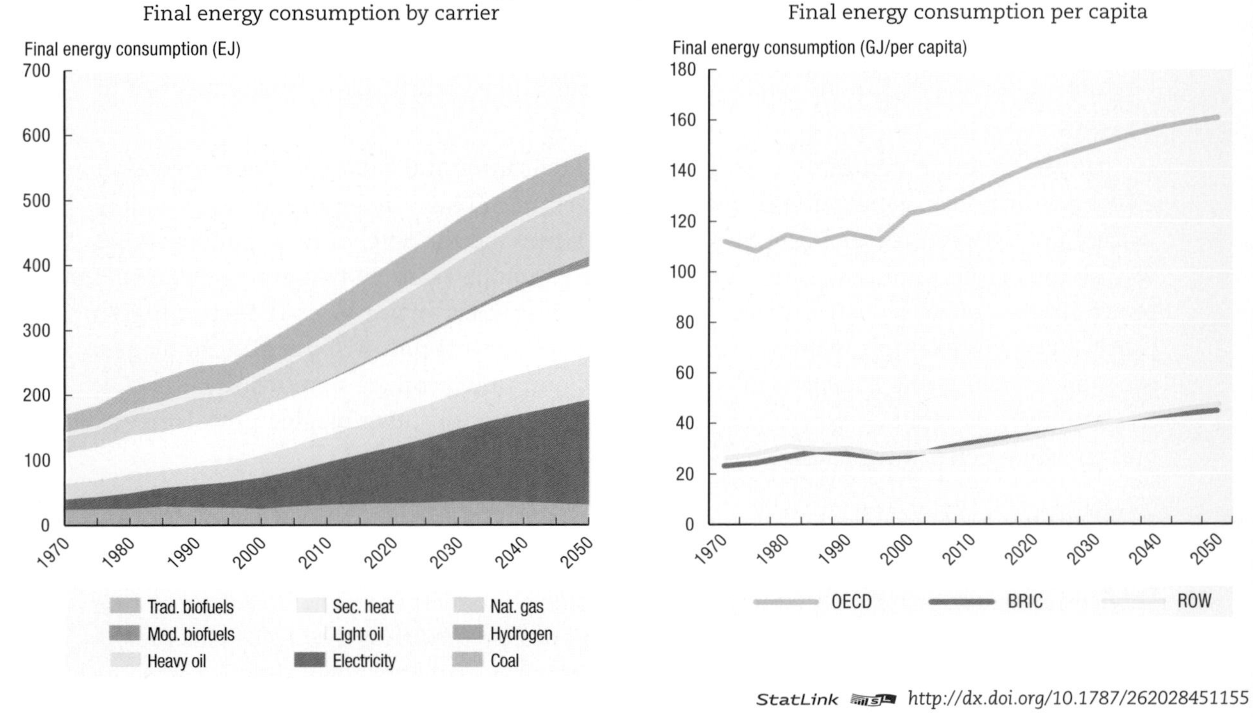

StatLink http://dx.doi.org/10.1787/262028451155

Source: OECD Environmental Outlook Baseline.

Policy implications

Based on current policies, rising consumption of fossil energy threatens to undermine the security of energy supply and exacerbate harmful environmental effects of energy use. Moreover, under current policies, much of the world's population will continue to have little or no access to modern energy services. In principle, establishing an environmentally sustainable energy system that is compatible with continued economic and social development is possible through changes in the fuel mix, more efficient use of energy, conservation and the use of new technologies such as carbon capture and storage (Box 17.4). But doing so will undoubtedly take several decades, given the slow pace of change in the physical energy infrastructure and in institutions, business practices and behaviour.

Technological developments

Technology and innovation are key to achieving energy and environmental goals and are a central goal of environmental policy. Environmental and energy policies affect both the use of existing energy-related technologies and the development and use of new technologies in the future. Pricing pollution emitted by energy use through environmental policies is an important driver for technological innovation and change (Jaffe *et al.*, 2003). The main technological avenues for mitigating the environmental effects of energy, by curbing the growth in energy use and/or related emissions of greenhouse gases or other pollutants, are:

- Improving end-use energy efficiency and conservation through a variety of process and technical innovations.
- Increasing reliance on non-fossil sources and carriers of energy, including renewables (notably hydro and wind power, photovoltaic and thermal solar energy,[5] liquid biofuels for transport and sustainable biomass technologies) and nuclear power.

- Improving efficiency of fossil-based power-generation technologies and switching to less carbon-intensive fuels (for example, from coal to gas).
- Using carbon capture and storage technology (Box 17.4).
- Using hydrogen technology.

Considerable progress could be made in making energy systems more sustainable by accelerating the uptake of state-of-the-art technologies that are already available (IEA, 2006a and b). These include cleaner and more efficient end-use equipment, vehicles and appliances, energy-efficient housing and less or zero carbon-intensive energy production. In many cases, such technologies are already competitive, yet are not widely used because of market barriers such as a lack of information or because they involve higher initial investment. In other cases, technologies are commercially proven, but may be more costly, even when the energy savings associated with their use are taken into account. The environmental, economic and social benefits of switching to such technologies may nonetheless outweigh the financial cost, justifying policy intervention. In the longer term, government action can accelerate the process of technology development (Box 17.4).

Box 17.4. **The outlook for energy technology**

A number of technologies currently being developed could improve energy efficiency significantly and mitigate the environmental effects of energy production and use. Prominent consumer demand-side technologies include plug-in hybrid vehicles, hydrogen fuel cells and zero-energy building designs. In power generation, research is focused on solar photovoltaics and concentrating solar power, in combination with long-distance electricity transportation, ocean energy, offshore wind turbines, hot dry rock geothermal, large-scale storage systems for intermittent power sources and distributed power generation. New nuclear reactor designs are also the subject of research. Analysis carried out by the IEA (see Box 17.5) suggests that efficiency improvements in end-uses could contribute up to half of the reduction in carbon dioxide emissions by 2050 in an accelerated technology scenario (IEA, 2006b) and these conclusions were recently confirmed by the IPCC (2007).

Carbon capture and storage (CCS) in geological formations is also a promising possibility. There are many different types of technologies that can be used to capture, transport and store CO_2, and these technologies are at different stages of development. The capture and transportation of carbon dioxide has been carried out for decades, albeit generally on a small scale and not with the purpose of ultimately storing it. There is a need to improve these technologies for use on a large scale and to lower the cost. At present, most CCS research and development is focused on post-combustion capture from burning fossil fuels in power plants. Much more work also needs to be done on carbon storage to demonstrate its viability and reduce the cost.

Economic instruments

Economic instruments, including taxes and subsidies (applied to the sale of fuels or to purchase energy-related equipment), mandatory emissions caps and trading, can be used to internalise the environmental externalities of energy production and use, and encourage the use of existing cleaner technologies and the development of new ones (see also Chapter 7 on climate change). Indeed, emissions trading schemes are already in place or planned in most OECD countries, as well as elsewhere: a cap and trade scheme is already in place in Norway and EU25 (to be extended to EU27 in January 2008); a small voluntary scheme is in place in

Box 17.5. **IEA technology scenarios**

The IEA scenarios lay out five different accelerated technology (ACT) energy futures and one TECH Plus Scenario. They are based on the same macroeconomic assumptions and underlying demand for energy services as used by the IEA in its Reference Scenario in the *World Energy Outlook 2005*. These scenarios do not consider the option of reducing the demand for energy services, such as restricting personal travel activity. Instead, they investigate the potential of energy technologies and best practices for reducing energy demand and emissions, and diversifying energy sources.

- Map Scenario (MAP) is optimistic in all technology areas: barriers to CCS are overcome, with costs reduced to USD 25/t CO_2 or less; cost reductions for renewable energy continue with increased deployment through learning effects; expansion of nuclear capacity occurs where it is economic to reduce CO_2 emissions and is acceptable; and progress in energy efficiency is accelerated as a result of successful policies. The other scenarios are mapped against the results of this scenario.
- Low Renewable Scenario (Low RenEn) assumes slower cost reductions for wind and solar.
- Low Nuclear Scenario reflects the limited growth potential of nuclear if public acceptance remains low, and nuclear waste non-proliferation issues remain significant.
- No CCS Scenario assumes that the technological issues facing CCS remain unsolved.
- Low Efficiency Scenario assumes less effective energy-efficiency policies.
- Technology Plus (TECH Plus) assumes faster technical progress.

Japan; and Switzerland, New Zealand, Australia, Canada, and several individual US states have all made proposals for emissions trading schemes that could be implemented within the next few years (Reinaud and Philibert, 2007; Ellis and Tirpak, 2006). A handful of OECD countries have also established – or are planning to implement – carbon taxes that are levied on selected uses of fossil fuels, electricity and/or heat.

Removing or reforming existing subsidies to energy production and use would also lead to price incentives to increase energy efficiency and switch to cleaner fuels. This has also been undertaken in some countries, *e.g.* Germany, which has provided tax relief for plants that generate both heat and power (IEA, 2007b).

Regulations and government ownership

A wide range of regulatory interventions in energy markets is in use today, including competition rules and environmental and technical standards. Minimum fuel-efficiency standards for equipment, appliances and vehicles, and labelling can be effective ways to encouraging the development and use of more efficient technology. Feed-in tariffs – the regulated price per unit of electricity that a utility or supplier has to pay for renewables-based electricity from private generators – have been used successfully in several countries. Other countries have preferred the use of renewable portfolio standards.

Direct government administration, management and control of energy resources and production are also common. Direct ownership can allow governments to dictate fuel and technology choices, such as the fuel mix in power generation. For example, the French government has been able to implement its preference for nuclear power through state ownership of the country's main electricity utility.

 OECD ENVIRONMENTAL OUTLOOK TO 2030 – ISBN 978-92-64-04048-9 – © OECD 2008

Research and development

Research and development (R&D) can be carried out directly by public organisations or indirectly through public financing of private-sector programmes. In practice, the level of commitment to R&D varies considerably across countries and over time, as well as among different fuels. In OECD countries, funding has increased since the mid-1990s. Despite some gains in funding for renewables, R&D funding remains heavily oriented to nuclear energy (Figure 17.5). In 2005, OECD country governments are estimated to have spent USD 9.6 billion in total to support R&D for energy, of which USD 1.1 billion was on energy efficiency and conservation, USD 1.1 billion on renewables, about USD 1 billion on fossil fuels and USD 3.9 billion on nuclear (IEA, 2007a; IEA, 2006c).

Figure 17.5. **Public energy research and development funding in IEA countries**

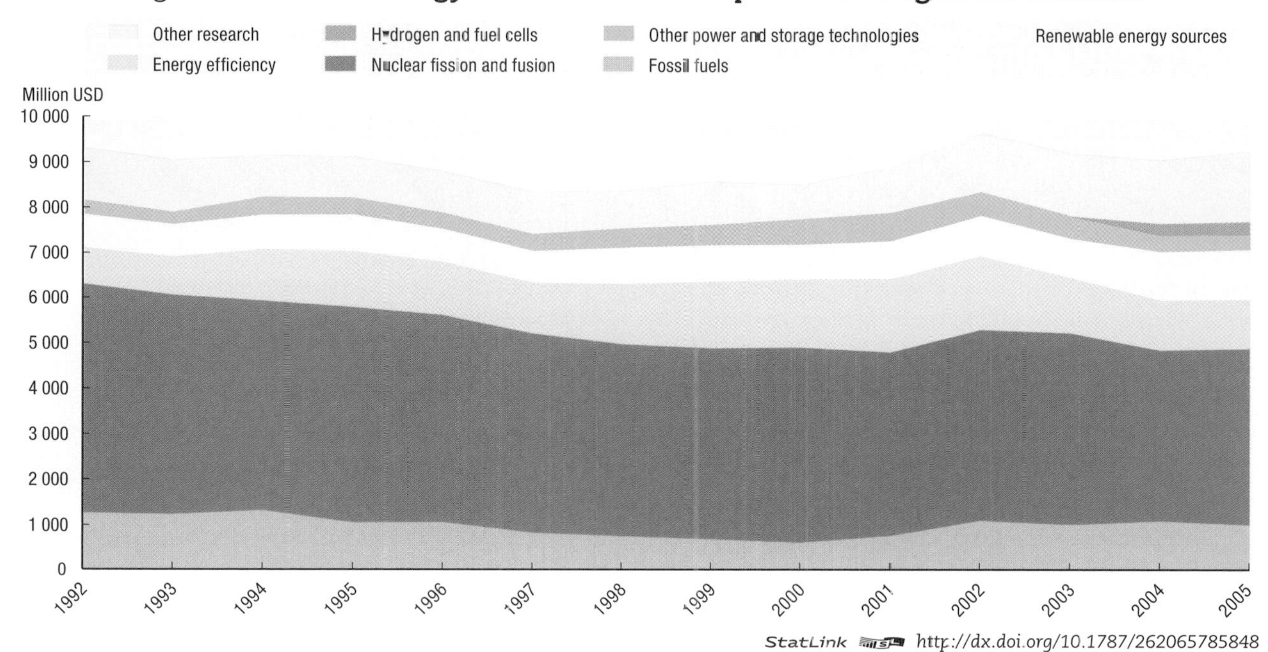

StatLink http://dx.doi.org/10.1787/262065785848

Note: Among OECD member states, only Iceland, Mexico, Poland and the Slovak Republic are not IEA members.

Source: IEA 2007, R&D database [accessed 4 July 2007].

Climate change policy simulations

Successfully addressing the threat of climate change will undoubtedly involve major changes in the pattern and level of energy use and production. Climate change policies will need to provide timely incentives to industry to shift the energy economy away from conventional fossil-based energy technologies towards more efficient and cleaner options. The OECD *Outlook* considers a variety of different GHG mitigation strategies, using GHG emission taxes as a proxy for climate policy action (see Chapter 7 for a discussion of climate change and more detail on these policy simulations, as well as Chapter 20 on environmental policy packages).

Figure 17.6 compares CO_2 emissions by energy sector for the OECD *Outlook* base year (2005) and Baseline projections (to 2050) with future emissions under mitigation policy scenarios in the 2050 timeframe. It presents results from the IEA Accelerated Technology Scenarios (IEA, 2006b and see Box 17.5) alongside the OECD *Outlook* policy cases. Interestingly all mitigation scenarios show significant emission reductions in the power generation

Figure 17.6. **IEA and OECD selected policy scenarios: CO_2 from energy in 2005 and 2050**

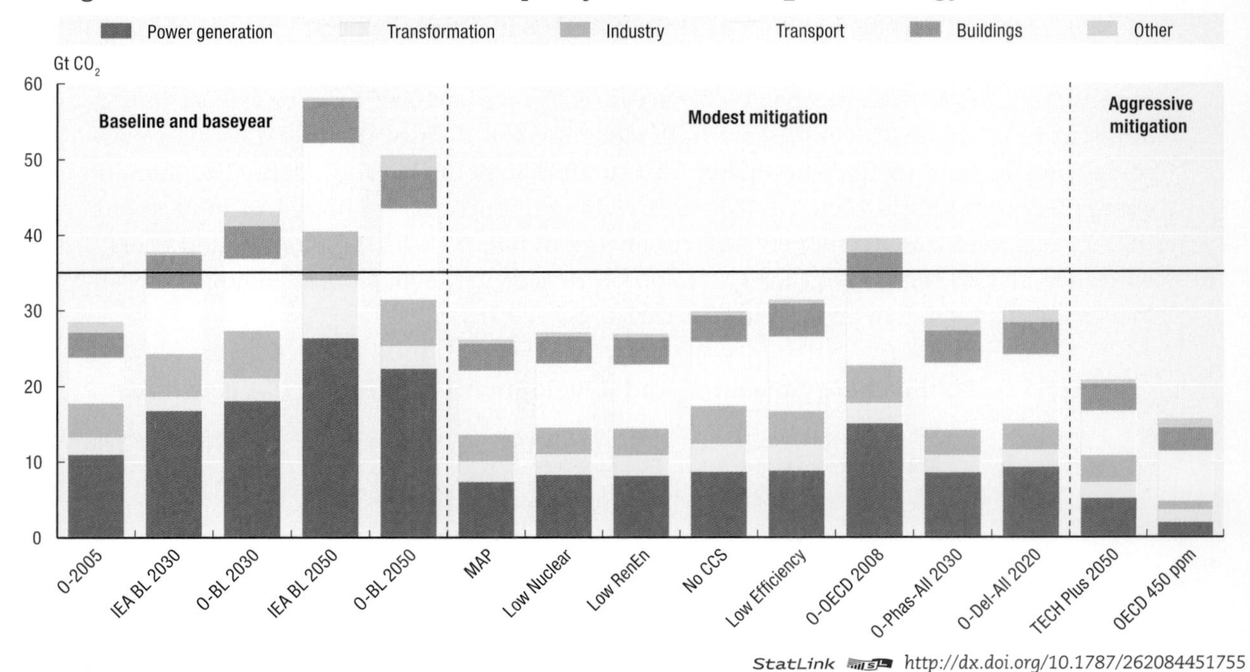

StatLink http://dx.doi.org/10.1787/262084451755

Note: OECD Outlook cases are indicated with O-OECD; all other cases are from IEA (2006b). See Box 17.5 for further detail.

Source: Adapted from IEA (2006b), *Energy Technology Perspectives 2006: Scenarios and Strategies to 2050*, OECD, Paris (Figure 2.1, p. 46).

sectors, highlighting the importance of the introduction of cleaner fuels and more efficient technologies. The most aggressive (and more costly) mitigation scenarios (TECH Plus, All 2008 and 450 PPM stabilisation) achieve even greater emission reductions from the power sector compared to the more modest mitigation scenarios. By comparison transport emissions become the largest source of energy CO_2 emissions in 2050 across all scenarios, replacing power generation as the dominant source today and in the 2050 Baseline.

Figure 17.7 shows the effect of policy scenarios on primary energy use in power generation between 2005 and 2030. More aggressive and earlier mitigation policy in the OECD and Brazil, India and China (BIC) lead to significant reductions in coal use in this sector by 2030, and an increase in natural gas and modern biofuels (especially in BIC). While the OECD *Outlook* shows an increase under Baseline conditions in the power sector's use of coal, this trend will reverse under a climate policy constraint; across all policy scenarios coal use is projected to grow less in absolute terms relative to 2005, except the Delayed case where climate policy is not imposed until 2020 (in the OECD and elsewhere). The 450 PPM case also drives reductions in coal use in all countries, including in Brazil, India and China, compared to the situation under less stringent or no mitigation policy (Baseline) conditions; what coal remains in the power sector is likely to be coupled with CCS in the last half of the simulation period to 2050.

Figure 17.8 shows the Baseline (top line) and the 450 PPM case (lower line). It highlights how, in the 450 PPM case, a wide variety of technologies and end-use changes will be required to reduce emissions to these very low levels to 2050 and beyond. Energy efficiency measures are central, as are low-cost measures that reduce non-CO_2 greenhouse gases, land use and forestry, all of which combine in the near to medium-term to keep mitigation costs low. Also essential by 2020 to achieve this objective is the use of advanced biofuels, carbon capture and storage (CCS) technologies and increased renewables on a worldwide basis.

Figure 17.7. **Change in primary energy use in power generation by fuel and region: policy scenarios compared with Baseline, 2005-2030**

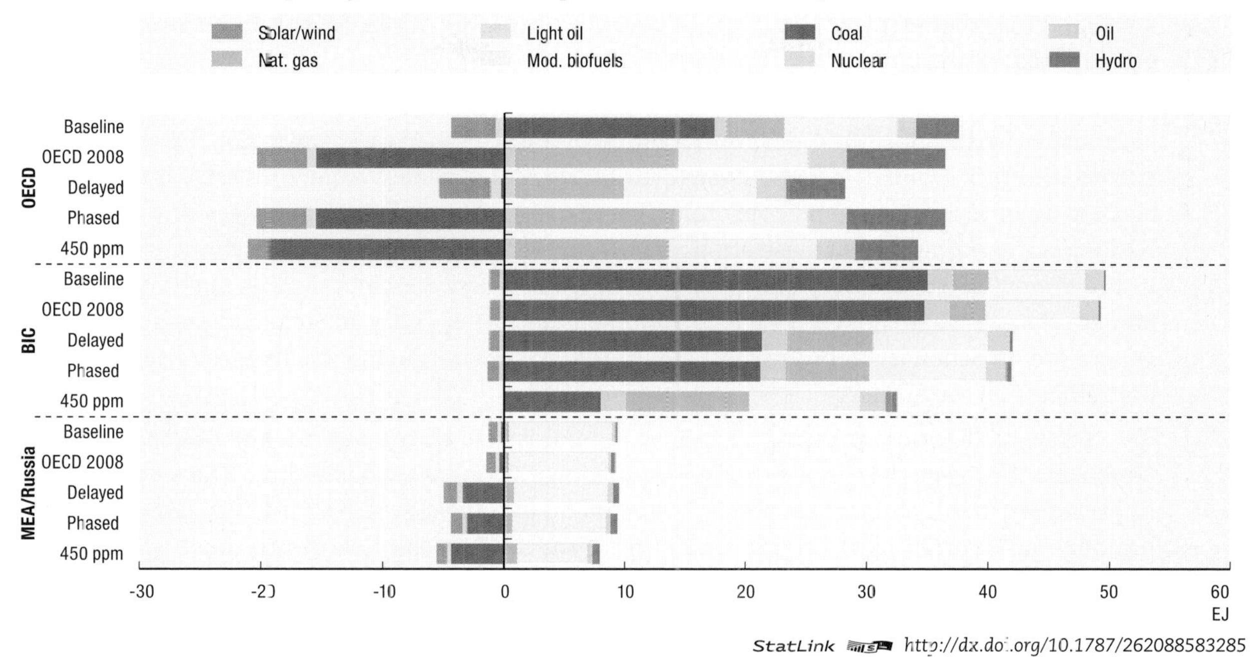

StatLink http://dx.doi.org/10.1787/262088583285

Source: OECD Environmental Outlook Baseline and policy simulations.

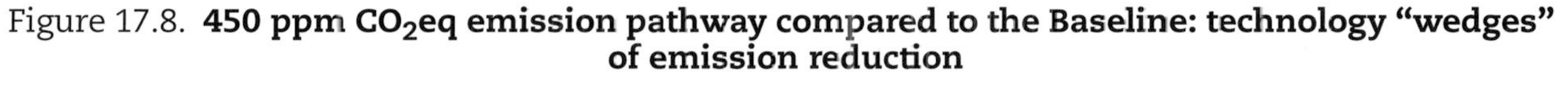

Figure 17.8. **450 ppm CO_2eq emission pathway compared to the Baseline: technology "wedges" of emission reduction**

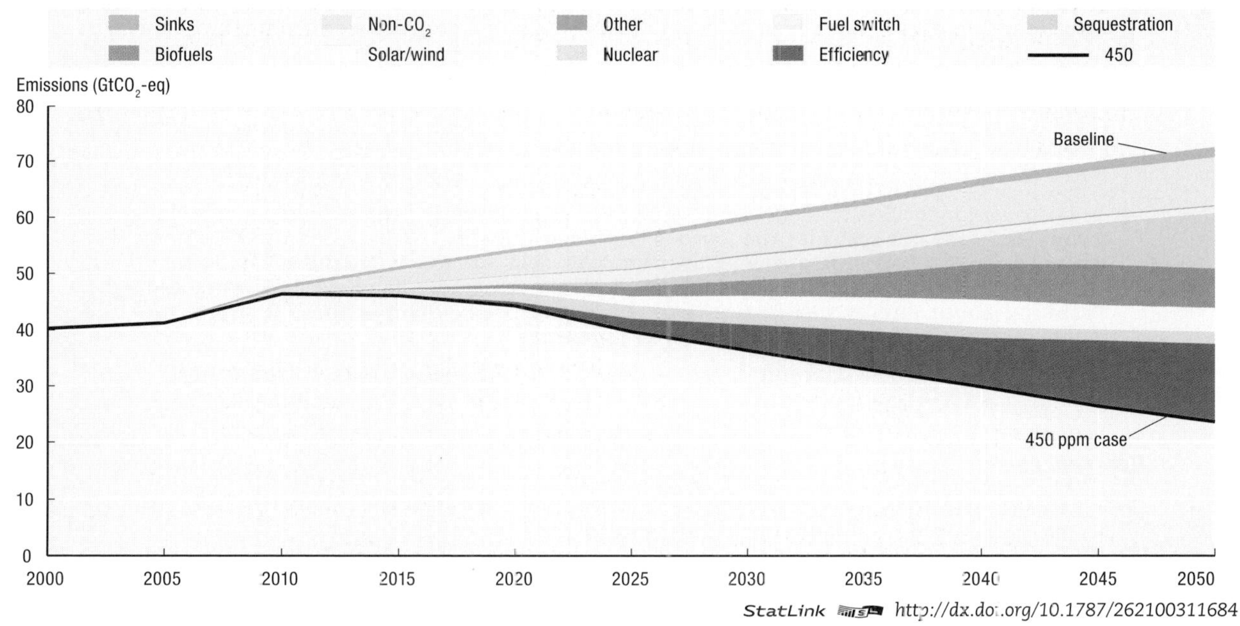

StatLink http://dx.doi.org/10.1787/262100311684

Source: OECD Environmental Outlook Baseline and policy simulations.

The costs of achieving substantial CO_2 emission reductions are significant but manageable. Even for the most stringent of cases, they are estimated to be less than a few percent of GDP by 2050 (see Chapter 7, Climate change). However without an explicit mechanism for burden-sharing, cost-effective implementation of climate change goals suggests that the costs will be greatest in non-OECD regions. Thus any comprehensive agreement on climate change must seek a means to redistribute the costs of mitigation amongst participants in a manner that is perceived to be fair, while not compromising the effectiveness of the outcome.

In conclusion, achieving an environmentally sustainable energy system that is compatible with continued economic and social development is possible. However, substantial new policies will be needed to redirect and generate new investment for cleaner ways of supplying and consuming energy. Major technological advances and cost reductions, underpinned by these stronger government policies, can accelerate innovation in the energy sector and beyond to achieve aggressive environmental goals such as for climate change. There is a growing recognition on the part of policy-makers and the public generally that action is needed urgently to address the environmental challenges raised by our reliance on fossil energy.

Amongst the most pressing of environmental issues of the day is climate change. Seriously tackling global climate change requires significant new policies in the near-term. In particular, it will require "early action" and broad participation across all large emitting nations and sources in the near to medium-term to limit the risk of the most severe consequences in the long-term.

Notes

1. Primary energy refers to energy in its initial form, after production or importation. World primary consumption includes international marine bunkers, which are excluded from the regional totals. Some primary energy is transformed, mainly in refineries, power stations and heat plants. Final consumption refers to consumption in end-use sectors, net of losses in transformation and distribution. The level and mix of fuels in primary energy use determine environmental effects.
2. China, South Asia and other Asia in the IMAGE model.
3. Electricity production is modelled here on the basis of electricity demand, which is met by fossil-fuel based power plants, biomass-based power plants, solar, wind, hydro and nuclear power. Fuel choice in each region is determined by a combination of the relative cost of each technology and government policy. Nuclear power and renewables inputs to generation are projected on a gross basis.
4. This discussion focuses on final energy delivered to end-users (as defined in IEA energy statistical terms) but does not consider the various end-uses and services provided by this energy. For example, roughly half of final energy uses provides heat (*e.g.* space heating of buildings, drying, washing or cooking, industrial process heat), with the remainder being used for work and light. Thinking about energy use in terms of these end-uses provides some insights into the requirements for alternative energy forms and systems.
5. The latter includes both concentrating solar power and solar thermal energy used as heat.

References

CIAB (Coal Industry Advisory Board) (2006), *Regional Trends in Energy-Efficient, Coal-Fired Power Generation Technologies*, November, International Energy Agency, Paris.

Ellis, J. and D. Tirpak (2006), *Linking GHG Emission Trading Systems and Markets*, OECD/IEA Information Paper, *www.oecd.org/env/cc/aixg*.

IEA (International Energy Agency) (2005a), *World Energy Outlook: Middle East and North Africa Insights*, OECD, Paris.

IEA (2006a), *World Energy Outlook 2006*, OECD, Paris.

IEA (2006b), *Energy Technology Perspectives 2006: Scenarios and Strategies to 2050*, OECD, Paris.

IEA (2006c), *Energy Policies of IEA Countries: 2006 Review*, OECD, Paris.

IEA (2007a), *World Energy Outlook 2007*, OECD, Paris.

IEA (2007b), *Energy Policies of IEA Countries – Germany: 2007 Review*, OECD, Paris.

IPCC (Intergovernmental Panel on Climate Change) (2007), *Summary for Policymakers*, in B. Metz *et al.* (eds.), *Climate Change 2007: Mitigation, Contribution of Working Group III to the Fourth Assessment Report of the Intergovernmental Panel on Climate Change*, Cambridge University Press, Cambridge, UK and New York.

Jaffe, A.B., R.G. Newell and R.N. Stavins (2003), "Technological Change and the Environment", in K.G. Mäler and J.R. Vincent (eds.), *Handbook of Environmental Economics*, Vol. 1, Elsevier Science, Amsterdam.

Reinaud, J. and C. Philibert (2007), *Emissions Trading: Trends and Prospects*, OECD/IEA Information Paper, *www.oecd.org/env/cc/aixg*.

ISBN 978-92-64-04048-9
OECD Environmental Outlook to 2030

Chapter 18

Chemicals

The chemicals industry is one of the largest sectors of the world economy, and nearly every man-made material contains one or more of the thousands of chemicals produced by the industry. While OECD countries have seen a reduction in releases from the production of chemicals, policies are needed to address releases from the use and disposal of products which include hazardous chemicals. Adopting a science-based risk assessment approach is among the policies reviewed in this chapter as a means to ensure that adverse impacts are avoided in the most cost effective manner. With the rapid increase of chemicals production in non-OECD countries, greater attention is needed to international co-operation with these governments to build capacity, share information and promote effective chemicals management globally.

KEY MESSAGES

Information is limited on the risks to health and environment posed by the production and use of many chemicals. While some progress has been made in collecting data and assessing the impacts from chemicals on the market, there is a need for better understanding of certain uses or sources of exposure (*e.g.*, chemicals used in products).

Little is known about releases of CO_2 (a greenhouse gas) from the chemicals industry in non-OECD countries, although given the lower energy efficiency of production in the BRIC countries than in most OECD countries, emissions are expected to rise as chemicals production increases in these countries.

New and emerging nanotechnologies may reduce energy use and pollution in the future, but their potential health and environmental effects need careful assessment.

The chemicals industry in OECD countries is continuing to make progress in reducing releases of pollutants during the manufacturing processes, and minimising emissions of CO_2.

Policy options

- Adopt a science-based risk assessment approach which takes into account Principle 15* of the Rio Declaration on Environment and Development and the costs and benefits of chemicals and chemical uses to ensure that adverse impacts are avoided in the most cost-effective manner.
- Conduct safety assessments of nanomaterials; this will require the development of new methodologies.
- Continue to co-operate in the development and implementation of international conventions; with the rapid increase of production in non-OECD countries, greater attention still needs to be given by OECD governments to international co-operation with governments in these countries to build capacity, share information and promote effective chemicals management globally.
- Implement the recently adopted Strategic Approach to International Chemicals Management. This will provide a good foundation for greater co-operation internationally on chemicals assessment and risk management.

Projected chemicals production by region, 2005-2030

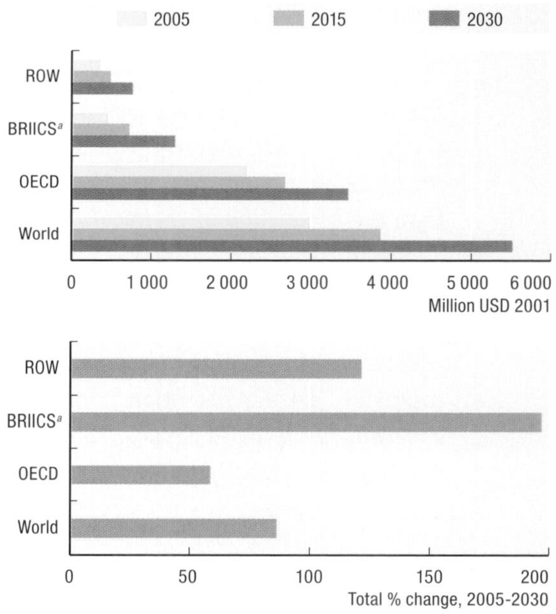

StatLink http://dx.doi.org/10.1787/262116088037

a) Includes Indonesia and South Africa.

Source: OECD Environmental Outlook Baseline.

Consequences of inaction

The release of chemical substances can cause serious damage to human health and the environment. While OECD countries have seen a reduction in releases from the production of chemicals, policies are needed to address releases from the use and disposal of products which include hazardous chemicals. Concerns have been raised about the reproductive and developmental effects of endocrine disrupting substances.

* Rio Declaration on Environment and Development, the United Nations Conference on Environment and Development (3 to 14 June 1992). Principle 15: In order to protect the environment, the precautionary approach shall be widely applied by states according to their capabilities. Where there are threats of serious or irreversible damage, lack of full scientific certainty shall not be used as a reason for postponing cost-effective measures to prevent environmental degradation.

Introduction

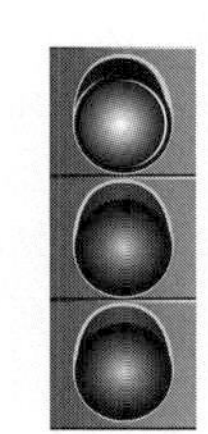

There is a need for better understanding of certain uses or sources of exposure (e.g., chemicals used in products).

The chemicals industry is one of the largest sectors of the world economy, and nearly every man-made material contains one or more of the thousands of chemicals produced by the industry. The world chemicals industry continues to grow steadily and is projected to increase by approximately 3.4% annually[1] to 2030. While production in OECD countries accounts for almost 75% of the world total, production in non-OECD countries, particularly Brazil, Russia, India and China (the BRIC countries[2]) is rapidly increasing and it is expected that the OECD's share of world production will drop to 63% by 2030.

Chemicals are very diverse, and their uses range from large volume commodity chemicals used as building blocks, to more specialised uses (*e.g.*, in coatings, electronics, additives, etc.); and from life science products (*e.g.*, pharmaceuticals, pesticides) to consumer care products. While these chemicals can improve the lives of people, their production and use can also have a negative impact on human health and the environment. The releases of certain substances can cause serious damage to human health and the environment, as has been seen in the past from harmful levels of exposure to PCBs, DDT and PBBs. Concern has been raised about the link between reproductive and developmental effects and endocrine disrupting substances in wildlife (*e.g.*, some alkylphenols used as raw materials in the production of a variety of industrial products, such as surfactants, detergents, phenolic resins, polymer additives and lubricants, can cause endocrine disruption in fish by interfering with oestrogen).

While the production and use of chemicals may pose risks to man and the environment, in general actual information on such impacts is not comprehensive. Many OECD countries have initiated significant voluntary and regulatory initiatives to provide such information, and good information exists on releases of pollutants from chemical factories in OECD countries. However, there is a lack of information on the health and environmental effects of many chemical substances on the market and on the products in which they are used. As both pollution (created during the production of chemicals) and products containing hazardous chemicals travel across borders, approaches to chemicals management should take these factors into account. To facilitate better management of chemicals around the world, in 2006 the International Conference on Chemicals Management adopted the Dubai Declaration on International Chemicals Management and the Overarching Policy Strategy. The conference also recommended the use and further development of the Global Plan of Action as a working tool and guidance document. Together, these three documents constitute the Strategic Approach to International Chemicals Management (SAICM; see section on "Policy implications" below for more detail).

Key trends and projections

Releases and use of hazardous chemicals

The emission of some pollutants to the environment from the chemicals industry in OECD countries continues to decrease.

Across OECD countries, emissions of hazardous substances from chemical plants have generally been steadily decreasing, as have overall releases of chlorofluorocarbons (CFCs). According to the Commission for Environmental Cooperation, total releases and transfers of the 152 chemicals that are common to the US and Canada and monitored by both countries, dropped 18% between 1995 and 2002 (CEC, 2005). In Japan, the chemicals industry reported a 54% reduction from 2000 to 2004 in the emissions of 354 substances listed in a national law on reporting releases of chemicals to the environment (JRCC, 2005). According to the European Commission, from 1990 to 2000, production of ozone-depleting substances had "almost stopped", emissions of acid rain precursors had dropped by 48%, ozone precursors by 38%, and non-methane volatile organic compounds by 26% (EC, 2003). The situation in non-OECD countries, for these, and for CFCs, is uncertain as past and current data are, in most cases, not available.

Releases of substances do not only occur during the chemical production phase. Use of chemical products can lead to the release of substances (*e.g.*, glues and adhesives from construction materials, chemicals in cleaning agents), as can final disposal. However, due to a lack of information, the risk posed by chemicals in products is unclear (see Box 18.1). With respect to pesticides, overall use in OECD countries has declined by 5% from 1990 to 2002, but the trends vary by country (OECD, 2007a). Even so, it is not possible to conclude that risks to man and the environment have been reduced accordingly, as the hazardous properties of pesticides used today are difficult to compare with those used in the past. (*e.g.*, if the potency of active ingredients increased over the years at the same time as the volume of pesticides used decreased, the risk may not have been reduced.)

Box 18.1. **Key uncertainties, choices and assumptions**

- There is no single definition of the chemicals industry for statistical purposes, and the industry sectors included in the various sources referenced in this report may not be strictly comparable; however, it is important to note that despite these differences, both the OECD and the chemicals industry project almost the same annual growth rate in the coming years.
- The economic model used in the *OECD Environmental Outlook* distinguishes the chemicals sector from other industries, but the models used to project environmental pressures and impacts do not. Hence, for this chapter all data on environmental impacts are derived from other sources.
- Production data cited in this chapter are based on sales, and may not correlate directly to production volume.
- Limited information on chemicals in products makes it difficult to document the extent of the risk posed to man and the environment due to releases from products.
- Determining the cost of inaction is an important consideration when suggesting policy choices, but sufficient data to do so were not available for this report.

 OECD ENVIRONMENTAL OUTLOOK TO 2030 – ISBN 978-92-64-04048-9 –

Use of fossil fuels

The chemicals industry uses a significant amount of coal, petroleum products and natural gas, both as a source of energy and as feedstocks for many of its products. As the BRIICS' share of world chemicals production grows, so does their share of total energy and feedstock use. In 1971, chemical companies in Brazil, India, Indonesia, China and South Africa consumed just 2.9% of the amount of energy and feedstocks used by companies in OECD countries; by 2003, this figure had grown to 39.4%[3] (IEA, 2005). In China, chemicals production is the second largest consumer of energy in the manufacturing sector after smelting of ferrous metals, and accounts for 18% of all energy consumed by manufacturers (National Statistics Bureau of China, 2004).

Emissions of CO_2 and hazardous pollutants from the chemicals sector in non-OECD countries are expected to rise.

The chemicals industry in OECD countries has made good progress in increasing energy efficiency and reducing or keeping CO_2 emissions constant. Energy consumed by the US chemical industry per unit of output declined from 65.9 to 57.4 (against an index of 100 for 1974) from 1990 to 2003, while CO_2 emissions have remained constant (ACC, 2004a). According to the European Commission (EC, 2003), from 1990 to 2000 there was a 50% fall in greenhouse gas emissions from chemical plants. Japan's Chemical Industry Association reported a drop in unit energy consumption from 100 in 1990 to 87 in 2004 (Joint Subcommittee for the Follow-up to the Nippon Keidanren Voluntary Action Plan on the Environment, 2005). While no energy efficiency data are readily available for the chemicals industry in the BRIICs, the growing production in these countries, coupled with their higher reliance on coal, could be cause for concern (OECD, 2001).

Production and use: historical trends

The world chemicals industry grew from sales of approximately USD 1 500 billion in 1998 to USD 2 245 billion[4] in 2004 (CEFIC, 2005; ACC, 2004a). While companies in OECD countries continue to account for the bulk of world production (74.5% in 2004), the OECD's share of world production has been dropping steadily, and now is 9% less than in 1970. Much of this shift has been to the major emerging economies, particularly the BRICs. In 2004, China accounted for most of the BRIC's production (47.7%), followed by Brazil and India (19.9% each) and Russia (12.6%). Since 1998, all of the BRICs, with the exception of Brazil, have far exceeded the world rate of growth in chemicals production. The chemicals sector in China has been growing at an annual rate of around 16.5% since 1987, several times the rate in most OECD countries, which has been around 1 to 4% over the last 10 years. As a result, it has recently surpassed Germany as the world's third largest producer of chemicals by turnover (ACC, 2004a and 2006).

Construction of new facilities and modification of existing facilities outside OECD countries has accelerated due to the high cost of natural gas in OECD countries and the need to be closer to customers (with high growth rates). In 2005, of the 120 large chemical plants being built (*i.e.*, plants costing USD 1 billion or more), 50 were being constructed in China, *versus* just one in the US (Arndt, 2005). From 2004 to 2009, the share of capital spending by American companies destined for the US is expected to drop from 71% to 59%, to remain the same for Western Europe (16.6% to 16.8%) and Japan (0.5% to 0.6%), whereas the amount destined for China is expected to jump three-fold (from 2.9% to 8.8%) (ACC, 2004b).

Production: long-term trends

The American Chemistry Council projects that world growth in chemicals over the next 10 years should average around 3.5%[5] per year, with the greatest growth in the Asia-Pacific region. The ACC estimates that during this period, China's chemicals industry will grow around 10.5% per year (ACC, 2006) and India's around 8% per year (ACC, 2004a). The economic Baseline for the *OECD Environmental Outlook* also projects similar world growth (3.4%) for 2005 to 2030 (see Figure 18.1). Annual growth in the BRICs is expected to be 7.9% during this period, *versus* 2.3% in OECD countries. This means that from 1998 to 2030, OECD countries' share of world production will have dropped from 77.5% to 62.7%, while the BRICs will have jumped from 10.8% to 23.5%.

Figure 18.1. **Projected chemicals production by region, 2005-2030**

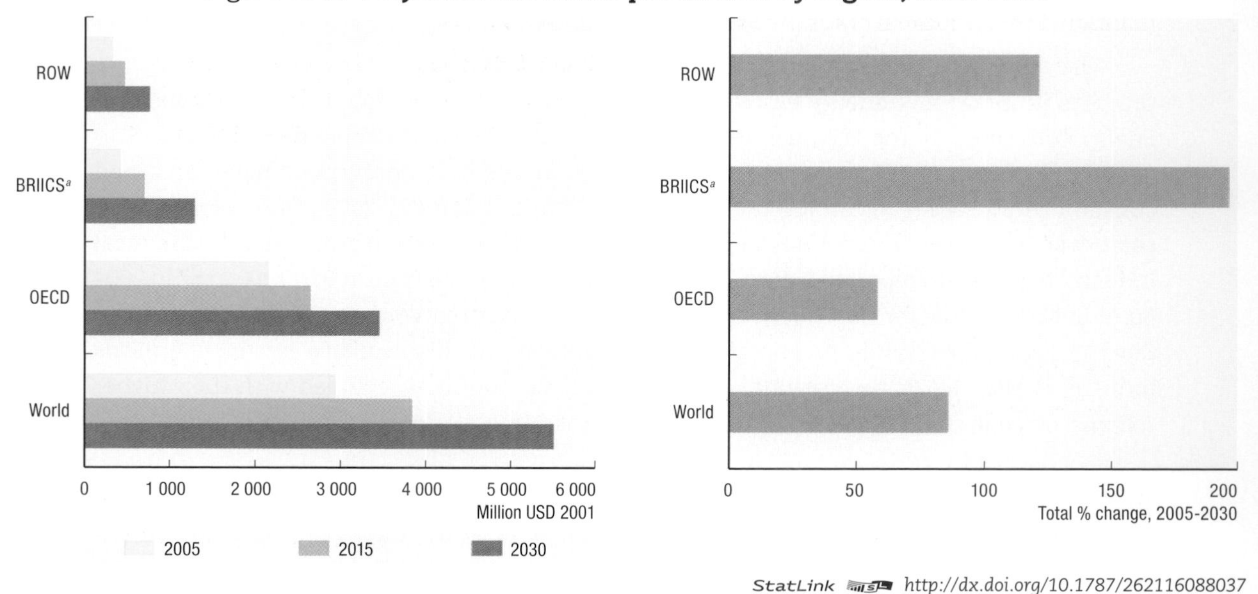

StatLink http://dx.doi.org/10.1787/262116088037

a) Includes Indonesia and South Africa.

Source: OECD Environmental Outlook Baseline.

Policy implications

In parallel to the tremendous growth in worldwide production and trade in chemical products over the last three decades, there has been a significant increase in activities by governments, international organisations, industry and environmental organisations to promote the safe use of chemicals (see Box 18.2).

Delegates to the 2002 World Summit on Sustainable Development agreed to work together to promote the sound management of chemicals throughout their life-cycle. The first international action which followed was the adoption of the Dubai Declaration and the Overarching Policy Strategy (OPS). The OPS includes important commitments to 2020 on the implementation of science-based risk assessment and management approaches. This also recognises and builds on the range of existing risk reduction tools from United Nations organisations and the OECD. A Global Plan of Action (GPA), which was developed to implement the Strategic Approach to International Chemicals Management (SAICM), "... lays out possible work areas and activities for addressing societal needs regarding chemical management." (UNEP, 2006a.) A number of activities aim to fill data gaps on

Box 18.2. **The OECD and chemicals**

In 1971, the OECD Chemicals Programme was launched to develop harmonised tools and policies for chemical safety which would allow countries and industry to increase efficiencies, reduce non-tariff barriers to trade, and improve the policies aimed at protecting man and the environment from the risks posed by chemicals.

Most significantly, in 1981 the OECD's Council Decision on the Mutual Acceptance of Data (MAD) was adopted, which requires OECD governments to accept test data developed for regulatory purposes in another country if these data were developed in accordance with the OECD Test Guidelines and GLP Principles. The yearly savings to governments and industry of avoiding duplicative testing through MAD are calculated to be EUR 60 million (OECD, 1998). Work at the OECD has also supported a recent agreement on globally accepted criteria for a complete classification system of the health and environmental hazards of chemicals – the Globally Harmonised System (GHS).

Over the years, a number of significant OECD Council Acts have been agreed to facilitate cost-effective chemicals management, including acts on: chemical accident prevention, preparedness and response; pollutant release and transfer registers; the systematic investigation of existing chemicals; the environmentally sound management of waste; and improving the environmental performance of public procurement.

Many important international conventions on environmental health and safety have also been agreed within the UN context. These include the Basel Convention on Transboundary Movements of Hazardous Wastes, the Montreal Protocol on Substances that Deplete the Ozone Layer, the Rotterdam Convention on the Prior Informed Consent Procedure for Certain Hazardous Chemicals and Pesticides in International Trade and the Stockholm Convention on Persistent Organic Pollutants.

chemicals, assess such information and make this information widely available and in a format which can be understood by the main parties. The OECD works on over 40% of the activities listed in the GPA and is considering how its work can further the implementation of SAICM worldwide. The OECD and its member countries are expected to play a large role in implementing SAICM.

Regulatory approaches

International co-operation for generating and assessing data on high production volume chemicals

Over the years, OECD governments and the chemicals industry have made and are continuing to make good progress in managing chemicals, through the establishment of a number of successful programmes to collect information, assess and manage risks posed by chemicals. In addition, since the beginning of the 1990s health and safety data on high volume chemicals produced in OECD countries have been generated and assessed as part of the OECD's Existing Chemicals High Production Volume (HPV) Programme. This programme focuses on chemicals produced or imported in quantities of greater than 1 000 tonnes in at least one OECD country or the European Union. While the number of high volume chemicals tends to be fairly small, approximately 4 800 in OECD countries (OECD, 2007c), when compared to all chemicals in commerce – estimated to be between 70 000 and 100 000 (UNEP, 2006b) – they do represent the bulk of total production. For example, of the total production volume of chemicals within the EU, 75% of this volume was for chemicals that are produced in volumes greater than 1 million tonnes/year (OECD, 2001). In the US, HPV chemicals[6] represent more than 93% of chemical volume[7] (US EPA, 2006). Under OECD's HPV Programme, for each chemical a minimum set of hazard data and available use/exposure information are collected and an initial assessment conducted. Currently, the majority of chemicals assessed in the OECD programme emanate from a voluntary commitment by industry to collect data and other information, carry out any needed testing, and provide draft assessments to OECD governments for review. Around 670 chemicals have currently been assessed and hundreds more are being reviewed. A web-based gateway to HPV-related data (eChemPortal) provides free access to such data (OECD, 2007b).

The work of the HPV Programme will result in the generation and assessment of hazard information on HPV chemicals in OECD countries, and, as this information is made publicly available, it will assist all countries wherever these chemicals are produced or used. However, there will still be a gap in knowledge about HPV chemicals which are unique to *non*-OECD countries. As more and more chemicals are being produced in non-OECD countries, particularly the BRICs, priority attention is needed to ensure that adequate health and safety information is available and assessed on these chemicals. However, the ability of such countries to meet this challenge is uncertain. International co-operation, information sharing and capacity-building can help provide a high level of support, share the burden of work and reduce duplication.

One such approach could be to invite those non-OECD countries with the largest (and fastest growing) chemicals industries (*i.e.*, the BRICs) to participate in OECD's HPV Programme. Participation could begin as a way to help build capacity in those countries for the assessment of HPV chemicals. As capacity is gained over time, the BRICs could be invited to take the lead on collecting data and assessing certain chemicals which are produced in high volumes both in an OECD country and a BRIC country. In the future, this capacity would also help the BRICs assess chemicals which are produced in high volumes only in their countries. Co-operating with OECD countries would not only build capacity in the BRICs, it would reduce potentially duplicative work if the same chemicals are assessed in the BRICs and also OECD.

Sharing the work on pesticides and new industrial chemicals

Since many pesticides used in OECD countries are the same, governments have recognised the substantial benefits that can be gained if the task of pesticide evaluations for registration and re-registration is shared, rather than duplicating each others' work. OECD governments have agreed to routinely share the work of such reviews. With the BRICs producing and using more pesticides each year, there would be value in inviting them to participate in such work sharing arrangements for agricultural chemicals. Similarly, the OECD's New Chemicals Programme has begun a pilot phase of a parallel process of notifications, in which a company can notify to multiple jurisdictions and authorise participating governments to share information when conducting their reviews. Inviting the BRICs to participate in the parallel process would be mutually beneficial to the companies and governments in both the OECD and the BRICs.

Mutual Acceptance of Data in non-OECD countries

Because of the OECD Council Decision on MAD (Box 18.2), OECD governments are able to work together to share information on new and existing industrial chemicals, as well as

 OECD ENVIRONMENTAL OUTLOOK TO 2030 – ISBN 978-92-64-04048-9 – © OECD 2008

pesticides. If OECD governments are to do the same with the BRICs, these countries must be adherents to MAD. Since 1997, the MAD system has been open to non-OECD economies as well, allowing them to participate with the same rights and obligations as member countries once they have implemented the two relevant Council Decisions. Currently, of the BRIICS, South Africa is a full member of the Council Decision on MAD, India and Brazil are provisional adherents to the MAD Council Decisions, and China is expected to ask for provisional adherence to the MAD Council Decisions in the short term. (Among other non-members, Israel and Slovenia are also full adherents and Singapore is a provisional adherent to the Council Decision). Efforts must continue to bring additional relevant non-OECD countries into the MAD system.

Economic instruments

Economic instruments such as taxes or charges could be used for certain products to reflect the costs of their health or environmental impacts. This would provide an incentive for consumers to reduce consumption of particularly damaging chemicals or to encourage them to select cheaper alternatives that are more environmentally friendly. For chemical products for which insufficient data are available on their environmental or health impacts, a charge could be levied to encourage and/or help to fund testing. Different European countries impose taxes on such products as fertilisers, pesticides, ozone-depleting substances, and chlorinated solvents (EEA/UNEP, 1998). Positive economic incentives could be considered to promote the development of new and innovative "sustainable chemicals". These could include tax deductions and the promotion of joint partnerships for research and development of "greener" chemicals and products. Reduced fees or a quicker review process for notifying new substances that pose a lower risk than other equivalent substances would be other possible approaches. Increased patent protection for low-risk products could also spur their development.

Voluntary approaches

Exchanging information across countries about chemicals throughout their life-cycle (both by governments and industry) is an important element in the management of chemical products. However, due to antitrust concerns, legal and cultural barriers, and language differences, sharing information is not always easy or possible. To facilitate the exchange of such information, information requirements and formats should be harmonised. There is also a need for a greater communication between manufacturers and users on the safe use and disposal of chemical products (OECD, 2004).

Voluntary approaches can be overseen by a government, but implemented by companies. For instance, under the US Environmental Protection Agency's 33/50 Program, companies agreed to voluntarily reduce releases and transfers of 17 priority chemicals by 33% by the end of 1992 and 50% by the end of 1995 based on a 1988 baseline. These goals were achieved one year ahead of schedule. More recently, under Korea's 30/50 programme, companies voluntarily agreed to reduce chemical releases by 30% by 2007 and 50% by 2009 from 2004 levels. Over 160 enterprises are participating.

Policies for "green procurement" can include supply and purchase policies that consider chemical safety as part of the procurement decision-making process (OECD, 2006).

Technological development and diffusion

There are a number of tools available to OECD governments to help facilitate the development of new and innovative, sustainable or more environmentally benign chemicals. Governments can recognise and reward sustainable chemistry accomplishments, promote the incorporation of sustainable chemistry principles into various levels of chemical education, help provide technical tools that can be used to design more benign chemicals, and promote the development of non-chemical alternatives for pesticides. (Quantitative) Structure-Activity Relationships [(Q)SARs] methods are a good example. These methods are used to estimate the properties of a chemical from its molecular structure (*i.e.*, without actual testing). Using (Q)SARs, companies can identify, during the design stage of new chemicals, products that may pose a risk to man or the environment and those that may be more benign. The OECD is developing a (Q)SAR application toolbox which will include a library of models which member countries, non-members and industry could use for various regulatory purposes.

Given the rapid development of nanomaterials, the assessment of the health and safety implications of such materials must keep pace.

With the growing demand for fossil fuels by the BRICs, improving energy efficiency should also be a focus of attention (see Chapter 17 on energy). The World Bank is exploring the opportunities to reduce greenhouse gas emissions from the chemicals industry in non-OECD countries through the use of energy efficient technologies. Energy efficiency measures in the manufacture of chemicals may also have the benefit of reducing chemical releases. Audits of chemical facilities in non-OECD countries could be supported; these could identify energy losses that could be reduced through improvements in technology

Box 18.3. **Nanotechnologies**

At the nanoscale – between 0.1 and 100 nanometers (1/1 000 000 mm) – the physical, chemical, and biological properties of materials differ in fundamental and often valuable ways from the properties of individual atoms and molecules or bulk matter. Research and development in nanotechnologies is directed toward understanding and creating improved materials, devices and systems that exploit these new properties. Such properties have been found to be very useful for an increasing number of commercial applications, for example: protective coatings, light-weight materials, inks and self-cleaning clothing material.

However, the special features of nanoparticles that may render them so useful in certain applications also mean they could prove hazardous to human health and the environment, and thus it is important that the health and environmental effects of nanomaterials are identified and assessed, as these products are now being placed on the market in increasing volumes. As the testing and assessment methods used to determine the safety of traditional chemicals are not necessarily (fully) applicable to nanomaterials, there should be a responsible and co-ordinated approach to ensure that potential safety issues are being addressed at the same time as the technology is developing. For this reason, many governments are working to address the safety implications, and the OECD has started a major new project to provide support for the development of methods for the testing and assessment of manufactured nanomaterials.

that may also reduce potential releases of hazardous chemicals (*e.g.*, the chlor-alkali industry). The US Department of Energy's "Bandwidth Study" conducted in 2004 looked at the most energy-intensive chemicals and related process technologies in the US, and identified 900 trillion British Thermal Units (Btus) of total potentially recoverable energy (US DOE, 2004). Co-operating with chemical industry associations who may already have conducted such studies and applying this type of approach could be a starting point for some of the large chemical facilities in the BRICs.

The ACC predicts that over the next couple of decades, bio-sciences will play a greater role in chemicals production. Biotechnology now accounts for 8% of overall shipments, as compared to less than 3% as recently as 1992 (ACC, 2004a). As such technologies are more energy-efficient, such a shift should result in less energy use and pollution. Nanotechnologies are also expected to play a greater role in the coming years in the production of chemical products (see Box 18.3). The shift in the types of chemicals being produced in OECD and non-OECD countries (from basic chemicals to life science, specialty and biotech products) has been more rapid than anticipated. The previous *OECD Environmental Outlook* (OECD, 2001) projected that life science chemicals would exceed basic chemical revenues by 2020, and specialty chemicals would rival basic chemicals by 2020. ACC now predicts that life sciences will exceed basic chemicals by 2010, and specialty chemicals will rival basic chemicals in 2010.

Notes

1. Rate is for real growth (*i.e.*, adjusted for inflation).
2. Note: This chapter also provides comparisons in places with the BRIICS countries (which also include Indonesia and South Africa), where data on those two additional countries are available.
3. 1971: 7 685 kilotonnes of oil equivalent (ktoe) for companies in the BRIICS countries *versus* 257 346 ktoe for companies in the OECD countries; 2003: 239 195 ktoe for the BRIICS companies *versus* 607 340 ktoe for the OECD companies (IEA, 2005).
4. Increase in sales from 1998 to 2004 is nominal growth (*i.e.*, figures are not adjusted for inflation).
5. Projected rate is for real growth (*i.e.*, adjusted for inflation).
6. Under the US Environmental Protection Agency's High Production Volume Challenge Program, HPV chemicals are classified as those chemicals produced or imported into the United States in quantities of 1 million pounds (approximately 450 metric tonnes) or more per year.
7. US data are for organic chemicals produced above approximately 5 metric tonnes/year.

References

ACC (American Chemistry Council) (2004a), *Guide to the Business of Chemistry*, (August, 2004): p. 122, ACC, Arlington, Virginia, US.

ACC (2004b), *American Chemistry Council: ACC's Year-End 2004 Situation and Outlook*, ACC, Arlington, Virginia, US.

ACC (2006), *American Chemistry Council: Business of Chemistry in China*, March 2006, ACC, Arlington, Virginia, US.

Arndt, M. (2005), "No Longer the Lab of the World", *Business Week*, 2 May, 2005.

CEC (Commission for Environmental Cooperation of North America), (2005), *Taking Stock: 2002 North American Pollutant Releases and Transfers* (May, 2005), CEC, Montreal.

CEFIC (The European Chemical Industry Council) (2005), *Facts and Figures 2005*, CEFIC Website, *www.cefic.org/factsandfigures*.

EC (European Commission) (2003), *Regulation of the European Parliament and of the Council Concerning the Registration, Evaluation, Authorisation and Restrictions of Chemicals, Establishing a European Chemicals Agency and Amending Directive 1999/45/EC and Regulation (EC) {on Persistent Organic Pollutants} Extended Impact Assessment,* Commission Staff Working Paper [SEC(2003)1171/3], EC, Brussels.

EEA/UNEP (European Environment Agency and United Nations Environment Programme), (1998) *Chemicals in the European Environment: Low Doses, High Stakes*, EEA/UNEP, Copenhagen.

IEA (International Energy Agency) (2005), *World Energy Statistics and Balance of OECD and non-OECD Countries*, Vol. 2005: release 01, IEA, Paris.

Joint Subcommittee for the Follow-up to the Nippon Keidanren Voluntary Action Plan on the Environment (2005), *The Follow-Up to the Nippon Keidanren Voluntary Action Plan on the Environment in Fiscal Year 2005*, Tokyo.

JRCC (Japan Responsible Care Council) (2005), *Responsible Care Report* 2005, JRCC, Tokyo.

National Statistics Bureau of China (2004), *China Statistical Yearbook*, National Statistics Bureau of China, Beijing, *www.stats.gov.cn/english/statisticaldata/yearlydata/*.

OECD (Organisation for Economic Co-operation and Development) (1998), *Savings to Governments and Industry Resulting from the Environmental Health and Safety Programme*, OECD, Paris.

OECD (2001), *OECD Environmental Outlook for the Chemicals Industry*, OECD, Paris.

OECD (2004), *OECD Report on the Workshop on Exchanging Information across a Chemical Product Chain* (Stockholm, Sweden; 15-16 June, 2004), OECD, Paris.

OECD (2006), *OECD Workshop Report on Consideration of Chemical Safety in Green Procurement* (Seoul, Korea; 8-10 November, 2005), OECD, Paris.

OECD (2007a), *OECD Environmental Indicators for Agriculture, Volume 4*, OECD, Paris, forthcoming.

OECD (2007b), *OECD eChemPortal, www.oecd.org/ehs/echemportal.*

OECD (2007c), *Description of OECD Work on Investigation of High Production Volume Chemicals*, Environment Directorate, OECD, Paris. *www.oecd.org/document/21/0,3343,en_2649_34379_1939669_1_1_1_1,00.html.*

UNEP (United Nations Environment Programme) (2006a), *Strategic Approach to International Chemicals Management Comprising the Dubai Declaration on International Chemicals Management, the Overarching Policy Strategy and the Global Plan of Action,www.chem.unep.ch/saicm/*, UNEP, Geneva.

UNEP (2006b), *New Global Chemicals Strategy Given Green Light by Governments*, 9th Special Session of the Governing Council of the United Nations Environment Programme/Global Ministerial Environment Forum. UNEP News Release, February 2006, UNEP, Geneva. *www.chem.unep.ch/ICCM/ICCM%20UNEP%20Press%20release.doc.*

US DOE (US Department of Energy) (2004), *Industrial Technologies Program; Chemical Bandwidth Study, Energy Analysis: A Powerful Tool for Identifying Process Inefficiencies in the US Chemical Industry*. US DOE, December, 2004, Washington, DC. *www.eere.energy.gov/industry/chemicals/pdfs/chemical_bandwidth_report.pdf.*

US EPA (US Environment Protection Agency) (2006), *Testimony of James B. Guilliford, Assistant Administrator, Office of Prevention, Pesticides and Toxic Substances*, Oversight Hearing on the Toxic Substances Control Act, before the Committee on Environment and Public Works, United States Senate; August 2, 2006. US EPA, Washington, DC.

ISBN 978-92-64-04048-9
OECD Environmental Outlook to 2030

Chapter 19

Selected Industries

- **STEEL AND CEMENT**
- **PULP AND PAPER**
- **TOURISM**
- **MINING**

This chapter outlines the projected growth, environmental impacts and policy implications for four other industries: steel (and cement), pulp and paper, tourism and mining. The steel sector, a major contributor to several environmental problems (e.g. air pollution and climate change), is projected to increase production significantly to 2030, especially in Brazil, Russia, India, Indonesia, China and South Africa (BRIICS countries). The pulp and paper sector is also expected to grow in the coming decades. Regulatory approaches, economic instruments, voluntary approaches, cleaner production and other instruments are explored as possible means to offset negative environmental impacts of this growth. Tourism has an impact on the environment in the destination country and at the global level (e.g. through air travel). This chapter reviews sustainable tourism policies and other initiatives to reduce the environmental impact of tourism activities. The rapid expansion of mining activity in developing countries constitutes an important challenge. Host governments will need to put in place policies to strengthen the capacity and institutional set-up to effectively manage the environmental risks associated with this development.

STEEL AND CEMENT

KEY MESSAGES

The steel sector is a major contributor to several environmental problems, including air pollution and climate change. It accounts for about 7% of anthropogenic emissions of CO_2. When mining and transportation of iron ore are included, the share may be as high as 10%. A strong increase in the production and use of steel is projected up to 2030, especially in the BRIICS countries.

Almost 60% of steel worldwide is produced using basic oxygen furnace (BOF) technology, which emits over four times as much CO_2 per unit of steel produced than standard electric arc furnace (EAF) technology.

Use of the heavily-polluting open hearth (OH) process to produce steel has declined significantly in recent decades worldwide, and now accounts for about 5% of total production.

Policy options

- Implementing a tax of USD 25 per tonne of CO_2 emitted from the sector would have a small impact on steel production in 2030, since demand for steel is relatively price inelastic, but would substantially reduce carbon emissions (see table).

Estimated changes in CO_2 emissions in the steel sector from a USD 25 tax
(% change in 2030 compared to the *Outlook* Baseline)

	The tax applies only to the steel sector in OECD	The tax applies to all sectors in OECD	The tax applies to all sectors globally
OECD	–34.0	–33.3	–31.4
BRIICS	0.5	1.4	–54.6
ROW	0.9	2.3	–46.4
WORLD	–7.4	–6.5	–48.0

StatLink http://dx.doi.org/10.1787/262664418128

- Applying such a tax globally would lead to a 15% reduction in SO_2 emissions from the sector in OECD countries, and a reduction of over 50% in non-OECD regions.
- Measures could be taken to reduce some of the competitiveness effects of such a tax (*e.g.* recycling the tax revenue back to the steel industry, or applying border tax adjustments), while maintaining at least some of the environmental benefits.
- Scaling back steel sector subsidies to close down unprofitable steel plants could be a low- or no-cost option for Annex I countries (the industrialised nations) to meet their CO_2 emission reduction targets under the Kyoto Protocol.

OECD ENVIRONMENTAL OUTLOOK TO 2030 – ISBN 978-92-64-04048-9 – © OECD 2008

Introduction

Steel production causes emissions of a number of pollutants, such as SO_2, NO_x, CO_2, particles, mercury, etc. According to the OECD (2003), the sector accounts for about 7% of anthropogenic emissions of CO_2. When mining and transportation of iron ore are included, the share may be as high as 10%. Cement production also has environmental consequences (explored in Box 19.2).

The CO_2 emissions associated with iron and steel production differ according to the different technologies used. The two main technologies are basic oxygen furnaces (BOF) and scrap-based, standard electric arc furnaces (EAF) (Table 19.1). In addition, some steel is produced in electric arc furnaces based on directly reduced iron (DRI). Steel is also produced in heavily-polluting open hearth (OH) processes, especially in certain Central and East European countries. Figure 19.1 illustrates the significance of BOF in particular, but EAF has also increased considerably over time, while OH and the now out-dated Bessemer furnaces (included in "Other Processes") have ceased to be of importance.

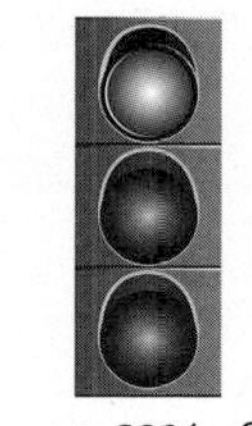

Almost 60% of steel worldwide is produced using BOF technology, which emits over four times as much CO_2 per unit of steel produced than EAF technology.

In general, the integrated BOFs are more energy-intensive than the EAFs, which rely on smelting of iron and steel scrap. The emissions produced by each also vary between countries and regions, depending on the energy-efficiency of the plants and the CO_2-intensity of the energy used. Globally, about 75% of the steel sector's CO_2 emissions are related to the use of coke and coal in iron making in the BOFs. Other important emission sources are the use of electricity, particularly in the EAFs, and the use of natural gas in the production of DRI.

Table 19.1. **Characteristics of different steel production technologies globally (2000)**

Technology	Share of world production (%)	Major inputs	Average CO_2 emissions per tonne steel (tonnes)
Basic oxygen furnace (BOF)	58	Ore, coal, scrap (10-30%)	2.5
Standard electric arc furnace (EAF)	27	Electricity, scrap (> 90%)	0.6
EAF based on directly reduced iron (DRI)	7	Ore, gas, electricity, scrap (20-50%)	1.2

StatLink http://dx.doi.org/10.1787/257574076351

Source: OECD (2003).

On the other hand, mercury emissions to the air are higher from EAF plants than from BOF plants in countries where mercury-containing switches in end-of-life motor vehicles are melted down in the EAF plants. The use of such switches has been phased out in many countries. Programmes to remove them before the vehicles are recycled are also in place in countries where these switches are still in use. Hence, remaining mercury emissions from EAF plants are likely to decrease significantly in the future.

Figure 19.1. **World crude steel production by process, 1970-2006**

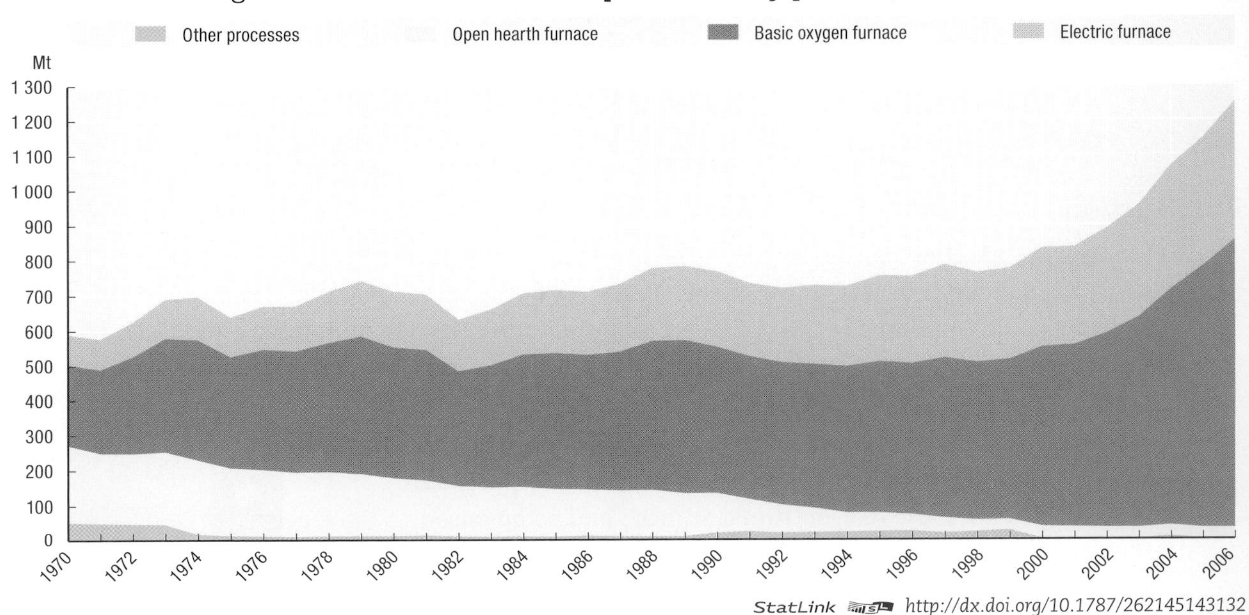

StatLink http://dx.doi.org/10.1787/262145143132

Source: Based on data from the International Iron and Steel Institute (IISI), (2008).

Key trends and projections

According to the International Iron and Steel Institute, global steel production increased from 562 million tonnes in 1980 to 1 106 million tonnes in 2005 (IISI, 2006). Growth in world steel production was 1.8% annually between 1980 and 1995, and 4.2% between 1995 and 2005. Within the OECD, the growth in steel production was much more modest: 0.8% a year between 1995 and 2005.

The Baseline produced for the *OECD Environmental Outlook* indicates a 3.4% annual growth in real value added[1] (at basic prices) in the iron and steel sector between 2006 and 2030 globally. This growth is estimated to be particularly strong in the BRIICS countries, which start out with an annual growth rate of 6.9% between 2006 and 2010. This is projected to decline gradually to 4.4% per year between 2020 and 2030, leading to an average growth rate in real value added of 5.1% per year between 2005 and 2030. In 2006, the share of the BRIICS countries in global steel production was just above 20%. This share is projected to increase to 32% by 2030 (Figure 19.2). A large part of this increase is estimated to take place in China, whose share of value added in the sector is projected to increase from 13 to 18% between 2006 and 2030.

The estimated growth rates in domestic demand for iron and steel are also particularly strong in the BRIICS countries. These are projected to average 5.1% annually for the period 2006-2030 as a whole; an annual growth of 7.2% per year between 2006 and 2010 is expected to gradually decline to 4.2% between 2020 and 2030. While steel demand in these countries represented 25% of the world total in 2006, this share is estimated to increase to 36% in 2030. The share of the rest of the world in total steel demand is estimated to increase from 13% in 2006 to 17% in 2030, while the share of OECD country demand decreases from more than 60% to below 50% (Figure 19.3).

In China, the demand for steel is estimated to grow particularly strongly. In 2006, China represented 17% of the global demand for steel. This share is estimated to increase to 26% in 2030. This strong demand increase in China is likely to lead to a large increase in

Figure 19.2. **Real value added in the iron and steel sector, 2006 and 2030**

At basic prices (2001), million USD

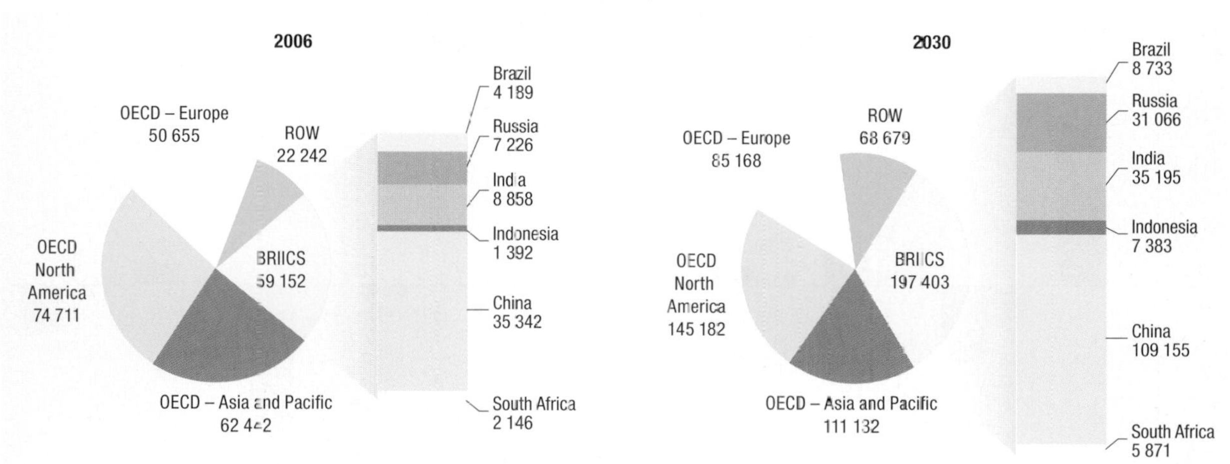

StatLink http://dx.doi.org/10.1787/262153137215

Note: OECD – A&P: Australia, Japan, Korea and New Zealand; OECD – NA: Canada, Mexico and the United States; OECD – E: OECD Europe.

Source: *OECD Environment Outlook* Baseline.

Figure 19.3. **Domestic demand for iron and steel, 2006 and 2030**

Million 2001 USD

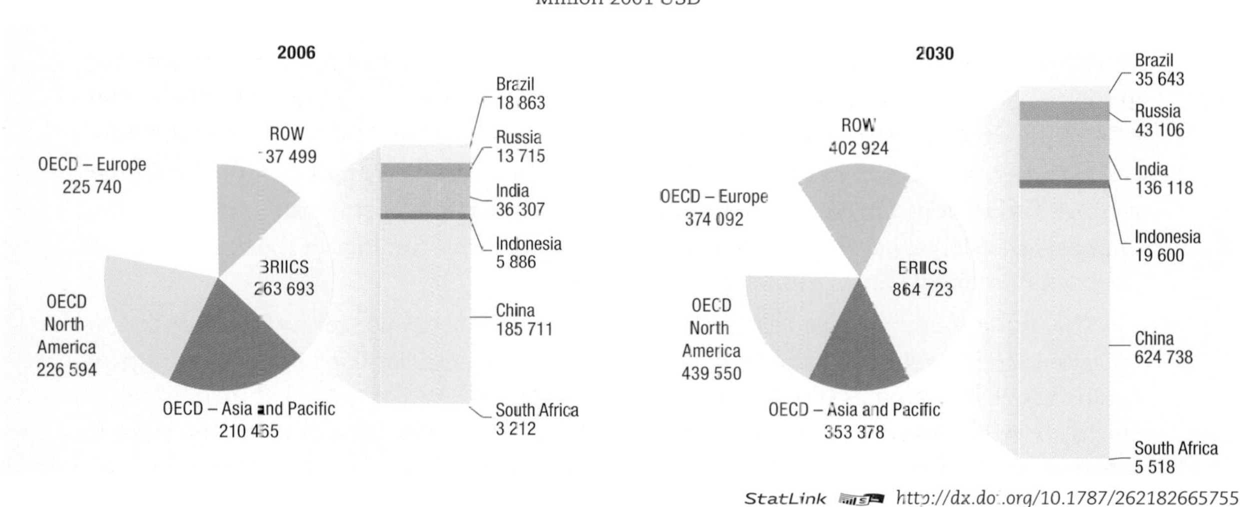

StatLink http://dx.doi.org/10.1787/262182665755

Note: Iron and steel products that are used as inputs in the iron and steel sector itself are double-counted in domestic demand. The totals are hence much higher for domestic demand than for value added. (Due to relatively low steel use, South Africa is not visible in the graph.)

Source: *OECD Environment Outlook* Baseline.

that country's trade deficit for iron and steel (Figure 19.4). The OECD countries of North America are also estimated to remain net importers of iron and steel products, whereas the export surpluses of the other OECD regions are estimated to increase significantly.

Policy simulations

Taxes or tradable permits

OECD (2003) explored the impacts of a hypothetical tax on CO_2 emissions levied on steel production and on electricity used in the sector. Within the context of that study, the

Figure 19.4. **Balance of trade in iron and steel products, 2006 and 2030**

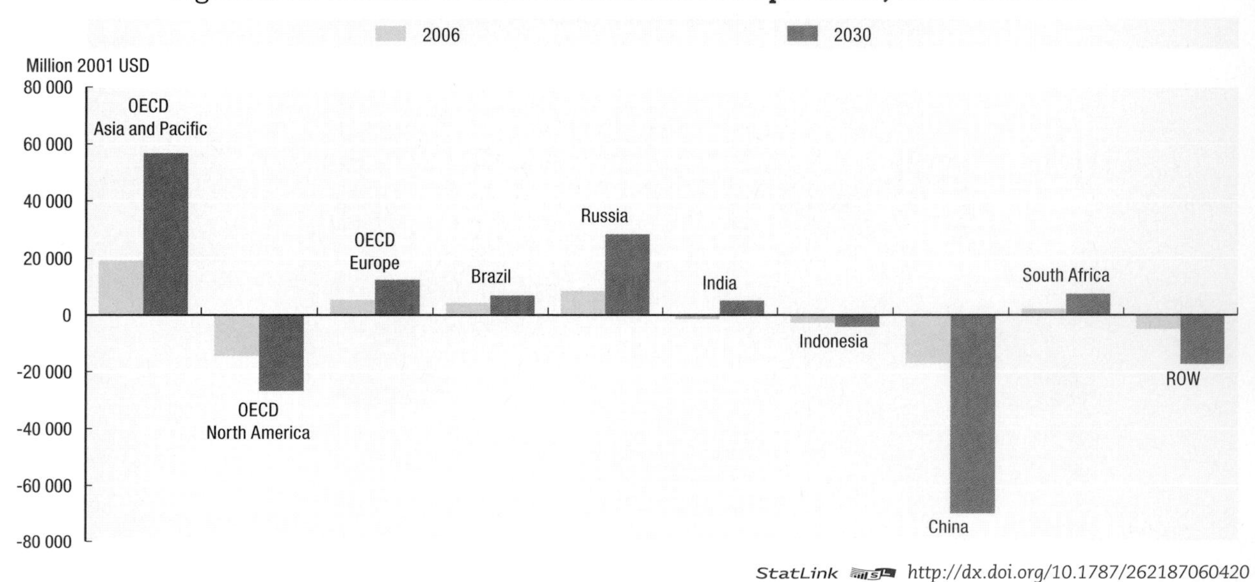

StatLink http://dx.doi.org/10.1787/262187060420

Source: *OECD Environment Outlook* Baseline.

simulated impacts would largely have been the same if an emission trading system had been applied instead of a tax. The purpose of the simulations was to examine the magnitude of the sectoral competitiveness problems related to environmental policy instruments, and to analyse possibilities for limiting such impacts, while still maintaining a positive impact on the environment (the goal was not to "single-out" the steel sector in particular for taxation).

Although significant changes have taken place in the steel sector since the base year of those simulations (1995), the study highlighted a number of points that are still valid. Emphasis should, however, be placed more on the qualitative findings of the study than on the exact numerical values estimated.

The study found that an OECD-wide carbon tax of USD 25 per tonne CO_2 (applied to emissions from steel plants and from the generation of electricity used in the steel sector) would reduce OECD country steel production by about 9%. While the exact magnitude of the total reduction in steel production in OECD countries in response to such a hypothetical tax is uncertain, the conclusion that the reduction in production would be much greater for the heavily polluting BOF plants (–12%) than for the scrap-based EAF plants (–2%) does seem robust. The simulations suggested that non-OECD production would increase by almost 5%, implying a fall in world steel production of 2%. The carbon tax would induce some substitution from the use of pig iron towards more use of scrap also in BOF steel-making. Due to limited supply, scrap prices would then rise, thereby – in isolation – weakening the relative competitiveness of the scrap-based EAF steel producers.

The study found that unilateral policies by single regions or countries could lead to quite dramatic cut-backs in the production of BOF steel (Figure 19.5), because unilateral approaches leave fewer opportunities to shift the tax burden over to suppliers or customers. For EAF steel producers, the net effect of unilateral policies was not found to differ much from an OECD-wide approach, partly because unilateral policies would lead to a smaller increase in scrap prices.

 OECD ENVIRONMENTAL OUTLOOK TO 2030 – ISBN 978-92-64-04048-9 – © OECD 2008

Figure 19.5. **Estimated changes in steel production in response to OECD-wide and unilateral taxes**

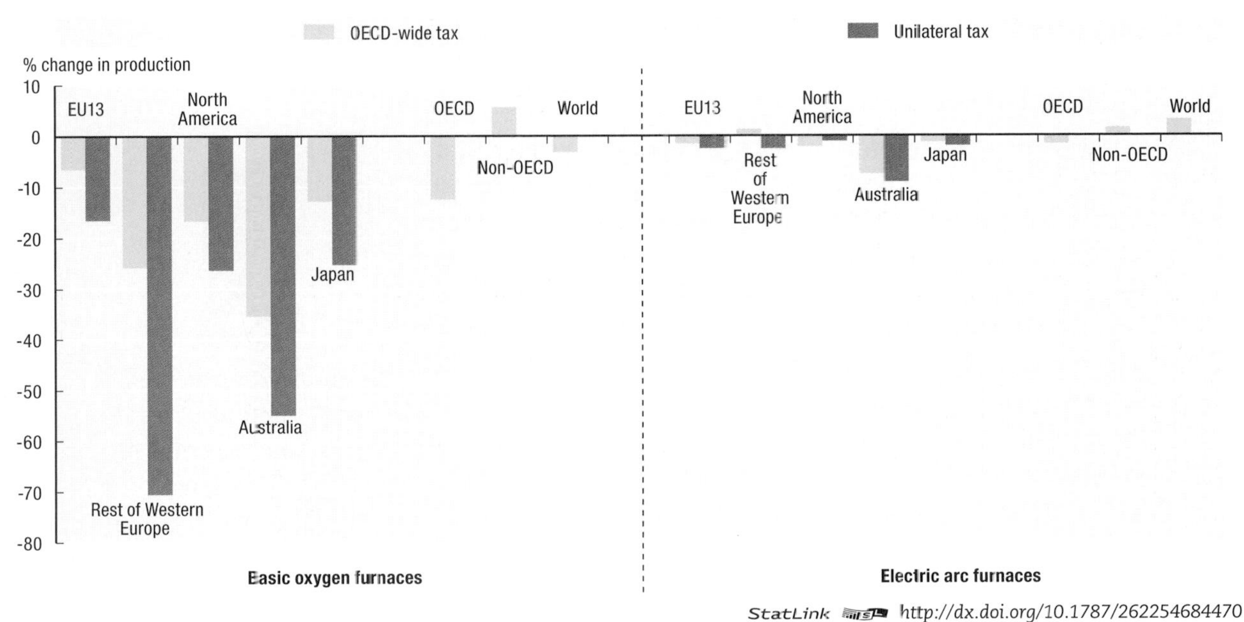

StatLink http://dx.doi.org/10.1787/262254684470

Note: EU13 comprises Austria, Belgium, Denmark, France, Germany, Greece, Ireland, Italy, Luxembourg, Netherlands, Portugal, Spain and United Kingdom.

Source: OECD, 2003.

An OECD-wide tax of USD 25 per tonne of CO_2 emitted from the steel sector and from the related electricity generation would, according to OECD (2003), reduce these emissions in OECD countries by about 19%. Despite relatively high emission intensities in non-OECD countries, global emissions from the sector were estimated to decline by 4.6%, *i.e.* more than twice the reduction in global steel production. This is due to substitution towards a cleaner input-mix and cleaner processes in the OECD area, in response to the simulated tax.

Because steel demand is relatively price-inelastic, and because steel produced in different ways has different qualities, a significant share of the gross tax burden was estimated to be carried by the steel users. The shift of the tax burden to steel users was found to be possible due to an increase in marginal production costs in non-OECD countries, as steel producers outside the OECD would be pushed closer to their capacity limits.

OECD (2003) also looked at various options to limit the negative impacts on the competitiveness of the steel sector of an OECD-wide tax. One possibility could be to recycle the tax revenue back to the steel industry, in accordance with current production levels (*i.e.* equivalent to an output subsidy). In this case, the decline in OECD steel production was estimated to be less than 1%. If the tax refund were uniform across processes, however, the OECD would see quite a significant restructuring towards the relatively clean EAF steel-making process. On the other hand, maintaining the competitiveness of the sector in this way would come at an environmental cost, as global CO_2 emission reductions in the sector would drop from an estimated 4.6%, to around 3%.

Another potential way of limiting the competitiveness impacts of the tax would be to apply border tax adjustments. The impacts of doing so would depend crucially on the scope and the design of the adopted scheme. If both import taxes and export subsidies were implemented, if these were differentiated between BOF and EAF steel-makers, and if the border tax rates were linked to emission levels in non-OECD countries, the decline in OECD steel production in response to an OECD-wide tax was estimated to be as small as 1%. At the same time, the reduction in global emissions was found to be slightly *larger* than without border tax adjustments. This was because border tax adjustments would keep a higher share of world steel production within the OECD area, thereby making more steel producers subject to the OECD-wide carbon tax.

To explore the long-term impacts of policies to limit carbon emissions from a sector like steel, a number of *hypothetical* taxes with a tax rate of 25 USD per tonne CO_2 were simulated for the *OECD Environmental Outlook*. These simulations indicate that by 2030, large reductions in CO_2 emissions (averaging roughly 30-35% in OECD countries) could be obtained for modest losses in output of iron and steel production (see Figures 19.6 and 19.7). Since the losses in output occur on a baseline that is increasing, the iron and steel industry in *all* regions would still be expected to be substantially larger than it is today.

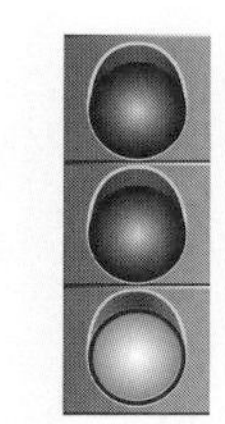

A tax on energy use in the steel sector would lead to a shift towards EAF production, with a reduction in the level of CO_2 emissions per unit of production and globally from the sector.

To the extent one can compare the simulations that were made (see Box 19.1), the simulations for the *OECD Environmental Outlook* seem to confirm some of the main findings made in OECD (2003). For example, whereas some "carbon leakage" could be expected if a tax was applied only in OECD countries, both sets of simulations indicate that *net* global emission reductions would occur. And while one should not place too much emphasis on the exact numerical production changes

Figure 19.6. **Effect of carbon tax on CO_2 emissions in the steel sector, 2010 and 2030**

% changes compared to the Baseline

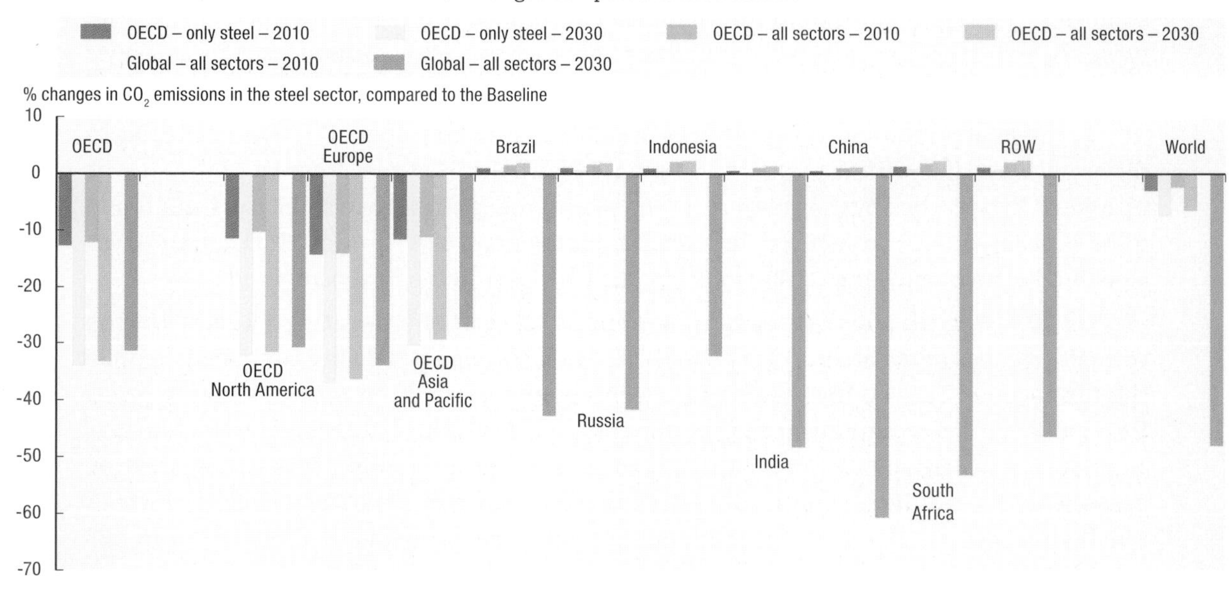

StatLink http://dx.doi.org/10.1787/262301558413

Note: In the simulations marked "OECD – Only steel", a tax of 25 USD per tonne CO_2 is applied only to emissions within the steel sector (*not* including emissions from the generation of electricity used in this sector). In the simulations marked "OECD – All Sectors", the tax is applied to all sectors within OECD, but this figure only shows changes in emissions in the steel sector. In the simulations marked "Global – All sectors", a 25 USD tax per tonne CO_2 is applied in all sectors in all regions; but, again, this figure only shows estimated impacts on emissions in the steel sector.

Source: OECD Environmental Outlook Baseline and policy simulations. Model used: OECD ENV-Linkages.

Figure 19.7. **Effect of carbon tax on production in the steel sector, 2010 and 2030**

% change compared to the Baseline

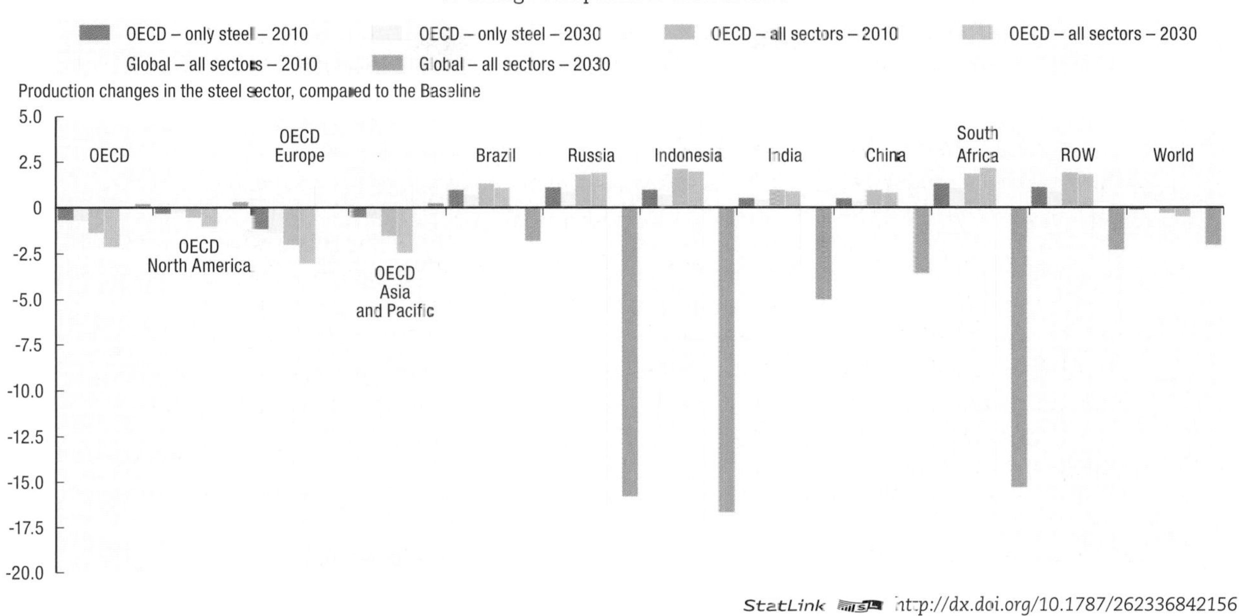

StatLink http://dx.doi.org/10.1787/262336842156

Note: For an explanation of the labels, see Figure 19.6.

Source: *OECD Environmental Outlook* Baseline and policy simulations.

that have been estimated, both sets of simulations seem to indicate that significant emission reductions can be achieved at a modest "cost" in terms of reduced production.

The reason why the estimated reduction in production in (and in emissions from) the steel sector is not much larger when a tax is applied *only* to that sector, compared to when a tax is applied to *all sectors* in the OECD, is linked to the assumed inflexibility in usage of materials-inputs in all production sectors that is built into the ENV-Linkages model used in the *Outlook* simulations (see Box 19.1).

A tax applied only to the steel sector in OECD countries would, according to these simulations, not have any macroeconomic impact. In 2030, real GDP in the OECD as a whole would be 0.0077% lower than in the Baseline; in the shorter term the impact would be even smaller. On the other hand, such a tax would only reduce total CO_2 emissions in the OECD countries by some 0.5%; globally, the emissions would fall by 0.3%. The macroeconomic impacts of an OECD-wide or a global carbon tax are discussed in Chapter 7 on climate change.

The estimated production reductions in the steel sector in the case of a hypothetical global carbon tax are much larger in *some* of the BRIICS countries than in the OECD regions and in the rest of the world. As can be seen from Figure 19.8, this is partly linked to the very high energy intensities of the iron and steel sector in these countries, and partly to their high reliance on fossil fuels for electricity generation.

The policy simulation also indicates that applying a 25 USD per tonne CO_2 tax to the steel sector would have a significant impact on SO_2 emissions from the sector. If the CO_2 tax were applied only in the steel sector in OECD countries, SO_2 emissions in these countries are estimated to decrease by almost 19% in 2030 compared to the Baseline (Table 19.2). If the tax were applied globally to all sectors, SO_2 emissions in OECD countries are estimated to decline by almost 15% in the sector, while these emissions in non-OECD countries are estimated to decrease more than 50%.

Box 19.1. **Model specifications and limitations**

The simulations in OECD (2003) focused on short to medium-term impacts of a hypothetical carbon tax, *i.e.* too short a time period to see any increases in production capacity in response to the simulated policies. The model used was a *static, partial* equilibrium model, focusing on the steel sector itself, and on sectors closely related to the steel sector (maritime transport, electricity generation, the scrap-iron market, etc.). The ENV-Linkages model used for this *Outlook* is, on the contrary, a *dynamic, general* equilibrium model, covering all sectors of the economy, but best suited to simulate longer-term impacts of a given policy shock. Whereas the model used in OECD (2003) distinguished between the main technologies for steel production, the ENV-Linkages model groups all iron and steel making into one sector. A direct comparison of all relevant results is thus not possible.

A drawback of the ENV-Linkages model (and many similar models) is that it assumes that the non-energy materials inputs into any production sector are *used in fixed proportions*. Hence, while changes in relative prices (*e.g.* due to the introduction of a tax) will trigger changes in households' demand for different products, they will in this model not trigger changes in the relative use of different materials for the production of a given final good or service. This means that the impacts of a given policy change could be underestimated. In reality, one would expect that an increase in the relative price of steel (compared to *e.g.* aluminium and other metals, plastics, cement, wood, etc.) would lead to a partial replacement of steel by other materials where possible. For example, in the building sector, wood or cement might replace steel in some applications – but wood could hardly replace steel in the car industry. In the car industry, substitution towards other metals or plastics would be more likely. Hence, the results of these simulations should be interpreted with caution.

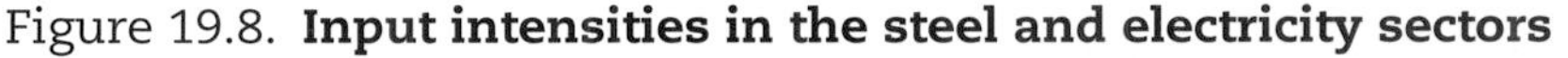

Figure 19.8. **Input intensities in the steel and electricity sectors**

Value of inputs as % of gross production value, 2005.

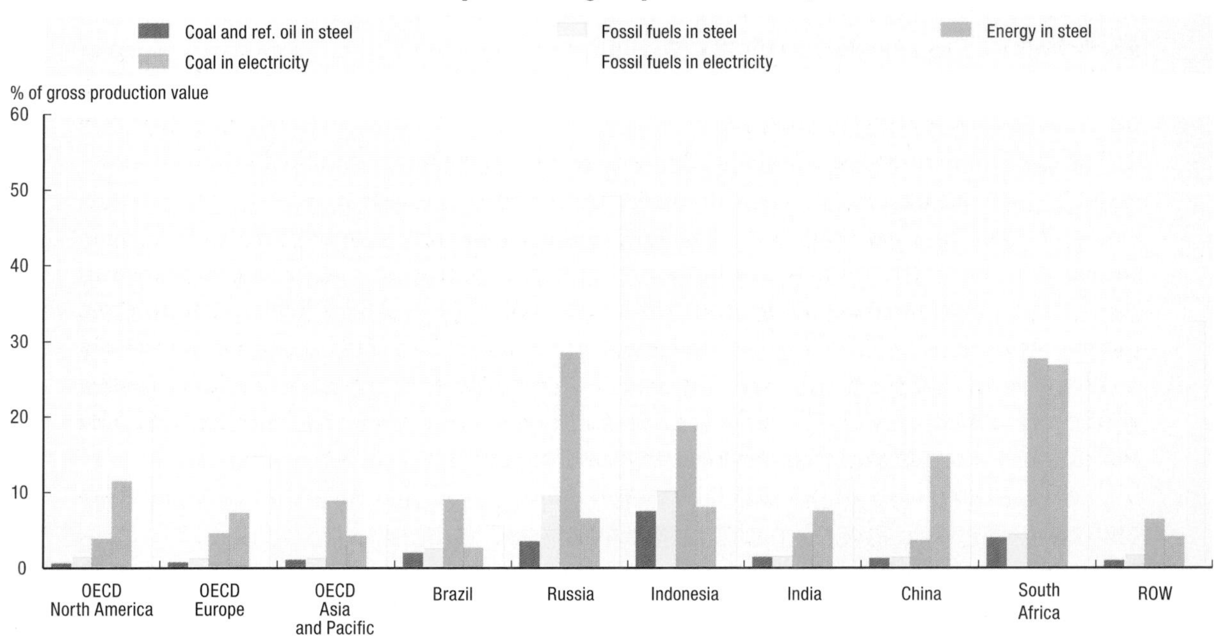

StatLink http://dx.doi.org/10.1787/262384265135

Note: These input intensities should be interpreted with caution. They are calculated as the value of certain inputs to either the "iron and steel" sector or to the "electricity" sector, as % of the gross production values of these sectors. In the GTAP database that is used as a basis for the model simulations, coke (which is largely used in basic oxygen furnaces) is classified in the sector "refined oil". "Fossil fuels" here includes the *value* of outputs from the sectors "coal", "crude oil", "natural gas", "gas distribution" and "refined oil" used as input in the iron and steel and the electricity generation sectors respectively, with no adjustment for the differences in carbon content of the different fuels. "Energy" includes "electricity" in addition to "fossil fuels".

Source: OECD Environment Outlook Baseline.

Table 19.2. **Estimated impacts on SO_2 emissions**

Changes in SO_2 emissions in the steel sector compared to the Baseline in 2030

	OECD – Only steel	OECD – All sectors	Global – All sectors
OECD	−18.7%	−17.1%	−14.5%
BRIICS	0.5%	1.0%	−52.6%
ROW	0.8%	2.0%	−54.0%
World	−1.9%	−1.2%	−48.5%

StatLink http://dx.doi.org/10.1787/257586111141

Source: *OECD Environmental Outlook* Baseline and policy simulations.

Reduction of environmentally harmful subsidies

For many years, there was very large over-capacity in the iron and steel sector. Hufbauer and Goodrich (2001) estimated that in 1998, the global over-capacity in the sector was at least 275 million tonnes, out of a production level of 775 million tonnes. This over-capacity put downward pressure on prices in the sector, which in turn stimulated demand, but nevertheless contributed to a very low profitability for many firms. Without any interventions in the market, unprofitable firms would gradually have been forced to close down, and many producers in the sector have historically therefore relied on large public subsidies for their survival. On top of various grants, preferential loans, loan guarantees, preferential tax provisions, etc., the sector has also benefited from a large number of trade-restricting practices (import quotas, anti-dumping measures, etc.)

In addition to examples of subsidies in the iron and steel sector, there are also instances of significant subsidies directed at their suppliers – such as coal mines and electricity generators – and to some of their customers, such as the shipbuilding sector. Likewise, some of the protection measures directed initially to the steel sector are passed backwards or forwards in the supply chain.

While there is broad agreement that subsidies to the iron and steel industry are still large and widespread, comprehensive quantitative estimates of their magnitude in different countries are hard to find. (A number of examples of subsidies and of trade-distorting practices are provided in UNCTAD, 2006.)

Two comments on the environmental impacts of a subsidy removal can nevertheless be noted. First, scaling-back subsidies in a way that would lead to the closure of unprofitable steel plants could be a low- or no-cost option for society as a whole in Annex I countries[2] to meet their CO_2 emission reduction targets under the Kyoto Protocol. In such a case, some of the financial resources that previously were provided as subsidies to steel producers could be used to help former employees find other employment and/or temporarily alleviate any social problems caused by the plant closures.

Second, the impacts on global CO_2 emissions of a reduction in subsidies to the iron and steel sector would depend on which subsidies were reduced and where, and on the emission intensities of the plants most affected by the subsidy reduction. Reducing subsidies to plants with high emission intensities would be particularly cost-efficient from an environmental point of view.

Box 19.2. **The cement sector**

OECD (2005) explored the impacts of a hypothetical OECD-wide tax carbon tax on the cement sector. This box briefly presents the main findings, which were largely similar to those made for the steel sector (OECD, 2003), which is an indication of the "robustness" of both studies.

Cement manufacturing consists of three main steps. Raw materials are first mined, ground and homogenised. Then they are burned at high temperatures: calcination and clinkerisation then take place to produce "clinker". Finally, the clinker is ground or milled and mixed with additional materials to produce cement. The calcium oxides required for the clinkerisation are provided by calcareous deposits, such as limestone, clay or chalk. These materials are the most common raw materials. The raw materials have to be homogenised, ground and crushed to the required fineness. Some other materials are required for the clinkerisation process: silica, iron oxide and alumina that are found in various ores and minerals, such as sand, shale, clay and iron ore. Power station ash, blast furnace slag, and other process residues can also be used as partial replacements for the natural raw materials.

The production of one tonne of clinker requires about two tonnes of raw materials and 25-30 kWh of energy, mostly electricity. The calcination of the calcium carbonate takes place at a temperature above 900°C, producing the calcium oxide required for the clinkerisation step. There are CO_2 emissions not only from the fuel combustion, but also from the process itself. Due to the CO_2 emissions, more than one third of the weight of the raw materials is lost.

The model used to study the cement sector was of a dynamic kind, and because of an embedded investment function, it was suited to analysis of long-term impacts. The cement model also took explicitly into account the high costs of transporting cement over long distances, especially inland – which tend to limit the impacts on production levels in OECD countries of potential climate policies in the sector.

The model's Baseline scenario projected an important increase in cement production (2% per year on average until 2030), entailing a strong rise in CO_2 emissions (1.5% per year). The CO_2 efficiency of cement production would thus rise by 0.5% per year, thanks to more intensive use of waste and wood fuels, and to an increasing share of modern and more energy-efficient technologies.

Several policy shocks were simulated:

- A CO_2 tax or an emission trading scheme, with auctioned allowances implemented in the countries that have ratified the Kyoto Protocol, assuming a CO_2 price of EUR 15 per tonne.
- The same policy implemented with border tax adjustments (BTA) (*i.e.* a rebate on cement exports and taxation of imported cement). There were two versions modelled: 1) exported production was completely exempted from the climate policy, and imports of cement from the rest of the world were taxed according to the CO_2 intensity of the cement production in the exporting country; 2) exports benefit from a rebate corresponding only to the least CO_2 intensive technology available at a large scale, and imports were taxed to the same level.

The implementation of a CO_2 tax was found to significantly decrease CO_2 emissions from the sector in these countries (around 20%), through more retrofitting toward energy-efficient technologies, a decrease in the rate of clinker (the CO_2 intensive input) in cement, a quicker switch to low-carbon fuels (gas, waste and wood fuels) and a decrease in cement consumption. The impact on cement production in these countries was significant (minus 7.5% in 2010), because of both a cut in their domestic consumption level, and because of a loss in competitiveness. For the latter reason, production and thus emissions in the rest of the world are projected to increase.

In the first BTA version, the loss of production in these countries was limited to 2% and the leakage was replaced by a spill-over, since emissions in the rest of the world also decrease. The decrease in world emissions is a bit higher than without BTA. The second BTA version was also found to prevent carbon leakage, with a leakage rate of around 4% in 2010.

PULP AND PAPER

KEY MESSAGES

Pollution from transportation in the pulp and paper sector remains a major environmental issue.

It is projected that the market for paper and board will continue to grow globally at 2.3% a year to 2030, with particularly rapid increases in developing and emerging economies. There will be significant differences among world regions, which will change the flows of trade.

There is room to improve paper recycling and to use more recovered paper in some regions of the world; this would reduce raw material and energy consumption in the industry.

The pulp and paper industry already generates approximately 50% of its own energy from biomass residues, and could eventually become a clean energy supplier.

New technologies have enabled a decoupling of environmental pressures from production. For instance, in Europe, Japan and North America, processes have been developed to eliminate the formation of chlorinated dioxins and furans, and to reduce formation of compounds containing organically bound chlorine. Further progress depends on the pace of diffusion of best available technologies around the world.

Policy options

- Ensure consistent government policies and industry action to maintain and further improve recent reductions in air and water emissions from the sector, and further increase the recycling of paper and board and the use of recovered paper.
- Design policy packages to reduce environmental pressures throughout the life-cycle of the products (from logging to recycling) whilst ensuring resource efficiency. These might include voluntary approaches, economic instruments, and command and control policies to stimulate R&D and the dissemination of innovation.
- Disseminate the best available techniques (*e.g.* to reduce the use and release of harmful chemicals in the bleaching process), especially within developing and emerging economies

Introduction

The whole life-cycle of paper production and consumption is significant from an environmental point of view, from harvesting of forest to conversion to paper and reuse. The type of feedstock (predominantly woodchips, but also recovered paper and to a lesser extent, rice and cotton), pulping process (mechanical or chemical) and final product determine the direct environmental impacts, which may include chemical, air and water pollution, deforestation and forest degradation. Progress has been made at each stage of the cycle, although at a different pace in different regions. At the same time, the industry can be the key enabler to meet biomass targets and produce biofuels, biodiesel and biochemicals.

The harvest of forests has an impact on ecosystems and biodiversity, depending on how forests are managed, which species are used for plantation, etc. (see also Chapter 9 on biodiversity). The industry contributes to the conversion of high conservation natural forests to managed forests (plantation or natural regeneration systems) in Southeast Asia (Indonesia) and Australia. Sustainable use of old growth forest is an issue in Canada and Russia. The environmental impact can be reduced and resource use efficiency increased through sustainable forest management. The industry indicates that only a portion of the timber harvested worldwide is used for paper-making; often, this is small dimension wood, sawmill waste and woodchips, as well as timber from forest thinning (extracted to enable remaining trees to grow).

The conversion of wood to pulp involves two preliminary methods: *i)* the mechanical grinding of woodchips, which is electricity intensive; and *ii)* the chemical separation of the wood fibres from the lignin which binds them; this produces a variety of atmospheric pollutants. The most polluting stage is the bleaching of the resulting pulp. This used to be done with chlorine-based bleaches, a process which is no longer used in Europe and which is also significantly reduced around the world. Outputs from this process include large volumes of water (for washing), which may contain highly toxic organic compounds, including furans, dioxins and other chlorinated organic compounds.

Cleaner technologies have radically improved the environmental performance of the industry in most regions of the world. Table 19.3 illustrates process evolution and performances: in the 1990s, a modern paper mill used about 85% less water than it did three decades ago; reductions in total suspended solids (TSS) and in five-days biological oxygen demand (BOD5) have also been considerable (FAO, 1996).

Recovered paper is an alternative input to timber in the production process (see Box 19.3); it helps lessen pressures on forests. Europe leads the way in the use of recovered paper, with Asia and North America trailing behind. Recovered paper is the most important raw material for the British paper and board industry, representing 68% of the fibre throughout this sector in 2004. The Confederation of European Paper Industries (CEPI) has set a target of 66% paper recovery by 2010 (see ERPC, 2006), up from roughly 55% in 2004 (CEPI, 2005). The US industry has set a paper recovery goal of 55% by 2012 for domestic use and export.

OECD ENVIRONMENTAL OUTLOOK TO 2030 – ISBN 978-92-64-04048-9 – © OECD 2008

Table 19.3. **Integrated kraft mill wastewater, TSS waste load and BOD5 waste load**

Technology	Waste water discharge (gal/tonne)		TSS waste load (lb/tonne)		BOD5 waste load (lb/tonne)	
	Bleached	Unbleached	Bleached	Unbleached	Bleached	Unbleached
1964 older	110 000	90 000	200	170	200	160
1964 current	45 000	27 000	170	130	120	90
1964 new	25 000	16 000	90	80	90	80
1990 design	16 000	8 000	50	45	60	50
% reduction 64/90	85%	91%	75%	73%	70%	69%

StatLink http://dx.doi.org/10.1787/257614204873

Source: K. Ferguson cited in FAO, 1996.

Box 19.3. **The procurement issue in perspective**

The industry has engaged in sustained efforts to diversify its fibre base and, in particular, to substitute virgin raw material with recovered paper. However, this policy has its limitations as, at least in Western Europe, most quality recovered paper sources are already tapped; the challenge now is to increase the quality of recovered paper (see CEPI, 2006a) and the recyclability of paper products, via an integrated environmental approach. Recovered paper is now increasingly traded around the world, in particular between the EU/US and China.

Another challenge is looming. Wood supply may well be influenced by the increasing demand for biofuels. A number of governments are enacting policies to support the development of bio-energies, including biomass, thus increasing competition for raw and recovered resources for the pulp and paper industry. According to a recent European Environment Agency report (EEA, 2006), increasing market values for bio-energy would lead to substantial mobilisation of wood biomass resources for bio-energy from other competing industries, including pulp and paper (for a more detailed discussion of the consequences of EU energy policy on forest-based industries, see EC-DG Environment, 2000). With a woodchip price of EUR 70/m^2, chemical pulp production in Europe might decline by around 10-15%. If the price increases to EUR 100/m^2, the reduction could be up to 50%. Since pulp and paper are produced globally and widely traded, higher production costs in Europe may not be reflected in pulp and paper prices, unless similar developments occur in the world market.

In Europe and in North America, at least, energy-efficiency has become a key issue for the industry, not least as a means to reduce energy costs (Jokinen, 2006). However, some energy savings achieved through improved production processes have been counteracted by demand for increasing quality paper products. However, the pulp and paper industry already generates approximately 50% of its own energy from biomass residues, and in the long term it could develop into a clean energy supplier.

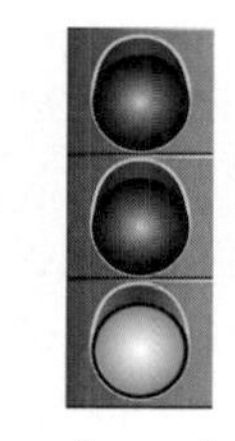

The pulp and paper industry already generates approximately 50% of its own energy from biomass residues, and could become a clean energy supplier.

It is notable that the European industry sets the world standards via the European reference document on best available technologies (BREF)[3] and the leading machine manufacturers are established in Europe. The BREF identifies new technologies which have not yet materialised but which are likely to have substantial environmental benefits over the *Outlook* period. For instance, black liquor gasification will yield higher returns on energy; new technologies for boilers will save energy; and biotechnologies will enable the production of by-products (ethanol), thus increasing the value-added from raw materials.

The diffusion of best available technologies (BAT) is indicated by the fact that in 2005, pulp manufactured without the use of any chlorine gas accounted for 85% of total world output (Alliance for Environmental Technology, 2005). The diffusion of BAT follows the investment cycle (including rebuilds): the industry is capital intensive and some equipment lasts over 20 years. Pulp and paper mills in many OECD countries will need to be replaced in the next 10-15 years, and this is an opportunity to install new technology.

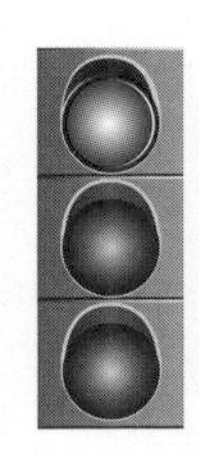

Pollution from transportation in the pulp and paper sector remains a major environmental issue.

Industry leaders consider that pollution from transportation in the pulp and paper sector remains a major environmental issue (see Ernst and Young, 2007); the modal choices vary from country to country, depending on distance, infrastructure and costs.

Key trends and projections

Demand for paper products comes from a number of sectors:

- Printing and publishing, though this is being challenged by electronic media.
- Packaging, though paper competes with alternative materials (aluminium, plastic, etc.).
- Sanitary and household sectors, which have a high demand for paper products.

Since 1990, the global market has been growing steadily, but it has been rocked by the rapidly changing demand and supply in Asia and Latin America. The consumption of paper and board in OECD countries increased by 27% (in volume) between 1990 and 2004. Over the same period, consumption in Southeast Asia has been booming: Chinese consumption increased by 213% and now amounts to one-quarter of the consumption of OECD countries;[4] over the same period, Indonesian consumption increased by 265%, though from a much lower level. Chinese imports of paper and board represented 10% of world trade in 2004 (6% in 1990); as for pulp, China's share of world imports rocketed from 3% in 1990 to 18% in 2004 (FAO data, at *http://faostat.fao.org*). World production has followed a parallel path with a 32% increase in OECD countries and a 207% increase in China for paper and board.

At the moment, the industry is characterised by fragmentation, overcapacity and low profitability (due to slow market growth, cost increases and capital intensity) (Ernst and Young, 2007). Some segments are in a better shape. In the value chain, the most profitable firms are those which create the most value; customers and suppliers of the industry are better off than the pulp and paper producers and merchants.

The *OECD Environmental Outlook* Baseline anticipates sustained growth of the market to 2030 (2.3% per annum – see Box 19.4 for key assumptions). The BRIC countries deserve particular attention:

- China[5] is already the second largest producer of pulp and paper worldwide (third, if EU member countries count as one). It is expected that rapidly growing demand (from

54.7 million tonnes in 2004 to 68.6 by 2010) will be mainly covered by domestic production. Over the same period, 50% of the demand for pulp and recovered paper will have to be imported, as China lacks domestic resources. Land use, energy and transportation issues are increasing and are driving up wood costs. The industry comprises old mills (small, family-owned, based on non-wood material, polluting) and recent ones (large, using primarily imported fibre, complying with international standards).

- The Indian market for paper and paperboard remains small, but it is expected to grow by 6% per annum until 2020 (Ernst and Young, 2007). Demand is led by printing/writing papers (India has become a hub for high quality printing at low cost) and containerboard. It is estimated that the industry will restructure heavily due to trade liberalisation and modernisation. The industry is expanding and investing in new technologies for cleaner and brighter paper. However, limited access to raw material prevents mills from growing and achieving economies of scale.
- Russia supplies raw material to Europe and China. Investment has remained minimal since 1990, but both production and consumption of newsprint are expected to increase by 7% per year until 2020 (see UNECE/FAO, 2005), allowing the Confederation of Independent States (CIS) countries to export newsprint to Europe and Asia. The Russian industry is highly consolidated, with five companies producing over 40% of all pulp and paper products. In the future, it will strive to retain the value-added (Russia is restricting its wood exports through additional export duties) and attract more investment from multinational companies.
- Pulp production is developing in Brazil (the annual growth rate for the 2002-2006 period is above 8%), based on large-scale plantations. Paper production is expanding less rapidly (data available at *www.bracelpa.org.br/eng*). The industry is expected to generate more value-added in the future.

Europe is expected to become a net importer of printing and writing paper by 2020 as a result of new demand in Eastern Europe and the reduction of growth in production in Western Europe (UNECE/FAO, 2005).

Box 19.4. **Key uncertainties, choices and assumptions**

The chapter relies on a number of assumptions. One is the continuation of the current trends in world production: it is assumed that *i)* no major upheaval will significantly change the way paper competes with other materials in the different markets, and *ii)* demand management will not dramatically impair demand for paper products.

Another set of assumptions relates to the environmental performance of the pulp and paper mills. We lack recent, comprehensive data on the environmental impacts of the pulp and paper industry, in particular as regards energy-efficiency (see Jokinen, 2006). The difference between the environmental performance of front runners and laggers in the sector is probably very large. There are key uncertainties about the pace of dissemination of best available technologies, and the role the multinational companies and supply chain management play in these dynamics.

The industry is taking initiatives to enforce compliance, such as wood tracking systems, codes of conduct and forest certification (see for instance the Confederation of European Paper Industries' Position Paper on Forest Certification; CEPI, 2005b).

In the future, the key location criteria for pulp and paper mills are expected to be access to the final market, access to the resource (but transport costs are still low and alternative, recycled resources are available worldwide) and energy costs. In this context, production capacity is expected to shift from North America and Western Europe to emerging markets (China, India, and Latin America). Additional criteria also gain importance, and will influence investment decisions, such as increasing transportation costs (due to high energy prices), non-tariff barriers to trade, and social and environmental requirements.

These dynamics have consequences for trade flows. In their review of forest product markets, UNECE/FAO (2005) claim that trade barriers continue to exist in the sector. These include technical barriers, such as antidumping measures and retaliatory tariffs, as well as uneven trade advantages that stem from divergent tax rules and labour standards. This is clearly not in line with GATT/WTO regulations which abolished paper and paperboard tariffs as of 2004, at least in trade between major industrialised countries.

Policy implications

The industry is subject to a wide array of policy instruments which aim to support cleaner production, more efficient processes (to save on energy, water and material usage) and end-of-pipe pollution abatement. Other options are being considered, such as demand management which encourages reduction of wasteful consumption of paper in OECD countries (see, for example, efforts by the Australian Conservation Foundation, 1992). The industry is also influenced by energy policies, which subsidise the use of renewable energy sources (including biomass), constraining the availability of raw materials and decreasing the incentives for separate collection of paper and recycling (see CEPI/WWF, 2006); energy-related taxes and emission trading schemes also affect the industry's competitiveness.

The impacts of these instruments have different time horizons: it can take decades before sustainable harvest of primary forests restores some biodiversity quality in a given territory, whereas changes in production processes can have an almost immediate impact on energy consumption.

Paper and pulp are traded globally and prices are set at the global level, whereas cost factors (some raw materials, energy, employment, compliance with environmental legislation) vary locally, which can affect competitiveness. Policy packages need to be designed and implemented in a way which addresses competitiveness concerns, at least at a regional level.

Regulatory approaches

Command and control instruments have been used widely in the industry. Typically, norms on emissions have been used to abate pollution.

The World Bank has issued *Environmental, Health, and Safety Guidelines* for pulp and paper mills, which guide proponents requiring financial assistance to set up mills; the initial document was drafted in 1998 and was under review in 2007.[6] Such approaches help to disseminate the best available technologies.

Economic instruments

Economic instruments at industry level are also appropriate. These include incentives for industry and households to recycle and use recovered materials; or green taxes targeting particular operations in the production process. As an illustration, a United

Nations Environment Programme (UNEP) report on economic instruments for environmental protection proposes an effluent charge to abate pollution in the pulp and paper industry in Indonesia (see UNEP, 2005). In Europe, pulp and paper mills that produce more than 20 tonnes per day are included in the EU Emission Trading System for greenhouse gas emissions. These instruments have to be regularly adapted to changes in the structure of markets and their impacts have to be monitored.

The Confederation of European Paper Industries (CEPI) assumes that green public procurement will become an incentive for environmental performance in the next five years or so. The public sector can represent up to 20% of the demand for some paper grades.

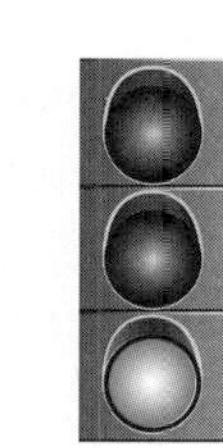

New technologies and further uptake of existing ones enable a decoupling of environmental pressures from production.

Voluntary approaches

The pulp and paper industry has been taking steps to reduce its environmental footprint. In France, the industry was among the first to sign a voluntary agreement to cut environmental impacts, back in 1972. Voluntary approaches at a global level are illustrated by the commitment by forest industry leaders to action on global sustainability (see ICFPA, 2006). In Europe, as in Canada, voluntary approaches are flourishing, for example on illegal logging (*e.g.* ICFPA's Statement of Support for WBCSD/WWF Certification and Illegal Logging Activities, ICFPA, 2006), on recycling (*e.g.* the European Declaration on Paper Recycling, ERPC, 2006), on biomass-based energy consumption (the European pulp and paper industry also committed to increase the use of biomass in on-site primary energy consumption to 56% by 2010), or on eradicating the emission of chlorinated organic and sulphur dioxide compounds.

Progress is monitored and reported by the Confederation of European Paper Industries, which publishes a set of performance indicators, most of them corresponding to Global Reporting Initiative indicators (see for instance CEPI, 2007).

R&D and cleaner production

In 2006 the industry spent an estimated 0.7% of its turnover on R&D, and industry leaders agree that this should increase strongly (Ernst and Young, 2007). The industry builds upon research all along the value chain. In 2004, the European industry reportedly invested EUR 560 million in environmental improvements (7% of its total capital expenditures) (Pöyry, quoted in CEPI, 2005). Suppliers, research institutes and other stakeholders invest in R&D as well.

In Europe, a platform has been established which brings together stakeholders in the forest and paper sectors around a long-term strategy and a vision of the industry for 2030 (see Forest-Based Sector Technology Platform, 2006). The platform emphasises research on modern timber breeding techniques (which will improve wood characteristics and fibre biomass, and reduce forest losses) and on "tailor-made" wood supply (adapting raw materials to customer demands and optimising the allocation of raw materials to different industrial applications).

The diffusion of best available technologies (BAT) can accelerate progress. According to the Worldwide Fund for Nature (WWF, 2006), most pulp mills meet BAT levels, but not all paper mills use the latest technologies. Action is needed by companies and governments, in particular in regions where the oldest part of the industry fails to meet the best industry standards (*e.g.* in some countries of continental Europe, parts of China and India). Resource use (procurement, resource efficient technologies) and cleaner production (closed loop processes, chlorine-free processes, elimination of persistent organic pollutants) are expected to attract attention in the future.

Other instruments

Other instruments include:

- Labelling, based on the life-cycle of the product, including reference to the raw material (the EU in particular has set up a partnership process with timber exporting countries, to support compliance enforcement in that sector).
- Extended producer responsibility (as an attempt to ensure coherence across the product life-cycle). Again, the industry can play a leading role, as is illustrated by the renewed commitment by CEPI in the field of recycling: it entails qualitative (and quantitative) targets throughout the value chain, and it passes responsibility to producers for ensuring waste prevention and better recycling.
- Monitoring and reporting. Reports from companies allow their performance to be assessed against the permits they hold, their commitments, or against best available technologies (see WWF, 2006); some reporting requirements are defined in permitting and certification schemes. Many companies also produce sustainability reports.

In addition, the UN Food and Agriculture Organization makes a strong case for environmental impact assessment and environmental auditing in the pulp and paper industry (FAO, 1996). According to the World Bank guidelines referred to earlier, large new pulp and paper mills and large expansions and projects located in or affecting a sensitive area require an environmental impact assessment to be submitted; the assessment includes a statement on the use of best available technologies.

TOURISM

KEY MESSAGES

According to some estimates, tourism contributes up to 5.3% of global anthropogenic greenhouse gas emissions, with transport accounting for 90% of this. Travel for tourism purposes is expected to grow significantly to 2030, with international tourism growing by over 4% per year, accompanied by increasing environmental pressures.

Tourism development can generate unsustainable pressures on the local environment, in particular if insufficient infrastructure is in place to cope in an environmentally sustainable way with large numbers of visitors and their activities.

Tourism and the environment can be mutually supportive. In a number of destinations, tourism is a driver of enhanced water quality and the protection of nature. In rural areas, it can contribute to the sustainable development of traditional activities (handicraft, agriculture, etc.). In urban settlements, it can generate additional resources to invest in environmental infrastructures and services.

Policy options

- Implement appropriate policies to support the development of sustainable tourism (including transport). These should involve an array of stakeholders (public and private), at international, domestic and local levels. Efficient mechanisms are needed to harness tourism for economic, environmental and social developments.
- Scale up innovative approaches to encourage environmental sustainability in the tourism sector. Certification and labelling schemes can help promote eco-tourism opportunities, for which there is a rapidly growing market.
- Increase the use of economic instruments to internalise the measurable externalities of tourism. Instruments can include price incentives, fees and subsidies to sustainable tourism activities.
- Adopt and promote the principles enshrined in declarations such as those on *Harnessing Tourism for the Millennium Development Goals* and *Action for More Sustainable European Tourism*.

Consequences of inaction

Tourism itself will be affected by changes in the environment. For example, climate change is expected to decrease the number of snow-reliable skiing days in the European Alps, and sea level rise will affect tourism operations in coastal areas and small islands.

Introduction

Tourism has an impact on the environment in the destination country and at the global level. According to a classification developed by the United Nations Environment Programme (see UNEP webpage: *www.unep.fr/pc/tourism*), the potential environmental pressures of tourism at the destination include:

- The depletion of natural resources: tourism often overuses water resources and requires oversized infrastructure (especially when demand fluctuates seasonally); it also affects local resources (*e.g.* energy, food stuffs), including land and scenic landscapes.
- Pollution: as with any other industry, tourism can generate air emissions, noise, solid waste, release of sewage, oil and chemical pollution, and visual pollution.
- Physical impacts: degradation of ecosystems, with coastal and mountain areas being particularly vulnerable.

At the global level, tourism can have an impact on biological diversity, the ozone layer and climate. It has been estimated that tourism contributes up to 5.3% of global anthropogenic greenhouse gas emissions, with transport accounting for 90% of this (Gössling, 2002 and Box 19.5).

Box 19.5. **Tourism, transport and the environment**

The OECD has recently explored the relationship between tourism and transport. Experts usually distinguish between tourism travel to and from destinations and tourism travel at destinations. The former usually, but not always, has the greater overall environmental impacts. Impacts of the latter may be lower when the quality of the destination facility is high.

Most tourism travel is by car. This is more the case in Europe than in North America, although the frequency and extent of all types of tourism travel in Europe is much lower. However, tourism travel is driven by the growth in availability of inexpensive air transport.

Tourism is estimated to account for about 75% of the demand for aviation, which is growing rapidly. Low-cost carriers have been moving passengers over longer distances for shorter and more frequent holidays, with 10-20 times the environmental impact per trip-day compared with tourism by road or rail. The low-cost-carrier phenomenon may be particularly evident in Europe, although North Americans continue to make many more longer-distance trips than Europeans. The tentative nature of these statements reflects the lack of reliable data, which in part reflects the lack of accepted definitions of leisure travel and tourism travel.

Current efforts towards reducing the environmental impacts of tourism travel include the marketing of packages involving both eco-tourism and eco-mobility. Several examples exist in Japan, Germany and Austria (*e.g.* Lake Neusiedl Region, and eco-mobility in the Alps).

Source: OECD, 2005a.

In turn, tourism can be affected by changes in the environment. For example, climate change will affect winter tourism in mountainous regions, and sea level rise will have potential consequences for tourism in coastal areas and small islands. These effects are ambivalent: they may generate new demands and change the geographical location of tourism supply and infrastructure. In many cases, tourism operators are beginning to adapt to these changes, for example by increasing the development of year-round tourism activities in skiing resorts and greater use of artificial snow-making machines (OECD, 2006). The degradation of the environment can diminish the capacity of a destination to attract tourists.

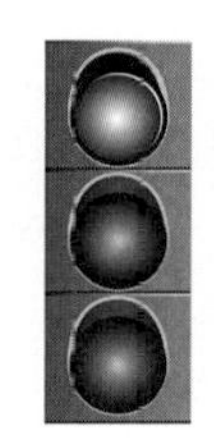

Climate change may adversely affect tourism opportunities – for example in coastal zones and ski areas.

Tourism and the environment can be mutually supportive: tourism is an opportunity to finance environmental infrastructure (water supply and sanitation, waste treatment) and it can contribute to the conservation of sensitive areas and habitat. Tourism can reduce poverty, depending on how revenues from the industry are shared and distributed along the value chain and among local communities. This is the objective of eco-tourism (see Box 19.9 below), which is emerging as a market segment.

Key trends and projections

By 2020, the United Nations World Tourism Organization (UNWTO) estimates that most industrialised countries will have come close to their upper limits in terms of supplying domestic tourism. Growth in that domain is expected to come from developing countries in Asia (in particular in China, see Box 19.6), Latin America, the Middle East and Africa.

Box 19.6. **Tourism in China**

China travel and tourism (encompassing transport, accommodation, catering, recreation and visitor services, for both domestic and international tourists) is estimated to have generated USD 265 billion of economic activity (total demand) in 2005, and is expected to grow (in nominal terms) to USD 875 bn by 2015. This represents an annual growth rate of 9.2%, in real terms, between 2006 and 2015 and would make China the second largest travel and tourism economy, after the US, by 2015. China travel and tourism capital investment for 2005 is estimated at USD 100 billion (9.9% of total investment); by 2015, this should reach USD 329 billion (10.7% of the total). The rise of disposable incomes has already fuelled domestic tourism. However, there is concern that these massive flows jeopardise the environment and generate excessive demand for environmental services (water, waste).

Source: World Travel and Tourism Council, 2005.

International tourism is characterised by steady growth in the recent past and in the foreseeable future. The UNWTO estimates that only 7% of the world's population currently capable of engaging in tourism has travelled abroad. Prospects for further growth are likely to be considerable. In OECD countries, the development of tourism is related to demographic changes, such as the ageing of populations and the resulting growth in the number of healthier, wealthier, car-driving older people. Outside the OECD, rising living standards are fuelling tourism. Growth is only temporarily and regionally sensitive to upsets (such as acts of terrorism or natural disasters).

Over the 1995-2004 period, international tourist arrivals grew by 3.8% annually. The receipts from international tourism rose even more steadily, from USD 405 billion to USD 622 billion. Table 19.4 shows how the different regions of the world have benefited from these trends. The focus of UNWTO data collection on international tourism provides a misleading impression that there is more tourist activity in Europe than in North America, where domestic tourism is particularly significant. Table 19.5 shows that tourists spend relatively more on tourism activities in the USA than in European countries.

Table 19.4. **International tourist arrivals by tourist receiving region (millions), 1995-2020**

	Base year	Forecasts		Market share (%)		Average annual growth rate (%)
	1995	2010	2020	1995	2020	1995-2020
World	565	1 006	1 561	100	100	4.1
Africa	20	47	77	3.6	5.0	5.5
Americas	110	190	282	19.3	18.1	3.8
East Asia, Pacific	81	195	397	14.4	25.4	6.5
Europe	336	527	717	59.8	45.9	3.1
Middle East	14	36	69	2.2	4.4	6.7
South Asia	4	11	19	0.7	1.2	6.2

StatLink http://dx.doi.org/10.1787/257658165751

Source: United Nations World Tourism Organization, 2001.

Table 19.5. **Trends for inbound tourism, 1995-2004**

	Number of tourist (overnight) arrivals			Tourism receipts		
	Rank 2004	Million	1995-2004	Rank 2004	USD billion	1995-2004
France	1	75.1	↘ –	3	40.8	↘ –
Spain	2	53.6	↗ +	2	45.2	↗ +
USA	3	46.1	↘ –	1	74.5	↘ –
China	4	41.8	↗ +	7	25.7	↗ +
Italy	5	37.1	↘ –	4	35.7	↘ –

StatLink http://dx.doi.org/10.1787/257672843555

Source: UNWTO, 2001.

The UNWTO anticipates that international tourism will continue to grow to 2020 (though see Box 19.7 for some methodological challenges to the projections). The number of international arrivals worldwide is expected to increase to almost 1.6 billion, 2.5 times the number recorded at the end of the 1990s; but the pace of growth will slow down to a forecasted 4% annually. Europe is expected to continue to be the most visited region (Table 19.4), but the anticipated growth rate is below the world average. East Asia and the Pacific will overtake the Americas as the second largest receiving region; China is likely to become the first destination country (in number of arrivals), ahead of France

Box 19.7. **Key uncertainties and assumptions**

The trends presented here are based on available data published by the UN World Tourism Organization, which regularly publishes market analyses and outlooks. Some major uncertainties remain. One major uncertainty is the pace of development of tourism from China.

Another example of uncertainty is the impact of climate change on the development of tourism. Recent work on the European Alps indicates that climate change can significantly affect the capacity of a region to sustain tourism (OECD, 2006). Adaptations are required, the magnitude of which are still unclear at the global level.

The qualitative shifts in tourism consumption may have severe impacts on long-haul travel, which could significantly change the environmental footprint of the industry.

Present data collection is inadequate for the kinds of analysis required, especially for domestic tourism. The data collected by the UNWTO focus on international tourism. Such a focus yields few insights for large OECD countries (*e.g.* United States) where most tourism is domestic.

and the USA. In the Americas, Northern America should perform less well than the other sub regions. Africa and the Middle East should perform above average, with a particularly notable increase in visitors to South Africa.

Annual receipts from international tourism (excluding transport) are projected to reach USD 2 trillion in 2020. The main driver of this expansion is expected to be rising incomes, spread across larger and new layers of the world's population, a growing share of which will be spent on travelling abroad.

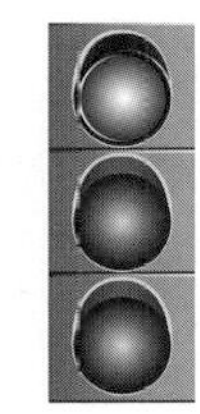

International arrivals are projected to reach almost 1.6 billion in 2020, increasing the environmental impacts of air travel.

According to UNWTO (2005a), tourism consumption is changing qualitatively; consumers are favouring destinations closer to home, taking a "wait-and-see" approach to travel plans and leaving bookings until the last minute. Tourists will travel more often, for shorter periods of time (see the multiplication of shorter holidays in Europe, North America and recently in Asia). The UNWTO notes that some products and sectors have benefited from these trends. These include non-hotel accommodation such as apartments and bed-and-breakfasts, and special interest trips with high motivation factors related to culture, sports or visits to family and friends. Long-haul destinations have been most affected by these trends. Short-haul travel is expected to enjoy comparatively stronger growth.

Policy implications

Tourism markets have not succeeded in systematically valuing the environment properly. There has been some progress, for example in valuing the contribution of Australia's biodiversity to the tourism industry (see Australian Government, 2004), but this remains slow and piecemeal. Active policies are needed to reverse unsustainable trends and market failures in the tourism industry.[7]

An agenda for sustainable tourism policies

The active role played by the United Nations Environment Program (UNEP), UNWTO and other international organisations such as the European Commission and the OECD is increasing recognition of the concept of sustainable tourism and clarifying actions required to support it. A representative group of government, industry, UN specialised agencies and civil society leaders met in New York, at the invitation of the UNWTO, on 13 September 2005 and adopted a declaration on *Harnessing Tourism for the Millennium Development Goals*. The declaration considers that tourism can make a substantially greater contribution than at present to poverty elimination, economic growth, sustainable development, environmental conservation, inter-cultural understanding and peace among nations.

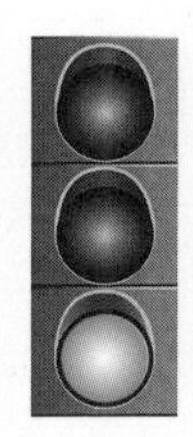

Eco-tourism is a rapidly growing industry, with potential benefits for the environment, economy and local communities.

UNEP/UNWTO (2005) outline the environmental agenda of sustainable tourism policies (see Box 19.8 for a social agenda):

- To maintain and enhance the quality of landscapes and avoid degradation of the environment.
- To support the conservation of natural areas, habitats and wildlife, and minimise damage to them.
- To minimise the use of scarce and non-renewable resources in the development and operation of tourism facilities and services.
- To minimise the pollution of air, water and land and the generation of waste by tourism enterprises and visitors.

Box 19.8. **The social agenda of sustainable tourism**

In addition to its environmental agenda, sustainable tourism also aims to avoid the potentially adverse social consequences of tourism (sexual exploitation, exploitation of women and local staff, long working hours and low seasonal wages, etc.) by:

- Creating jobs, creating capacity and generating income for local staff.
- Bringing benefits for local communities.
- Respecting and supporting regional cultures and habits.
- Ensuring the informed participation of all stakeholders, etc.

Compliance with domestic and international regulations (*e.g.* International Labour Organization regulations) is an issue. Codes of ethics and corporate social sustainability have been developed to address these issues, but their dissemination needs to be supported and monitored.

The European Commission has engaged in similar work: the Tourism Sustainability Group (TSG) was established in 2004 to support sustainability in European tourism. Its conclusions and recommendations were published in February 2007 and will serve as a basis for communicating an agenda for the sustainability of European tourism (see Tourism Sustainability Group, 2007).

The multi-layered governance structure

The UNEP/UNWTO agenda requires structures and institutions. The trend towards decentralisation, as well as the treatment of broader issues related to development, employment or the environment, means that the relevant institutions have to adapt their governance. The industry structure should be taken into account when considering policies to make the most of the economic and social impact of tourism on business and the local community, while at the same time minimising the adverse effects on the environment.

The rapidly changing structure of tourism-related industries and the dual nature of the industry (comprising some large multinational enterprises but a majority of small and medium-sized enterprises, SMEs) have paved the way for new modes of co-operation and participation in supply chains and distribution networks. Co-operation between companies and destination governments is playing a growing part. Thus national, regional and local authorities can play an important role in enhancing the development and diffusion of tourism best practices and innovation, *e.g.* in the areas of environment, education, information and communication technologies, notably in small enterprises. An OECD report on trends in innovation and tourism policies (OECD, 2005b) illustrates the need to push the diffusion of innovative practices, and the part played by competition and co-operation to stimulate structural change and innovation in a fragmented industry.

Sustainability at the destination relies on the capacity of those involved to work together. Stakeholders include the state, local jurisdictions and communities, as well as the business sector, whether international tour operators or SMEs. Agenda 21 and local charters, although they are non-binding documents, can foster forward-thinking dialogues among these groups.

A consistent set of policy instruments

Measurement instruments

Indicators and data are used to measure the environmental impacts of tourism and to support outlooks and anticipations. They play a key part in designing, implementing and enforcing sustainable tourism policies at the destinations. UNWTO (2004) has published a guidebook on this subject.

However, a major constraint to sustainable tourism is the inadequate knowledge of the interactions between tourism and the environment. Considerable research is required to develop the knowledge base needed to underpin a sustainable tourism industry.

Regulatory instruments

Typical regulatory instruments in the industry are licensing for tourism enterprises (*e.g.* tour operators), land use planning and development control. These will remain important for governments' capacity to control operations by the business community. Land use planning can be used to take account of the value added by the environment to the activity and the environmental impact of the activity.

Economic instruments

Economic instruments can be used to internalise the measurable externalities of tourism. However, the sector does not make sufficient use of such incentives.[8] Relevant instruments could include:

- Reviewing capital investment programmes for tourism development and tourism-related infrastructure; in particular, public investment in infrastructure (such as transport, water supply and sanitation) can be used to make tourism more sustainable, if costs are shared according to the externalities of the public good.

- Price incentives, to ensure that the cost of an activity includes the positive/negative externalities. In some cases, admission fees would make it possible to reduce impacts on protected and/or sensitive areas and to generate revenues which could be redistributed to protect the environment.
- Fines for illegal activities in protected zones (*e.g.* illegal camping or picking flowers).
- Subsidies in the field of tourism development; they too often fail to take the environmental and social dimension into account, or do so insufficiently.

Voluntary agreements and eco-labelling schemes

Tourism firms, and especially the international tourism industry, have come to realise that the environment is an essential resource for the growth of the industry. As a result, major international investors have exerted considerable pressure on destination countries to make their tourism products greener in response to demand.

The International Tourism Partnership is an example of a voluntary initiative in the industry. It claims that the industry as a whole needs to design, develop, refurbish and operate a new generation of tourism destinations which have a minimal ecological footprint and which support and strengthen the communities in which they operate. Such initiatives, however, often fail to reach the less elaborate tour operations, accommodation and services which are most frequently used worldwide.

Dissemination of technologies (*e.g.* sun and wind generated energy, co-generation, wastewater and sewage treatment plants and buildings designed for recycling) can contribute to the development of sustainable tourism. Case studies from Australia and around the world demonstrate that initial upfront costs can often be recovered from savings in reduced energy, water, waste disposal costs and improved staff morale and productivity (see UNEP/UNWTO, 2005). Promoting the financial benefits of implementing sustainable tourism is therefore an integral part of sustainable tourism programmes.

Certification schemes (such as Eco Management and Audit Scheme, ISO 14001) and eco-labelling (based on initiatives such as Global Reporting Initiative, Corporate Social Responsibility, or on more industry-specific codes of conduct) can help consumers choose sustainable tourism options and provide incentives to tour operators to ensure sustainability. A contribution to this is the creation of an international task force on sustainable tourism development (UNDESA/UNEP/UNWTO), chaired by France, within the framework of the UN Marrakech process on sustainable production and consumption patterns.

Transport and sustainable tourism policies

Addressing the environmental impacts of tourism requires giving appropriate attention to travel for tourism purposes (including impacts at tourism destinations) and, where necessary, to the need for co-ordination between travel service providers, tourism providers, tour operator associations, hotel operators, municipalities and public and private transport enterprises (bus, rail, car-sharing, taxis, etc.). In Germany, sustainable transport systems are part of a policy on sustainable tourism at all levels (federal, Land, local).

Instruments to mitigate the impacts of tourism travel include internalising environmental costs of all transport modes, including aviation, and increasing the availability and convenience of more environmentally friendly transport modes (see also

Box 19.9. **The potential of ecotourism**

According to the Quebec Declaration on Ecotourism,* ecotourism "embraces the principles of sustainable tourism... and.... contributes actively to the conservation of natural and cultural heritage, includes local and indigenous communities in its planning, development and operation, contributing to their well-being, interprets the natural and cultural heritage of the destination to visitors, lends itself better to independent travellers, as well as to organized tours for small size groups".

In a joint publication, UNEP and WTO (UNEP/UNWTO 2005) note that, as a development tool, ecotourism contributes to the three main goals of the Convention on Biological Diversity: conserve biological and cultural diversity, promote the sustainable use of biodiversity, and share the benefits equitably with local communities and indigenous people.

Ecotourism is a field for experiment and innovation. It is a growing niche market, but its elusive and multifaceted nature makes it difficult to measure its size and market share. An extremely rough estimate of the world's international ecotourism arrivals (notwithstanding the domestic visitors to natural areas) would be 7% of all tourism arrivals (Lindberg, quoted in UNEP/UNWTO, 2005), which is expected to amount to 70 million visitors in 2010.

Ecotourism raises a number of expectations. It generates risks as well, that the fragile ecosystems it is based on will be threatened by its very development, if not properly managed. This is one reason why ecotourism certification is a fundamental tool to ensure businesses are meeting ecotourism standards. Efforts in this direction have been led by Australia, which launched the first ecotourism specific certification programme in 1996.

* See *www.world-tourism.org/sustainable/IYE/quebec/anglais/declaration.html.*

Chapter 16 on Transport). Innovative projects have been identified, *e.g.* at the European Expert Conference on Environmentally Friendly Travelling in Europe, and recommendations have been drafted for future action by the transport sector, the tourism industry, the destinations, and policy-makers, to scale up these experiences (see European Expert Conference, 2006).

Sustainability certification schemes for the travel and tourism industry could make an important contribution towards improving tourism's environmental performance, but only if they also consider transport to, from and in the destination. Such schemes can include provision for travellers' participation in carbon-offset programmes, whereby revenues support projects in non-Kyoto countries – to avoid double-counting of emissions reductions – which reduce greenhouse gas emissions (*e.g.* through reforestation or installation of household biogas digesters).

UNWTO recommends *a)* providing incentives for tourists to use local public transport in tourism cities instead of personal cars; *b)* developing rail networks able to compete with air transport for short and medium distances; *c)* raising awareness about the consequences of travel; *d)* encouraging the further development of environmental voluntary initiatives and certification in the passenger transport sector (including transport to, at and from destinations); *e)* developing a set of indicators to monitor the impacts of tourism transport; and *f)* including transport in general tourism plans (see OECD, 2005a).

MINING

KEY MESSAGES

Small and medium-sized mining operators, particularly in developing countries, often lack the know-how and resources to apply sufficient health and environmental safeguards. As most of the additional production of mined materials to 2030 will originate in non-OECD countries, which often also have weak environmental policies, it is likely that the environmental impacts of mining may increase on average across countries.

Worldwide consumption of key mined commodities has been increasing steadily in recent years, and is expected to continue, due to strong demand in emerging economies. Production of mined metal commodities is expected to increase by about 250% to 2030.

There is potential for considerable environmental impact from most exploration, mining, and mineral processing, although significant progress has been achieved in developing ways of avoiding or reducing these impacts. Most of these impacts are local, but they also include climate change and loss of biodiversity.

The mineral and metal-intensity of OECD economies are continuing to decrease, reflecting a decoupling in the material intensity of the economy.

Policy options

- Implement policies to encourage more efficient use of minerals and metals, greater recycling and reuse of scrap metals, and substitution by other materials to further decrease the mineral and metal-intensity of economies.
- Address environmental impacts through national mining and environmental policies, as most impacts are local.
- Spread international best practices in the operation of mines more widely across the industry.
- Strengthen and support initiatives by the industry to develop and apply corporate governance approaches to the mining sector internationally.
- Work together to strengthen the capacity and institutional set-up for managing the environmental risks of rapidly expanding mining activity in developing countries. OECD countries can provide technical assistance and financial support where necessary.

Consequences of inaction

Without new policies, the environmental impacts of global mining activity per unit of output are likely to increase in the coming years. This is because the additional mining activity is expected to take place in countries with relatively lower environmental protection practices. To address this challenge, countries hosting the new mining activity might develop and implement best practice environmental and mining policies, and/or encourage use of corporate environmental best practices.

OECD ENVIRONMENTAL OUTLOOK TO 2030 – ISBN 978-92-64-04048-9 – © OECD 2008

Introduction

Mining operations worldwide are responsible for providing vast quantities of coal, metals and industrial materials for use in industrial processes, energy production and in consumer products. Without appropriate policies and precautions, mining operations can lead to negative impacts on the environment and on human health. Many of these impacts are local, while a few are global (such as climate change and loss of biodiversity).

While mining may have many potential environmental impacts (Box 19.10), it does not follow that all of these will occur – industry performance varies from responsible operations, concerned to minimise impacts as far as possible, to those that exhibit no concern at all.[9] With modern practices many of these effects can be avoided, or at least greatly reduced. Much of the damaging impact can be minimised through careful project planning, choice of appropriate mining technologies, and careful ongoing operation (UNEP, 1993).

Box 19.10. **Potential environmental impacts of mining**

Environmental impacts:	*Pollution impacts:*
• Destruction of natural habitat at the mining site and at waste disposal sites • Destruction of adjacent habitats as a result of emissions and discharges • Destruction of adjacent habitats arising from influx of settlers • Changes in river regime and ecology due to siltation and flow modification • Alteration in watertables • Change in landform • Land degradation due to inadequate rehabilitation after closure • Land instability • Danger from failure of structures and dams • Abandoned equipment, plant and buildings	• Drainage from mining sites, including acid mine drainage and pumped mine water • Sediment runoff from mining sites • Pollution from mining operations in riverbeds • Effluent from minerals processing operations • Sewage effluent from the site • Oil and fuel spills • Soil contamination from treatment residues and spillage of chemicals • Leaching of pollutants from tailings and disposal areas and contaminated soils • Air emissions from minerals processing operations • Dust emissions from sites close to living areas or habitats • Release of methane from mines

Source: UNEP (1993).

There are a number of phases in a mining operation which affect the environment in different ways (UNEP, 1993):

- **Exploration:** including surveys, field surveys, drilling and exploratory excavations. Some pollution can already be produced at this stage from land disturbance and from the wastes produced.
- **Project development:** includes development of the site by construction of roads and buildings, underground work on access tunnels, erection of treatment plants, overburden stripping and placing, preparation of disposal areas, construction of service

infrastructure such as power lines or generating plants, railways, water supplies and sewerage, laboratories and amenities.

- **Mine operation:** operations can be extremely varied, including underground mining, surface mining from open-pits or placer deposits, hydraulic mining in or near river beds. Newer processes may also include heap-leaching of tailing dumps, bio-leaching of surface heaps or deposits, and solution mining of buried deposits.
- **Beneficiation:** on-site processing may include comminution to reduce particle size, flotation using selected chemicals, gravity separation or magnetic, electrical or optical sorting and ore leaching with a variety of chemical solutions.
- **Associated transport and storage of ore and concentrates:** these may be a handling risk and can result in localised site contamination.
- **Mine closure:** this is an important and sometimes neglected aspect of mine operation. Rehabilitation is best done progressively rather than at the end of life of the mine, and accordingly needs to be a part of ongoing operation. While closure and rehabilitation are intended to mitigate environmental and social impacts, it is important that they do not create secondary effects such as excessive fertiliser use, spread of weeds, siltation and incompatible landscape features. Ongoing monitoring and maintenance may be important in some situations.

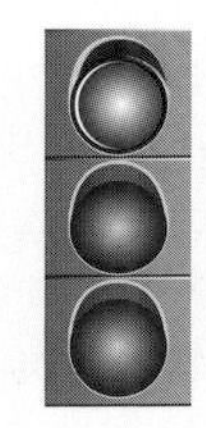

Small-scale mining operations, particularly in developing countries, often lack the know-how and resources to apply sufficient health and environmental safeguards.

The large multinational mining companies have been making significant progress in applying management methods and technologies that minimise the environmental impacts of mining. These corporations explore, mine, smelt, refine, and sell metals on world markets. About 30-40 companies are in this category. However, there are many actors in the industry that do not perform to these best practices, in particular some of the "junior" and "small-scale" miners.

In many instances, junior companies find new ore bodies and sell them on to larger companies. Intermediates offer growth potential through mergers between themselves or by being taken over by the largest corporations. Junior companies now spend more than 50% of the global exploration budget and their importance looks set to continue to grow. Artisanal and small-scale mining plays an important role in some minerals, especially gold and gemstones. These actors often lack know-how and resources to apply sufficient environmental and social safeguards.

Key trends and projections

Most of the increase in mining activity to 2030 is expected to take place in developing regions, due to the rapidly increasing demand in these economies and because of decreasing ore grades of marketable commodities in more mature mining regions (see Box 19.11). Already, China has become the world's largest miner or refiner of a number of metals (World Bank, 2006).

Global trends and the demand for mining products

There is enormous diversity in minerals, with mining commodities grouped into three broad categories: coal, metals and industrial minerals.[10] The production volumes and

Box 19.11. **Key uncertainties and assumptions**

The chapter relies on a number of assumptions. A key one is the continuation of current trends in the production and consumption of mined commodities. It is assumed that there are not going to be any major technological innovations leading to the massive substitution of mined commodities by other materials. This assumption is relatively robust, given the broad array of minerals that this chapter is considering. It is therefore assumed that the demand for mined minerals would grow in parallel with growth of GDP.

Another assumption is that the shift of minerals production away from the OECD towards less developed countries is going to result in a deterioration on average of the environmental performance of mine operators compared to today, as generally environmental standards in those countries tend to be lower for mining activities. This assumption could be proven wrong if large mining corporations in emerging markets embrace international corporate social responsibility standards more rapidly than expected.

dollar values of these minerals vary widely (see Table 19.6). It is estimated that the production of aggregates and construction materials exceeds 15 billion tonnes per year (2000). This is followed by coal mining, with 4.973 billion tonnes in 2005. Of the metalliferous ores, iron (used mainly in the form of steel) is the largest volume.

Table 19.6. **Production and prices of some major mineral commodities, 2000-2005**

Mineral commodity	2000 production[a] (thousand tonnes)	Price 2000[b] (USD/tonne)	2005 production[c] (thousand tonnes)	Price 2005 (USD/tonne)	Annual value (USD million)
Finished steel	762 612	300	1 012 000[d]	n.a.	n.a.
Coal	3 400 000	40	4 973 000[e]	99[f]	492 327
Primary aluminium	24 461	1 458	31 900	2 007.52[g]	64 039
Refined copper	14 676	1 813	15 000	3 681.72[h]	55 225
Gold	2 574	8 677 877	2 470	12 979 166.67[i]	32 058
Refined zinc	8 922	1 155	9 800	1 383.91[j]	13 611
Primary nickel	1 107	8 642	1 490	14 744[k]	21 968
Phosphate rock	141 589	40	147 000	27.76[l] (2004 price)	4 108
Molybdenum	543	5 732	185	71 672.28[m]	13 259
Platinum	0.162	16 920 304	0.239	21 145 333[n]	5 053
Primary lead	3 038	454	3 270	976[o]	3 191
Titanium minerals	6 580	222	5 200	n.a.	n.a.
Fluorspar	4 520	125	5 260	n.a.	n.a.

StatLink http://dx.doi.org/10.1787/257703582223

a) Source: CRU International (2001), *Precious Metals Market Outlook*, CRU International, London.
b) Source: CRU International (2001), *Precious Metals Market Outlook*, CRU International, London.
c) US Geological Survey, *Mineral Commodity Summaries*, January 2007 unless otherwise noted.
d) *www.unctad.org/infocomm/*.
e) *www.worldcoal.org/pages/content/index.asp?PageID=188*.
f) Teck Cominco Limited (2005), Annual Report, Vancouver, *www.teckcominco.com*.
g) *http://minerals.usgs.gov/minerals/pubs/commodity/aluminum/alumimyb05.pdf*
h) Teck Cominco Limited (2005), Annual Report, Vancouver, *www.teckcominco.com*.
i) Teck Cominco Limited (2005), Annual Report, Vancouver, *www.teckcominco.com*.
j) Teck Cominco Limited (2005), Annual Report, Vancouver, *www.teckcominco.com*.
k) *www.outokumpu.com/29679.epibrw*.
l) *http://minerals.usgs.gov/minerals/pubs/commodity/phosphate_rock/phospmyb04.pdf*.
m) *www.outokumpu.com/29679.epibrw*.
n) *www.kitco.com/scripts/hist_charts/yearly_graphs.plx*.
o) *www.xstrata.com/annualreport/2005/review/page67*.

The prices of minerals and metals vary widely, affecting demand for their production and substitution between materials (Table 19.6). Platinum prices averaged nearly USD 26 million per tonne in 2005, while coal averaged USD 99 per tonne. While low value minerals (per unit of weight), such as sand, gravel and stone, are mainly marketed locally, high value minerals are sold in the global market.[11] Finished steel is the largest mineral commodity traded in sales value, followed by coal. These are the only minerals or metals for which the value of sales exceeded USD 100 billion in 2005. Copper, aluminium, zinc and gold were all in the USD 10-100 billion range, while fluorspar, at the low end, was well below USD 1 billion in value (Table 19.6). International metals prices have risen substantially in the last three years and are at all time highs in nominal terms, and in some cases match or exceed the highest real levels seen in the last 30 years. Prices have been driven up by strong global economic growth, and particularly strong metal demand in China that caught the industry by surprise (World Bank, 2006).

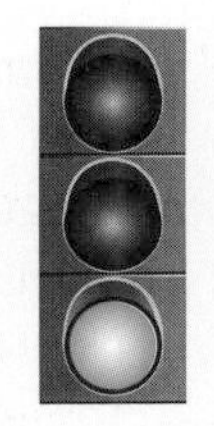

The mineral-intensity of OECD economies is continuing to improve.

In the last four decades, production of the six major industrial metals[12] grew on average by about three-and-a-half times. More recently growth has varied from 2.1% to 3.9% per annum. This growth is expected to continue in the future, despite consumption growing more slowly or levelling off in most OECD countries. This is because most of the increase in demand for metals in the future will come from rapidly industrialising developing countries, continuing recent trends (World Bank, 2006). For instance, over the 15 years since 1990, Chinese metals demand growth has averaged 10% a year and in the last five years it has accelerated to 17% a year. For a number of metals, China accounted for 70% or more of global demand growth in the last five years, and the country is now the world's single largest user of almost all metals (World Bank, 2006).

Several studies have suggested that the intensity of use of a mineral (the use of a mineral commodity divided by GDP) depends on the level of economic development as measured by GDP per capita, and that the pattern of intensity of use follows an inverse U-shape as economies develop (Malenbaum, 1975; Altenpohl, 1980; Tilton, 1990). As development takes place, countries focus on building infrastructure (such as rails, roads, and bridges, housing and other buildings and water supply and electricity transmission) and people buy more durable goods, which rapidly increases the demand for mineral commodities. As economies mature, all other things being equal, they move to a less materials-intensive phase, spending more on education and other services, which reduces the intensity of minerals use. Other factors that affect the intensity of use include government policies, shifts in demographics, materials substitution and new technologies.

Empirical research in resource economics has shown that metal use intensity (defined as metal consumption per unit of GDP) is also a function of per capita income. This function varies across countries and materials, but again often follows an inverse U-shaped curve. Metal requirements change in different phases of economic development – from agriculture-based economies (low intensity), to manufacturing-based economies (high intensity), to services-based economies (low intensity) (Tilton, 1986). They also change following substitution with other materials, or changes in metal requirements as a result of technological development, leading to more efficient raw materials use in the production of final goods (Bernardini and Galli, 1993). As a result of these trends, the inputs in metals required for the production of one unit of GDP have been constantly decreasing in the OECD, reflecting a decoupling in the material intensity of the economy.[13]

At the same time, many developing countries with large populations have recently been accelerating their economic growth, moving from agriculture to more manufacturing-based activities, which has led to a strong increase in the demand for metals. For instance, the annual per capita usage of refined copper was less than a kilo in India in 2003, while it was about 10 kg in Japan and other OECD countries. In India, copper-intensive applications such as the telecommunications industry are predicted to grow by a factor of 10 over 2000 levels (Mining Minerals and Sustainable Development, 2002). Another example is aluminium, of which Africans currently use only 0.7 kg per capita per year, compared with 22.3 kg in the US. Expectations are similar for many other mined commodities. Over the next 25 years, the World Bank expects that China's demand for metals could grow to two to four times current levels depending on the metal, implying annual increases in demand of about 2.5% to 4.8% (World Bank, 2006).

Hence, if current trends continue (Table 19.7), global extractive activity is expected to increase by a factor of 2 5 to 2030 under the *OECD Environmental Outlook* Baseline, roughly in line with projected growth in world GDP. Growth in trade of metals and minerals will be strongest in the BRIICS[14] countries, with imports increasing by a factor of six by 2050, while imports in the OECD are predicted to "only" double.[15]

Table 19.7. **Trends in the production of metals, 1995 to 2005**

	Production 1995[a] (thousand tonnes)	Production 2005[b] (thousand tonnes)	Average annual growth in production 1995-2005 (%)
Copper	10 000	15 000	4.14
Aluminium	19 400	31 900	5.10
Iron ore	1 000 000	1 540 000	4.41
Lead	2 710	3 270	1.90
Nickel	1 040	1 490	3.66
Silver	14.6	19.3	2.83
Tin	194	290	4.10
Zinc	7 120	9 800	3.25

StatLink http://dx.doi.org/10.1787/257774142575

a) US Geological Survey *Commodity Statistics and Information*, *http://minerals.usgs.gov/minerals/pubs/commodity/* 1997 statistics.

b) US Geological Survey *Commodity Statistics and Information* *http://minerals.usgs.gov/minerals/pubs/commodity/* 2007 statistics.

Source: US Geological Survey US Geological Survey *Commodity Statistics and Information*, *http://minerals.usgs.gov/minerals/pubs/commodity/*.

Global trends and the environment

As the *OECD Environmental Outlook* estimates that the demand for metals and other mined commodities will more than double over the next 25 years, significant additional pressures on the environment are to be expected from the sector, due to the simple expansion of the scale of mining operations that will be needed to meet steeply increasing world demand.

The location of future mine production will be determined by the economic geological resource base, but other factors, such as investors' capacity to access resources, government policies and so on will also play a part. As high grade ore deposits in the OECD are being depleted, and environmental regulation becomes more stringent, mineral deposits in developing and transition countries become more competitive (IIED, 2002).

While traditional mining centres in Australia and North America – which currently account for 30-40% of mine production and exploration – will continue to play an important role, other parts of the world are likely to gain in importance. Already, China has become one of the largest metals producers, with 17% of world production in 2005. Africa's share of production is also likely to expand significantly, as suggested by data on planned projects and exploration spending (World Bank, 2006).

Increases in the production of industrial minerals – such as sand, stone and gravel, which are too expensive to transport over long distances – are expected to occur mostly in rapidly developing economies, where most of the demand will be generated. Hence, the environmental impacts that are linked to mining activity will increasingly occur outside of OECD countries.[16]

Some mineral commodities can be recycled. Recycling reduces the demand for primary metals and requires considerably less energy than producing primary metal (see also Chapter 11 on waste and material flows). For example, scrap aluminium requires about 5% and scrap steel about 25% of the energy required to produce primary metals. Already, about 50% of total steel use is being derived from recycled material and the situation is similar for other metals. Overall recovery of lead in the US stands at about 55%. But for most minerals, at least in the medium term, while the overall demand for mineral products continues to rise, the effect on primary production of increased recycling is likely to be minimal due to limited supply of the secondary materials, and hence the potential for avoiding mining related environmental impacts through improved recycling policies is also limited (IIED, 2002).

The environmental impacts of global mining activity are likely to be more than proportional to growth in production, unless countries hosting the additional mining activity develop and implement best practice environmental and mining policies and/or corporate environmental best practices spread to a much wider array of mining companies.

Policy implications

Most environmental impacts of mining are local and need to be addressed in the framework of national mining and environmental policies in the host countries. The rapid expansion of mining activity in developing countries constitutes an important challenge. Host governments will need to put in place policies to strengthen their capacity and institutional set-up to effectively manage the environmental risks associated with this development. OECD countries can support them by providing technical assistance and financial support where necessary.

It is also to be expected that an increasing share of operations will be overseen by companies headquartered in the countries generating much of the additional demand for mined commodities (*i.e.* China and India). Those involved in mining from these regions should be included in voluntary approaches, such as the OECD's Guidelines for Multinational Enterprises or the UN's Global Compact.[17] This would be a useful complement to efforts to improve national mining and environmental policies in host countries. This is particularly relevant when mining takes place in "weak governance zones", where national mining and environmental policies are non-existent or not properly implemented.

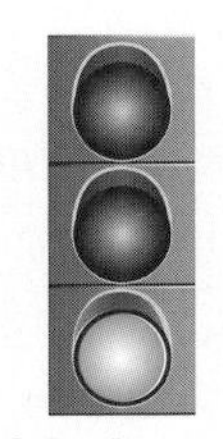

Some of the larger mining companies are working together towards better environmental practices in mining operations.

In addition to policies governing activities in the mining sector within countries, a number of large mining companies internationally are working to strengthen corporate governance in the sector, including environmental management (Box 19.12).

Box 19.12. **Corporate governance in the mining sector**

Many large mining corporations are acknowledging the fact that successful company strategies need to integrate the concepts of sustainable development into core business practice. In 2000, nine of the largest mining companies decided to initiate a project to examine the role of the minerals sector in contributing to sustainable development and explore how that contribution could be increased. Through the World Business Council for Sustainable Development, they contracted with the International Institute for Environment and Development to undertake a two-year independent process of research and consultation: the Mining, Minerals and Sustainable Development Project (MMSD; IIED, 2002). The work was presented at the World Summit on Sustainable Development in Johannesburg in 2002. It lays out a vision of how the sector would look if it were to maximise its contribution to sustainable development. In 2001, the International Council for Mining and Metals (ICMM), representing 16 of the largest corporations and associations, was formed to take forward the agenda identified in this report.

Through 10 mandatory sustainable development principles (*www.icmm.com/icmm_principles.php*), ICMM members are committed to continual improvement of their performance. Members have been given numerous awards by government agencies and other independent bodies. ICMM has published guidance to assist members to improve their performance in several areas, including most recently, biodiversity. This guidance was prepared with the assistance of and through a dialogue with the World Conservation Union (IUCN).

The member companies of ICMM have committed themselves to implementing a sustainable development framework comprising 10 principles, reporting in accordance with the Global Reporting Initiative framework (including a Mining and Metals Sector Supplement, which was developed jointly by ICMM and the GRI), and independent assurance.

Notes

1. Value added equals gross production in the sector, minus the use of intermediate products.
2. The industrialised nations who have specific emission limits agreed under the Kyoto Protocol (see Chapter 7 on climate change).
3. Available on the European IPPC webpage, at *http://eippcb.jrc.es/pages/FActivities.htm*; the initial document was adopted by the European Commission in 2001.
4. Part of Chinese consumption is due to the boom in manufacturing and export of goods to the US and Europe.
5. This paragraph relies on documents presented by the FAO Advisory Committee on Paper and Wood Products, 47th Session, 6 June 2006.
6. See World Bank, 1998 and latest information at *www.ifc.org/ifcext/enviro.nsf/Content/EnvironmentalGuidelines#note*.
7. Currently exotic but potentially high-impact concepts such as space-tourism are not discussed here, although they may become relevant by 2030.
8. According to most chapters on tourism in the OECD country Environmental Performance Reviews.
9. *www.mineralresourcesforum.org/aboute.htm#Overview*.
10. Iron, copper, lead and zinc, gold and silver are metals. Potash, soda ash, borates, phosphate rock, limestone, and other crushed rock are grouped into industrial minerals. See: *www.eere.energy.gov/industry/mining/pdfs/overview.pdf*.
11. This section focuses on mined commodities that are traded internationally, due to data limitations on minerals that are mostly produced and used locally, such as construction materials.

12. Aluminum, copper, lead, nickel, tin and zinc.
13. There are questions, however, as to what extent the reduction in metals intensity is linked to shifts in production and environmental burden to less developed countries, from where manufactured products are exported to the OECD.
14. Brazil, Russia, India, Indonesia, China and South Africa.
15. It should be noted though that predicting future demand for mined commodities over such a long time frame is very difficult, since technological innovations and substitution of materials are impossible to predict (see Box 19.11 for more).
16. It should be noted that while mining generates environmental impacts, it also creates opportunities for economic growth, and therefore might have an overall positive effect on social welfare in these countries (see ICMM's Resource Endowment Initiative, *www.icmm.com*).
17. *www.oecd.org/daf/investment/guidelines.*

References

Alliance for Environmental Technology (2005), *Trends in World Bleached Chemical Pulp Production: 1990-2005* (available at *http://aet.org/science_of_ecf/eco_risk/2005_pulp.html*), AET, Melbourne.

Altenpohl, D.G. (1980), *Materials and World Perspective. Assessment of Resources, Technologies and Trends for Key Materials Industries*, Springer Verlag, Berlin.

Australian Conservation Foundation (1992), *Pulp and Paper Mills for Australia*, Policy Statement No. 50, ACF, Melbourne (available at *www.acfonline.org.au*).

Australian Government (2004), "Two Way Track. Biodiversity Conservation and Ecotourism: Investigation of Linkages, Mutual Benefits and Future Opportunities", *Biodiversity Series*, Paper No. 5, Department of the Environment and Water Resources, Canberra.

Bernardini, O. and R. Galli (1993), "Dematerialisation: Long-term Trends in the Intensity of Use of Materials and Energy", *Futures*, 25, 431-447.

CEPI (Confederation of European Paper Industries) (2005), *Sustainability Report*, CEPI, Brussels.

CEPI (2006a), *Recovered Paper Quality Control*, CEPI, Brussels.

CEPI (2006b), *Position Paper on Forest Certification*, CEPI, Brussels.

CEPI (2007), *Sustainability Newsletter*, CEPI, Brussels.

CEPI/WWF (2006), *WWF and CEPI Recommendations for an Effective Implementation of European Renewable Energy Sources* (RES) Policies, CEPI, Brussels.

EC-DG Environment (2000), *EU Energy Policy Impacts on the Forest-based Industries*, Nangis, Wageningen, Netherlands.

EEA (European Environment Agency) (2006), *How Much Bioenergy Can Europe Produce Without Harming the Environment?* EEA, Copenhagen.

Ernst and Young (2007), "At the Crossroads", *Global Pulp and Paper Report 2007*, Ernst and Young, Helsinki.

ERPC (European Recovered Paper Council) (2006), *European Declaration on Paper Recycling*, European Recovered Paper Council, Brussels.

European Expert Conference (2006), *Environmentally Friendly Travelling in Europe – Challenges and Innovations Facing the Environment*, Vienna, 30-31 January 2006; documentation available at *www.eco-travel.at/english/kongress.php*.

FAO (1996), "Environmental Impact Assessment and Environmental Auditing in the Pulp and Paper Industry", *FAO Forestry Paper*, No. 129, FAO, Rome.

Forest-Based Sector Technology Platform (2006), FTP, Brussels (available at www.forestplatform.org/).

French Directorate for Tourism (2006), *Tourisme Info Stat*, n°2006-4, French Directorate for Tourism, Paris, information available at *www.veilleinfotourisme.fr*.

Gössling S. (2002), "Global Environmental Consequences of Tourism", *Global Environmental Change*, 12, 283-302.

Hufbauer, G.C. and B. Goodrich (2001), *Steel: Big Problems, Better Solutions*, Policy Brief 01-9, Institute for International Economics, Washington, DC. Available at *www.iie.com/publications/pb/pb.cfm?ResearchID=77.*

ICFPA (International Council of Forest and Paper Associations) (2006), *Forest Industry Leaders Commit to Action on Global Sustainability* (available at *www.icfpa.org*).

IIED (International Institute for Environment and Development) (2002), *Breaking New Ground – Mining Minerals, and Sustainable Development*, IIED, London.

IISI (International Iron and Steel Institute) (2008), personal communication, 18 January 2008.

Jokinen, J. (2006), *Energy-efficiency Developments in the North and South American and European Pulp and Paper Industry*, presentation at an IEA Workshop, October 9, Pöyry Forest Industry Consulting, Montreal, Canada.

Malenbaum, W. (1975), *WorldDemand for Raw Materials in 1985 and 2000*, McGraw-Hill, New York.

Mining Minerals and Sustainable Development (2002), *Breaking New Ground – Mining Minerals, and Sustainable Development*, London.

OECD (2003), *Environmental Policy in the Steel Industry: Using Economic Instruments*, OECD, Paris, available at *www.oecd.org/dataoecd/58/20/33709359.pdf*.

OECD (2004), *Tourism Policies and Environmental Integration*, OECD, Paris.

OECD (2005a), *Leisure Travel, Tourism Travel and the Environment*, ENV/EPOC/WPNEP/T(2005)1, ENV/EPOC/WPNEP/T(2005)2, OECD, Paris.

OECD (2005b), *Trends in Innovation and Tourism Policies*, OECD, Paris.

OECD (2005c), *The Competitiveness Impact of CO_2 Emissions Reduction in the Cement Sector*, OECD, Paris, available at *http://appli1.oecd.org/olis/2004doc.nsf/linkto/com-env-epoc-ctpa-cfa(2004)68-final*.

OECD (2006), *Climate Change in the European Alps: Adapting Winter Tourism and Natural Hazards Management*, OECD, Paris.

Tilton, J.E. (1986), "Atrophy in Metal Demand", *Materials and Society*, 10. 241-243.

Tilton, J.E. (1990), *WorldMetal Demand, Resources for the Future*, Washington DC.

Tourism Sustainability Group (2007), *Action for More Sustainable European Tourism*, Tourism Sustainability Group, Brussels. *http://ec.europa.eu/enterprise/services/tourism/doc/tsg/TSG_Final_Report.pdf*.

UNCTAD (United Nations Conference on Trade and Development) (2006), *Dealing with Trade Distortions in the Steel Industry*, UNCTAD, India Programme, New Delhi. Available at www.unctadindia.org/displaymore.asp?subitemkey=421&itemid=310&subchnm=59&subchkey=59&chname=Other.

UNECE/FAO (United Nations Economic Commission for Europe and UN Food and Agriculture Organization) (2005), *European Forest Sector Outlook Study 1960-2000-2020*, United Nations Economic Commission for Europe, Geneva.

UNEP (United Nations Environment Programme) (1993), *Pollution Prevention and Abatement Guidelines for the Mining Industry*, UNEP, IE/PAC, Paris. *www.mineralresourcesforum.org/docs/pdfs/minguides.pdf*.

UNEP (2005), *Sustainable Use of Natural Resources in the Context of Trade Liberalization and Export Growth in Indonesia*. A Study on the Use of Economic Instruments in the Pulp and Paper Industry, United Nations Environment Programme, Geneva.

UNEP/UNWTO (2005), *Making Tourism More Sustainable. A Guide for Policy Makers*, United Nations Environment Programme and the United Nations World Tourism Organization, UNWTO, Paris, France and Madrid, Spain.

UNWTO (United Nations World Tourism Organization) (2001), *Tourism 2020 Vision. Global Forecasts and Profiles of Market Segments*, UNWTO, Madrid, Spain.

UNWTO (2004), *Indicators of Sustainable Development for Tourism Destinations*, UNWTO, Madrid, Spain.

UNWTO (2005a) *Tourism Market Trends 2004*, UNWTO, Madrid, Spain.

UNWTO (2005b), *Harnessing Tourism for the Millennium Development Goals*, UNWTO, Madrid, Spain, available at *www.world-tourism.org/sustainable/doc/decla-ny-mdg-en.pdf*.

World Bank (1998), *Pollution Prevention and Abatement Handbook*, The World Bank, Washington, DC.

World Bank (2006), *The Outlook for Metals Markets*, Background paper prepared for G20 Deputies Meeting in Sydney, September 2006, World Bank, Washington DC. *http://siteresources.worldbank.org/INTOGMC/Resources/outlook_for_metals_market.pdf*.

World Travel and Tourism Council (2005), "China Travel and Tourism. Sowing the Seeds of Growth", *The 2005 Travel and Tourism Economic Research, WTTC, London*.

WWF (Worldwide Fund for Nature) (2006), *The Ideal Corporate Responsibility Report*, WWF International, Gland, Switzerland.

IV. PUTTING THE POLICIES TOGETHER

20. Environmental Policy Packages

21. Institutions and Approaches for Policy Implementation

22. Global Environmental Co-operation

ISBN 978-92-64-04048-9
OECD Environmental Outlook to 2030

Chapter 20

Environmental Policy Packages

This chapter examines how different types of policy instruments can be combined into an instrument mix to tackle environmental problems. It examines the benefits of combining instruments in a mix, and some of the challenges to avoid in terms of potentially overlapping or conflicting policy instruments. The chapter also examines a broad policy package to address a number of the key environmental challenges outlined in the Outlook report. It finds that significant environmental improvement can be achieved at relatively low cost to the economy, if the right mix of policies is used.

KEY MESSAGES

- Significant environmental improvement can be achieved at relatively low cost to the economy and with little negative social impact if the right mix of policies is used. The necessary policies and technological solutions to tackle the key environmental challenges are both available and affordable.
- There is no silver bullet – a package of policies will be needed to tackle the environmental challenges identified in the first part of this *Outlook*. Even for a single environmental problem, an instrument mix may be needed given the often complex and inter-connected nature of many environmental challenges, the often large number and variety of sources exerting pressure on the environment, and the many market and information failures.
- Instrument mixes need to be carefully constructed to ensure that they achieve a given environmental goal in an effective and economically efficient manner, while providing consumers and producers with flexibility in how they meet the targets, so as to enable innovation. Social or equity impacts should be addressed. Instrument mixes should provide clear, short- and long-term policy signals to support appropriate investment decisions.
- The policy instruments used in a mix should be complementary and reinforcing, rather than duplicative or conflicting. The net effect of an instrument mix needs to be considered, not just the effect of individual instruments. This can identify positive ancillary benefits from policy instruments, or whether an instrument has instead simply shifted the problem or exacerbated other environmental problems.
- Environmental policy evaluations should be integrated into the cycle of policy design, implementation and reform.

What would be the environmental and economic impacts of a global policy package?

This chapter simulates the impacts of an *OECD Environmental Outlook* (EO) policy package, designed to tackle some of the key environmental challenges identified in the *Outlook*. Its impacts would include:

- Global emissions of nitrogen and sulphur oxides that are 31% and 37% less in 2030 respectively than under the Baseline (and about one-third less than 2005 levels).
- Greenhouse gas (GHG) emissions that are 38% less in 2050 than under the Baseline.

Difference in key environmental variables between the Baseline and the EO policy package in 2030 (2050 for GHG)

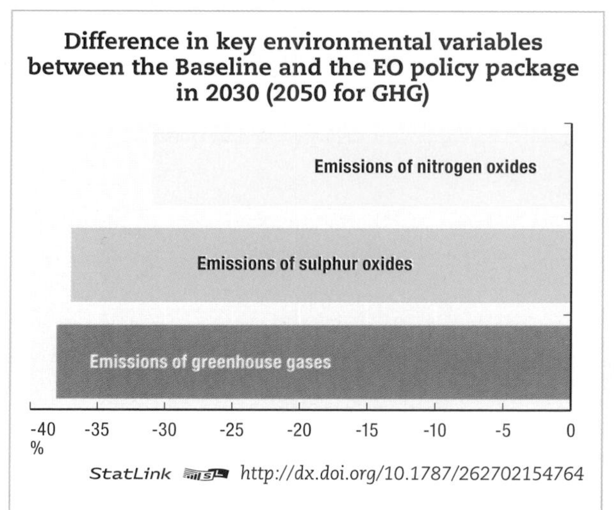

StatLink http://dx.doi.org/10.1787/262702154764

- A reduction in projected GDP growth to 2030 of only about 0.03 percentage points per year.

Introduction

Preceding chapters of the *Outlook* have shown that without more ambitious policy action, increasing pressures on the environment could cause irreversible damage within the next few decades. The consequences and costs of environmental policy inaction are high (see also Chapter 13).

The often complex nature of the environmental problems identified in this *Outlook*, the "red light" issues that require urgent attention, and the large number of actors that contribute to these across the production and consumption chain, mean that a single policy instrument is unlikely to tackle them successfully. Instead, some form of mix of policy instruments will be required to deal with most environmental problems. The first part of this chapter examines how mixes of policy instruments can be designed and implemented to effectively tackle environmental problems.

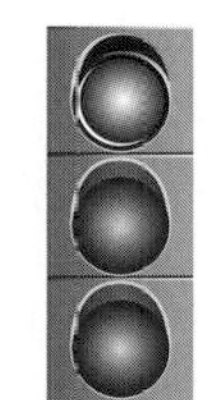

Ambitious policy action is needed to prevent irreversible environmental damage. The complexity of many environmental challenges means that a mix of policy instruments will be needed.

At the same time, given the inter-connected nature of many environmental pressures, the mixes of policy instruments developed to tackle individual environmental problems will often interact with each other. In some cases they can be mutually supportive, possibly providing co-benefits of enhanced or less costly achievement of a given goal; in other cases they can counteract or duplicate each other. The second part of this chapter examines how a broad package of policy instruments, designed to address a number of the main environmental challenges identified in this *Outlook*, might work together.

Designing and implementing effective mixes of policy instruments

A range of policy instruments is already in use, often in combination with each other (see Box 20.1). While some of these instruments are under the purview of environment ministries, others are applied under the responsibility of sectoral or economic ministries.

A well-designed instrument mix can be both environmentally effective and economically efficient. For instance, for environmental challenges where there are information failures, environmental taxes can be effectively combined with eco-labelling schemes and other information-based measures, for example to provide information on the fuel efficiency of different vehicles. Similarly, while a "cap and trade" system of tradable permits can achieve an environmental goal with fairly low compliance costs, if there is considerable uncertainty about abatement costs, it may be useful to combine it with a "tax" which sets an upper limit on permit prices. In situations where environmental impacts differ greatly depending upon the location of the emission, tradable permits combined with location-specific performance standards can be a useful policy mix. The

Box 20.1. **Policy instruments for environmental management**

- *Regulatory instruments: e.g.* bans on certain products or practices, emission standards, ambient quality standards, technology standards, requirements for the application of certain ("best available") technologies, obligations for operational permits, land use planning and zoning, etc. These are used to address a broad range of environmental problems and can make environmental outcomes more certain, but they can be less economically efficient than market based instruments, do not always provide incentives for technological innovation and often have significant information requirements for their design. Regulations that target an environmental outcome, rather than, for example, specifying the technologies to be used, give firms more flexibility in seeking low-cost abatement policies.
- *Environmentally-related taxes:* help to ensure that prices reflect the negative environmental externalities of various products and processes. In the short term they provide incentives for polluters and resource users to change their behaviour, while in the longer term they encourage innovation in the development of new production methods and new products that meet consumer demand while reducing damage to the environment. They can be economically efficient and are less demanding in terms of information needs than regulations, but have a lower degree of certainty on environmental outcomes. There is high potential for wider use of environmental taxes, although they need to be well-designed and their potential impacts on international competitiveness and income distribution identified and addressed, as appropriate. Exemptions or reduced tax rates should be scaled back to increase the effectiveness and economic efficiency of existing environmental taxes. Closely linked to environmental taxes are prices, fees and charges for various environmentally related services (*e.g.* waste collection, water supply, waste water treatment, energy supply).
- *Tradable permit systems:* these set limits, either as a maximum ceiling for "cap and trade" schemes, or as a minimum performance commitment for "baseline and credit" schemes. The limits can be either in absolute terms or in relative terms, and the permits can be denominated either in terms of rights to emit pollutants (*e.g.* greenhouse gas emissions), or rights to access natural resources (*e.g.* water, fish stocks). Cap and trade systems can combine a high degree of environmental certainty with economic efficiency – their flexibility helps to reduce the costs of abatement, while their environmental effectiveness is high because the environmental objective is explicitly reflected in the number of permits that are issued. But the transaction costs can also be high, and decisions are needed on politically-sensitive issues such as which activities or sectors will be covered, and the initial allocation of permits.
- *Voluntary approaches:* such approaches include environmental agreements negotiated between industry and public authorities, as well as voluntary programmes developed by public authorities in which individual firms are invited to participate. However, while the environmental targets of most existing voluntary approaches do seem to have been met, there is little evidence of situations where such approaches have contributed to environmental improvements that are significantly different from what would have happened anyway. They can be useful in raising awareness and in getting buy-in from business and industry on the need for action, but their environmental effectiveness has to be carefully assessed, and they need careful monitoring and reporting. They are most useful when used in combination with other policy instruments, or during a phase-in period for the use of another instrument.
- *Subsidies for environmental improvement:* A large number of subsidies are used for environmental policy purposes, *e.g.* to promote the diffusion of environmentally benign products, to reward environmentally friendly behaviour, to finance environmental infrastructure investments – *e.g.* in water supply and waste-water treatment – or to stimulate research and development of environmentally friendly technologies. But if not time-bound, subsidies can get locked-in, along with the (potentially inefficient) practices or technologies that they support. Providing subsidies to encourage compliance with direct regulations can result in significant economic distortions and strategic behaviour among firms. In general, it is better to tax environmental "bads", rather than to subsidise environmental "goods". However, public funding can be justified to support basic research and development.

Box 20.1. **Policy instruments for environmental management** *(cont.)*

- *Removal or reform of environmentally harmful subsidies:* Subsidies for various purposes are pervasive worldwide, with OECD countries currently transferring at least USD 400 billion to different economic sectors each year. Subsidies are costly to taxpayers or consumers. Subsidies distort prices and resource allocation decisions, altering the pattern of production and consumption, and as a result many subsidies can have unintentional negative effects on the environment. While scaling back subsidies has proven a difficult and long-term process in many countries, greater progress is being made in identifying and reforming the subsidies that are particularly environmentally harmful, trade distorting, and/or ineffective at achieving given social aims. For example, agricultural subsidies are moving away from the more environmentally damaging forms (which support over-production), towards subsidies which require farmers to undertake environmental practices.
- *Information-based instruments:* Good quality information is essential for identifying environmental challenges, better designing and monitoring the impacts of environmental policies, building support for these policies, and providing relevant information to inform consumption and production decisions. This includes a range of activities, such as environmental data collection and dissemination, development of indicators, environmental valuation, education and training, eco-labelling or certification schemes, Pollutant Release and Transfer Registers (PRTRs), etc. If designed properly, these can complement and strengthen the effectiveness of other policy instruments, such as environmental taxes.

environmental effectiveness of voluntary approaches can be enhanced if they are combined with the threat of a tax or regulatory measure if sufficient improvements are not realised.

Most environmental instrument mixes in use to date have evolved as the result of a succession of *ad hoc* decisions to adapt to evolving challenges and political demands. Only in a few cases have combinations of policy instruments been used in a fully articulated and coherent manner. If they are not carefully designed, however, instrument mixes can result in inefficiencies, redundancy (for instance, by targeting the same externality twice), and high administrative costs and complexities. There can be conflicts between policy objectives in an instrument mix as well, for example between policies to support agricultural production or input use, and taxes on fertilisers to encourage reductions in input use and nitrogen loading.

Most environmental policy mixes in place today have developed in an ad hoc *manner, and may contain overlapping or even conflicting policy instruments.*

Lessons learned from the implementation of instrument mixes

Recent work in the OECD has identified a number of key lessons for the successful implementation of environmental instrument mixes (OECD, 2007). From the perspectives of both environmental effectiveness and economic efficiency, policy instruments used to address a given environmental problem should be applied as broadly as possible (*e.g.* covering all relevant sectors of the economy and all countries). They should provide similar incentives at the margin to all sources of the environmental problem. Economic instruments (*e.g.* emission trading systems and taxes) can "automatically" provide equal marginal abatement incentives, while this is generally more difficult to achieve with regulatory instruments.

To maintain flexibility in the design of environmental policy, and to promote economic efficiency, policy-makers should set long-term targets (as opposed to short-term annual targets). These can provide clear signals for long-term investment decisions by business and consumers, for example in more efficient processing technologies, low-energy buildings, hybrid vehicles, insulation in housing, water saving appliances, etc.

For environmental problems that have many diffuse or varied sources (*e.g.* water pollution from agriculture), it can be appropriate to supplement instruments that address the total amount of pollution with instruments that address the way a certain product is used, when it is used, where it is used, etc. Regulatory instruments, information instruments, training, etc., can be better suited to address these latter dimensions than a tax or an emission trading system.

In some cases, combining two instruments can enhance the effectiveness and efficiency of both instruments. For example, a well-designed system for separate collection of recyclables can enhance the net environmental benefits of a variable waste collection charge, including by limiting the danger of illegal dumping of waste. The charge will make households more inclined to sort out recyclables that they can dispose of for free. In order to exploit such possibilities for mutual strengthening, instruments that provide as much flexibility as possible to the targeted groups should be used.

In some cases, policies used to address one environmental problem can have ancillary benefits in another environmental area at the same time. For example, some policies to tackle GHG emissions can also lead to ancillary benefits in terms of reduced air pollution emissions, and *vice versa*. In other cases, environmental policies can instead shift the environmental problem to another area or even exacerbate other environmental challenges.

A well-designed instrument mix can help to address some of the obstacles to successful implementation of environmental policies, such as concerns about their impact on low-income households, employment and industrial competitiveness (see Chapter 21). Care needs to be taken, however, to ensure that instrument mixes are not "over-burdened" with too many competing policy objectives.

Except for situations where mutual reinforcement between instruments is likely, or when the instruments address different "dimensions" of a given problem, policy-makers should generally avoid introducing overlapping instruments as such overlaps can reduce flexibility and create unnecessary administrative costs. It is often preferable to address non-environmental market failures (*e.g.* incomplete information, unclear property rights over natural resources, split incentives between landlords and tenants) with non-environmental instruments, such as competition policy instruments, improvements to patenting systems, etc., rather than adjusting environmental policy instruments to address these problems.

Assessing the impacts of policy mixes

The precise role of each instrument used, and its relationship with the other instruments which "target" the same environmental problem, need to be evaluated with care. Furthermore, to ensure policy coherence, existing market and intervention failures that exacerbate the environmental problem – such as environmentally harmful subsidies, distortionary tax provisions and inefficient, costly and contradicting policy instruments – should be removed.

Systematic evaluation is required to better plan, monitor and improve environmental policies over time. It encourages transparency and accountability within public administration, as well as being an important element of performance management. This includes both policy evaluations before the policy is put in place, as well as evaluations to verify the effects of the policy after it has been implemented in order to adjust it as necessary.

Policy decisions should be taken where possible based on information on the marginal benefits, costs and effects of alternative options for addressing a particular environmental problem. The costs and benefits of not taking policy action – *i.e.* policy inaction or delaying policy action – should also be considered (see also Chapter 13). As is clear from the first part of this *Environmental Outlook*, some of the environmental problems identified are worsening rapidly, and may be approaching thresholds beyond which there will be irreversible damage. In such cases, delaying policy action could be very costly to society as a whole.

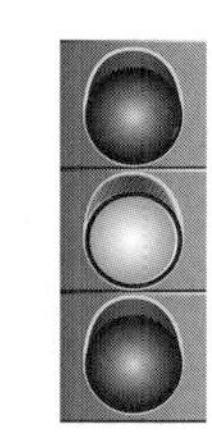

While cost-benefit assessments are increasingly used, most environmental policies are developed without their benefit.

The impacts of policy reform on specific groups in society – for example, low income households or workers in affected industries – are important to identify and address early in the policy-design process, in order to gain support for the policy measures (see also Chapter 21, Institutions and approaches for policy implementation).

OECD countries are progressively developing and using cost-benefit assessments, although in most cases environmental policies are fixed without their use (Pearce *et al.*, 2006). After the policy is in place, assessments of its actual impacts should be used to guide decisions to amend the policy, and can provide valuable information for future policy decisions. However, systematic and independent policy evaluations – particularly *ex post* evaluations – remain relatively rare. The systematic use of environmental policy evaluations should be integrated into the full policy design, implementation and reform process, as a means to ensure good governance.

Policy packages to address the key environmental issues of the OECD Outlook

While a mix of instruments may be used to tackle a single environmental problem, instrument mixes are rarely applied in a policy vacuum. Instead, they are likely to interact with other instruments in the broader package of environment-related policy measures in place. This is particularly true of policies targeted to the agriculture/land use/biodiversity nexus, or those addressing the climate change/air pollution/energy/transport nexus. A co-ordinated approach is needed to ensure policy coherence, to identify policies that are mutually reinforcing and to avoid those that are potentially conflicting or duplicative.

A number of specific policies and policy packages were examined using the modelling framework for this *OECD Environmental Outlook*, and the resulting impacts on environmental conditions, the economy and competitiveness assessed. The impacts on the environment and on GDP of specific policy simulations are described in the relevant chapters of this *Outlook*.[1]

This chapter does not reproduce that analysis, but instead simulates the impacts of combining some of these specific policies and some additional ones into a global policy package to address the key environmental challenges identified in this *Outlook*. Some of the relevant results of this EO policy package simulation are also reflected in other chapters, in particular Chapters 8 on air pollution, 10 on freshwater, and 12 on health and environment (see the Introduction to this report for a mapping by chapter of the descriptions of the policy simulations in the *Environmental Outlook*).

The global co-operation *Environmental Outlook* (EO) policy package includes:

- A 50% reduction of agricultural subsidies and tariffs worldwide, phased-in between 2010 and 2030 by a 3% decrease per year.[2]
- Application of a price on carbon across all sectors, via a carbon tax starting at USD 25 per tonne of CO_2eq, which increases in real terms by 2.4% per year. The carbon price was phased-in by region, starting in OECD countries in 2012, BRIC countries in 2020 and the rest of the world (ROW) in 2030.[3]
- Policies to bring forward the introduction and uptake of second generation biofuels, *i.e.* those using agricultural waste material or woody inputs developed on abandoned or marginal soils, rather than competing with agricultural land use.
- Regulatory policies that would move towards – but not necessarily reach – "maximum feasible reductions" (MFR) in air pollutant emissions, differentiated by region and sector (transport, power, refineries, industry). Measures are introduced in those sectors in which action will become cost-effective during the period to 2030, such as marine shipping. The policies are phased-in linked to economic growth; thus some low-income countries reach the target level long after the 2030 timeline for this *Outlook*.
- An increase in the rate of connections to public sewerage systems to close the gap by 50% between the level of connections in 2000 and connecting all urban dwellers with improved sanitation in 2030. For existing sewage treatment, treatment is upgraded to the next best level in terms of removal of nitrogen compounds.

This EO policy package does not attempt to reflect an "ideal" or "comprehensive" package of environmental policies. Instead it reflects a combination of a limited set of policies that: *a)* cut across many of the main environmental challenges identified in the *Outlook*; and *b)* can be simulated in the modelling framework used for the *Outlook*. It does not, however, include any policies explicitly aimed at protecting biodiversity[4] or enhancing agricultural technology uptake, for example. The EO policy package has been designed to be reasonably politically and practically realistic in terms of its scope, phasing the policies in over time, and with some consideration of regional capacity.

Environmental impacts of the EO policy package

Some of the most notable impacts of the EO policy package scenario compared with the Baseline are on air pollution (see Table 20.1). Given recent trends and existing policies, emissions of nitrogen oxides are projected to decrease slightly worldwide from 2005 to 2030 under the Baseline, but under the EO policy package they would decrease by almost one-third compared to 2005 levels. Similarly, the Baseline projects that emissions of sulphur oxides would increase by 5% from 2005 to 2030 worldwide; but under the EO policy package they would decrease by about one-third over this period. Particularly notable reductions in air pollution could be expected to be achieved with the EO policy package in North America (predominantly in Mexico), Eastern Europe, Russia, and China (see Figure 20.1). In some regions, little difference is expected between the air pollution emissions under the Baseline and those under the policy package. This in part reflects the time needed to replace more polluting technologies and infrastructure – decreasing emissions as a result of the policy package may only be seen after the period in this *Outlook* as capital stock is turned-over.

Table 20.1. **Change in selected environmental variables under the Baseline and EO policy package scenario**

Environmental variables	Baseline scenario % change 2005-2030	EO policy package % change 2005-2030	Projected difference in 2030 between Baseline and the EO policy package (difference in percentage points in 2030)
Agricultural land use	10%	11%	1
of which:			
food crops	16%	15%	–1
grass and fodder	6%	6%	1
biofuel crops[a]	242%	775%	139
Emissions of nitrogen oxides	–0.6%	–32%	–31
Emissions of sulphur oxides	4.5%	–34%	–37
Natural forest areas	–8%	–9%	–1
Greenhouse gas emissions	37%	13%	–18

StatLink http://dx.doi.org/10.1787/257781513106

a) Although land use for biofuels shows strong growth over the Baseline period, biofuels are still expected to account for well under 1% of agricultural lands by 2030.

Source: OECD Environmental Outlook Baseline and policy simulations.

As a result of the EO policy package, water quality would be expected to be better in 2030 than under the Baseline. The proportion of the population without connection to public sewerage would reduce to 55% worldwide in 2030 (compared with the 67% projected under the Baseline). As a result, nitrogen loading of waterways is projected to increase only slightly compared with 2000 levels (from 9 to 10 million tonnes reactive nitrogen per year worldwide).

Figure 20.1. **Change in emissions of sulphur and nitrogen oxides under the Baseline and EO policy package scenarios, 1980-2030**

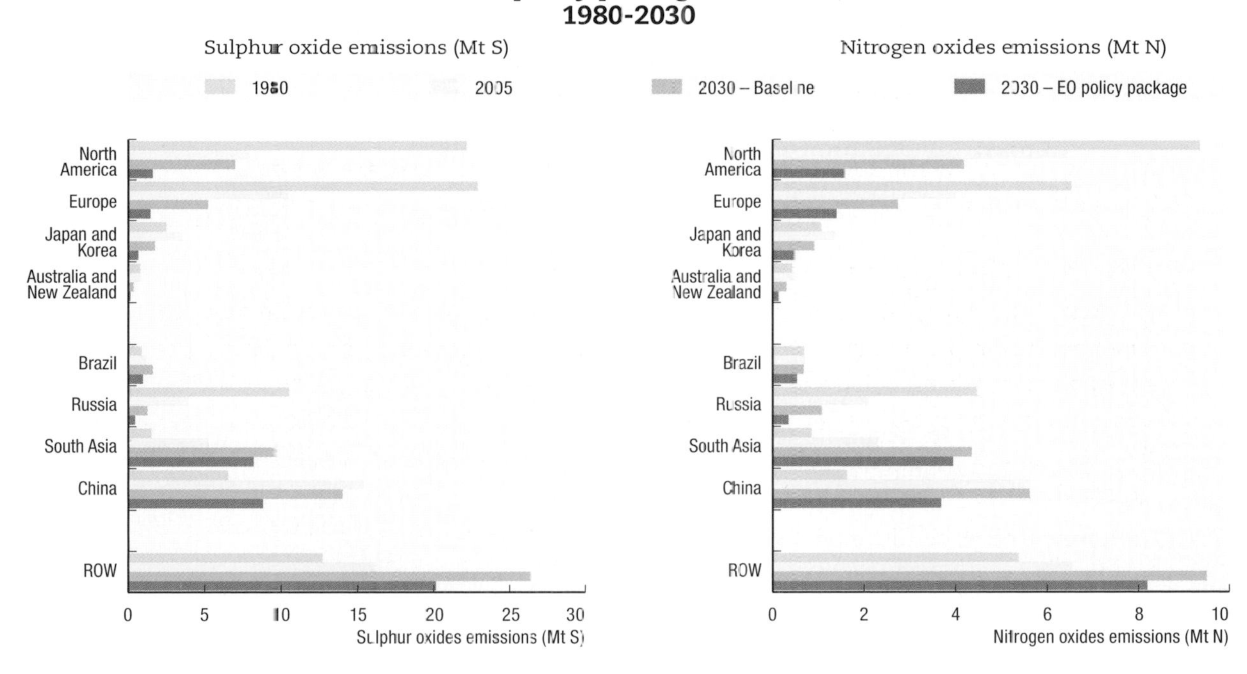

StatLink http://dx.doi.org/10.1787/262412673482

Source: OECD Environmental Outlook Baseline and EO policy package simulations.

The land used for agriculture could be expected to increase very slightly by 2030 under the EO policy package compared with the increase seen in the Baseline (Table 20.1). This reflects the combination of reducing food crops and increasing the land used for grass, fodder and biofuel production (see Box 20.2 for a discussion on the environmental importance of compact agriculture, and the impact of modelling simulations to reflect policies to keep it compact). The impacts on biodiversity of the EO policy package – as measured by the area under natural forest and the mean species abundance (MSA, see Chapter 9) – would be almost negligible in 2030 as compared with the Baseline. As indicated above, this policy package does not include any policies explicitly aimed at protecting biodiversity.

Under the Baseline, human land use in most regions increases, reflecting urban sprawl and increasing agricultural land use. The EO policy package projects that land use for human purposes in regions such as Brazil and other Latin American countries would expand even faster. This reflects the effects of liberalisation of agricultural production and trade, causing agricultural production to increase in low-income countries where land is cheap.

Global GHG emissions would be expected to grow in the EO policy package by only 13% from 2005 to 2030, compared with the projected 37% growth in the Baseline. In part, this reflects the relatively low ambition of the policies simulated, and in part the fact that the climate policies in the EO package are phased-in, with OECD countries starting action in 2012, BRIC countries (Brazil, Russia, India and China) in 2020, and the rest of the world (ROW) only in 2030. Thus, a more noticeable impact of these policies can be seen after 2030, with a net *reduction* in global GHG emissions of 5% from 2005 levels projected under the EO policy package in 2050 compared with an *increase* of 52% under the Baseline. Such a phasing-in of involvement by BRIC and developing country regions may be politically more realistic; however, delaying action will lock-in a much higher level of emissions than could have been realised with earlier action, for example because investments will be made in long-lived energy infrastructure and buildings which are not designed to minimise GHG emissions. The impacts of more ambitious climate policies, for example to move onto a pathway to achieve a relatively stringent (450 ppm CO_2eq) stabilisation of greenhouse gases in the atmosphere, are discussed in Chapter 7 on climate change.

Combining air pollution policies with climate change policies is likely to lead to mutually reinforcing benefits, as fossil fuel burning for energy use is one of the main sources of both environmental problems. This suggests clear co-benefits for regional air pollution from climate change policies. The 450 ppm stabilisation pathway simulation found that, in addition to reducing GHG emissions, the ambitious climate change policies would also lead to a range of environmental and health co-benefits, with for example reductions in sulphur oxides of 20-30% by 2030 and in nitrogen oxides of 30-40% (see Chapter 7).

Ambitious policy packages, if based on complementary and efficient policy instruments, can achieve substantial environmental improvements at relatively low cost.

Economic and social impacts of the EO policy package

The overall cost of the EO policy package would be relatively low in economic terms. It is expected that it would reduce global GDP growth by about 0.03 percentage points a year on average to 2030. This would be about a 1.2% loss in gross GDP in 2030. In other words, instead of realising about a 99% growth in the economy as projected under the Baseline, there would be roughly a 97% growth in GDP between 2005 and 2030. In reality,

Box 20.2. **More compact agriculture**

Despite ongoing improvements in land productivity and the shift towards mixed and landless livestock systems, the land required for agriculture continues to grow in both the Baseline and EO policy package cases. In all cases, the amount of food crops produced grows at a projected rate of 1.6% per year over the period 2005-2030, declining from around 2% a year now to 1% a year by 2030.

The area required to grow these food crops expands far less over the period. While the average yield (in tonnes per hectare) is assumed to increase by around 1% a year over the time horizon, the agricultural land area used grows by 0.6% per year between 2005 and 2030. This means that in total agricultural land used for food crops expands by 16% from 2005-2030, *i.e.* 2.7 million km^2 of land is converted to farming.

In light of the negative impacts associated with such an expansion, notably carbon emissions and loss of ecosystems and biodiversity, an alternative "compact agriculture" policy scenario was simulated. Based upon preliminary results from the *International Assessment of Agricultural Science and Technology for Development* (World Bank *et al.*, 2008) it was assumed that land productivity increases start from 1.6% per year in 2010 (instead of 1% as in the Baseline), and then slow to 1% per annum by 2030. Thus, the yield improvements match the growth in output while the global food crop area remains at the current level (Figure 20.2). The modelling simulation was done in a simple and partial way, assuming uniform yield improvements across all regions and crops, and ignoring possible feedbacks to commodity prices, consumption and trade volumes.

For grass and fodder, the growth in land use is already much less than for food crops in the Baseline, due to changes in grazing intensity and the aforementioned shift to mixed and landless livestock production systems. On average, the grass and fodder area expands by 6% until 2030 (0.22% a year); assuming a similar improvement in land productivity as for food crops above, in the compact agriculture simulation this expansion of land use of 6% is reversed to a decline of 5%.

Total agricultural land (excluding land use for biofuels) grows by 9% in the Baseline between 2005 and 2030, but ends up 2% below the 2005 level in the compact agriculture case.

Figure 20.2. **Global agricultural land area changes under Baseline and compact agriculture scenarios, 2000-2030**

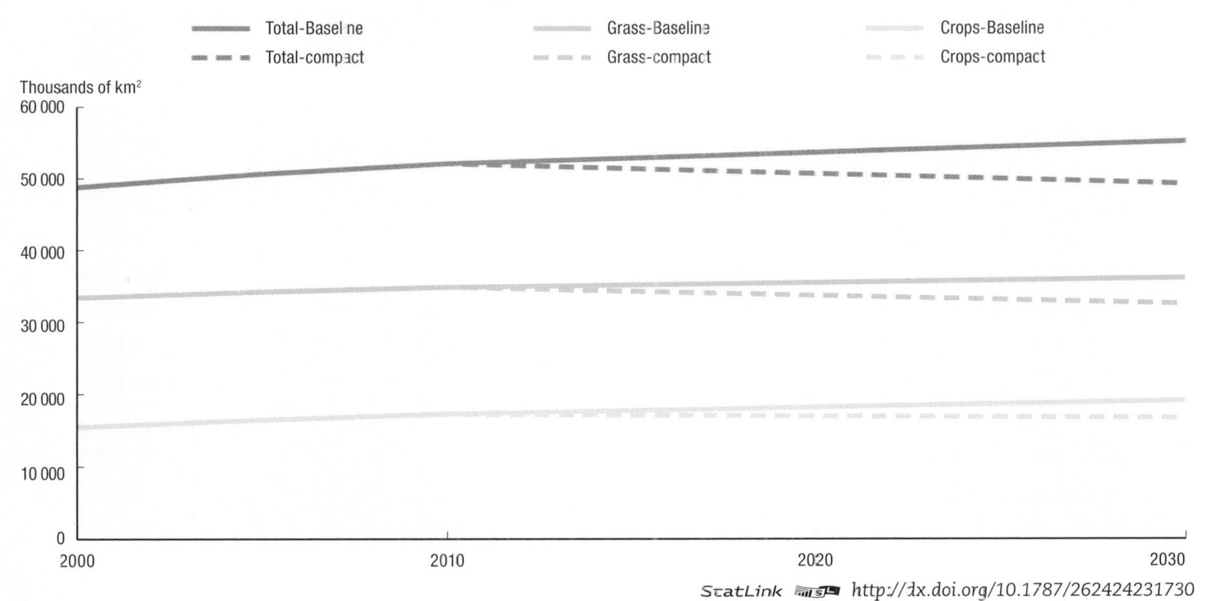

StatLink http://dx.doi.org/10.1787/262424231730

Source: *OECD Environmental Outlook* Baseline and policy simulations.

these losses might be counter-balanced to some extent by the social welfare improvements arising from the policy package, for example from improved environmental and health conditions, but these effects were not captured in the modelling framework.

The economic impacts of the policy package would not be the same for all regions or for all sectors. The expected impacts would be largest in Russia, with average GDP growth increasing by 0.2 percentage points less annually under the EO policy package than under the Baseline (see Figure 20.3). These impacts reflect the high energy-dependency of the Russian economy, and the relatively low energy prices for consumers, as a result of which any globally applied tax on carbon would affect Russian consumers relatively more than consumers already paying higher energy prices.

Figure 20.3. **Average annual GDP growth by region for the Baseline and the EO policy package, 2005-2030**

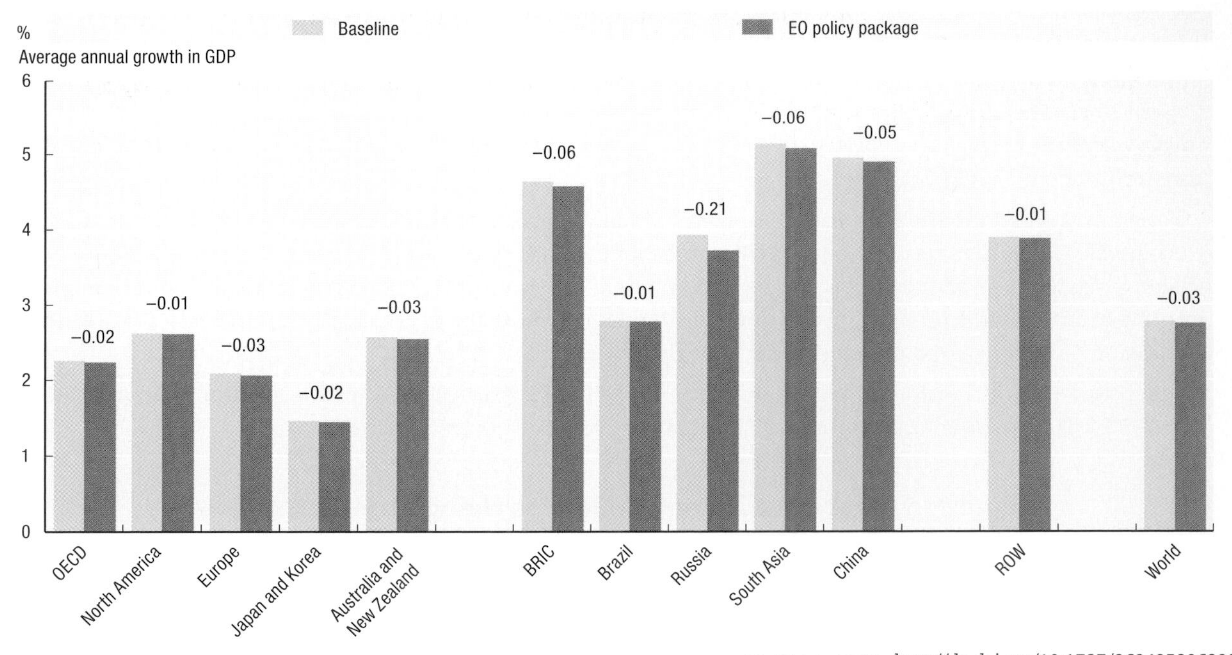

StatLink http://dx.doi.org/10.1787/262425206332

Source: *OECD Environmental Outlook* Baseline and policy simulations.

The relatively low economic costs of the EO policy package in part reflect the "political realism" of the policies simulated. More ambitious policies might well cost more. For example, the policy simulation to stabilise the concentration of greenhouse gas emissions in the atmosphere at around 450 ppm (Chapter 7 on climate change) would reduce GDP growth by about 0.1 percentage points on average each year over the 2005 to 2030 period.

However, the EO policy package would also be expected to reduce the impacts of environmental damage on human health, societies and the economy compared with the Baseline, although these are not directly reflected in the economic results above. The significantly lower levels of air pollution emissions compared with the Baseline would result in less environment-related health problems and their associated economic and social costs (see also Chapter 12 on health and environment). Similarly, reductions in GHG emissions compared with the Baseline would lessen the projected impacts of climate change on infrastructure and communities (see Chapter 13 for some information on the costs of environmental impacts).

Notes

1. See, for example, Chapters 6, Key variations to the standard expectation to 2030; 7, Climate change; 8, Air pollution; 9, Biodiversity; 12, Health and environment; 14, Agriculture; 15, Fisheries and aquaculture; 17, Energy; 19, Selected industries – steel and cement.
2. This combines policy simulations that were reflected in Chapter 9 on biodiversity and Chapter 14 on agriculture.
3. This is similar to policy simulations that were reflected in Chapter 7 on climate change
4. In part because of the difficulties in reflecting such policies in the modelling framework.

References

OECD (2006), *The Political Economy of Environmentally Related Taxes*, OECD, Paris.

OECD (2007), *Instrument Mixes for Environmental Policy*, OECD, Paris.

Pearce, D., G. Atkinson and S. Mourato (2006), *Cost-Benefit Analysis and the Environment: Recent Developments*, OECD, Paris.

World Bank, FAO and UNEP (2008), *International Assessment of Agricultural Science and Technology for Development*, World Bank, Washington DC.

ISBN 978-92-64-04048-9
OECD Environmental Outlook to 2030

Chapter 21

Institutions and Approaches for Policy Implementation

Government environmental institutions initiate and support the policy-making process, facilitate the development and implementation of environmental policies, and ensure compliance with environmental requirements. Several governments are moving away from the direct provision of services (e.g. water supply and sanitation, waste management) and towards regulating private markets for service provision. Although most OECD environment ministries have cabinet status, they often struggle to get approval for sufficiently ambitious environmental policies. Environment ministries need to work closely with other ministries, private sector and civil society for the development and implementation of environmental policies. This chapter examines recent trends and possible future developments in the institutions for developing and implementing environmental policies at the national and sub-national level. It identifies some of the main obstacles to successful environmental policy reform, and suggests how these can be addressed to build acceptance for ambitious environmental policies and to enhance the benefits of reform.

KEY MESSAGES

What are the key institutional and policy challenges?

Although most OECD environment ministries have cabinet status, lack of political commitment at the highest level often denies them the power to implement sufficiently ambitious environmental policies.

Further efforts are needed to integrate environmental concerns into sectoral policies.

One of the greatest challenges to ambitious environmental policy implementation stems from concerns that the short-term costs of these policies will be too high, thereby harming economic competitiveness, or that low-income households will be disadvantaged. However, it is often possible to design effective environmental policies in such a way that competitiveness or distributive impacts can be reduced to politically acceptable levels.

What is the main role for institutions?

- Government environmental institutions are increasingly becoming the catalysts and facilitators of environmental policy development, as well as the providers of environmental services and protection.
- Enforcement agencies play an increasingly important role in responding to serious breaches of regulations and in establishing a widespread perception that violations will not be tolerated.
- Decentralisation of environmental governance to local authorities is continuing, following the principle of "subsidiarity". Central governments will play a greater role in co-ordinating and bridging between the local level and international efforts, in particular for transboundary or global environmental problems.
- Mechanisms for consensus-building with stakeholders – such as industry, trade unions, NGOs and the media – will continue to play an important role in environmental policy-making.

Policy options

OECD analysis shows that well-designed environmental policies are unlikely to have significant negative impacts on income distribution or on a country's net competitiveness. Any distributive or competitiveness effects can be overcome by designing environmental policies or accompanying measures that address these concerns, while still providing an incentive for better environmental behaviour. Environmental and policy impact assessments can identify any potential problems and help to design better environmental policies. Some additional steps will aid policy implementation:

- Phase-in policies gradually, according to a pre-announced timetable and following stakeholder consultations, to build acceptance for more ambitious environmental policies, while providing time for affected individuals and industries to adjust.
- Further integrate environmental concerns into sectoral policies, either through inter-ministerial co-ordination mechanisms, or directly through sectoral ministries (*e.g.* agriculture, transport, energy). This will continue to develop internal capacities for dealing with environmental issues and promoting sustainable development.
- Strengthen the economic and integrated analysis skills of environment ministries, and at the same time the environmental expertise of sectoral ministries, to enhance their ability to work with specialised research and development institutions for a sound scientific basis for decision-making.
- Strengthen enforcement and other facilitative systems to monitor and ensure compliance with environmental regulations in both OECD and non-OECD countries.

OECD ENVIRONMENTAL OUTLOOK TO 2030 – ISBN 978-92-64-04048-9 – © OECD 2008

Introduction

An effective and efficient public institutional framework[1] is an inherent part of an environmental management system that promotes environmental improvement and/or changes environmentally harmful behaviour in the context of sustainable development. The institutions required to implement the regulatory policy agenda are numerous and varied, and their evolution in different countries varies according to their history, culture and particular stage of development, among other things. The main pillars of an environmental institutional framework are government authorities (administration), appointed and authorised by elected officials to carry out tasks at the national (federal) and sub-national (regional and lower) levels. These include regulatory management and oversight bodies within cabinets and executive government and, increasingly, within parliaments as well. The institutional framework also entails a web of formal and informal organisations and arrangements with established rules and communication patterns, responsible for regulating selected policy areas, ensuring dialogue and consultation, monitoring, analysis, dissemination of information and awareness-raising.

However, it is not always easy to implement environmental policies, sometimes due to mal-adapted institutions, but mainly for political economy reasons. In general, implementation of policies creates both "winners" and "losers", and each of these groups can influence how policies are actually designed and implemented. While the environment itself benefits, these policies can have negative impacts on the competitiveness of certain industries, or on the distributive burden facing certain groups of households. As a result, some environmental policy reforms face resistance from affected industries and the general public. However, it is possible to overcome these obstacles to environmental policies, achieve efficient and equitable outcomes, and even create opportunities to gain competitive advantage. *Ex ante* impact assessments of new policies can identify any possible problems, and flanking measures can be designed to address them.

This chapter examines recent trends and future pressures on the institutions for developing and implementing environmental policies at the national and sub-national level.[2] It then identifies some of the main obstacles to successful environmental policy reform and suggests how to address these and enhance the benefits of reform.

Institutions for policy development and implementation

Evolution of environmental administrations

Environmental ministries or departments initiate and support the policy-making process (including identifying key areas for priority attention, goals and objectives), facilitate the development and implementation of environmental policies, and ensure compliance with environmental requirements to guarantee legal protection of environmental endowments and human health. Government environmental institutions also work towards co-ordinating policies with other sectoral agencies to ensure integrated and coherent policies. Within the institutional framework, the achievement of policy objectives, goals and targets is often evaluated, and the performance of various elements of the regulatory cycle assessed.

Environmental institutions, like other public agencies, evolve following changes in political and management approaches to deal with public policy problems. Governments have faced pressures for change from growing demands stemming from globalisation and international competition, and such pressures are likely to increase in the coming years. Pressures also come from citizens' expectations for more openness, higher levels of service quality delivery and the increasing need for solutions to more complex environmental problems. As a result OECD public administrations are increasingly applying good governance principles[3] and are becoming more efficient, more transparent and customer-oriented, more flexible, and more focused on performance. These approaches enable institutions to develop and implement better policies and to generate political and public support.

After decades in which new government initiatives were funded by extra revenue, growing fiscal constraints now mean that public institutions must increase their effectiveness and efficiency, and prioritise among competing policy objectives (OECD, 2005a).

Thus, several governments are moving away from the direct provision of services – including environmental services, such as water supply and sanitation, sewerage or waste management – towards regulating markets for service provision, leaving a greater role for other entities, both private and non-profit. Government's role is increasingly to ensure fair competition and the introduction of market-based mechanisms (Box 21.1). Outsourcing has been growing significantly over the past two decades and has become a mainstream element of modern public administration in many OECD countries. This trend is likely to continue in the future, reflecting the demands on government and the further development of private sector services. It will be important for governments to oversee the quality and pricing of services provided by non-government entities to limit abuses of market power. Where outsourcing or privatisation entails government entities pulling out of these activities, public mistrust or concerns about the displacement of public sector workers will need to be carefully considered.

Most OECD countries have a fully fledged environmental ministry or department with cabinet status. This gives environmental issues higher prominence and increases the weight accorded to the department's views in interdepartmental discussions. However, environmental authorities often still lack real power to fully implement environmental policies or to oversee the implementation of environmentally-related policies in other sectors, generally because of lack of political commitment at the highest level. If the environment is not a major and uncontested concern of the government and the country, the environment minister has limited influence within the government, regardless of the status of the ministry.

Although most OECD environment ministries have cabinet status, they often are not given the power needed to implement sufficiently ambitious environmental policies.

Historically, the main government environmental authority had primary responsibilities for the regulation of air and water pollution, municipal and industrial waste disposal, noise, nature and biodiversity protection, and environmental impact assessment (EIA) procedures. Today, in several countries the environment ministries are also responsible for

Box 21.1. **The changing skills base of environment authorities**

Environmental authorities periodically review their structure to identify and fill in possible gaps in professional expertise as priorities evolve. Such restructuring is also related to the fact that environmental policy-making is a complex task which requires a mix of political, economic and legal skills on the one hand, and scientific and engineering skills on the other. The staff of environmental authorities has been dominated by environmental professionals, mostly engineers and scientists. The increasing complexity of environmental problems, combined with the growing costs of environmental policies, is beginning to force environment ministries to broaden their mandates to include the economic (and more recently, the social) aspects of environmental issues. With the gradual development of integrated environmental policy-making, environment ministries have begun to expand their internal capacities for economic analysis to support stronger cross-sectoral and cross-medium integration.

Increasingly, flexible teams ("clusters") of technical and managerial staff from different parts of the environmental authority are formed to focus on problems related to specific sources (*e.g.* petroleum refining, chemical industry), pollutants (*e.g.* particles, lead, CO_2), environmental resources (*e.g.* groundwater) or other groupings of activities (*e.g.* children's health). This trend is likely to continue as the cross-channels of communication and responsibility fostered by the teams/clusters can give the ministry the character of a matrix organisation, enhancing integration at the policy level.

Policy implementation also increasingly requires diplomatic and negotiation skills to facilitate consultations at the national level (between those responsible for drafting regulations and the regulated community) and at the international level (when international commitments and agreements are discussed). Many environmental authorities will need to acquire such skills in the future, especially to play an active and constructive role in negotiating international environmental agreements. In addition to scientists, engineers, lawyers, economists, and various support staff, a modern environment authority will need staff trained in information management, public relations and project management.

managing natural resources, such as water (Australia, Portugal, Slovakia), forestry (Poland, Turkey), or special planning and territory (Italy, the Netherlands). In Austria, Germany, France, the Netherlands and the United Kingdom environment is part of a larger operating ministry with budgetary and management responsibilities for other sectoral issues, such as energy, nuclear safety, housing or food and rural development (OECD, 1990-2007). Some countries (*e.g.* Sweden and France) have experimented with creating ministries for sustainable development where environmental management has been linked closely with priority sectors such as energy and transport. This approach may be replicated by other countries in coming years.

While the environment ministry is responsible for setting environmental policies and managing inter-sectoral decision-making, several countries also have a separate environmental protection agency and/or research institutes for technical and analytical support. These agencies/institutes are usually responsible for monitoring and assessing the implementation of environmental policies and the overall state of the environment; they also develop technical proposals for rules and guidelines, provide environmental information services and public outreach. They employ hundreds of highly qualified experts from a wide range of disciplines in a number of environmentally relevant departments working in close co-operation with scientific institutions at home and abroad. They provide technical support for policy development and will continue to play an important role in establishing links between science, politics, the authorities and practical application.

Environment ministries and agencies have been developing extensive monitoring systems to gather data on emissions and the ambient environment. Monitoring is often delegated to the sub-national level and only supplemented by national networks. However, monitoring and information systems are often fragmented and dispersed among agencies, reflecting the historical distribution of policy and implementation responsibilities. Efforts have been made to focus, streamline and increase the cost-effectiveness of the institutional frameworks for data collection and processing in many OECD countries, as well as to supplement them with other monitoring methods that reduce costs. In the future, governments are likely to increasingly use complementary industrial self-monitoring, supported by citizens' monitoring/non-compliance detection. Self-monitoring, which uses operators' measurements of process conditions and releases, as well as environmental conditions, can lead to clearer understanding for industry of their compliance status and provides easier data gathering to support regulatory reform. Citizens' monitoring (including "whistle blowers" in the case of illegal activities) can also complement extensive state monitoring systems. In some cases, monitoring is supported by modelling as the basis for strategic decisions. The combination of state monitoring systems with self-monitoring, citizens' monitoring and computer modelling can help governments to refocus their environmental research efforts towards better environmental priority setting and integration of environmental concerns into sectoral policies.

Compliance assurance

Enforcement or compliance assurance programmes are important policy tools used by environmental enforcement agencies (environment inspectorates or other specialised units within environmental agencies – Box 21.2). Their role is likely to continue to increase in importance as they are the first to respond to non-compliance in the short term and help to continuously prevent breaches of environmental and other requirements.

Sectoral policy integration

Many environmental issues cut across the mandates of several ministries. Over time, previously compartmentalised agencies have begun closer co-operation for addressing cross-sectoral issues. Many sectoral ministries are establishing environmental units within their organisational structures. The main reason is to carry out routine analysis of environmental impacts in order to strengthen internal capacity and address increasing environmental concerns and negative impacts within their sector (OECD, 2001a).

Further efforts are needed to integrate environmental concerns into sectoral policies.

At the same time, integrating environmental concerns into sectoral policies is also increasingly discussed through inter-ministerial working groups or cabinet-level committees, commissions of enquiry, task forces, etc. In several countries inter-agency integration of environmental concerns is carried out at the highest level, within the president or prime ministers' offices. In some countries, institutions have been established to carry out independent audits of

Box 21.2. **Compliance assurance**

Environment inspectorates play an important role by establishing a deterrence atmosphere, *i.e.* creating the widespread perception that violations will not be tolerated. This can be achieved in several ways: by providing strict and timely reaction to non-compliance; by establishing social disapproval of violators by "naming and shaming" poor environmental performers; by publicising successful enforcement actions; by addressing with perseverance minor but widespread violations; and by creating incentives to improve compliance. Deterrence can stimulate voluntary compliance, eliminating the need for a physically omnipresent enforcement agency, and thus reducing enforcement costs.

The role of inspectors has often been under-appreciated or left outside the reform agendas even though the inspectors are in many cases those who face the regulated community on a daily basis and can be critical in assisting with compliance, putting additional pressure to comply or applying actions against non-compliance.

The trend of enforcement agencies being at arm's-length from policy-makers and elected authorities is likely to continue, as this ensures that once regulations are set, political considerations do not interfere with enforcement. This is especially important at the sub-national level where local administrations responsible for economic development and job creation may have a tendency to protect offenders from enforcement responses. At the same time, however, several safeguards such as appeal mechanisms or supervisory boards are needed to ensure that quality, fairness and integrity are maintained during the enforcement processes. Another important aspect has been the establishment of mechanisms for feedback between the inspectors and decision-makers to inform the latter about compliance problems and the need for regulatory revisions.

Source: OECD (2000); OECD (2005c).

government actions in the field of the environment, such as Canada's Commissioner of the Environment and Sustainable Development, the UK House of Commons' Environmental Audit Committee, or Korea's Presidential Commission on Sustainable Development. The main functions are to monitor and report on progress made by departments or agencies in implementing environmental or sustainable development strategies.

In the 1990s, administrations and agencies, parliamentary commissions, the scientific community and non-government organisations (NGOs) devoted major efforts to clarifying the concept of, and designing strategies for, sustainable development. Subsequently, several countries adopted strategies and set up inter-agency or inter-ministerial co-ordinating committees for sustainable development to provide an overarching body and framework for integrated action (Box 21.3). Achieving sustainable development depends a great deal on high-level political commitment, well-functioning government institutions and overcoming co-ordination failures in public policies. Therefore, involving and co-ordinating a wide range of government departments allows sustainable development strategies to take a broad view of issues, give voice to a range of dispersed interests and address trade-offs across policy areas.

Subsidiarity in environmental management

In many countries, constitutions reserve international relations, basic legislation and national planning for the central government, while granting legislative and managerial jurisdiction over environmental issues to the autonomous regions. Regions are then

Box 21.3. **Good governance for national sustainable development**

Although many countries now have considerable experience with the governance aspects of national sustainable development strategies (NSDS), approaches differ in terms of whether the NSDS is: *1)* top-down or bottom-up; *2)* horizontal or embedded in a single department; *3)* underpinned by legislation; *4)* linked to budget processes; *5)* fully open to stakeholders; and *6)* linked to sub-national levels.

In most OECD countries, overall responsibility for sustainable development strategy implementation is housed in the ministry of environment, either directly or indirectly through a co-ordinating committee which it oversees. This is true for countries such as Austria, Denmark, Greece, Ireland, Luxembourg, the Netherlands and the United Kingdom. Although the United Kingdom replaced its Green Cabinet with a Sustainable Development Cabinet, the Department of Environment, Food and Rural Affairs (DEFRA) leads the preparation of sustainable development strategies and manages implementation across the government.

One good practice is to assign overall co-ordination to a prime minister's office or equivalent; these have greater authority to demand inputs and resolve conflicts than line ministries. France, Portugal and Germany have placed responsibility for their national sustainable development strategies directly under the prime minister's office to achieve maximum coherence. In the Belgian federal government, the responsibility for strategy implementation is under the State Secretary for Sustainable Development who chairs the Interdepartmental Committee for Sustainable Development, which includes all federal departments. In countries where responsibility for implementing the NSDS is assigned to the prime minister's office, the presence of a sustainable development minister or equivalent tends to improve results.

Another approach is to assign responsibility for national sustainability strategies to finance ministries, which can ensure that strategic management is linked to fiscal priority setting, national expenditure and revenue generation. Thus, Norway has placed responsibility for its sustainable development plan within the Ministry of Finance, while in the Czech Republic, the Governmental Council for Sustainable Development is chaired by the Prime Minister. In Italy, the Interministerial Committee for Economic Planning, which is responsible for sustainable development, is chaired by the Minister of Economy and Finance. Sustainable governance in practice would require that departmental expenditures are justified through their contributions to the goals and priorities of the NSDS.

NSDS, to be fully effective, should involve a range of ministries, departments and agencies. Preferably, they should be top-down in that government bodies design their SD programmes in accordance with an overarching strategy (*e.g.* as in the United Kingdom). However, some countries (*e.g.* Canada) have a bottom-up approach where individual departments formulate their SD programmes in the absence of an overall strategy. Different types of strategies may be needed in federal countries (*e.g.* Belgium, Canada) than in those where the government is more centralised.

Stakeholder involvement is a fundamental part of NSDS. Recognising that transparency is central to sustainable development, most countries have included stakeholders in strategy development and implementation. But approaches differ. Some countries (*e.g.* Austria, Czech Republic) include stakeholders in the government bodies responsible for NSDS implementation and oversight. Others (*e.g.* France, Germany and the United Kingdom) have separate stakeholder councils which advise the government.

Lastly, links should be established to sub-national authorities in order to catalyse action, leverage their involvement, and manage the interdependency between different levels of government. But degrees of co-ordination with local governments in the context of NSDS vary from high (*e.g.* France, South Korea) to medium (*e.g.* Sweden, Finland) to low (*e.g.* Germany, Portugal).

Source: OECD (2005b); Swanson and Pintér (2006).

responsible for certain matters that directly concern the local community and individual citizens, including, for example, water supply, waste water treatment and waste disposal. The regions often confer to the municipalities the power to enforce local aspects of federal or regional legislation. Responsibility for decision-making that affects the environment is being increasingly decentralised to local governments. In line with the principle of "subsidiarity", functions are being devolved to the most appropriate level to address problems in the most effective and efficient way, *e.g.* river basin management. However, some functions are likely to continue to be carried out at the national level, for example when legal protection and uniformity are the most important factors or when matters require international environmental co-operation (see Chapter 22 on global environmental co-operation). It is important that responsibilities are clearly allocated between levels of government to avoid confusion, frictions, ineffectiveness or lengthy procedures and that devolved responsibilities also come with adequate resources.

A centralised approach to addressing environmental problems has the advantage of co-ordination and the ability to provide for integrated development with internal human and material resources. The main disadvantages of centralisation are inadequate knowledge of local conditions and slow response. Conversely, decentralised institutions can provide more flexibility and are usually more specialised. Their disadvantages can include poor co-ordination and redundancy among several different institutions working in a single area, and the overriding of environmental problems by economic or social priorities. There is also a tendency to delegate functions to institutions before they have a relevant mandate, or without providing adequate resources (including financial and human) for implementing these mandates.

Many countries are taking steps to clarify responsibilities and strengthen vertical co-ordination, following a phase of decentralisation. In federal countries, federal-provincial-territorial co-ordinating councils are being created in several policy areas (including the environment, energy, agriculture, fisheries, forestry, wildlife and protected areas) to promote consultation and co-operation at the federal level and to develop coherent strategies and guidelines addressing issues of common concern. In some countries, regional environmental centres or agencies are being established to serve as co-ordination nodes between central and local governments in implementing national environmental policies.

Stakeholder involvement

Much discussion of modern environmental problems concerns the need to engage the public more directly in solving these problems. In several countries, consultative bodies to the national governments, which include representatives from civil society, industry and trade unions, have been established to provide strategic advice on environmental and sustainable development issues (see Box 21.3). This is part of a trend towards more horizontal institutional structures that are capable of quick reaction, consultation and exchange of information, multidisciplinary analysis, and a significant degree of structural flexibility. Stakeholder involvement is a crucial component of formulating environmental policies if they are to be implemented successfully.

Political economy of environmental policies

This chapter has explored some of the institutional approaches that can help support the development of environmental policies. But beyond these, there remain other obstacles to the successful implementation of policies to address environmental challenges. These "political economy" factors include concerns about competitiveness, reconciling social objectives and policy integration across sectors. Each is discussed now in turn and a range of solutions proposed (OECD, 2006; OECD, 2005d; OECD, 2002b; OECD, 2001b).

Winning support for environmental policies

A major obstacle to successful implementation of environmental policy reform is uncertainty or insufficient information about the severity or causes of the environmental problem it is meant to address. While it is widely recognised that environmental policies should be science-based, policy action is sometimes needed, even with remaining uncertainties, in order to minimise potential risks. Differences in points of view will always remain between those who are more risk-adverse and those more risk-accepting. However, continued improvement in our knowledge base is important, particularly for issues that are currently poorly understood – such as the impacts of the build-up of chemicals[4] in the environment – or for global challenges for which there is not yet consensus on policy action. Scientific understanding of some key global issues has moved forward recently, notably on biodiversity with the 2005 Millennium Ecosystem Assessment, and for climate change with the 4th Assessment Report of the Intergovernmental Panel on Climate Change (IPCC, 2007).

At the international level, another important "political economy" obstacle is the uneven distribution of costs and benefits of policy action across regions and countries. This has typically been the case for global environmental problems, such as climate change, transboundary movements of persistent organic pollutants (POPs), and biodiversity loss (see Chapter 22 on global environmental co-operation). For these problems, the location of the damage can be a long distance away or unrelated to the location of the causes of the damage, resulting in a mismatch between the costs and benefits of action. This problem can be addressed by financial mechanisms to redistribute costs which would otherwise be distributed "unfairly" between developed and developing countries, based on the notion of "common but differentiated responsibility". Thus, for example, financial transfers have been made in a number of instances between countries to help overcome differences in environmental protection priorities. Some regional trade agreements between countries with common borders also include clauses addressing trans-boundary environmental issues through environmental co-operation and capacity-building (see also Chapter 4 on globalisation).

From competitiveness concerns to competitive advantage

The greatest challenge to environmental policy implementation often stems from concerns that the costs of these policies will be too high, thereby harming economic competitiveness. As world economies are expected to become ever more integrated over the *Outlook* period, the competitiveness issue is likely to remain on the agenda in the years to 2030 and beyond.

It is very important to distinguish between competitiveness impacts at the level of the national economy, and those at the level of the individual firm (or sector). Within a country, the "winners" from a given policy initiative may partly or fully outweigh the "losers". It is therefore the net impacts on competitiveness at the national level that should be taken into account. Here, the following important points need to be emphasised:

- Environmental policies – if properly designed and implemented – will generate positive welfare benefits for society as a whole. Even in cases where these policies lead to negative economic results in the short term, these effects will be more than offset in the long term by the environmental benefits the policies engender. For example, even though policies that target heavily polluting industries may cause some firms to go out of business, the positive public health benefits that result from the reduced pollution may outweigh the negative consequences for individual firms.

- Most available evidence suggests that national economies do not generally suffer losses of overall competitiveness as a result of existing environmental policies. Losses in one area of the economy tend to be offset by gains in another part of the economy. In addition, the costs associated with meeting environmental objectives are currently a relatively small share of total costs (most OECD economies currently spend only 0.5-2.1% of their national GDP on environmental goals (OECD, 2007).
- It is possible that very ambitious environmental policies could generate more significant national competitiveness problems for some countries in the short to medium-term future (*e.g.* energy-intensive economies which face stringent climate change mitigation policies). But even in such cases it is not clear that the overall results would be politically unsustainable, provided that some of the flanking measures described below or some form of transfers between countries are used. The long-term environmental benefits may outweigh the short-term challenges.
- The competitiveness problems facing individual firms will depend on more than just the stringency of environmental policies. In particular, the technology being used by the firm, the location of the factory, and the market power being exercised by the firm will all affect its overall competitiveness. It would therefore not be appropriate to attribute all of the responsibility for a given competitiveness problem to environmental policies.

In some cases, environmental policies may actually enhance the economic competitiveness of targeted sectors or industries. Firms that are forced to "clean up" can gain market advantages, either by being the first to exploit "green" markets or by moving into new technologies that can in turn be marketed or lead to efficiency savings. Although such opportunities may be realised by individual firms in particular circumstances, it is unlikely that they will apply to economies as a whole (or for very long). Again, the main reason is that gains in one part of the economy will tend to be offset by losses in another part.

Still, governments often face opposition from firms that envisage that they will suffer from particular environmental policies. This opposition is likely to be particularly intense where the firm(s) involved face strong international competition. Broadly, these firms fear that environmental policy measures (*e.g.* taxes, emissions trading or regulatory standards) will undermine their competitiveness because competitors in other countries (or jurisdictions) do not face similar requirements. For example, a number of OECD governments have decided not to introduce energy or carbon taxes in recent years following intense discussions and strong opposition from energy-intensive industries. Before its introduction in 2007, the new EU regulatory framework on the Registration, Evaluation and Authorisation of Chemicals (REACH) also faced significant resistance from the chemical industry, because of the expected impact of higher compliance costs on sectoral competitiveness. Close consultations with the industry and other stakeholders, as well as extensive *ex ante* impact assessments during the policy formulation phase (now a standard feature in the EU), were crucial for the eventual adoption of REACH. Such consultative efforts to address competitiveness concerns about environmental policies will become increasingly important in the coming years.

This is one of several approaches that governments can take to address competitiveness concerns of individual producers, while still ensuring that the policies are effective in inducing firms to change their environmentally damaging behaviours. Such approaches include, for example, providing better information on the actual competitiveness impacts of policies, and through a well-planned and transparent phasing-in of the policy to allow for adjustments. Recycling environmental tax revenues back into affected sectors, or co-operating internationally to "level the playing field", can also help to reduce competitiveness impacts. The following sections provide more details about some of these options.

Better information

Industry resistance to environmental policies can sometimes be overcome by creating a common understanding of the problem at hand (including its causes and impacts and the effects of possible instruments for addressing the underlying environmental problem). One way to build such common understanding is to involve relevant "stakeholders" in policy formulation, for example through broad formal consultations when new policy instruments are being proposed. Green Tax Commissions, with participation from relevant ministries, affected industries, trade unions, environmental organisations, etc., can be a useful way to build communication between stakeholders. Perceived negative competitiveness impacts on individual firms, sectors or industries should also be carefully assessed, as they may be exaggerated by special interest groups.

Timing

Timing is crucial – new environmental regulations or other policy tools that seem impossible to implement at one period in time may become feasible later, when circumstances are more favourable. Phasing-in new policy tools gradually, and according to an agreed timetable, can also help. This allows industries time to adjust (*e.g.* to replace capital stocks gradually with cleaner technologies) and offers industry more certainty for long-term planning.

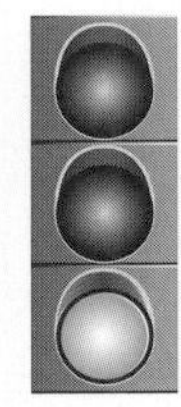

It is often possible to design effective environmental policies in such a way that competitiveness or distributive impacts can be reduced to politically acceptable levels.

Proactive employment adjustment policies – such as gradual phasing-in of policies coupled with targeted and transitional support for re-training or job search support – can help workers to gradually adjust to other forms of employment. Governments wishing to address employment concerns related to negative competitiveness impacts can best do so via "flanking" measures that do not reduce the environmental effectiveness of the original environmental policy.

Broader fiscal reform

In the case of new environmental taxes, introducing these as part of broader fiscal reforms can also make it easier to win political acceptance. For example, a Norwegian CO_2 tax introduced in 1991 was combined with reduced taxes on incomes in "remote" areas. Similarly, a 1999 ecological tax reform in Germany increased mineral oil duties and electricity taxes, using additional revenues to reduce pension insurance contributions.

Levelling the playing field

One way to limit the competitiveness impacts associated with environmental policies is through broader international co-operation to "level the playing field" among

income or to support tied to environmental cross-compliance (see Chapter 14 on agriculture). OECD analysis has found that these latter measures can be even more efficient in transferring support to farmers than those linked to production or inputs (OECD, 2005d).

In the field of water pricing, OECD countries have experimented with a range of measures to ensure affordable access by all segments of society to water supply and sanitation services. Examples include progressive tariffs (*e.g.* charges increase with each additional unit of water that is used); income measures (*e.g.* direct subsidies to low-income consumers or those with large water requirements, such as for dialysis purposes); and policies limiting the disconnection from basic water services (*i.e.* in the case of overdue bills due to affordability problems).

Policy co-ordination for win-win solutions

Most of today's key environmental challenges cannot be solved by environment ministries alone, but require coherent government-wide policy action, both horizontal and vertical. Often, the most effective and efficient policies to meet environmental objectives – such as energy taxes, cleaner fuel choices, increased public transport provision, or reform of agricultural or fisheries subsidies – lie well outside the responsibility of environment ministries. Thus, it will become even more important in the coming years for finance and economy ministries, as well as those regulating relevant economic sectors (energy, agriculture, transport, industry), public health and development co-operation to integrate environmental concerns into their sectoral policies. A mix of environmental policies can achieve synergistic "win-win" solutions for the economy, the environment and human health (see Chapter 20 on environmental policy packages). Because competencies for different aspects of environmental management rest with different levels of government, it is important to work across regional, state and local government levels.

Governments may also find creative win-win solutions to environmental challenges by working in partnership with the business community, research institutions, civil society organisations and trade unions which can play important roles both in the development and implementation of policies. Governments should therefore seek to promote this kind of dialogue and set consistent policy frameworks. For example, governments should encourage environmentally responsible corporate behaviour, including the wider use and further development of environmental technologies, not by picking "winners" but by setting long-term policies that send stable economic signals and enable the private sector to make long-term business plans (see Chapter 1 on consumption, production and technology).

Notes

1. The term "institutional framework" or "institutions" is used in a very broad sense here. In particular, it refers to "organisations" which are regimes with staff and a physical location. They generally involve some form of charter or formal governance agreement, setting out objectives and identifying institutional means to achieve them. This broad interpretation includes institutional frameworks at the sub-national levels.
2. For discussion on international environmental governance, including multilateral and international environmental institutions operating within the framework of global and regional environmental co-operation, see Chapter 22 on global environmental co-operation.

3. The OECD Public Management Committee (OECD, 2002a) has adopted a set of principles that explain the key components of good public governance: *i)* rule of law, *ii)* accountability, *iii)* transparency, *iv)* efficiency and effectiveness, *v)* responsiveness and *vi)* forward vision.
4. See also Chapter 18 on chemicals.
5. See *www.oecd.org/env/policies/database*.

References

IPCC (Intergovernmental Panel on Climate Change) (2007), *Fourth Assessment Report*, IPCC, Geneva, forthcoming.

Millennium Ecosystem Assessment (2005), *Ecosystems and Human Well-Being*, Island Press, Washington DC.

OECD (1990-2007), *OECD Environmental Performance Reviews,* OECD, Paris.

OECD (2000), *Guiding Principles for Reform of Environmental Enforcement Agencies in Eastern Europe, Caucasus and Central Asia*, EAP Task Force, OECD, Paris.

OECD (2001a), *OECD Environmental Outlook*, OECD, Paris.

OECD (2001b), *Environmentally Related Taxes in OECD Countries: Issues and Strategies*, OECD, Paris.

OECD (2002a) *OECD Reviews of Regulatory Reform: Regulatory Policies in OECD Countries – From Interventionism to Regulatory Governance*, OECD, Paris.

OECD (2002b), *Implementing Domestic Tradable Permits: Recent Developments and Future Challenges*, OECD, Paris.

OECD (2005a), *Modernising Government: The Way Forward*, OECD, Paris.

OECD (2005b), *National Sustainable Development Strategies: Good Practices in OECD Countries*, Sustainable Development, OECD, Paris.

OECD (2005c), *Economic Aspects of Environmental Compliance Assurance Proceedings from the OECD Global Forum on Sustainable Development,* 2-3 December 2004, OECD, Paris.

OECD (2005d), *Environmentally Harmful Subsidies: Challenges for Reform*, OECD, Paris.

OECD (2006), *The Political Economy of Environmentally Related Taxes*, OECD, Paris.

OECD (2007), *Pollution Abatement and Control Expenditure in OECD Countries*, [ENV/EPOC/SE(2007)1], *www.oecd.org/env*, OECD, Paris.

Swanson, D. and L. Pintér (2006), *Governance Structures for National Sustainable Development Strategies Study of Good Practice Examples*, International Institute for Sustainable Development (IISD), Winipeg, Canada.

ISBN 978-92-64-04048-9
OECD Environmental Outlook to 2030

Chapter 22

Global Environmental Co-operation

Many environmental challenges are inherently global: there is only one common atmosphere and many ecosystems provide global public goods. Watersheds cross national borders, and some pollutants travel across continents and oceans. Responding to global environmental challenges requires global solutions and international co-operation. This chapter summarises the key emerging trends in global and regional environmental co-operation. The chapter focuses primarily on the traditional means of government-to-government co-operation: multilateral environmental agreements on the "environment ministries track" and environmental aid on the "development ministries track". Alternative co-operation mechanisms – such as intra-industry technology transfer, community-to-community de-centralised co-operation and sustainable development partnerships – are increasingly important, and are discussed briefly.

KEY MESSAGES

What is the outlook for global co-operation?

It will be increasingly important to 2030 for developing countries to share the burden of solving global environmental challenges. But the distribution of the responsibility for action amongst countries is likely to prove increasingly problematic and, if unresolved, may prevent major advances in environmental co-operation.

Environmental aid has been decreasing since 1996 as a share of donor country GDP and as a share of total aid.

While many countries are working to address environmental issues through international means and instruments, a coherent and effective system at the international level is still lacking. Improvements in the international environmental governance system are taking place, but at a slow pace.

Environmental issues are becoming more prominent in the international economic governance framework – such as in Regional Trade Agreements. However, the number of trade and investment agreements with commitments to co-operate on environmental matters are still comparatively few.

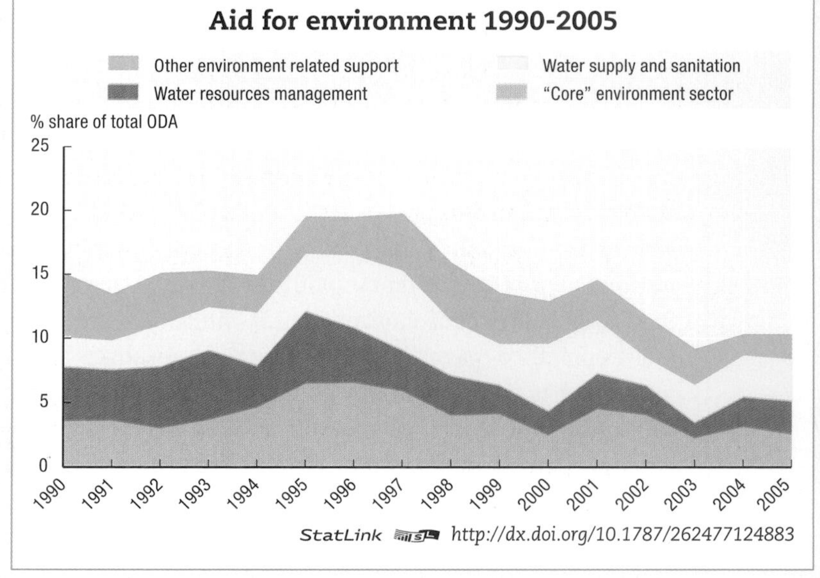

StatLink http://dx.doi.org/10.1787/262477124883

Why is global co-operation essential?

- Many environmental challenges are inherently global: we share one atmosphere and many ecosystems provide global public goods. Watersheds cross national borders, and some pollutants travel across continents and oceans. Responding to global environmental challenges requires global solutions and international co-operation.
- Globalisation reinforces the need for global environmental co-operation as the environmental impact of developing countries grows and competitiveness concerns slow down the implementation of more ambitious environmental policies by individual countries.
- Policy coherence in the development agenda requires environment and development co-operation policies to support each other.

What can be done?

- Commit greater efforts to streamlining and strengthening the global environmental governance system, building on the experience of successful multilateral environmental agreements, including through enforcement mechanisms and stable and predictable funding.
- Take advantage of new development co-operation mechanisms to bring sustainable development into the policy discussions with developing countries, and develop specific environmental co-operation mechanisms (including with middle-income countries) to complement budget support.
- Continue to nurture newly emerging forms of environmental co-operation which go beyond binding multilateral agreements and traditional project-based development co-operation, such as policy dialogues and partnerships with the private sector and civil society.

Introduction

The first *OECD Environmental Outlook* (OECD, 2001) suggested that OECD countries needed to strengthen their dialogue with developing countries about environmental problems that were increasingly becoming a mutual concern. Since then, the arguments for strengthening co-operation have become even more powerful.

This chapter summarises the key emerging trends in global environmental co-operation, defined as environmental co-operation on a worldwide scale. Rather than attempt to analyse all possible co-operation mechanisms, the chapter focuses largely on the traditional means of government-to-government co-operation: multilateral environmental agreements on the "environment ministries track" and environmental aid on the "development ministries track". Alternative co-operation mechanisms – such as intra-industry technology transfer, community-to-community de-centralised co-operation and sustainable development partnerships[1] are increasingly important, and are mentioned briefly, but their analysis falls largely beyond the scope of this chapter.

One of the main arguments for international environmental co-operation is to ensure the provision of global public goods (such as climate stability and biodiversity conservation) and the internalisation of environmental externalities. Globalisation has an influence in at least three ways: *i)* accelerated economic growth without corrective environmental policies brings ever-increasing environmental degradation; *ii)* relocation of industrial production, often from OECD to non-OECD countries, reduces the effectiveness of traditional OECD environmental policies aimed at protecting global public goods; and *iii)* the growing economic weight of emerging economies makes their participation increasingly important to address global environmental problems effectively and efficiently (see Chapter 4 on globalisation for more detail)

A second argument relates to the interest of OECD (and other) countries in pursuing stricter environmental policies without facing competitiveness impacts as a result. Although some analysts observe an environmental "race-to-the-top" (particularly within the European Union), issues of competitiveness are increasingly prominent in the dialogue between environmental regulators and the private sector. OECD countries would find it easier to reach the optimal level of environmental regulation if stricter environmental policies were better co-ordinated among themselves and with emerging non-OECD countries with whom they compete on international markets (see Box 22.1). Increased co-ordination on environmental rules would, in turn, reduce the cost for industry to comply with them (see Chapter 4 on globalisation and Chapter 21 on institutions and approaches for policy implementation). Co-operation with BRICs, which is particularly important, is increasingly taking place (see Box 22.2 for the case of China).

A third major argument for increased environmental co-operation is linked to the policy coherence agenda. Contributing to the socio-economic development of developing countries, particularly least developed countries (LDCs) in Sub-Saharan Africa and other regions, is a well established policy objective of most OECD countries. As the 2006 Meeting

Box 22.1. **Reaping mutual benefits from co-operation: The OECD MAD system**

The OECD Mutual Acceptance of Data (MAD) is an example of how countries can derive mutual benefits from environmental co-operation. The OECD is a forum for discussion where governments express their points of view, share their experiences and search for common ground. When member countries consider it appropriate, an accord can be embodied in a formal OECD Council Act. The testing of chemicals is labour intensive and expensive and often the same chemical is to be tested and assessed in several countries. Because of the need to reduce some of this burden, the OECD Council adopted a decision in 1981 stating that data generated in a member country in accordance with OECD Test Guidelines and Principles of Good Laboratory Practice shall be accepted in other member countries for assessment purposes and other uses relating to the protection of human health and the environment. Non-member countries have also been able to benefit from the MAD system since 1997, when another Council Decision set out a stepwise procedure for non-OECD countries with a significant chemical industry to take part as full members in the system. It is estimated that the MAD system saves governments and industry about EUR 60 million per year by avoiding duplicative testing (see also Chapter 18 on chemicals).

of OECD Ministers of Environment and Development recognised (OECD, 2006), achieving environmental sustainability is critical if the gains of development are not to be short-lived. Thus, both OECD and non-OECD countries can gain from co-operation aimed at ensuring sustainable development in the developing world.

The case for environmental co-operation, however, is not always straightforward. The scope for environmental co-operation varies across environmental issues and regions. Moreover, there are significant barriers working against stronger environmental co-operation:

- The very nature of global environmental public goods. Global (and regional) environmental problems are often characterised by asymmetries in the distribution of costs and benefits of co-operation and the "free rider" problem.[2] Mechanisms to deal with those features – such as compensatory payments and enforcement clauses in multilateral environmental agreements (MEAs) – are rarely used.
- Multilateral governance limitations. These include: *i)* the dynamics of international environmental negotiations (often characterised by lack of trust and a high level of complexity); *ii)* lack of coherence in the current international governance system; and *iii)* the often low profile of environmental issues on the foreign policy agenda.
- Political and capacity constraints in non-OECD countries. The low status of the environment on domestic political agendas – often due to the combination of more pressing issues, a weak understanding of poverty and environment links and a low level of public awareness – and fragile environmental institutions (see Chapter 21 on institutions and approaches for policy implementation) prevent many non-OECD countries from engaging in mutually beneficial co-operation.
- Weak analytical base. Uncertainty about the underlying data and the analysis of environmental problems and policy options prevents some countries from fully engaging in global co-operation.

Box 22.2. **China and international co-operation**

Given its strong global economic role and large population, China has emerged as a major contributor to environmental pressures. China is the world's largest producer and consumer of ozone depleting substances, is likely to already be the largest contributor of greenhouse gases, is a major source of acid rain in Northeast Asia and is responsible for a large part of the land-based pollution of the East Asian regional seas.

At the same time, the last decade has seen a dramatic increase in China's engagement with other countries in addressing environmental challenges. China is now an active, constructive participant in a broad array of regional and global environmental conventions, institutions and programmes, and it is drawing heavily on international financial institutions and special mechanisms to augment its own resources so as to ensure that its international commitments are met. In addition to substantive efforts to tackle a range of transboundary environmental issues, the Chinese government has examined how its trade and investment policies can work to support environmental management goals as a first step to ensure that Chinese corporations operating overseas contribute to sustainable development.

But lack of strong monitoring, inspection and enforcement capabilities and associated penalties in China are limiting the effectiveness of otherwise sound policies, laws and regulations. Funding limitations and inadequate institutional co-ordination also hinder progress. To achieve success with its ambitious international environmental agenda will require increased financial efforts from China, as well as major technical support and targeted financial assistance to China from OECD countries and international financial institutions.

Source: OECD (2007a).

Delivering better international environmental governance

Environmental co-operation takes place to a large extent through the implementation of negotiated international legal instruments. Multilateral environmental agreements (MEAs) thus constitute the basis of the global environmental governance system. But as this *Outlook* shows, the global environment is not improving. This, together with the expected new challenges associated with globalisation, provides a strong rationale for continuing to improve the system of global environmental governance.

International environmental governance has been strengthened by the entry into force of a number of important MEAs in recent years, although many other MEAs have not been sufficiently ratified. There are more than 500 environment-related international treaties and other agreements, of which 323 are regional and 302 date from the period starting in 1972 (UNEP, 2006).[3] The emergence of regional integration bodies concerned with the environment, such as in Central America and Europe, has contributed to this trend. The largest cluster of MEAs is related to the marine environment, accounting for over 40% of the total. Biodiversity-related conventions form a second important but smaller cluster. Since 1972, two new important clusters of conventions have emerged, governing: *i)* chemicals and hazardous waste, primarily of a global nature; and *ii)* the atmosphere and climate change. As the number of MEAs in force has continued to expand, their implementation has become more demanding.

The rate of signature and ratification of new MEAs has decreased in the last few years, and this is likely to continue in the immediate future. The rate of ratification of MEAs picks up around major global conferences, such as the 1972 UN Conference on the Human Environment, the 1992 UN Conference on Environment and Development, and, to a lesser extent, the 2002 World Summit on Sustainable Development (WSSD). In the last decade the focus has shifted from getting new agreements signed (see Figure 22.1), to implementing existing agreements. This can also be seen at the regional level, for example, in the preparations for the 2007 Environment for Europe Ministerial Conference.

Figure 22.1. **Multilateral environmental agreements, 1960-2004**

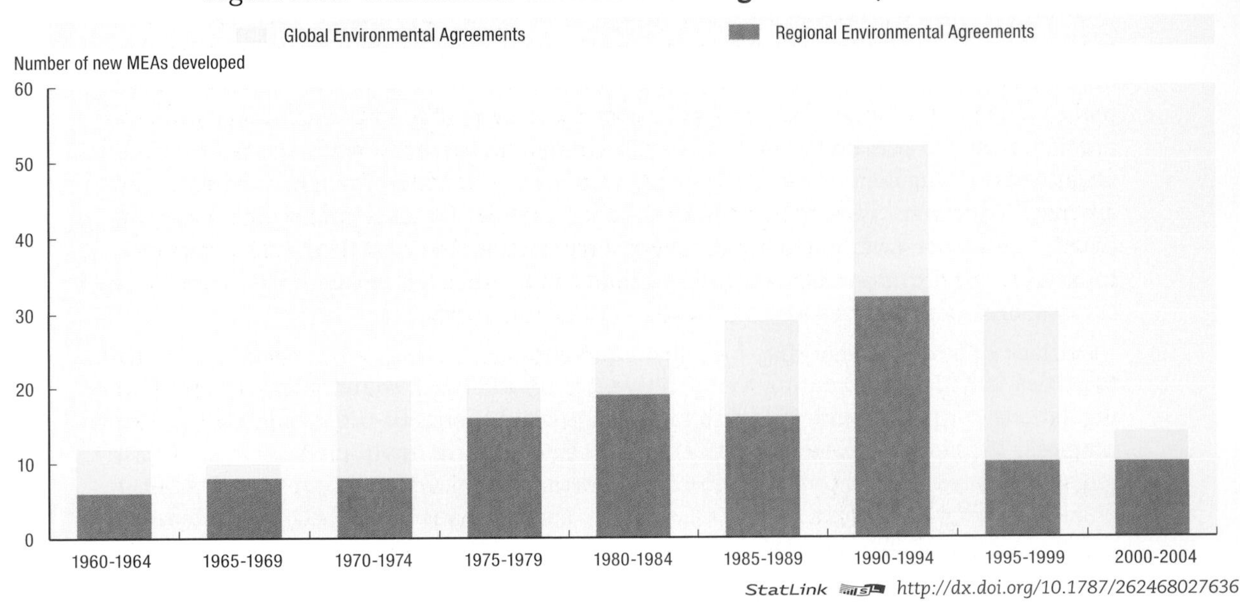

StatLink http://dx.doi.org/10.1787/262468027636

Source: UNEP Environmental Law Instruments Website, *www.unep.org/dpdl/Law/Law_instruments/multilateral_instruments.asp*

While many countries have displayed genuine interest in addressing environmental issues through international means and instruments, the international community has not been able to create a coherent system to support those countries. The system of MEAs has evolved largely in a piecemeal fashion, and as the number of MEAs has increased, problems of multiplicity, overlap and conflict have become more evident. The next decades are likely to witness significant efforts to streamline the system of MEAs and strengthen its coherence. The launching of the Strategic Approach to Integrated Chemicals Management (SAICM, see Chapter 18 on chemicals) and discussions about a World Environment Organisation (see Box 22.3) are evidence of the trend towards increased coherence. In addition, bottom-up approaches are also emerging, such as the novel decision of the Basel, Rotterdam and Stockholm conventions on waste and chemicals management to prepare joint recommendations for enhancing co-operation and co-ordination.

Many developing and transition countries have signed MEAs in the hope that significant support for implementation would be provided, but the result may have been counter-productive. The proliferation of international processes has placed a particularly heavy burden on developing countries which often lack the capacity to engage meaningfully and consistently in the lengthy negotiations for the development of international environmental policy. When MEA-related funding arrives, activities linked to MEAs may be given priority over more pressing national environmental priorities that do not receive adequate attention by national governments. A possible solution is greater co-ordination of capacity-building across MEAs, such as that spearheaded by the Global Environment Facility/United Nations Development Programme projects on capacity-building for implementing the Rio Conventions. Beyond that, future trends may include

OECD ENVIRONMENTAL OUTLOOK TO 2030 – ISBN 978-92-64-04048-9 – © OECD 2008

Box 22.3. **Towards a world environment organisation?**

Proposals to create a world environment organisation have been made for over 30 years. The initial response of the international community was to set up the United Nations Environment Programme (UNEP) in 1972.* Originally, UNEP was to evolve into an "environmental conscience" within the UN system that would act as a catalyst triggering environmental projects in other bodies and helping co-ordinate UN environmental policies. The debate about a larger, more powerful agency for global environmental policy has resurfaced at several points in time. Reasons advanced historically for such a body include improving the effectiveness of UNEP, strengthening co-ordination among MEAs, securing stable financing and having an environmental counterweight to the World Trade Organisation (WTO). But critics maintain that while the benefits of a world environment organisation remain uncertain and questionable, political attention and scarce resources should not be diverted to experiments with organisational reform. The WSSD gave new impetus to the debate, and by 2004 no less than 17 proposals for a new intergovernmental organisation had been put forward (Bauer and Biermann, 2004).

Options fall into three categories: *i)* upgrading UNEP to a UN specialised agency (with the World Health Organization and International Labour Organization as role models); *ii)* integrating the multiple existing agencies and programmes dealing with environmental issues into one all-encompassing world environment organisation outside the UN system (with the WTO as a role model); or *iii)* creating a hierarchical intergovernmental organisation equipped with majority decision-making as well as enforcement powers for states that fail to comply with international environmental agreements. Although it is often argued that the third option is the only one that would overcome the free rider problem that has traditionally undermined the effectiveness of MEAs, support for this model remains scarce. One alternative to a world environment organisation would be to cluster MEAs. This would help to address issues of institutional overlap and fragmentation amongst MEAs by allowing individual governments to champion well designed clusters that address environmental macro issues such as the atmosphere, hazardous substances, the marine environment and extractive resources (von Moltke, 2005).

While there seems to be an emerging convergence of views towards the first option, the outlook remains uncertain. The next decade is likely to witness the strengthening of UNEP in one way or another, with the current discussion in international deliberations focusing on the options for an enlarged mandate and a more predictable financial basis. The future of UNEP is currently being debated in the UN General Assembly. The UN Secretary General's High-Level Panel for UN Reform has recommended that "UNEP should be upgraded and have real authority as the environmental policy pillar of the UN system" (UN, 2006). The UN Informal Consultative Process on the Institutional Framework for the UN's Environmental Activities concluded that "there is wide recognition that the efforts to create a more coherent institutional framework for the UN's environmental activities should start by strengthening and building upon existing structures and better implementing past agreements" (Berruga and Maurer, 2006). A recent common EU proposal is to transform UNEP into a UN specialised agency (to be known as the United Nations Environment Organisation, UNEO) that would exercise cross-cutting functions for MEAs such as information exchange and centralisation, regional and global co-ordination of activities, and streamlining of the international agenda of MEA meetings. But there is no common position on a possible UNEO, even among OECD countries.

* Not as a specialised UN organisation, but as a subsidiary body of the General Assembly supported by a "small secretariat".

stronger work for aligning global and national agendas and provision of support for general institutional strengthening that would contribute to the management of both national and global environmental issues.

A major set of MEA challenges includes enforcement, financing and developing burden-sharing amongst countries. Control or review mechanisms on the implementation of a number of existing conventions have recently been strengthened – such as the 1979 Geneva Convention on long-range transboundary air pollution – but enforcement of MEAs remains a major issue. Currently MEAs are ill-equipped to fight the "free rider" problem as hardly any MEA has enforcement provisions. The Kyoto Protocol to the UN Framework Convention on Climate Change is an important exception, with a new compliance mechanism launched in March 2006. This is possibly the fore-runner of a new trend making enforcement mechanisms part and parcel of MEAs. Compliance with MEAs among developing and transition countries is often promoted through financial mechanisms, such as the Multilateral Fund for the Implementation of the Montreal Protocol on ozone-depleting substances or the Global Environment Facility (see Box 22.4). However, complaints about insufficiency of donor resources to support MEA implementation are constantly voiced in MEA discussions and these are likely to continue.

Improvements in the international environmental governance system are taking place, but at a slow pace.

As developing countries increasingly become the drivers of environmental degradation, and as their economic power strengthens, pressures on them to help share the burden of implementing MEAs can be expected to increase significantly in the period covered by this *Outlook*. However, distribution issues (such as access and benefit-sharing under the Convention on Biological Diversity or the role of developing countries in a post-2012 international climate change framework) are likely to prove increasingly problematic and, if unresolved, may prevent major advances in environmental co-operation.

In addition to multilateral environmental agreements, environmental issues are also considered in international and regional trade agreements (see Chapter 4 on globalisation). The World Trade Organization (WTO) has included environmental elements in the trade

Box 22.4. **The Global Environment Facility (GEF)**

The Global Environment Facility (GEF) was established in 1991 to help developing countries fund projects and programmes that protect the environment, initially in the fields of biodiversity, climate change, ozone depletion and international waters. It emerged from the 1992 UN Conference on Environment and Development as the main multilateral financial mechanism for distributing funds for global environmental goals, reflecting a recognition that developed and developing countries have shared, but differentiated, responsibilities in achieving these goals. After a 1994 re-structuring, the GEF is replenished every four years. Between 1994 and 1998, 34 countries contributed USD 2 billion; 1998 to 2002 saw 36 countries donating USD 2.75 billion, and from 2002 to 2006, USD 3 billion was contributed by 32 countries. For the latest period, 2006-2010, 32 countries have pledged USD 3.13 billion. In recent years, the GEF's remit has expanded to cover new environmental issues, such as land degradation and persistent organic pollutants. It is not clear, however, whether budget replenishments have kept up with the expanded agenda.

negotiations, but progress on the environmental component is unlikely to be fast given the slow progress with the Doha Development Agenda. Given the growing importance of regional trade agreements (RTAs) in advancing international trade, the on-going inclusion of environmental elements in RTAs is encouraging. Most RTAs dealing with environmental issues do so in the form of commitments by parties to co-operate on environmental matters. The scope and depth of these commitments vary, and range from co-operation in one specific technology area to fully-fledged co-operation programmes (see Chapter 4 on globalisation).

Aid for environment in a changing development co-operation context

In the international arena, OECD countries are often proponents of progressive solutions for environmental problems, but they also co-operate with developing countries on environmental issues through traditional development co-operation channels. Indeed, a significant part of environmental co-operation takes place within the broader development co-operation agenda (the development track of environmental co-operation). In recent years, this agenda has been evolving in ways that pose both challenges and opportunities for strengthening environmental co-operation.

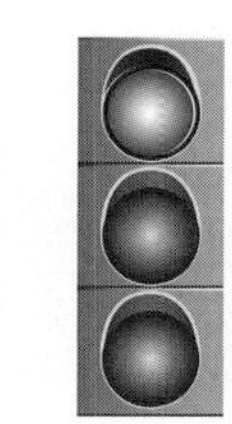

Environmental aid has been decreasing since 1996 as a share of donor country GDP and total aid.

Official development assistance to developing countries from member countries of the OECD's Development Assistance Committee (DAC) has increased rapidly in recent years. The 2002 Monterey Conference on Financing for Development set the target of doubling official development assistance (ODA) from a base of USD 50 billion. From 2004 to 2005 alone, ODA rose 31% to a record high of USD 106 billion, or 0.33% of DAC members' combined gross national income (GNI). This rapid growth in ODA has mainly been fuelled by debt relief,[4] and so it is unlikely to continue as debt relief scales back. Indeed, between 2005 and 2006 aid fell by 5.1%.

The environment has not benefited from the increased availability of aid money. In real terms, aid for environment has been relatively stable over the last 15 years when defined in broad terms, but declining when defined narrowly.[5] The decline in "core" environmental aid can be attributed to a 17% reduction in support from bilateral donors (who have traditionally provided over 80% of this aid) between 1996 and 2005. A recent upsurge in "extended" environmental aid (which peaked in 2005 at over USD 12 billion) is explained by much stronger support of bilateral donors for water-related programmes – it more than doubled between 2003 and 2005. By any definition, however, environmental aid has been declining as a share of donor country GDP and of total aid (see Figure 22.2). This is not just the case for the "environment sector" and it can be partly explained by an increase in non-sector specific ODA, such as debt relief and aid for emergencies and reconstruction.

The composition of environmental aid is changing. The water sub-sector accounts for the lion's share of "extended" environmental aid, at about 40% of total environmental aid since 1990. This is expected to continue at least for the next decade, given the visibility of water issues within the Millennium Development Goal framework (MDG, see below). Donor support for biodiversity and solid waste has increased by some 50% in real terms in the same period, but it remains relatively small, at less than 2% of total environmental aid for

Figure 22.2. **Aid for environment, 1990-2005**

Environment-related official development assistance (ODA) as % of total ODA

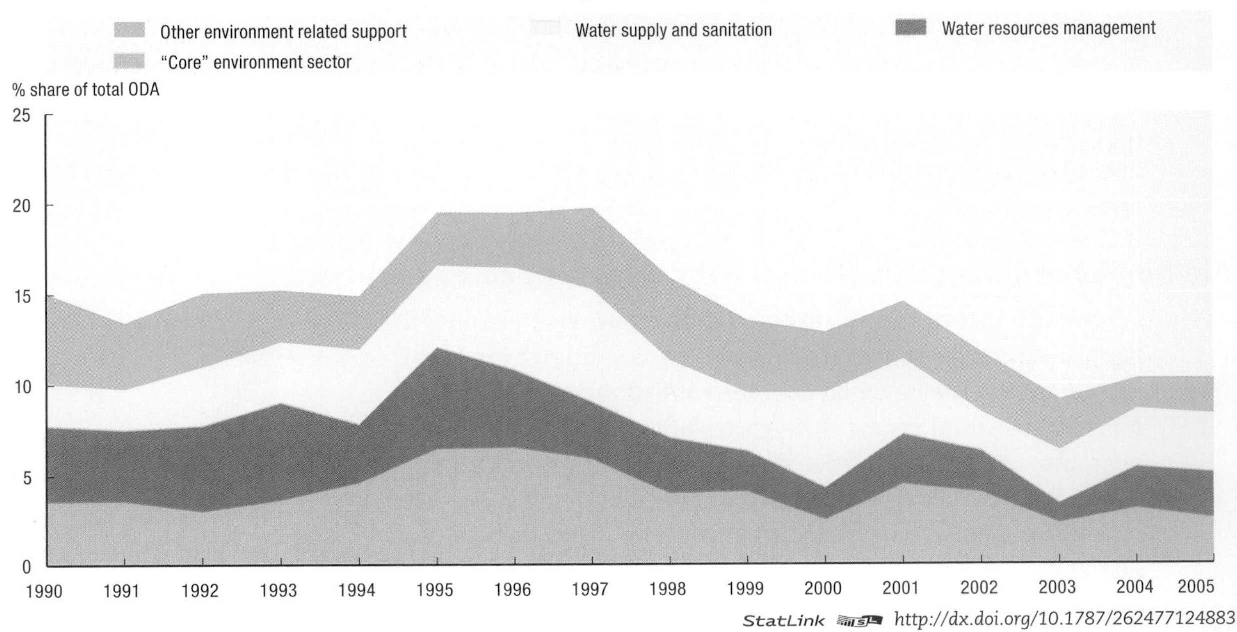

StatLink http://dx.doi.org/10.1787/262477124883

Note: Data refer to bilateral and multilateral ODA. For the purpose of this analysis the environment-related sector includes the following activities:

- "Core" *environment sector*: general environmental protection (environmental policy, biosphere protection, biodiversity, environmental education/research), waste management, renewable energy and agricultural land resources.
- *Water resources management*: water resources protection, flood prevention/control, river development, agricultural water resources.
- *Water supply and sanitation*: basic drinking water supply and sanitation, large systems water supply and sanitation.
- *Other environment-related support*: urban and rural development, forestry and fisheries development.

Source: OECD CRS Aid Activity Database (Creditor Reporting System) at *www.oecd.org/dac/stats/idsonline*.

each sub-sector. Aid for renewable energy had decreased significantly until 2003 but it is picking up – largely due to the changing fortunes of hydro-power projects, which represented 93% of aid for renewable energy in 1990, 32% in 2003 and 43% in 2005. Despite its importance for agricultural productivity, aid for land management has also decreased significantly since 1997, from 3.2% to 2% of total environmental aid in 1997.

Environmental aid is not evenly distributed across regions.[6] At the same time, the geographical distribution of environmental aid has been evolving. For example, Eastern Europe and Central Asia have gained a significant share of environmental aid over the last 15 years (see Table 22.1). Projections of how environmental aid allocations will be distributed in the future are difficult to substantiate. Total aid resources to Sub-Saharan Africa are rapidly increasing and it is likely that environment-related flows would also increase, although at a slower pace. In addition, given South Asia's current low income levels and low allocations of environmental aid on a per capita basis (Table 22.1), it would seem that this region is poised to receive an increasing share of global environmental aid.

The Millennium Development Goals (MDGs) are an integrated target-based framework to guide development co-operation around which development partners have rallied. The accepted approach for achieving the MDGs is to support developing country-owned broad-based growth and poverty reduction strategies, as reinforced by the Paris Declaration on Aid Effectiveness. Support for those strategies will increasingly be channelled through general and sector-specific budget-support instruments, rather than through specific investment projects.

Table 22.1. **Environmental aid to developing regions, 1990-2005**

	GNI per capita in 2005 (thousand USD)	Environmental aid to each region (as % of total environmental aid)			Environmental aid in 2005 (USD per capita)
		1990-1994	1995-1999	2000-2005	
Europe and Central Asia	4.1	2.6	4.4	6.5	2.3
Sub-Saharan Africa	0.7	16.9	14.1	17.4	4.0
Latin America and the Caribbean	4.0	23.4	21.9	16.7	4.1
East Asia and Pacific	1.6	32.3	29.2	27.1	2.4
Middle East and North Africa	2.2	9.4	10.4	13.9	8.8
South Asia	0.7	14.5	18.5	15.2	2.0
Unallocated/unspecified		0.9	1.6	3.1	

StatLink http://dx.doi.org/10.1787/257811010326

Note: Data include official development assistance (ODA) and other official flows (OOF).
Source: DAC Creditor Reporting System database and authors' calculations.

This new development context poses major challenges for environmental co-operation. So far, the treatment of environment in poverty reduction strategies (formulated in what are known as Poverty Reduction Strategy Papers – PRSPs) is not encouraging. World Bank reviews (see Bojo *et al.*, 2004) find that while there is considerable variation among countries, the level of environmental mainstreaming is generally low. With the growing importance of general budget support, it is becoming more difficult to ensure that a given share of aid gets devoted to supporting environmental sustainability. And as the share of aid for investment projects decreases, the role of traditional tools for integrating environmental considerations into development assistance, such as environmental impact assessments, is reduced. There is now a clear need to enhance the capacity of environmental staff in developing countries to interact with their finance ministry colleagues and make convincing arguments to steer budget-support resources towards achieving environmental objectives. As conditioning aid to environmental performance becomes less and less feasible, strengthening recipient-country environmental constituencies so that those constituencies can get environment higher on the national co-operation agenda will become increasingly important.

At the same time, with the expansion of sector-specific budget support, the importance of inter-sectoral policy dialogue and strategic environmental assessment will also increase. These developments may also open opportunities to influence sectoral policies and put the sectors (whether agriculture, transport or energy) on a more sustainable path. Seizing this opportunity will not prove easy, as shown by the fate of the environmental MDGs (see Box 22.5) and the problems faced by many developing countries to take advantage of the Kyoto mechanisms (see Box 22.6). On the positive side, awareness about the impact of environmental quality on development is increasing and may continue to do so as better knowledge is brought to the fore. For example, information like the fact that 25% of total wealth in developing countries is environment-based (compared to less than 4% in OECD countries) and that about 24% of the global burden of disease is attributable to environmental causes is becoming more widely known (see Chapter 12 on health and environment).

Another emerging feature is the risk of depriving middle-income countries of environmental aid at a critical moment. Environmental co-operation has traditionally been facilitated by grant money. As grant money increasingly gets diverted to the poorest countries – which have many pressing needs that are higher on the agenda than the environment – middle-income developing countries are finding it harder to access grant

Box 22.5. **The environment and the Millennium Development Goals**

Environment has a place in the MDG framework,* both through a dedicated goal (MDG7: Ensure environmental sustainability) and through the linkages to the other goals. Promoting off farm sources of income and technological improvement will be key for reducing income poverty in rural areas so as to achieve MDG1 (Eradicate extreme poverty and hunger). But it is difficult to imagine achieving this where land is degraded and water absent. Reductions in child mortality (MDG4) will be more likely if households have access to adequate water supply, sanitation facilities, and modern fuels. Ready access to fuel and water lessens the time demands on women and girls, allowing them to engage more in productive activities (MDG3: Promote gender equality and empower women) and attend school (MDG2: Achieve universal primary education). Climate change will favour the spread of vector-borne diseases (undermining MDG6: Combat HIV/AIDS, malaria and other diseases) and increase the likelihood of natural disasters. Those disasters, in turn, reduce income and destroy the infrastructure for education and health.

Implementing MDG7 (Ensure environmental sustainability) poses a major challenge, however. This is especially true for Target 9 on integrating the principles of sustainable development into country policies and programmes and reversing the loss of environmental resources. This is the only non-quantitative target in the MDG framework, and as a result it often gets pushed aside in the programmes of bilateral donors and international financial institutions. By contrast, Target 10 on halving the proportion of people without access to safe water, is proving more successful in attracting the attention of the development community (see Chapter 10 on freshwater).

* See *http://www.un.org/millenniumgoals/* for the full list of goals.

resources for the environment. Ironically, this is occurring precisely at the moment in their development trajectory when environmental issues tend to gain attention in national agendas. At the same time, as domestic financial resources become increasingly available for environmental protection, expertise is becoming a more critical constraint for emerging economies to improve their environmental performance. Indeed, countries like Brazil and China ask for international financial institution loans primarily to access knowledge. The coming years will witness increased needs and opportunities for OECD countries to co-operate with influential emerging economies over knowledge transfer.

Box 22.6. **Who is benefiting from the Clean Development Mechanism?**

The Clean Development Mechanism, established under the Kyoto Protocol to the UN Framework Convention on Climate Change, has emerged as an important source of additional environmental finance. By early 2007, more than 500 CDM projects had been registered and 1 000 were undergoing evaluation. The majority of credits are targeted to projects to reduce hydrofluorocarbon (HFC) and nitrous oxide (N_20) gases from industrial production, or to support development of renewable energy. However, contrary to what some expected when the CDM was established, the poorest countries, and in particular African ones, are benefiting little from this new mechanism. Eighty-four per cent of expected credits from registered projects will come from just China, India, Brazil, Mexico and Korea, and 51% from China alone.

Source: Ellis and Kamel (2007).

 OECD ENVIRONMENTAL OUTLOOK TO 2030 – ISBN 978-92-64-04048-9 – © OECD 2008

Responses to those challenges are starting to emerge. The 2006 Meeting of OECD Development and Environment Ministers launched an alliance for sustainable development aimed at combining the expertise and resources of OECD environment and development ministries to support developing country needs through better designed development programmes and targeted capacity-building. Given the scarcity of available resources and the focus on country-ownership and aid effectiveness, pressures for improved donor co-ordination will continue to increase.

The emergence of alternative forms of co-operation

Beyond binding multilateral agreements and traditional project-based development co-operation, alternative forms of environmental co-operation are emerging. Policy dialogues are one of them. While they are different initiatives, policy dialogues tend to complement traditional forms of government-to-government co-operation by helping to provide a framework, guide traditional initiatives and, in some cases, help to promote donor co-ordination. They are implemented through training workshops and policy seminars. Examples include the UK Sustainable Development Dialogues[7] (with China, Brazil and India), which aim to make sustainability a core principle in bilateral relationships and provide a coherent framework for co-operation using a cross-governmental, multi-level approach. The EU Water Initiative Policy Dialogues are another example.

Beyond government-to-government co-operation, the importance of other actors in environmental co-operation is growing. In looking for genuine opportunities to create value within the framework conditions laid down by governments, the private sector can be an effective instrument of environmental co-operation. Responding to environmental demands from investors, consumers and employees, leading OECD multinational enterprises are finding a competitive advantage in pursuing sustainability (Esty and Winston, 2006). They are a vector of technology transfer through direct investments, and a driver of enhanced performance through demands on their suppliers (see Chapter 4 on globalisation). Analysis of the business role in implementing MEAs shows that the private sector is actively contributing to the achievement of environmental objectives in certain domains, although not in others (see Box 22.7).

Box 22.7. **Business and the environment: trends in MEA implementation**

Business action that contributes to addressing the goals of the UN Framework Convention on Climate Change has increased significantly, especially since the entry into force of the Kyoto Protocol in 2005. Private sector action in addressing the goals of the UN Convention on Biological Diversity is much lower, but on the rise, although very limited business action is occurring to meet the goals of the UN Convention to Combat Desertification. The financial sector is increasing its involvement, especially in the area of climate change. Its involvement includes developing standards to incorporate social and environmental criteria in their lending practices, investing in clean technology, especially renewable energy; and offering metrics and benchmarks to assess the effect of environmental issues on risk management.

Source: OECD, 2007c.

Additional environmental co-operation actors include local jurisdictions ("decentralised co-operation") and civil society organisations (CSOs). In addition to taking part in implementing environmental projects and monitoring environmental policies, CSOs also influence co-operation by governments and business through their impact on public opinion. The involvement of stakeholders in environmental policy design and implementation is likely also to continue to increase (see also Chapter 21 on institutions and approaches for policy implementation).

Partnerships for sustainable development, which made a strong appearance at the 2002 World Summit on Sustainable Development (WSSD), allow different stakeholders to work together to achieve sustainable development outcomes. The UN Commission on Sustainable Development (CSD) maintains a database that includes over 300 such initiatives, but in addition to the CSD-registered partnerships many other partnerships aimed at promoting environmental sustainability do exist. While partnerships come in all shapes and sizes, they are potentially an important instrument for environmental co-operation between OECD and non-OECD governments and other actors. Recent analyses of UN CSD-registered partnerships by the OECD and by the World Bank on global and regional programmes stress the need for further efforts in ensuring and evaluating the effectiveness and efficiency of partnerships (see Box 22.8). Partnerships are likely to become an increasingly important complement to government commitments and multilateral environmental agreements.

Box 22.8. **Effectiveness and efficiency of partnerships involving OECD governments**

Partnerships can be defined as voluntary arrangements that share risks and benefits among partners and combine and leverage the financial and non-financial resources of partners to achieve specific goals. The use of the partnership approach is growing worldwide. Partnerships are often seen as a complement to traditional governmental approaches to environmental protection and sustainable development, and as having an important role to play in leveraging funding from different sources, supporting dissemination of technology, and bringing together expertise from governments, universities, the business community, environmental organisations and others.

Comparatively little work has been done on evaluating partnerships, perhaps because insufficient time has elapsed since partnerships were launched at the WSSD. A survey of CSD-registered partnerships conducted by the OECD revealed that only 28% of the partnerships that responded had completed an evaluation. At the same time, there is an increasing interest in methodological aspects of partnership evaluation, and several organisations have developed assessment frameworks and methodologies to evaluate partnerships. Existing evaluations of partnerships have revealed that success factors relate both to good project management (such as clear objectives, detailed plans, good leadership, sufficient resources and accountability) and the dynamics of partnerships (such as understanding the needs of different partners, shared ownership and flexibility). Examination of the major costs and benefits generated by partnerships, however, has been rare.

In addition to procedural aspects, evaluations of partnerships involving governments could address: policy rationale for the partnership, effectiveness, efficiency (including transaction, operational and opportunity costs), benefits, financial leverage, policy consistency and sustainability. Two aspects that risk hindering the development of robust evaluations are the absence of clear objectives and the failure to analyse the costs and benefits of non-partnering solutions.

 OECD ENVIRONMENTAL OUTLOOK TO 2030 – ISBN 978-92-64-04048-9 – © OECD 2008

In a globalised world, there is a compelling rationale for OECD countries to increase environmental co-operation with non-OECD countries. Improving international environmental governance and supporting developing countries to steer their economies onto a more sustainable path through evolving co-operation modalities pose major challenges. But they also provide new opportunities to improve the world's environmental outlook.

Notes

1. Sometimes known as Type II Partnerships, these multi-stakeholder partnerships were launched at or after the 2002 World Summit on Sustainable Development.
2. The free rider problem refers to a situation where agents that do not contribute to providing a benefit cannot be excluded from enjoying it.
3. For a more detailed treatment of particular MEAs (such as the United Nations Framework Convention on Climate Change, Convention on Biological Diversity or the United Nations Convention to Combat Desertification), please see relevant chapters earlier in this report.
4. In 2005, debt relief for Iraq and Nigeria amounted to USD 19 billion.
5. The level of environmental aid depends on its definition. For the "narrowest" definition, the level of environmental aid since the mid-90s has fluctuated around USD 3 billion (in constant terms), roughly 80% of which has been provided by bilateral donors. For the "broadest" definition (that also includes water resources management, water supply and sanitation as well as urban, rural, forestry and fisheries development) the level has fluctuated at around USD 10 billion, with bilateral donors accounting for two-thirds of the total.
6. There is no reason why it should be – allocation of environmental aid should in principle be guided by recipient country characteristics such as endowment of environmental resources of global importance, the role of environmental resources in fighting poverty, and the ability of the recipient country to transform aid resources into effective environmental protection. But environmental aid does not seem to be always allocated on those grounds. For example, Acharya *et al.* (2004) showed that within the World Bank, environmental aid is highly correlated with the size of the Bank's country programmes.
7. Led by the UK Department of Environment, Food and Rural Affairs, in close collaboration with a range of other government departments including the Foreign and Commonwealth Office and Department for International Development.

References

Acharya, A. *et al.* (2004), "How Has Environment Mattered? An Analysis of World Bank Resource Allocation", *World Bank Policy Research Working Paper* No. 3269, World Bank, Washington, DC.

Bauer, S. and F. Biermann (2004), "Does Effective International Environmental Governance Require a World Environment Organisation? '*Global Governance Working Paper* No. 13, "The Global Governance Project", Amsterdam, Berlin, Oldenburg, Postdam: The Global Governance Project.

Berruga, E. and P. Maurer (2006), *Co-Chairs' Summary of the Informal Consultative Process on the Institutional Framework for the UN's Environmental Activities*, United Nations, New York.

Bojo, J., K. Green, S. Kishore, S. Pilapitiya and R. Reddy (2004), "Environment in Poverty Reduction Strategies and Poverty Reduction Support Credits", *World Bank Environment Department Paper* No. 102 World Bank, Washington, DC.

Ellis, J. and S. Kamel (2007), *Overcoming Barriers to Clean Development Mechanism Projects*, [*www.oecd.org/env/cc/aixg*], OECD/IEA, Paris.

Esty, D. and A. Winston (2006), *Green to Gold: How Smart Companies Use Environmental Strategy to Innovate, Create Value, and Build Competitive Advantage*, Yale University Press, New Haven, CT.

Gupta J. (2005), "Global Environmental Governance: Challenges for the South from a Theoretical Perspective", in F. Biermann and S. Bauer (eds.), *A World Environment Organization: Solution or Threat for Effective International Environmental Governance?* Ashgate, Aldershot (UK).

Moltke von K. (2005), "Clustering International Environmental Agreements as an Alternative to a World Environment Organization", in F. Biermann and S. Bauer (eds.), *A World Environment Organization: Solution or Threat for Effective International Environmental Governance?* Ashgate, Aldershot (UK).

OECD (2001), *OECD Environmental Outlook*, OECD, Paris.

OECD (2006), *Meeting of the OECD Development Assistance Committee and the Environment Policy Committee at Ministerial Level – Co-Chairs Summary*, OECD, Paris.

OECD (2007a), *Environmental Performance Review of China,* OECD, Paris.

OECD (2007b), *Environment and Regional Trade Agreements*, OECD, Paris.

OECD (2007c), *Business contribution to MEAs: Suggestions for Further Action*, ENV/EPOC/GSP(2007)1/FINAL, [*www.oecd.org/env*] OECD, Paris.

OECD (2008), *Evaluating the Effectiveness and Efficiency of Partnerships Involving Governments from OECD Countries*, ENV/EPOC(2006)15/FINAL, [*www.oecd.org/env*], OECD, Paris.

UN (2006), *Delivering as One – Report of the Secretary-General's High Level Panel,* United Nations, New York.

UNEP (United Nations Environment Programme) (2006), *Multilateral Environmental Agreements* (webpage). *httwww.unep.org/dpdl/Law/Law_instruments/multilateral_instruments.asp* (accessed 6th July 2006).

UNEP (2006), *UN Reform – Implications for the Environment Pillar,* Issue paper by the Deputy Executive Director UNEP/DED/040506, UNEP, Nairobi.

ISBN 978-92-64-04048-9
OECD Environmental Outlook to 2030

ANNEX A

Regional Environmental Implications

This annex summarises the Outlook's *main Baseline developments for a number of world regions, including the economic and social drivers of environmental change, and the main environmental developments to 2030. The key projections for each region are highlighted, and global indicators allow regional performance to be compared with global averages.*

Introduction

Country clusters are an appropriate level of aggregation for a number of environmental issues. World regions are increasingly more integrated; for instance, intra-regional trade has grown in all regions (with the exception of Central and Eastern Europe), and will be a major driver of economic integration to 2030 (see Chapter 4 on globalisation). The vulnerability of different regions to environmental damage will vary to 2030. In combination with variations in environmental pressures – including from climate change which will have the greatest impacts on developing countries – uneven capacities to respond will cause region-specific physical, economic and social impacts.

This annex summarises the *Outlook*'s main Baseline developments by region, including economic and social drivers, and environmental developments (see Box A.1 for some key assumptions and limitations). Because the level of information is uneven between regions, some of the 13 regional clusters used in the *Outlook* (Table A.1) have been merged.

Each section covers one region: it summarises the main data of note for that particular region,[1] including regional data developed by the Intergovernmental Panel on Climate Change (IPCC, 2007). Table A.15 contains all the key global indicators, so that the performance of each region can be compared with the global averages. Data for Brazil, Russia, India and China (the BRICs countries) are emphasised in particular, as the emerging influence of these rapidly industrialising countries is of particular interest in this *Outlook*.

This annex also highlights differences in region-specific environmental agendas, information that would be relevant for the development of regional environmental co-operation. Interdependencies within regions make regional environmental co-operation particularly relevant. Regions are unevenly governed in this domain.

Box A.1. **Assumptions and key uncertainties**

The limitations of many of the regional projections lie in the limited data and uncertainty about the underlying economic, demographic, technological and other factors. For example:

- The spatial distribution of changes in temperature and precipitation are subject to large uncertainties.
- Future developments of irrigated areas and volumes are very uncertain. This has consequences for the projections of water withdrawals, and hence projections of the availability of fresh water.
- In Eastern Europe, the Caucasus and Russia, the database to gauge future growth potential is limited.
- A high economic growth scenario, based on the most recent performance of the regions in terms of productivity growth, generates more optimistic forecasts for GDP growth than those presented in this chapter, but higher environmental pressures, especially in Latin America and Africa (see also Chapter 6).

Table A.1. **The 13 regional clusters used in the *Outlook***

OECD	BRIC	Rest of the world
OECD North America	**Brazil**	**Middle East**
• Canada	• Brazil	• Middle East countries
• USA		
• Mexico		
OECD Europe	**Russia and the Caucasus**	**Other Asian countries**
• Western Europe	• Russia	• Indonesia
• Central Europe	• Caucasus	• Rest of South East Asia
• Turkey		
OECD Asia	**South Asia**	**Eastern Europe and Central Asia**
• Japan	• India	• Ukraine region
• Korea	• Other South Asian countries	• Central Asia
OECD Pacific	**China**	**Other Latin American and Caribbean Countries**
• Australia	• China region	• Central America and the Caribbean
• New Zealand		• Rest of South America
• Rest of Oceania		
		Africa
		• All countries in the African continent

Regional environmental profiles

OECD North America

The population in the US is projected to grow faster than in most OECD countries, and to receive almost half of the annual flow of international migrants. North America's share of world population (Table A.2) will remain stable to 2030 (above 6%). This increase in labour force is expected to be a major driver of the region's economic performance to 2030. The level of GDP per capita will remain significantly higher than in other regions.

It is expected that North America will represent 21% of world energy consumption in 2030, down from 25% in 2005. This reduction is driven by the increase of the service sector in the economy. Final[2] energy use per capita remains high, however, and is still

Table A.2. **North America: Key figures, 1980-2030**

	1980	2005	2030	% changes 1980-2005	% changes 2005-2030
Population (million)	**322**	**429**	**522**	**33.0**	**21.9**
% of world total	7.2	6.6	6.3		
GDP per capita (USD)	–	30 253	47 495		57.0
Primary energy consumption					
Total (% of world total)	**27**	**25**	**21**	**39.4**	**29.8**
Final energy use					
Total (% of world total)	**27**	**25**	**21**	**32.4**	**32.4**
Climate change					
GHG basket emissions (% of world total)	**22**	**20**	**18**	**30.2**	**25.0**
Energy related CO_2 emissions (Gt CO_2)	**5.27**	**7.23**	**9.14**	**37.2**	**26.3**
Energy related CO_2 emissions per capita (t CO_2)	16.35	16.87	17.49	3.2	3.7
Nitrogen emission (% world)	30.7	21.8	14.2	–31.1	–34.9
Sulphur emission (% world)	27.5	12.3	10.4	–64.3	–11.5
Land use					
Food crops (% of world total)	18.5	16.7	15.8	2.5	9.9
Natural forested area (%)	19.3	20.2	21.5	–4.6	–2.4
Population living in areas under severe water stress (% of population)		**40.6**	**39.4**		**18.2**
Biodiversity	1970	2000	2030	1970-2000	2000-2030
Remaining species abundance (% of potential)	**78.4**	**74.5**	**68.8**	**–3.9**	**–5.7**
Loss due to crop area (%)	11.2	11.5	13.1	0.4	1.6

StatLink http://dx.doi.org/10.1787/257828287161

Note: In the regional tables, reference years for water stress and biodiversity are not 1980-2005-2030, because of features of the IMAGE model.

Source: OECD *Environmental Outlook* Baseline.

expected to be 26% higher than the OECD average by 2030 (down from 55% in 1980). The fuel mix in the region is characterised by a high share of oil in the primary energy mix, and of natural gas and light oil in final energy use.

By 2030, the region is projected to generate 18% of global GHG emissions (down from 20% in 2005). The growth of energy-related CO_2 emissions per capita is expected to stabilise over the 2005-2030 period. Transport will represent a relatively higher share of the region's energy-related CO_2 emissions by 2030 (33%, compared with 22% for the world).

Nitrogen surplus from agriculture is projected to stabilise in North America by 2030 as a result of cross-compliance policies in agriculture. However, nitrogen from urban sewerage is expected to increase, as population growth and urbanisation expand faster to 2030 than the construction of sanitation and waste water treatment facilities, in particular in Mexico.

Agriculture is projected to be the major pressure on biodiversity, as food crop area expands to 2030, particular in the US and in Canada. The North American region will remain a leading producer of food crops (with East Asia) and of animal products (with South Asia and Western Europe). Together Brazil and the United States are likely to be the leading exporters of oilseeds, the fastest growing agricultural product.

The IPCC indicates that annual mean warming is likely to exceed the global mean warming in most parts of North America. At the same time, annual mean precipitation is likely to decrease in the southwest (IPCC, 2007). Such changes are likely to make droughts more persistent in western areas of North America. Large areas where temperate cereals are grown are likely to be negatively affected by climate change, with a resulting decrease in the potential crop yield.

OECD Europe

OECD Europe's share of the world's population will decrease from 9% in 2005 to less than 8% in 2030 (Table A.3). While in some countries (including Germany and Italy) the population is expected to be lower in 2050 than in 2005, international migration will compensate for this trend in Germany and, to a lesser extent, in Italy. People will tend to migrate within the region, from Central and Eastern Europe towards Western Europe.

By 2030 it is expected that OECD Europe will represent about 15% of total energy consumption (down from 18% in 2005). Compared to the world average, the region consumes a relatively higher share of natural gas for primary energy consumption and final energy use.

Europe's share of global GHG emissions is expected to drop from 20% of the world total in 1980 to 12% in 2030. Energy-related CO_2 emissions per capita (largely from power generation) will increase comparatively slowly. Nitrogen and sulphur emissions are expected to be significantly reduced by 2030 (by 34 and 50% respectively). Deaths from particulate matter pollution are likely to be particularly high in Europe compared to other OECD countries, mainly from the relatively high use of diesel fuel in transport (see Chapter 8 on air pollution).

It is anticipated that increased urbanisation, sanitation and food production will raise river nitrogen levels by 21% from 2000 to 2030 in Europe. This would lead to an increased incidence of problems associated with eutrophication of coastal seas (see Chapter 15, Fisheries and aquaculture).

Table A.3. **OECD Europe: Key figures, 1980-2030**

	1980	2005	2030	% changes	
				1980-2005	2005-2030
Population (million)	**537**	**598**	**621**	**11.5**	**3.7**
% of world total	12.0	9.2	7.5		
GDP per capita (USD)	–	16 034	25 951		61.9
Primary energy consumption					
Total (% of world total)	**22**	**17**	**15**	**17.3**	**31.9**
Final energy use					
Total (% of world total)	**22**	**18**	**15**	**19.0**	**32.0**
Climate change					
GHG basket emissions (% of world total)	**20**	**13**	**12**	**–5.7**	**23.5**
Energy related CO_2 emissions (Gt CO_2)	**4.78**	**4.92**	**6.02**	**2.9**	**22.3**
Energy related CO_2 emissions per capita (t CO_2)	8.90	8.22	9.70	–7.7	18.0
Nitrogen emission (% world)	21.5	13.5	9.3	–38.7	–31.4
Sulphur emission (% world)	28.5	16.3	7.8	–54.3	–50.1
Land use					
Food crops (% of world total)	12.2	10.5	9.6	–2.2	7.1
Natural forested area (%)	4.5	4.7	4.9	–4.6	–3.3
Population living in areas under severe water stress (% of population)		**36.3**	**42.3**		**20.9**
Biodiversity	1970	2000	2030	1970-2000	2000-2030
Remaining species abundance (% of potential)	**50.5**	**47.8**	**39.7**	**–2.7**	**–8.1**
Loss due to crop area (%)	28.2	27.8	29.4	–0.5	1.6

StatLink http://dx.doi.org/10.1787/257887873880

Source: *OECD Environmental Outlook* Baseline.

Pressure on biodiversity will remain, especially in central Europe, where biodiversity levels are already low. Pressures are expected to increase to the point that only 40% of pristine state ecosystems are likely to remain in Europe by 2030. This decline results from the expansion of agricultural land and human infrastructure, particularly in the new member states of the European Community.

Water-stressed southern Europe is vulnerable to further precipitation decline and drought in the coming decades due to climate change. The IPCC anticipates that the risk of summer drought is likely to increase in central Europe and in the Mediterranean area (IPCC, 2007). These trends may have negative impacts on agriculture and human settlements.

OECD Asia and Pacific

Japan will face the impact of an ageing population and lower rates of population replacement. The resulting fall in labour force participation will pull down aggregate GDP growth (Table A.4). Ageing will be particularly stark in both Japan and Korea, and the grey dependency ratio[3] will reach record highs (70% in Japan by 2050, from 28% in 2005).

The share of OECD Asia in world primary energy consumption in 2030 is expected to be around 5% (down from 7% in 2005). The share of nuclear energy in the energy mix is relatively high (17% of the world total). It is expected that, by 2030, 3% of final energy use in OECD Asia will come from modern biofuels (twice the world's average, and 13% of the world total).

Table A.4. **OECD Asia: Key figures, 1980-2030**

	1980	2005	2030	% changes	
				1980-2005	2005-2030
Population (million)	**172**	**198**	**194**	**15.0**	**–1.8**
% of world total	3.9	3.0	2.4		
GDP per capita (USD)	–	25 233	36 951		46.4
Primary energy consumption					
Total (% of world total)	**6**	**7**	**5**	**74.8**	**16.8**
Final energy use					
Total (% of world total)	**1**	**2**	**2**	**124.1**	**45.7**
Climate change					
GHG basket emissions (% of world total)	**5**	**5**	**4**	**56.8**	**10.6**
Energy related CO_2 emissions (Gt CO_2)	**1.23**	**1.98**	**2.18**	**61.2**	**9.8**
Energy related CO_2 emissions per capita (t CO_2)	7.14	10.02	11.20	40.2	11.8
Nitrogen emission (% world)	3.5	4.7	3.1	28.1	–33.9
Sulphur emission (% world)	3.1	5.4	2.6	40.9	–50.1
Land use					
Food crops (% of world total)	0.7	0.6	0.4	–11.5	–20.5
Natural forested area (%)	0.9	1.0	1.1	–0.9	–3.3
Population living in areas under severe water stress (% of population)		**20.7**	**25.5**		**20.9**
Biodiversity	1970	2000	2030	1970-2000	2000-2030
Remaining species abundance (% of potential)	**60.2**	**56.5**	**46.4**	**–3.8**	**–10.1**
Loss due to crop area (%)	18.8	18.0	14.7	–0.8	–3.3

StatLink http://dx.doi.org/10.1787/258003748064

Source: *OECD Environmental Outlook* Baseline.

The Baseline projects that GHG emissions will increase at a relatively moderate pace over the 2005-2030 period.

Land degradation is an issue in the region. OECD Asia (Japan and Korea) already has high levels of human encroachment on nature, and further biodiversity is expected to be lost by 2030.

UNEP (2007) signals that waste, in particular the illegal traffic in electronic and hazardous waste, is a new challenge in this region.

In the OECD Pacific region (Australia, New Zealand and rest of Oceania; Table A.5), Australia tends to be a major destination for migrants, in particular from Asian countries (which account for 50% of annual flows of migrants in the region). It is noteworthy that the value added in the agriculture sector in the region is expected to out-perform the rest of the economy and the world average for this sector.

Climate change is expected to increase the frequency of extreme high daily temperatures in Australia and New Zealand, and to reduce precipitation in southern and south-western Australia. These changes are likely to affect large areas where temperate cereals are grown; the potential yield is likely to decrease as a result. The IPCC (2007) concludes that increased risks of drought in southern areas of Australia are very likely.

The region is rich in biodiversity, and pressures on this biodiversity are expected to be lower than in the rest of the world. However, changes in land use and conversion of vast natural areas for agriculture are expected to result in additional biodiversity loss.

Table A.5. **OECD Pacific: Key figures, 1980-2030**

	1980	2005	2030	% changes	
				1980-2005	2005-2030
Population (million)	**19**	**25**	**31**	**28.2**	**23.5**
% of world total	0.4	0.4	0.4		
GDP per capita (USD)	–	19 004	29 073		53.0
Primary energy consumption					
Total (% of world total)	**1**	**1**	**1**	**73.7**	**43.6**
Final energy use					
Total (% of world total)	**1**	**1**	**1**	**71.4**	**46.1**
Climate change					
GHG basket emissions (% of world total)	**1**	**2**	**2**	**86.8**	**28.1**
Energy related CO_2 emissions (G CO_2)	**0.24**	**0.41**	**0.56**	**72.0**	**37.0**
Energy related CO_2 emissions per capita (t CO_2)	12.14	16.28	18.06	34.1	10.9
Nitrogen emission (% world)	1.4	1.5	1.1	6.4	–32.1
Sulphur emission (% world)	1.0	1.2	0.5	1.3	–56.6
Land use					
Food crops (% of world total)	3.3	3.5	3.5	20.1	15.2
Natural forested area (%)	2.2	2.3	2.0	–3.8	–20.3
Population living in areas under severe water stress (% of population)		**22.6**	**23.0**		**25.7**
Biodiversity	1970	2000	2030	1970-2000	2000-2030
Remaining species abundance (% of potential)	**80.9**	**78.0**	**72.9**	**–2.8**	**–5.1**
Loss due to crop area (%)	5.0	6.7	7.8	1.7	1.1

StatLink http://dx.doi.org/10.1787/258042315056

Source: *OECD Environmental Outlook* Baseline.

Russia and the Caucasus

According to UN projections, the population in this region is expected to be lower in 2050 than in 2005, reflecting the degradation of social and sanitary services which has increased mortality rates. The Russian Federation, in particular, will face a shorter life expectancy than in the 1960s. Transition economies and particularly Russia will outperform the world for average GDP growth. GDP per capita is projected to multiply by three in this region to 2030 (compared to less than two as the world average).

The economies of Russia and the Caucasus are energy intensive. In 2005, the region represented less than 3% of the world population, but some 7% of total energy consumption; final energy use per capita is higher than the OECD average and is expected to remain so over the *Outlook* period. However, the energy intensity of the Russian economy is expected to decrease as a result of energy price reforms and the introduction of energy-efficient technologies. By 2030 the region is likely to be consuming some 5% of the world's total energy, essentially natural gas; natural gas is expected to comprise 53% of the primary inputs to produce energy/power by 2030, compared with the world average of 27%.

Energy-related CO_2 emissions are expected to remain remarkably stable over the period. Power generation is responsible for half of energy-related CO_2 emissions in the region. Nitrogen and sulphur emissions are expected to be divided by two over the *Outlook* period.

Vast natural and sparsely populated areas in Russia are rich in biodiversity. Russia hosts almost one-third of the world's natural forest areas. A slight loss of biodiversity is expected over the 2005-2030 period, resulting from some conversion from grasslands or forests to crop lands.

Table A.6. **Russia and the Caucasus: Key figures, 1980-2030**

	1980	2005	2030	% changes	
				1980-2005	2005-2030
Population (million)	**153**	**164**	**143**	**7.0**	**–12.5**
% of world total	3.4	2.5	1.7		
GDP per capita (USD)	–	2 464	7 380		199.4
Primary energy consumption					
Total (% of world total)	**14**	**7**	**5**	**–18.9**	**10.2**
Final energy use					
Total (% of world total)	**11**	**7**	**5**	**–9.3**	**13.3**
Climate change					
GHG basket emissions (% of world total)	**12**	**6**	**5**	**–22.6**	**15.1**
Energy related CO_2 emissions (Gt CO_2)	**2.90**	**2.22**	**2.24**	**–23.2**	**0.5**
Energy related CO_2 emissions per capita (t CO_2)	18.93	13.58	15.61	–28.3	14.9
Nitrogen emission (% world)	14.7	7.1	3.7	–52.9	–48.4
Sulphur emission (% world)	13.1	6.1	1.9	–62.9	–67.8
Land use					
Food crops (% of world total)	9.7	8.6	8.7	0.5	18.1
Natural forested area (%)	26.0	27.9	29.7	–2.1	–2.1
Population living in areas under severe water stress (% of population)		**23.4**	**25.7**		**–3.8**
Biodiversity	1970	2000	2030	1970-2000	2000-2030
Remaining species abundance (% of potential)	**85.4**	**83.1**	**77.8**	**–2.2**	**–5.3**
Loss due to crop area (%)	8.0	7.7	9.1	–0.3	1.4

StatLink http://dx.doi.org/10.1787/258075835463

Source: OECD Environmental Outlook Baseline.

South Asia (including India)

The population in the South Asian region will be one of the fastest growing in the world, and is projected to reach 2 billion by 2030 (Table A.7). From 2005 to 2030, per capita income will multiply by 2.5 in the region and will grow twice as fast as the world average. However, per capita income is expected to remain below half of the world average by 2030. The service sector will perform better than other sectors, but industrial growth will be robust as well.

A consequence of this rapid economic and population growth will be a doubling of the region's share in the world consumption of energy between 1980 and 2030, bringing it to 9% by 2030. The region relies more heavily than the rest of the world on coal and traditional biofuels for primary energy consumption and for final energy use. Over the *Outlook* period the consumption of coal in the region as a primary energy source will grow three times faster than the world average, and is expected to represent 11% of the world total by 2030 (up from 3% in 1980). The share of the total consumption of traditional biofuels is expected to remain stable, but high (26%) for both primary and final energy uses.

India and China are projected to account for half of the total increase in residential energy use in non-OECD countries through 2030; in these two countries, residential energy use will be nearly 30% higher than the OECD total for this sector by the end of the period. Growth rates in passenger transport activity in the region are expected to be about 2% per year (compared with 1% growth in the OECD). Despite this growth, per capita use of energy is expected to remain roughly 20 GJ/year, one-fifth of per capita energy consumption in OECD countries; and energy-related CO_2 emissions per capita will be less than one-third of the world average.

Table A.7. **South Asia (including India): Key figures, 1980-2030**

	1980	2005	2030	% changes	
				1980-2005	2005-2030
Population (million)	**909**	**1 483**	**2 035**	**63.1**	**37.2**
% of world total	20.4	22.8	24.7		
GDP per capita (USD)	–	559	1 426		155.0
Primary energy consumption					
Total (% of world total)	**4**	**7**	**9**	**172.2**	**99.4**
Final energy use					
Total (% of world total)	**5**	**7**	**9**	**129.3**	**82.9**
Climate change					
GHG basket emissions (% of world total)	**4**	**8**	**10**	**160.6**	**63.4**
Energy related CO_2 emissions (Gt CO_2)	**0.30**	**1.41**	**3.37**	**368.3**	**139.2**
Energy related CO_2 emissions per capita (t CO_2)	0.33	0.95	1.65	187.1	74.3
Nitrogen emission (% world)	2.8	7.8	14.8	167.9	88.7
Sulphur emission (% world)	1.9	8.1	14.4	244.7	85.9
Land use					
Food crops (% of world total)	14.5	14.8	16.5	15.7	30.0
Natural forested area (%)	1.9	1.5	0.5	–27.3	–68.0
Population living in areas under severe water stress (% of population)		**79.0**	**83.2**		**44.6**
Biodiversity	1970	2000	2030	1970-2000	2000-2030
Remaining species abundance (% of potential)	**60.6**	**50.0**	**29.8**	**–10.6**	**–20.1**
Loss due to crop area (%)	31.7	37.6	53.0	6.0	15.4

StatLink http://dx.doi.org/10.1787/258088606772

Source: *OECD Environmental Outlook* Baseline.

By 2030 GHG emissions from South Asia will represent 10% of the world total (up from 8% in 2005). India is likely to surpass the US in energy-related carbon dioxide emissions around 2040 (almost 50% of which will arise from power generation), and the use of coal will increase.

It is projected that nitrogen and total sulphur emissions (energy- and industry-related emissions) will continue to increase from 2005 to 2030, by which time they will represent 14% of the world total (more than North America). Nitrogen surplus from agriculture is projected to increase in India. Nitrogen from urban sewerage is also expected to increase, as population growth and urbanisation will expand faster than the construction of sanitation and waste water treatment facilities.

Food crop production in South Asia doubled between 1980 and 2005 and is expected to further multiply by 1.8 between 2005 and 2030 (amounting to 15% of the world total in 2030). Animal products tripled over the same period and are expected to almost double by 2030. The increase in food crop production is projected to cause natural forested areas in South Asia to be significantly reduced over the *Outlook* period.

It is projected that agriculture will be responsible for 53% of the loss of species abundance.

The region will be particularly vulnerable to climate change. Changes in temperature regimes and in precipitation are likely to affect large areas where temperate cereals and rice are grown; the potential yield is likely to decrease as a result. The IPCC expects precipitation in the summer to increase in the region; intense precipitation events will be more frequent.

The population in South Asia experiencing medium to severe water stress is expected to expand to 2030 by half a billion, mostly in India. This reflects the higher use of water resulting from increasing population and income per capita. In a region plagued with international security issues, the management of international river basins is particularly sensitive (see UNEP, 2007).

China

China will remain one of the fastest growing economies in the world. Population increase has been a major driver of this growth over recent decades. In the future (2005-2030), however, the pace of population growth is projected to be reduced by a factor of three and the ageing population will begin to become more dependent on younger generations. Nevertheless, the Chinese population is expected to grow by more than 130 million people to 2030, and is likely to become more concentrated in urban settlements. Per capita income is expected to multiply by more than 3.5 in China between 2005 and 2030, and will be above the world average at the end of the period (Table A.8).

China is expected to consume some 16% of the world's energy in 2030, up from about 14% in 2005. Its share of the world use of coal for final energy use is projected to amount to 57%. Coal represents roughly 85% of power inputs; this percentage is projected to remain unchanged in 2030. Passenger transport activity in China is expected to grow by about 3% a year (compared with 1% in the OECD). However, high density Asian cities result in lower fuel consumption for private transportation (see also Chapter 5 on urbanisation).

GHG emissions from China are expected to increase by two-thirds between 1980 and 2030 (compared to the world average of one-half), and the country is projected to be

Table A.8. **China region: Key figures, 1980-2030**

	1980	2005	2030	% changes 1980-2005	% changes 2005-2030
Population (million)	**1 024**	**1 326**	**1 457**	**29.5**	**9.9**
% of world total	22.9	20.4	17.7		
GDP per capita (USD)	–	1 671	5 088		204.5
Primary energy consumption					
Total (% of world total)	**7**	**15**	**18**	**215.6**	**88.3**
Final energy use					
Total (% of world total)	**8**	**14**	**16**	**142.0**	**72.7**
Climate change					
GHG basket emissions (% of world total)	**8**	**17**	**19**	**185.6**	**56.9**
Energy related CO_2 emissions (Gt CO_2)	**1.13**	**4.92**	**9.10**	**333.9**	**85.0**
Energy related CO_2 emissions per capita (t CO_2)	1.11	3.71	6.25	235.1	68.4
Nitrogen emission (% world)	5.3	19.1	19.1	247.2	–0.3
Sulphur emission (% world)	8.1	23.9	20.8	135.7	–9.0
Land use					
Food crops (% of world total)	7.0	10.0	9.4	61.4	9.5
Natural forested area (%)	5.1	2.6	2.1	–53.8	–25.9
Population living in areas under severe water stress (% of population)		**37.1**	**39.4**		**16.5**
Biodiversity	1970	2000	2030	1970-2000	2000-2030
Remaining species abundance (% of potential)	**75.2**	**64.0**	**57.5**	**–11.2**	**–6.4**
Loss due to crop area (%)	14.3	16.8	17.5	2.5	0.7

StatLink http://dx.doi.org/10.1787/258114713842

Source: *OECD Environmental Outlook* Baseline.

emitting 19% of the world's GHGs in 2030 (more than North America). Power generation will be responsible for half the energy-related CO_2 emissions.

China's energy policy is changing rapidly, with China's first law on renewable energy coming into force in 2006, and new targets set in 2007 for energy consumption and emission abatement (OECD, forthcoming). The challenge is how to implement these targets. Recent environmental policy initiatives have resulted in new power plants being equipped with sulphur scrubbers to tackle local and regional air pollution. This has significantly reduced sulphur emissions in the region.

China is expected to remain an important producer of food crops and will emerge as a major producer of animal products (which it was not in the 1970s), although it will also become the biggest net importer of meat. Climate change will change the prospects for agricultural yield in the region. The northern part of the country will suffer from warming well above the global mean, whereas precipitation will increase in the whole region. Heat waves are likely to last longer, while extreme cyclonic rainfall is expected to increase (IPCC, 2007).

By 2030, almost 600 million people in China, or roughly 40% of the population, are expected to live in regions under severe water stress.

The expansion in livestock production, coupled with an increased demand for rice from growing populations, is expected to increase methane emissions from agriculture to 44% to 2030. However, more efficient, intensive production and dietary changes are likely to lower emissions per kilogramme of food produced.

Nitrogen overloading – from both agriculture and untreated sewage – is already one of the main drivers of biodiversity loss in East Asia. Urbanisation, sanitation and increased food production will increase river nitrogen levels by over 40% to 2030.

The Middle East

Population in this region is among the fastest growing in the world today, and is expected to grow two times faster than the world population to 2030 (Table A.9). This trend will be accompanied by rapid urbanisation. Economies in the region have not performed steadily in the past. Even in Israel, per capita economic growth is mediocre and immigration has fuelled GDP growth. In other countries, it is not clear how the current oil-commodity boom will benefit long-term growth.

Energy consumption is projected to grow in the region at roughly the world average rate. It is expected to amount to 83 gigajoules per capita by 2030. Natural gas and oil play a major part in the fuel mix.

The region's impact on climate change has significantly worsened since 1980. Over the *Outlook* period, regional emissions are expected to grow faster than the world average. This average masks high internal discrepancies: per capita carbon dioxide emissions in the UAE are almost double those of the US, while per capita emissions in Kuwait are slightly less than those in the US (see Esty *et al.*, 2007).

The region will become a leading importer of rice to feed the fast growing population. Rapid growth in animal production will drive high coarse grain imports.

Table A.9. **The Middle East: Key figures, 1980-2030**

	1980	2005	2030	% changes	
				1980-2005	2005-2030
Population (million)	**91**	**195**	**302**	**114.4**	**54.6**
% of world total	2.0	3.0	3.7		
GDP per capita (USD)	–	4 209	7 130		69.4
Primary energy consumption					
Total (% of world total)	**2**	**4**	**6**	**222.8**	**110.4**
Final energy use					
Total (% of world total)	**2**	**4**	**5**	**203.9**	**111.4**
Climate change					
GHG basket emissions (% of world total)	**2**	**4**	**5**	**141.3**	**86.4**
Energy related CO_2 emissions (Gt CO_2)	**0.52**	**1.23**	**2.49**	**138.8**	**102.0**
Energy related CO_2 emissions per capita (t CO_2)	5.67	6.32	8.26	11.4	30.7
Nitrogen emission (% world)	2.2	4.6	6.1	102.9	31.3
Sulphur emission (% world)	1.4	3.7	5.1	108.1	44.6
Land use					
Food crops (% of world total)	1.9	1.8	1.7	6.0	6.4
Natural forested area (%)	0.0	0.0	0.0	–100.0	–
Population living in areas under severe water stress (% of population)		**95.6**	**96.3**		**55.7**
Biodiversity	1970	2000	2030	1970-2000	2000-2030
Remaining species abundance (% of potential)	**84.3**	**80.7**	**77.6**	**–3.5**	**–3.1**
Loss due to crop area (%)	6.5	7.3	8.0	0.8	0.7

StatLink http://dx.doi.org/10.1787/258116406850

Source: OECD Environmental Outlook Baseline.

It is expected that biodiversity losses in the region will be lower than in the rest of the world, due to the widespread arid and desert biomes that are not easily converted to human activities. Nitrogen loading of marine coastal areas is expected to increase, however, as population growth and urbanisation will expand faster than the construction of sanitation and waste water treatment facilities.

The population living in water stressed areas will almost double in the region, because of climate change and increased water demand for urban and agriculture use. It is expected that almost 300 million people will face severe or medium water stress by 2030. The fastest population growth rates will be in the most arid areas.

Brazil and other Latin American and Caribbean countries

The cycles of economic growth and contraction make long-term growth projections difficult in the region, especially in Brazil and Argentina.

The population of Latin America and the Caribbean is expected to grow roughly at the average world rate to 2030 (Table A.11), though internal migration will change the population distribution. Urbanisation is expected to be a driver of environmental degradation. Brazil is projected to have 226 million inhabitants in 2030, nearly double its 1980 population (Table A.10).

Energy consumption will grow, but is still expected to be below 60 GJ per capita in 2030. Brazil's share of world energy consumption will be roughly 3% at the end of the *Outlook*

Table A.10. **Brazil: Key figures, 1980-2030**

	1980	2005	2030	% changes	
				1980-2005	2005-2030
Population (million)	**121**	**179**	**226**	**48.3**	**26.3**
% of world total	2.7	2.8	2.7		
GDP per capita (USD)	–	3 162	4 980		57.5
Primary energy consumption					
Total (% of world total)	**2**	**2**	**2**	**31.1**	**85.0**
Final energy use					
Total (% of world total)	**3**	**2**	**3**	**21.7**	**77.6**
Climate change					
GHG basket emissions (% of world total)	**4**	**3**	**3**	**30.4**	**9.7**
Energy related CO_2 emissions (Gt CO_2)	**0.22**	**0.36**	**0.73**	**66.8**	**100.8**
Energy related CO_2 emissions per capita (t CO_2)	1.81	2.03	3.23	12.5	58.9
Nitrogen emission (% world)	2.3	2.4	2.3	–0.2	–1.6
Sulphur emission (% world)	1.1	1.8	2.4	31.8	40.3
Land use					
Food crops (% of world total)	3.7	4.1	3.8	27.1	6.3
Natural forested area (%)	11.0	11.1	11.8	–7.7	–2.2
Population living in areas under severe water stress (% of population)		**8.9**	**11.1**		**57.3**
Biodiversity	1970	2000	2030	1970-2000	2000-2030
Remaining species abundance (% of potential)	**79.6**	**74.6**	**68.8**	**–5.0**	**–5.8**
Loss due to crop area (%)	9.2	9.3	9.6	0.1	0.4

StatLink http://dx.doi.org/10.1787/258243421451

Source: *OECD Environmental Outlook* Baseline.

Table A.11. **Other Latin America and the Caribbean: Key figures, 1980-2030**

	1980	2005	2030	% changes	
				1980-2005	2005-2030
Population (million)	**171**	**264**	**349**	**54.3**	**32.0**
% of world total	3.8	4.1	4.2		
GDP per capita (USD)	–	3 831	6 322		65.0
Primary energy consumption					
Total (% of world total)	**3**	**3**	**3**	**51.2**	**91.3**
Final energy use					
Total (% of world total)	**3**	**3**	**4**	**57.7**	**92.8**
Climate change					
GHG basket emissions (% of world total)	**5**	**5**	**4**	**49.3**	**6.4**
Energy related CO_2 emissions (Gt CO_2)	**0.48**	**0.67**	**1.29**	**39.1**	**93.0**
Energy related CO_2 emissions per capita (t CO_2)	2.80	2.52	3.69	–9.8	46.2
Nitrogen emission (% world)	2.7	3.3	3.7	17.6	11.2
Sulphur emission (% world)	3.3	5.6	8.2	36.7	53.1
Land use					
Food crops (% of world total)	4.7	4.9	4.8	18.6	13.7
Natural forested area (%)	8.9	8.8	8.8	–9.3	–8.3
Population living in areas under severe water stress (% of population)		**23.0**	**25.8**		**48.5**
Biodiversity	1970	2000	2030	1970-2000	2000-2030
Remaining species abundance (% of potential)	**76.6**	**71.6**	**64.5**	**–5.0**	**–7.1**
Loss due to crop area (%)	9.1	10.0	10.3	0.8	0.4

StatLink http://dx.doi.org/10.1787/258253337436

Source: *OECD Environmental Outlook* Baseline.

period. The region is projected to rely on a relatively high proportion of modern biofuels and renewable energy sources for primary energy consumption, and of modern biofuels for final energy use (11% of the world total) in 2030.

Over the period, Brazil is projected to contribute 3% of world GHG emissions and the rest of the region will account for an additional 4% of global emissions. Brazil's energy-related carbon dioxide emissions will surpass Germany's in 2010 and Japan's around 2015. It is noteworthy that power generation is responsible for roughly 20% of energy-related GHG emissions in the region (compared with 42% as the world average). Transport and, in Brazil, industry, are the region's major sources of energy-related CO_2 emissions.

Latin America is one of the regions where the forest area losses are the greatest, and deforestation is projected to continue, albeit at a slower rate. Climate change, especially temperature rise in the Amazon region, is a major threat to forests. Grassland area is expected to expand significantly, particularly in South America. Along with the United States, Brazil is projected to be a leading exporter of oilseeds, its fastest growing agricultural product. Agriculture and habitat fragmentation caused by urban sprawl are expected to be responsible for losses in species abundance in the region to 2030.

Africa

Africa's population is the fastest growing in the world (Table A.12). It has doubled over the last 25 years and is expected to increase by 61% between 2005 and 2030. However, this masks intra-regional differences. Regional migrations within sub-Saharan Africa will change the distribution of populations among countries. Population pressure is likely to

Table A.12. **Africa: Key figures, 1980-2030**

	1980	2005	2030	% changes	
				1980-2005	2005-2030
Population (million)	**476**	**946**	**1 525**	**98.7**	**61.3**
% of world total	10.7	14.6	18.5		
GDP per capita (USD)	–	740	1 391		87.9
Primary energy consumption					
Total (% of world total)	**4**	**5**	**7**	**75.6**	**120.1**
Final energy use					
Total (% of world total)	**5**	**5**	**7**	**62.4**	**115.0**
Climate change					
GHG basket emissions (% of world total)	**9**	**8**	**10**	**37.7**	**62.2**
Energy related CO_2 emissions (Gt CO_2)	**0.56**	**0.99**	**2.42**	**75.7**	**145.3**
Energy related CO_2 emissions per capita (t CO_2)	1.18	1.04	1.58	–11.6	52.1
Nitrogen emission (% world)	3.8	5.6	11.0	43.4	94.9
Sulphur emission (% world)	3.4	5.4	10.2	26.5	97.1
Land use					
Food crops (% of world total)	12.1	13.6	14.9	27.8	27.4
Natural forested area (%)	12.3	11.9	9.9	–11.2	–24.0
Population living in areas under severe water stress (% of population)		**24.5**	**22.7**		**49.3**
Biodiversity	1970	2000	2030	1970-2000	2000-2030
Remaining species abundance (% of potential)	**81.9**	**77.2**	**68.0**	**–4.7**	**–9.2**
Loss due to crop area (%)	4.9	6.1	8.0	1.2	2.0

StatLink http://dx.doi.org/10.1787/258268650040

Source: *OECD Environmental Outlook* Baseline.

increase the flows of sub-Saharan migrants to OECD countries (Spain, Northern America). Because parts of the region are particularly vulnerable, environmentally stressed areas are expected to lead to out-migrations. Rapid population growth and migrations will add pressure on the environment, driving land degradation and land use change (through both rapid urbanisation and desertification). South Africa is projected to have one of the lowest population growth rates in the region.

Despite improved economic performance, GDP per capita is projected to remain low, in both absolute and relative terms (below one-third of the world average). Per capita use of energy is expected to remain below 24 GJ a year in 2030 (one-third of the OECD's per capita energy consumption). Traditional biofuels are expected to remain an important component of primary energy consumption (26%), with associated high health impacts and costs, particularly for women and children (UNEP, 2006). The share of oil in the primary energy mix will grow faster, with oil expected to become the major source of primary energy by 2030.

GHG emissions are expected to double over the *Outlook* period to reach 10% of world emissions. Energy-related CO_2 emissions per capita will remain relatively low, less than half of the world average. Nitrogen and sulphur emissions are also expected to double between 2005 and 2030.

Demand for fuel and agricultural land, and to a lesser extent infrastructure development, are major causes of deforestation, habitat and biodiversity loss in the continent. At the same time, land degradation threatens a number of ecosystems (see UNEP, 2007).

Crop and animal production are expected to grow respectively two and three times faster than the world average. Food crop area will expand significantly in the region. The productivity of land resources is closely related to other environmental goods and services; on the one hand, desertification hampers land fertility and agricultural productivity; on the other hand, the use of genetically-modified technology is very controversial in the region, but is expected to gain ground in the coming decades (UNEP, 2006).

Medium to severely water stressed areas will expand in the region, with 400 million people expected to live in such areas by 2030. Population growth will lead to more water withdrawals. The consequences of climate change will add to this burden, as IPCC (2007) anticipates above average temperatures throughout the continent and in all seasons. Rainfall is likely to decrease in much of the continent, except in East Africa where it is expected to increase. Water availability will shape development opportunities.

Eastern Europe and Central Asia[4]

Most countries in this region lack the strong drivers for environmental improvement that exist in western countries (*e.g.* public demand, price signals) and Central European countries (EU accession requirements).

As stated in the Fourth Environmental Assessment report (EEA, 2007), the economies of Eastern Europe and Central Asia are gradually moving away from reliance on agricultural output towards service industries. Nonetheless, the region is still relatively more dependent on mineral extraction and agriculture. Resource use efficiency is poor. This often results in major environmental pressures and high volumes of waste; increasingly so now that countries are recovering from the economic and financial crisis that plagued their economies at the turn of the century.

Table A.13. **Eastern Europe and Central Asia: Key figures, 1980-2030**

	1980	2005	2030	% changes	
				1980-2005	2005-2030
Population (million)	**105**	**124**	**125**	**18.5**	**0.7**
% of world total	2.4	1.9	1.5		
GDP per capita (USD)	–	1 131	2 814		148.8
Primary energy consumption					
Total (% of world total)	**6**	**3**	**3**	**–18.6**	**20.7**
Final energy use					
Total (% of world total)	**6**	**3**	**3**	**–19.2**	**38.3**
Climate change					
GHG basket emissions (% of world total)	**6%**	**3**	**2**	**–28.4**	**21.1**
Energy related CO_2 emissions (Gt CO_2)	**1.41**	**1.00**	**1.04**	**–29.2**	**4.8**
Energy related CO_2 emissions per capita (t CO_2)	13.38	8.00	8.33	–40.2	4.1
Nitrogen emission (% world)	7.1	3.2	2.6	–56.3	–19.5
Sulphur emission (% world)	6.4	3.6	2.3	–55.3	–33.7
Land use					
Food crops (% of world total)	6.1	4.5	4.0	–15.5	2.7
Natural forested area (%)	0.7	0.6	0.5	–9.3	–21.5
Population living in areas under severe water stress (% of population)		**84.2**	**85.5**		**2.4**
Biodiversity	1970	2000	2030	1970-2000	2000-2030
Remaining species abundance (% of potential)	**63.8**	**61.6**	**54.9**	**–2.2**	**–6.7**
Loss due to crop area (%)	20.3	19.9	20.6	–0.4	0.7

StatLink http://dx.doi.org/10.1787/258273071233

Source: OECD Environmental Outlook Baseline.

The quality of the region's water supply and sanitation services has deteriorated continuously over the past 15 years, with the rural population being most affected. It is unlikely that the water-related Millennium Development Goal will be met. In its recent report, the European Environment Agency (2007) notes that high leakage losses in water distribution systems, poor management and maintenance of irrigation systems, and unsustainable cropping patterns exacerbate the impacts of droughts and water scarcity.

No country in the region has a national strategy on invasive alien species, nor is any developing such a strategy.

Other Asian countries

This region is facing rapid population and economic growth (Table A.14). However, at the end of the period, the GDP per capita will remain below two-thirds of the world average.

The region is relatively more dependent on oil and traditional biofuels for energy. At the end of the *Outlook* period, the region is projected to be responsible for 6% of global GHG emissions (compared to 8.6% of the world population and 6% of global energy use).

The region's development path relies on land use patterns which are detrimental for natural forested areas and biodiversity. Discussions are under way on the possibility of fully incorporating the economic value of ecosystem goods and services into national policies. International co-operation will be needed to share the costs of preserving these goods and services.

Table A.14. **Other Asian countries: Key figures, 1980-2030**

	1980	2005	2030	% changes	
				1980-2005	2005-2030
Population (million)	**363**	**563**	**706**	**55.0**	**25.3**
% of world total	8.1	8.7	8.6		
GDP per capita (USD)	–	1 455	3 178		118.4
Primary energy consumption					
Total (% of world total)	**2**	**4**	**6**	**186.7**	**105.3**
Final energy use					
Total (% of world total)	**3**	**4**	**6**	**145.5**	**98.2**
Climate change					
GHG basket emissions (% of world total)	**3**	**6**	**6**	**167.6**	**38.2**
Energy related CO_2 emissions (Gt CO_2)	**0.27**	**1.06**	**2.48**	**287.6**	**132.7**
Energy related CO_2 emissions per capita (t CO_2)	0.76	1.89	3.51	150.1	85.6
Nitrogen emission (% world)	1.9	5.4	9.0	183.6	64.3
Sulphur emission (% world)	1.3	6.6	13.4	315.1	111.8
Land use					
Food crops (% of world total)	5.5	6.4	6.9	33.2	25.7
Natural forested area (%)	7.3	7.3	7.2	–8.3	–10.0
Population living in areas under severe water stress (% of population)		**17.2**	**30.2**		**120.4**
Biodiversity	1970	2000	2030	1970-2000	2000-2030
Remaining species abundance (% of potential)	**72.4**	**64.2**	**50.8**	**–8.1**	**–13.4**
Loss due to crop area (%)	19.0	22.6	26.4	3.6	3.8

StatLink http://dx.doi.org/10.1787/258281777327

Source: *OECD Environmental Outlook* Baseline.

Table A.15. **The world: Key figures, 1980-2030**

Assumptions

		1980	2005	2030	Total % change 1980-2005	Total % change 2005-2030
Population (million inhabitants)		4 464	6 494	8 236	45	27
GDP per capita (USD)			5 488	8 606		57
Value added, per sector (million USD$)	Agriculture		1 316 026	2 517 590		91
	Industry		9 863 188	19 694 210		100
	Services		24 509 329	50 175 246		105

Energy consumption

		1980	(%)	2005	(%)	2030	(%)	Total % change 1980-2005	Total % change 2005-2030
Primary energy consumption (EJ)	Coal	75	25	129	28	198	28	71	54
	Oil	132	43	168	36	239	33	27	42
	Nat. gas		18	98	21	175	24	77	78
	Mod. biofuels	0	0.2	2	0.5	16	2	334	658
	Trad. biofuels	34	11	44	10	53	7	32	19
	Nuclear	3	1	9	2	13	2	271	38
	Solar/wind/hydro	6	2	11	2	20	3	83	81
	Total	306	100	462	100	714	100	51	54
Primary inputs to produce electric power (EJ)	Coal	40	46	87	55	148	55	117	71
	Oil	18	20	10	7	4	1	–42	–64
	Light oil	1	2	3	2	1	0	99	–69
	Nat. gas	18	21	35	22	72	27	92	106
	Mod. biofuels	0	0.4	2	1.1	11	4	381	485
	Nuclear	3	3	9	6	13	5	271	38
	Solar/wind/hydro	6	7	11	7	20	7	83	81
	Total	86	100	157	100	268	100	82	71
Final energy use (EJ)	Coal	26	12	29	9	37	8	13	26
	Heavy oil	28	13	38	12	55	12	32	47
	Light oil	59	28	78	25	115	24	33	47
	Nat. gas	30	14	52	17	83	18	71	59
	Mod. biofuels	0.11	0.1	0.32	0.1	6	1	187	1637
	Trad. biofuels	34	16	44	14	53	11	32	19
	Hydrogen	0	0	0	0	0	0.0	–	–
	Sec. heat	9	4.5	12	3.9	13	2.7	26	4
	Electricity	24	11	55	18	111	23	128	101
	Total	211	100	309	100	472	100	46	53
Final energy use (GJ/per capita)		47		48		57		1	20

Water

	2005		2030		Total % change 2005-2030
Population living in areas under water stress (million)					
Severe	2 837	44%	3 901	47%	38
Medium	794	12%	1 368	17%	72
Low	835	13%	866	11%	4
No	2 028	31%	2 101	26%	4
Total	6 494	100%	8 236	100%	27

Table A.15. **The world: Key figures, 1980-2030** *(cont.)*

Climate change

		1980		2005		2030		Total % change 1980-2005	Total % change 2005-2030
GHG emissions (Gt CO_2eq)		32.9		46.9		64.1		43	37
CO_2 emissions from energy (Gt CO_2)	Industry and other[a]	7.6	39%	9.0	32%	12.5	29%	19	39
	Power generation	6.2	32%	11.0	39%	18.0	42%	78	65
	Residential	2.0	11%	2.3	8%	2.8	7%	14	22
	Transport	3.5	18%	6.1	21%	9.6	22%	73	58
	Total	19.3	100%	28.4	100%	43.0	100%	47	52
CO_2 emissions from energy (t CO_2/per capita)		4.3		4.4		5.2		1	19
CO_2 concentration (ppmv)		339		383		465		13	21
Mean global temperature change (°C)		0.21		0.69		1.34			

Air pollution

	1980	2005	2030	Total % change 1980-2005	Total % change 2005-2030
Nitrogen emission (Mt)[b]	30.5	29.6	29.4	–3	–1
Sulphur emission (Mt)[b]	80.5	64.4	67.3	–20	5

	2000	2030	Total % change
Loss of health (per million inh.)[c]	1 632	3 507	115%
Mortality (deaths/per million inh).[d]	164	412	150%

Biodiversity

		1970 (%)	2000 (%)	2030 (%)	Total % change 1970-2000	Total % change 2000-2030
Species abundance	Remaining	77.7	72.9	65.6	–4.8	–7.4
	Loss to crop area	10.7	11.8	13.6	1.0	1.9
	Loss to infrastructure	4.4	6.0	8.8	1.6	2.8
	Loss to woody fuels	0.0	0.04	0.1	0.04	0.04
	Loss to pasture area	4.4	4.7	5.1	0.3	0.5
	Loss to climate change	0.5	1.6	3.2	1.1	1.7
	Loss to forestry	1.1	1.2	1.4	0.1	0.3
	Loss to nitrogen deposition	0.5	0.7	0.8	0.2	0.1
	Loss to fragmentation	0.7	1.1	1.3	0.4	0.2

Agriculture

	1980	2005	2030	Total % change 1980-2005	Total % change 2005-2030
Food crop production (million t)	2 346	3 471	5 151	48	48
Animal products (million t)	621	951	1 386	53	46

OECD ENVIRONMENTAL OUTLOOK TO 2030 – ISBN 978-92-64-04048-9 – © OECD 2008

Table A.15. **The world: Key figures, 1980-2030** *(cont.)*

Land use

		1980		2005		2030		Total % change	
								1980-2005	2005-2030
Natural forested area (000 km^2)		46 274		42 254		38 826		–9	–8
Crop area (000 km^2)	Biofuel crops	33	0%	102	0.2%	349	1%	214	242
	Food crops	14 447	31%	16 420	32%	19 098	34%	14	16
	Grass and fodder	32 176	69%	34 222	67%	36 137	65%	6	6
	Total	46 655	100%	50 745	100%	55 585	100%	9	10

StatLink http://dx.doi.org/10.1787/258301648536

Note: Totals may not add up due to number rounding.

a) The term "other' includes energy-related emissions of CO_2 from: services bunkers, energy transformation, losses and leakages, and other.

b) Total nitrogen and sulphur dioxide emissions consider both industry-related and energy-related emissions.

c) The figures for loss of health were obtained by adding up loss of health attributable to outdoor exposure to ozone plus loss of health attributable to particulate matter, per million inhabitants.

d) Mortality was defined as the sum of deaths related to outdoor exposure to ozone plus deaths attributable to particulate matter, per million inhabitants.

Source: *OECD Environmental Outlook* Baseline.

Notes

1. This is the reason why tables do not all show the same indicators.
2. Primary energy refers to energy in its initial form, after production or importation. Final consumption refers to consumption in end-use sectors, net of losses in transformation and distribution.
3. The ratio of adults aged 65 and above to the working-age population, i.e. those aged 15 to 65.
4. For more details on the progress made in environmental policies in the region see OECD, 2007.

References

EEA (European Environment Agency) (2007), *Europe's Environment The Fourth Assessment*, European Environment Agency, Copenhagen, Denmark.

Esty, D.C., M.A. Levy, and A. Winstonne, (2007), "Environmental Sustainability in the Arab World", in P. Cornelius (ed.), *The Arab World Competitiveness Report 2002-2003*, New York, Oxford University Press.

IPCC (Intergovernmental Panel on Climate Change) (2007), "Regional Climate Projections", in *Climate Change 2007: The Physical Science Basis*, Contribution of Working Group 1 to the Fourth Assessment Report of the Intergovernmental Panel on Climate Change, Geneva.

OECD (2007), *Policies for a Better Environment. Progress in Eastern Europe, Caucasus and Central Asia*, OECD, Paris, France.

OECD (forthcoming), *Global Forum on Sustainable Development: The OECD Environmental Outlook to 2030. A BRIC Perspective*, OECD, Paris, forthcoming.

UNEP (United Nations Environment Programme) (2006), *Africa Environment Outlook. Our Environment, Our Wealth*, UNEP, Nairobi, Kenya.

UNEP (2007), *Global Environmental Outlook. Chapter 6: Regional Perspectives*, UNEP, Nairobi, Kenya.

ANNEX B

Modelling Framework

The analysis for the OECD Environmental Outlook has been supported by two modelling frameworks that have been coupled: i) the ENV-Linkages computable general equilibrium model for the economic variables; and ii) a set of environmental models linked to the Integrated Model to Assess the Global Environment (IMAGE). This annex provides information about the models, and the main assumptions used in developing the Outlook Baseline and policy simulations. Particular attention is given to the way these models have been connected together for use in the OECD Environmental Outlook. *The annex includes a tabular overview of which environmental estimates were produced with what model. It also outlines some specific sources of model-related uncertainty.*

Introduction

The analyses for the *OECD Environmental Outlook* have been supported by two modelling frameworks that have been coupled: *i)* the ENV-Linkages economic model; and *ii)* a set of mostly environmental models linked up to the Integrated Model to Assess the Global Environment (IMAGE) integrated assessment framework. This annex summarises the models and refers to more in-depth descriptions. Particular attention is given to the way these models have been connected specifically for the *OECD Environmental Outlook*. The section on environmental modelling contains a tabular overview of which environmental estimates were produced with what model. This annex also outlines some specific sources of model-related uncertainty. The analysis methods and tools of the *OECD Environmental Outlook* are more fully described in a background report (MNP and OECD, 2008), along with detailed results and a broader discussion of uncertainty issues.

The ENV-Linkages macroeconomic framework

The ENV-Linkages model continues an OECD tradition of quantitative simulation analysis. For environmental policy the work with the GREEN model (*e.g.* Burniaux, *et al.*, 1992) established a line of analyses that has continued to the present. GREEN was originally used for studying climate change policy, and culminated in Burniaux (2002). It was developed into the Linkages model, and subsequently became the JOBS modelling

Figure B.1. **Structure of production in ENV-Linkages**

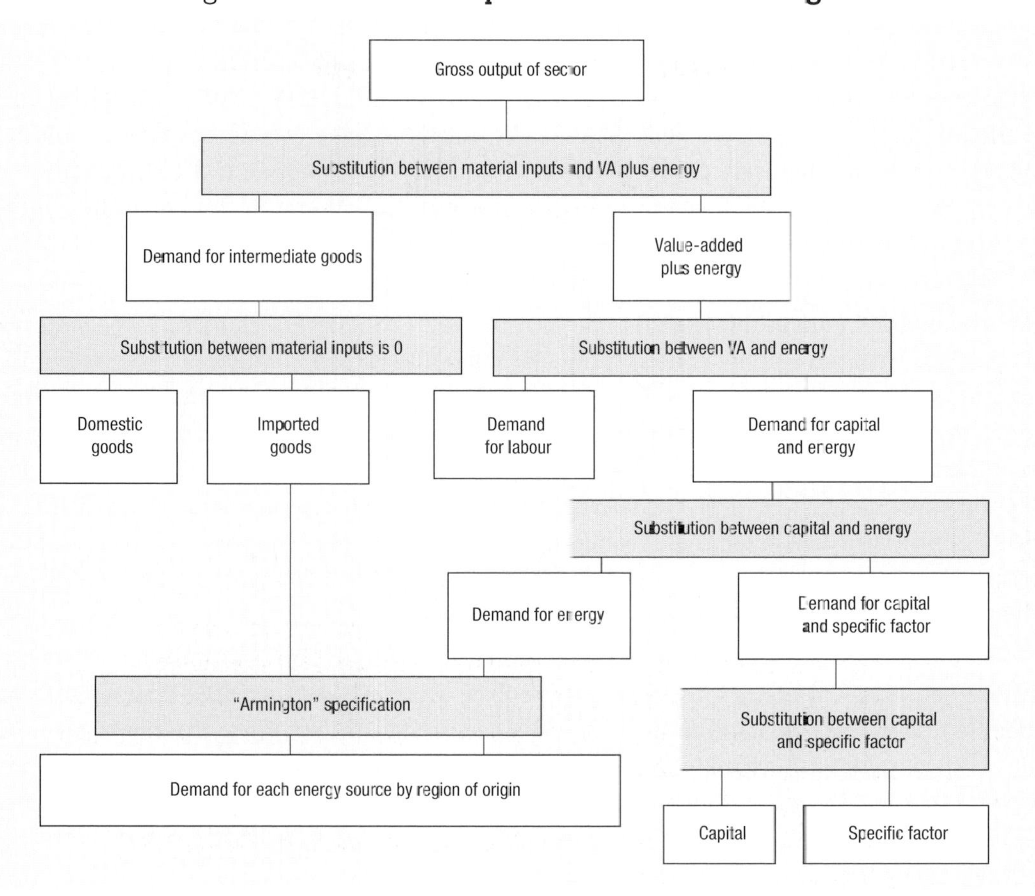

Source: OECD Environmental Outlook.

platform. This was used to help underpin the *OECD Environmental Outlook* to 2020 (OECD, 2001). Subsequent versions of the Linkages model are also in use at the World Bank for research into global economic development issues. Further developments have been incorporated, and the ENV-Linkages model is now in use in the OECD's Environment Directorate.

The ENV-Linkages model is a global economic model built primarily on a database of national economies. The model represents the world economy in 34 countries/regions (Table B.2), each with 26 economic sectors. Each of the 34 regions is underpinned by an economic input-output table (usually published by a national statistical agency). These tables identify all the inputs into an industry and identify all the industries that buy specific products. Some industries explicitly use land, while others, such as fisheries and forestry, also have a "natural resource" input – *e.g.* fish and trees.

Since it is an economic model, ENV-Linkages does not represent physical processes. Instead, physical processes are summarised from empirically derived relationships between inputs and outputs. That is, industries (rather than individual firms) are observed over time to be able to vary the use of inputs such as labour, capital, energy and materials. When prices for the inputs or outputs change, individual firms adjust, but the industry as a whole adjusts more strongly by favouring the firms that gain advantage as a result of the price changes. In the real world, even firms that produce the same product are very heterogeneous. Such responsiveness can be represented mathematically and tested for robustness (see *e.g.* Hertel *et al.*, 2003; Valenzuela *et al.*, 2007). Inputs and outputs are measured in the constant currency of a base year – inflation is thus removed from the value of output. Moreover, output can be calculated in either the real price of a given year, or the initial price of the base year. Calculating output in base year price gives a "volume" measure that would closely parallel physical quantities in that year. If the composition of output in any given sector does not change much over time, then the change in the volume of output is equal to the change in the physical quantities.

Income generated by economic activity ultimately reflects demand for goods and services by final consumers. ENV-Linkages represents consumers as being largely similar at a very aggregated level of consumption. As such, the model postulates a representative consumer who allocates disposable income according to preferences: among consumer goods and savings. More formally, household consumption demand is the result of static maximisation behaviour which is formally implemented as an "Extended Linear Expenditure System". A representative consumer in each region – who takes prices as given – optimally allocates disposable income among the full set of consumption commodities and savings. Saving is considered as a standard good and therefore does not rely on a forward-looking behaviour by the consumer.

In the model, the technological representation of production is accomplished using a nested sequence of constant elasticity of substitution (CES) functions. Four factors are specified: land, labour, capital and a sector-specific natural resource. Energy is an input that is combined with capital. There is a parameterisation of the substitutability between inputs, so the intensity of using capital, energy, labour and land changes when their relative price changes: as labour becomes more expensive, less of it is used relative to capital, energy and land.

All production is assumed to operate under cost minimisation, in well-functioning markets and with constant returns to scale technology. Changes to these assumptions are possible, but were not used for the *OECD Environmental Outlook*. The production technology is specified as nested CES production functions in a branching hierarchy. The top node thus represents an output, using intermediate goods combined with value-added. This structure is replicated for each output, where the parameterisation of the CES functions may differ across sectors. Figure B.1 illustrates this hierarchy.

As is illustrated, the valued-added bundle is itself specified as a CES combination of labour and a capital/energy input. In turn, the capital/energy bundle is a CES combination of energy and a broad concept of capital. The definition of capital is broad because in some sectors capital will have been combined with a resource input (*e.g.* land, fish or trees) before it is combined with energy. In the "crop" and "livestock" sectors, there are different structures that also incorporate fertiliser and feed. In the "crop" production sector, the broad capital is itself a CES combination of fertiliser and another bundle of capital-land-energy. The intention of this specification is to reflect the possibility of substitution between extensive and intensive agriculture. In the "livestock" sectors, substitution possibilities are between bundles of land and feed on the one hand, and of a capital-energy-labour bundle on the other hand. This reflects a similar choice between intensive and extensive livestock production. Production in other sectors is characterised by substitution between labour and a bundle of capital-energy (and possibly a sector-specific factor for primary resources).

 OECD ENVIRONMENTAL OUTLOOK TO 2030 – ISBN 978-92-64-04048-9 – © OECD 2008

Total output for a sector is actually the sum of two different production streams: resulting from the distinction between production with an "old" capital vintage, and production with a "new" capital vintage. The substitution possibilities among factors are assumed to be higher with new capital than with old capital. In other words, technologies have putty/semi-putty specifications. This will imply longer adjustment of quantities to prices changes. Capital accumulation is modelled as in traditional Solow/Swan growth models.

This version of the model does not include an investment schedule that relates investment to interest rates. Investment is equal to domestic saving in each period; *i.e.* investment is equal to the sum of government savings, consumer savings and net capital flows from abroad induced by trade imbalances. The differences in sectoral rates of return determine the allocation of investment. The model features two vintages of capital, but investment adds only to new, more flexible capital. Sectors with higher investment, therefore, are more able to adapt to changes than are sectors with low levels of investment. Indeed, declining sectors whose old capital is less productive begin to sell capital to other firms (which they can use after incurring some cost for modifications).

A full range of market policy instruments (tax, etc.) is also specified. The government in each region collects various kinds of taxes in order to finance a given sequence of government expenditures. For simplicity it is assumed in the Baseline that these expenditures grow at the same rate as the real GDP of the previous period. Since predicting corrective government policy is not an easy task, the real government deficit is exogenous. Closure of the model to ensure reasonable long-term properties therefore implies that some fiscal instrument is endogenous – in order to anchor the given government deficit. The fiscal closure rule in ENV-Linkages is that the marginal income tax rate adjusts to offset changes that may arise in government expenditures, or as a result of other taxes. For example, a reduction or elimination of tariff rates is compensated for by an increase in household direct taxation, other things being equal. If change in the long-term deficit is desired as a result of the tariff change, the deficit can be changed exogenously by the amount of the decreased revenues – so there is no offsetting change in income taxes.

World trade in ENV-Linkages is based on a set of regional bilateral flows for the model's 24 sectors. The basic assumption is that imports originating in different regions are imperfect substitutes; *i.e.* different countries may produce similar goods, but they are never identical (though some goods, such as crude oil, are very similar). At a 24-sector level, this assumption is tenable since each sector will be composed of different goods and services in each country. Therefore in each region, total import demand for each good is allocated across trading partners according to the relationship between their export prices. This specification of imports – commonly referred to as the Armington specification – formally implies that each region faces a reduction in demand for its exports if domestic prices increase. The Armington specification is implemented using two CES nests. At the top nest, domestic agents choose the optimal combination of the domestic good and an aggregate import good consistent with the agent's preference function. At the second nest, agents optimally allocate demand for the aggregate import good across the range of trading partners. The bilateral supply of exports is specified in parallel using a nesting of constant-elasticity-of-transformation (CET) functions. At the top nest, domestic suppliers optimally allocate aggregate supply across the domestic market and the aggregate export market. At the second nest, aggregate export supply is optimally allocated across each trading region as a function of relative prices.

Each region runs a current-account surplus (or deficit), which is fixed (in terms of the model numéraire basket of goods). Closure on the international side of each economy is achieved by having a counterpart of these imbalances result in a net outflow (or inflow) of capital, which is subtracted from (added to) the domestic flow of saving. In each period, the model equates gross investment to net saving (which is equal to the sum of saving by households, the net budget position of the government and foreign capital inflows). Given the rules for government and international closure, this final particular closure rule implies that investment is driven by saving.

Trade measures are fully bilateral and can include both export and import taxes/subsidies. Trade and transport margins can also be included; in which case world prices would reflect the difference between free on board (FOB) and cartage, insurance and freight (CIF) pricing.

A technical description is available for the original World Bank Linkages model in van der Mensbrugghe (2003).

Integrated assessment and environmental models coupled to the IMAGE framework

The Integrated Model to Assess the Global Environment (IMAGE) is the central tool for the environmental analysis reported here. IMAGE is a dynamic integrated assessment framework to model global change. It was developed at the National Institute for Public Health and the Environment (RIVM) initially to assess the impact of anthropogenic climate change (Rotmans, 1990). During the 1990s IMAGE was extended to include a more comprehensive coverage of global change issues (IMAGE team, 2001a and b).

IMAGE is a "medium complexity" model, falling between stylised macro models and true earth system models. It operates at a resolution of 24/26 world regions (for most socio-economic parameters) and a geographical 0.5 – 0.5 degree grid (for land use and environmental parameters). The medium complexity character of IMAGE allows analyses that take into account key characteristics of the physical world (*e.g.* local soil and climate characteristics of technology detail) without excessive calculation times. Figure B.2 provides an overview of the IMAGE modelling framework used for this *OECD Environmental Outlook*. This is IMAGE version 2.4, documented in Bouwman *et al.* (2006).

For the *OECD Environment Outlook*, a wider IMAGE framework has been used. This framework includes tools that have been described in the literature as models or databases in their own right, such as the TIMER global energy model (de Vries, *et al.*, 2001), the FAIR model to analyse environmental and cost implications of future commitment regimes (den Elzen and Lucas, 2003) and the GLOBIO 3 framework for the assessment of global terrestrial biodiversity. In addition, two regular collaborative model links were used, namely with the LEITAP model for agricultural economy (collaboration with the LEI Institute) and with WaterGAP for water quantity issues (collaboration with the University of Kassel).

For air pollution, results of a number of tools were kindly made available by the Joint Research Centre of the European Commission and by the World Bank. They fed into the analysis as described briefly below. A fuller description can be found in the background report (MNP and OECD, 2008) as well as a separate OECD paper on air pollution work for the *OECD Environmental Outlook* (de Leeuw *et al.*, forthcoming).

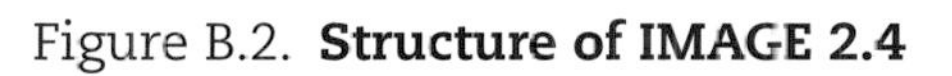

Figure B.2. **Structure of IMAGE 2.4**

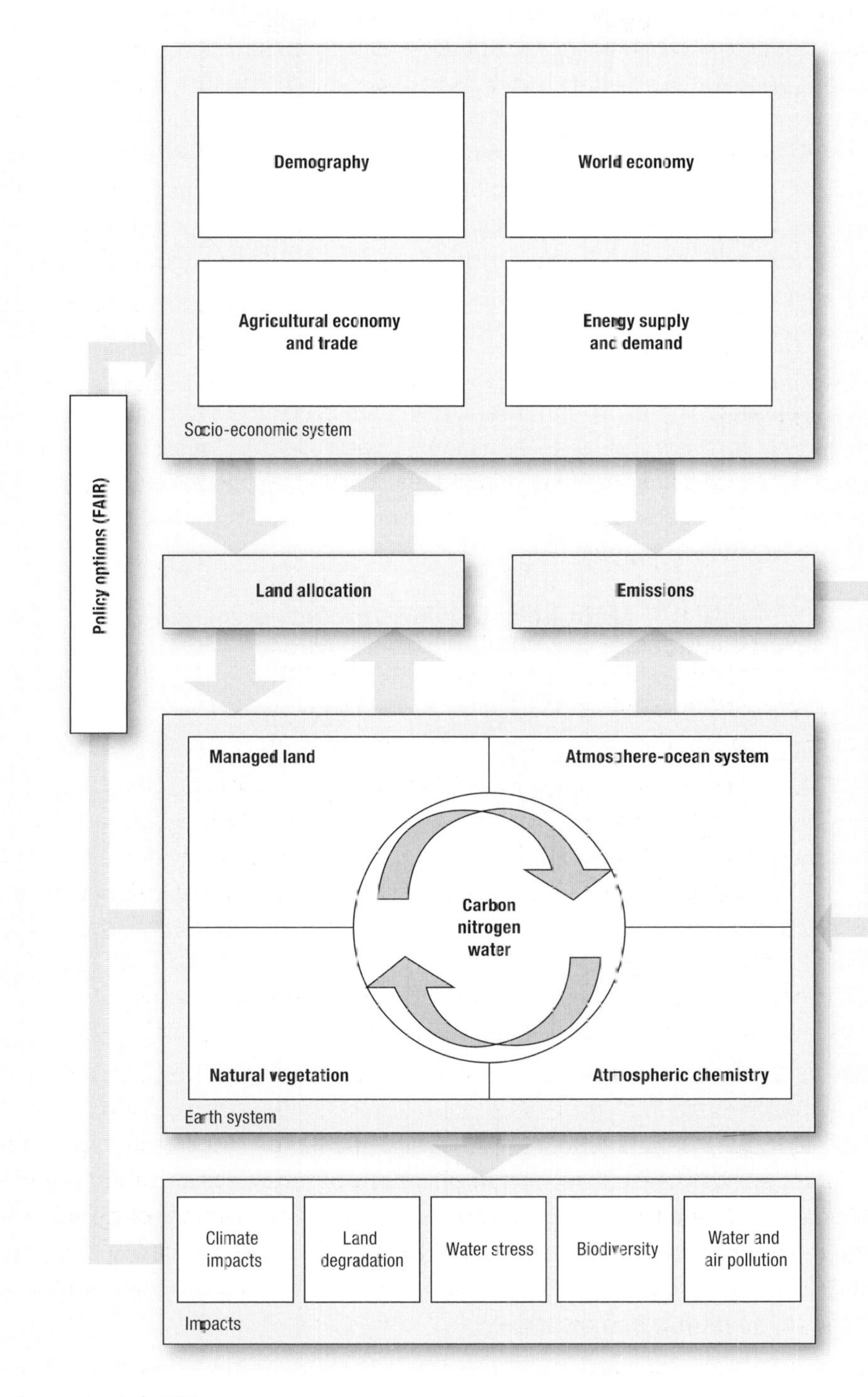

Source: Bouwman *et al.*, 2006.

From the point of view of the *OECD Environmental Outlook*, the models coupled to the IMAGE framework can be subdivided into two broad categories:

- Models that help specify important socio-economic drivers of environmental change (energy and the agricultural system) with the necessary detail.
- Models with a predominantly environmental focus.

Models that describe socio-economic drivers of environmental change

Economic models, such as ENV-linkages, describe socio-economic activities in accounting units that allow aggregation: *e.g.* monetary units or utility indexes. While this facilitates a description of shifts between the deployment of production factors in very broad terms (labour, energy, land), it does not permit insight into the changes in more physical parameters as in energy technology or the technology used for crop growing or animal farming in different regions.

However, assessing the environmental consequences of the Baseline and simulated policy requires this sort of physical, technical and spatial detail. Therefore, as depicted in Figure B.3, part of the IMAGE framework applied for the *OECD Environmental Outlook* more or less operates as a bridge between the macroeconomic description of the Baseline and the environmental systems modelling.

Figure B.3. **Main links between models deployed for the OECD *Environmental Outlook***

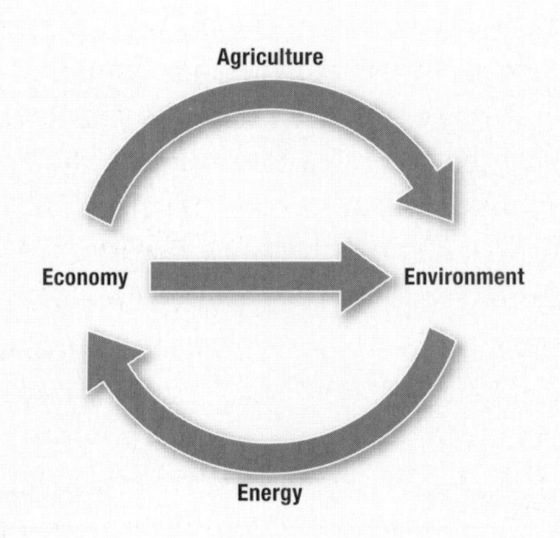

Source: MNP and OECD, 2008.

The two main models for this function in the *OECD Environmental Outlook* are the LEITAP model on agricultural economy and the TIMER model of energy supply and demand. Both can be found in the literature as models in their own right, but are here applied as part of the IMAGE framework.

Agricultural land supply and use

The LEITAP model, named after the LEI Agricultural Economics Institute that developed and applies it, is an extended version of the GTAP model developed at Purdue University. A more detailed description of LEITAP is included in the background report to this *OECD Environmental Outlook* (MNP and OECD, 2008); an example of a stand-alone application can be found in Francois *et al.* (2005).

OECD ENVIRONMENTAL OUTLOOK TO 2030 – ISBN 978-92-64-04048-9 – © OECD 2008

The base version of GTAP (to which also ENV-Linkages is related) represents land allocation in a structure of constant elasticities of transformation, assuming that the various types of land use are imperfectly substitutable, but the substitutability is equal among all land use types. LEITAP extends the land use allocation structure by taking into account the fact that the degree of substitutability of types of land differs between types (Huang *et al.*, 2004). It uses the more detailed OECD's Policy Evaluation Model (OECD, 2003) structure. This structure reflects the fact that it is easier to shift land between producing crops like wheat, coarse grains and oilseeds, than between land uses like pasture, sugarcane or, even more so, horticulture. The values of the elasticities are taken from OECD (2003).

In the standard GTAP model the total land supply is exogenous. In LEITAP the total agricultural land supply is modelled using a land supply curve which specifies the relationship between land supply and a land rental rate in each region. Land supply to agriculture can be adjusted as a result of idling of agricultural land, conversion of non-agricultural land to agriculture, conversion of agricultural land to urban use and agricultural land abandonment. The concept of a land supply curve has been based on Abler (2003).

The general idea underlying the land supply curve specification is that the most productive land is first taken into production. However, the potential for bringing additional land into agriculture is limited. If the gap between potentially available agricultural land and land used in the agricultural sector is large, the increase in demand for agricultural land will lead to land conversion to agricultural land and a modest increase in rental rates to compensate for the cost of bringing this land into production.

The land supply curve is derived using biophysical data from the IMAGE modelling framework. In the IMAGE model, climate and soil conditions determine the crop productivity on a grid scale of 0.5 by 0.5 degrees longitude-latitude. This allows spatially heterogeneous information on land productivity to be fed into the agro-economic model with LEITAP. In practice, land use change projections are iterated between LEITAP and the IMAGE until a stable solution is reached – typically one iteration is enough. Land supply functions differ between region according to survey results on land type supply constraints.

For the *OECD Environmental Outlook*, LEITAP calculations take projected changes in GDP factor productivity as input from ENV-Linkages. The projections of crop production and yield changes are coupled to IMAGE and form the key driver in the calculation of many environmental variables. In the results reported within the *OECD Environmental Outlook*, this is done for the Baseline and the comprehensive policy packages (including the 450 PPM case), but not for the simulations run solely with ENV-Linkages.

Energy supply and demand (IMAGE/TIMER)

The IMAGE/TIMER global energy model describes long-term trends in the world energy system, based on an interplay between dynamic factors such as development of energy demand, depletion and technology development of various energy sources and technologies, cost-based substitution and the development of climate policy. The TIMER model has been described in various documents (de Vries *et al.*, 2001; van Vuuren, 2007).

In the *Outlook*, demand for energy services in TIMER is modelled on the basis of general economic projections by ENV-Linkages in terms of GDP, household consumption and value added in industry, services and agriculture. The activity indicators are combined with assumptions about technology development for end-use technologies, assumptions about autonomous energy efficiency improvements and structural change. These factors have all been calibrated so that the TIMER model more-or-less follows the IEA's 2006 *World Energy Outlook* baseline in terms of the relationship between economic drivers, energy supply and use.

Energy demand is met by a large set of energy carriers, including coal, oil and natural gas, traditional and modern biomass, electricity, hydrogen and heat. These energy carriers are selected on the basis of relative costs via a multinomial logit distribution function (de Vries *et al.*, 2001). Most of the final energy carriers are produced from a range of primary energy carriers that in turn compete for market share on the basis of costs (*e.g.* electricity can be produced from fossil fuels, biomass and nuclear, solar, wind and hydropower).

Throughout the model, inertia is introduced by an explicit treatment of vintages of capital stock. The costs of the primary energy carriers are determined in the long-term by learning-by-doing (*i.e.*, technologies improve with their cumulative build-up of installed capacity) and resource depletion (driving up costs for extraction of exhaustible energy resources with their cumulative production; and of renewable resources with annual production). The main outputs from TIMER in this study are primary and final energy consumption by energy type, sector and region; cost indicators; and greenhouse gas and other emissions.

Emissions of air pollutants are determined by multiplying exogenously set emission factors (corresponding to the assumptions in each scenario) with different energy consumption and production indicators. An important technology in TIMER in the context of climate policy is carbon capture-and-storage. This technology can be applied in combination with fossil-fuel and biomass fired power plants, in the industry end-use sector and in the production of hydrogen. Its use is cost driven, which is a function of capture costs (that decline over time) and storage costs (that increase along with depletion of storage capacity).

Models that focus on environmental change

Land use and land cover (IMAGE)

An important aspect of the IMAGE model is the geographically explicit description of land-use and land-cover change. The model distinguishes 14 natural and forest land-cover types and 6 man-made land-cover types.

The land use model describes both crop and livestock systems on the basis of agricultural demand, demand for food and feed crops, animal products and energy crops. A crop module based on the FAO agro-ecological zones approach (FAO, 1978-1981) computes the spatially explicit yields of the different crop groups and the grass, and the areas used for their production, as determined by climate and soil quality. Where expansion of agricultural land is required, a rule-based "suitability map" determines the grid cells selected (on the basis of the grid cell's potential crop yield, its proximity to other agricultural areas and to water bodies). An initial land-use map for 1970 is incorporated on the basis of satellite observations combined with statistical information. For the period 1970-2000, the model is calibrated to be fully consistent with FAO statistics. From 2000 onwards, agricultural production is driven by the production of agricultural products as determined by LEITAP and demand for bio-energy crops from the TIMER model.

Changes in natural vegetation cover are simulated in IMAGE 2.4 on the basis of a modified version of the BIOME natural vegetation model (BIOME, Prentice *et al.*, 1992). This model computes changes in potential vegetation for 14 biome types on the basis of climate characteristics. The potential vegetation is the equilibrium vegetation that should eventually develop under a given climate.

 OECD ENVIRONMENTAL OUTLOOK TO 2030 – ISBN 978-92-64-04048-9 – © OECD 2008

Carbon cycle (IMAGE)

The consequences of land use and land-cover changes for the carbon cycle are simulated by a geographically explicit terrestrial carbon cycle model. The terrestrial carbon cycle model is suitable for simulating global and regional carbon pools and fluxes (pools include the living vegetation, and several stocks for carbon stored in soils). The model accounts for important feedback mechanisms related to changing climate (*e.g.* by different growth characteristics), carbon dioxide concentrations (carbon fertilisation) and land use (*e.g.* conversion of natural vegetation into agricultural land or *vice versa*). In addition, it allows for an evaluation of the potential for carbon sequestration of natural vegetation and carbon plantations.

In addition to the terrestrial system, the carbon cycle model also describes the carbon included in the atmospheric and ocean systems, the fluxes between these systems, and their subsequent effect on greenhouse gas concentrations and climate change (van Minnen *et al.*, 2000).

Nitrogen cycle (IMAGE)

IMAGE 2.4 includes a module for assessing the consequences of changing population, economy, land use and technological developments for surface-nutrient balances and reactive nitrogen emissions from point sources and non-point sources. These surface balances are the basis for describing the major fluxes in the global and regional nitrogen cycle, as well as the effects on water and air quality.

Processes that are accounted for in this module are human emissions, wastewater treatment, surface nitrogen and phosphorous balances for terrestrial systems, ammonia emissions, denitrification and emissions of dinitrogen oxide and nitrogen oxide from soils, nitrate leaching, and transport and retention of nitrogen in groundwater and surface water.

For the *OECD Environment Outlook*, nitrogen loadings have been estimated, but not phosphorus.

Air pollution

The *OECD Environmental Outlook* addresses several aspects of conventional air pollution including emissions of sulphur oxides, nitrogen oxides, airborne particulate matter and ground level ozone. The focus is on what the Baseline and policy measures mean for urban air quality worldwide.

Ambient particulate matter is partly directly emitted into the atmosphere (dominant sources are fossil fuel use, wood burning and road transport); partly it is formed in the atmosphere from precursor gases (sulphur dioxide, nitrogen oxides, ammonia and, to a lesser extent, volatile organic compounds). Ground-level ozone is a secondary pollutant: it is not directly emitted but formed in the atmosphere. Important precursors of ozone are nitrogen oxides, volatile organic compounds, methane and carbon monoxide.

Future emissions of sulphur oxides, nitrogen oxides, methane and carbon monoxide from the energy system are calculated by IMAGE/TIMER using a system of sector/region/substance specific emission coefficients (based on the EDGAR database), calibrated to historic trends and reflecting assumptions of the policy packages. Land use related emissions are calculated in a similar manner based on the land use and agricultural parameters included in IMAGE.

The policy simulations focused further on sulphur dioxide emissions. Based on published cost curves, a default long-term ambition level was set relative to maximum feasible reductions (Cofala *et al.*, 2005). The pathway towards this long-term ambition was differentiated by region, in function of the regional GDP per capita as projected with ENV-Linkages (and interpreted to be equivalent to purchasing power parity). A further differentiation was applied by sector. Emissions from international shipping were addressed separately.

For the *OECD Environment Outlook* Baseline, hemispheric transport of air pollution (eastward from one continent to the next) has been brought into the picture using results from the TM3 model of the European Commission's Joint Research Centre (JRC/IES at Ispra, Italy) for ground level ozone (Dentener *et al.*, 2005; 2006).

A uniform set of urban air concentrations of airborne particulate matter for 3 265 urban agglomerations worldwide for 1995 and 2000 was obtained using the World Bank's GMAPS model (Pandey *et al.*, 2006). The exposure of future urban populations to air pollution by particulate matter was estimated by scaling historic pollution levels using TIMER emission projections and disaggregated UN 2004 medium projections of urban population growth (UN, 2004).

Health impacts of the change in exposure to particulate matter were estimated on the basis of these exposure estimates and regionalised projections of the overall health status resulting from demographic trends, health care expectations, etc. These were evaluated using the World Health Organization's system of comparative risk assessment in terms of premature mortality and loss of healthy life expectancy (Ezzati *et al.*, 2004).

For aggregated presentations, in terms of regions, the results were weighted by the population concerned.

Climate change (IMAGE)

The emissions of greenhouse gases and air pollutants are used in IMAGE to calculate changes in the concentrations of greenhouse gases, ozone precursors and species involved in aerosol formation at a global scale. These calculations, except for carbon dioxide (see carbon cycle), are based directly on those described in the Intergovernmental Panel on Climate Change (IPCC)'s Third Assessment Report (IPCC, 2001). Next, changes in climate are calculated as global mean changes using a slightly adapted version of the MAGICC[1] model which has also been extensively used by the IPCC. Finally, changes in temperature and precipitation are estimated at a scale of 0.5 – 0.5 degrees using the standard IPCC approach to pattern scaling (including the revisions proposed by Schlesinger *et al.*, 2000, for the impact of sulfate aerosols) and using the *HadCM2* pattern (data obtained from the IPCC distribution centre).

An important factor in these calculations is the so-called climate sensitivity, which is the increase in global mean equilibrium temperature for a doubling of the greenhouse gas radiative forcing. The IMAGE parameter settings are consistent with the IPCC's Third Assessment Report, which calculated the value of this parameter to be between 1.5 and 4.5°C, with a medium estimate of 2.5°C. The recently published IPCC Fourth Assessment Report has re-calculated the most likely value of climate sensitivity as 3.0°C, implying that the IMAGE climate calculation is somewhat conservative (IPCC, 2001 and 2007).

Terrestrial Biodiversity (GLOBIO 3)

For the *OECD Environment Outlook*, the significance of the projections for terrestrial biodiversity was evaluated using GLOBIO 3. This takes into account the impacts of climate and land use change, ecosystem fragmentation, expansion of infrastructure such as roads and built-up areas, deposition of acidity and reactive nitrogen. A detailed description of the model structure can be found in Alkemade *et al.* (2006) and a sample application in CBD and MNP (2007).

For projections into the future, the underlying assumption is that the higher the pressure on biodiversity the lower the probability of a high mean species abundance. The GLOBIO model contains global cause-effect relationships between the pressure factors considered and mean species abundance, based on more than 700 publications. These are applied in a spatially explicit fashion, namely grid cells of 0.5 – 0.5 degree longitude-latitude, with a frequency distribution representing the occurrence of various biomes within each cell. The considered pressure values are calculated and combined per grid cell. The mean species abundance per region or for the world is the uniformly weighted sum over the underlying grid cells. In other words, each square kilometre of every biome is weighted equal (ten Brink, 2000).

The GLOBIO model is a joint venture between the Netherlands Environmental Assessment Agency, the UNEP World Conservation Monitoring Centre at Cambridge, UK, and the UNEP Global Resources Information Database centre at Arendal, Norway, in conjunction with others (details in CBD and MNP, 2007).

Climate policy options (FAIR, IMAGE/TIMER)

Climate policies are described within the IMAGE framework using the closely coupled models FAIR (to describe climate policy), TIMER (to describe the energy system) and the IMAGE land use system. FAIR adds an explicit description of assumed climate policies (such as burden sharing), but also a relatively simple framework to optimise the costs of reducing energy-related greenhouse gas emissions (as described in TIMER) against other forms of emissions (den Elzen and Lucas, 2003). The FAIR model also links long-term climate targets and global reduction objectives with regional emission allowances and abatement costs, accounting for the Kyoto Mechanisms such as Emissions Trading, the Clean Development Mechanism and Joint Implementation. IMAGE provides information on the potential for bio-energy use, adds the ability to evaluate environmental and land-use impacts of different energy scenarios and, finally, describes other sectors that are relevant for climate change.

In principle, in all simulations for the *OECD Environmental Outlook* the climate policies are based on a selection of low-cost reduction options by introducing a greenhouse gas permit price (see also van Vuuren *et al.*, 2007).

Water stress (University of Kassel)

The water stress variable in the *OECD Environmental Outlook* brings together information on future water availability and water withdrawals per river basin. Both variables are computed by the WaterGAP model (Alcamo *et al.*, 2003 and 2003b), soft-linked with IMAGE via co-operation with the University of Kassel. Taking the river basin as the basic unit is essential, because this is where availability and demand physically meet. National water balances have limited or no significance, especially for large countries which, for example, may have plentiful water in the north but demand concentrated in the south.

"Water availability" is defined here as total river discharge: the combined surface runoff and groundwater recharge. Long-term average annual water availability for the current and future situation is calculated on the basis of monthly climate input data from the climate normal period (1961 to 1990 time series).

WaterGAP has two main components: a Global Hydrology Model and a Global Water Use Model. The Global Hydrology Model simulates the macroscale behaviour of the terrestrial water cycle to estimate water resources, while the Global Water Use Model computes water use for the domestic, industrial, irrigation and livestock sectors. Both water availability and water use computations cover the entire land surface of the globe (except the Antarctic) and are performed for cells on a 0.5° by 0.5° spatial resolution.

Total water withdrawals are the sum of water withdrawals for the three main water-use sectors, *i.e.* households, industry and irrigation. "Current" water withdrawals per watershed for the household and industrial sectors reflect the 1995 country-specific water-use data taken from Shiklomanov (2000) and WRI (2000). Future annual water use in these sectors is computed based on the proxy driving-forces of population, electricity production and structural changes based on income. Current and future water withdrawals for irrigation are calculated for average climate conditions (*i.e.* 1961 to 1990 time series) using the 1995 distribution of irrigated areas. Water use for irrigation modelling incorporates medium assumptions of technology development (Döll *et al.*, 2003).

Simulation results are expressed in terms of water stress. This is the long-term average of the annual withdrawal-to-availability ratio. The concept of "water stress" is often used for assessing the world's water status. It indicates the intensity of pressure on water resources. In principle, the higher the ratio, the more intensively the waters in a river basin are used; this reduces either water quantity or water quality, or both, for downstream users. The *OECD Environmental Outlook* presents water stress projections in a number of severity classes. On the basis of experience and expert judgement, it is assumed that if the long-term average withdrawal-to-availability ratio in a river basin exceeds 40%, the river basin (its management, the ecosystem and the local economy) experiences severe water stress.

Policies in the Baseline

While population, GDP/sectoral value-added, land and energy are the key quantitative drivers of environmental developments over the next decades, the *Outlook* Baseline contains another set of important assumptions for modelling the environmental impacts of these drivers under "no new policies" conditions. Of course, the projections for the macroeconomy, land and energy are meant to reflect this no new policies assumption.

However, the modelling of many environmental impacts is also directly influenced by assumptions about what policies are still active during the Baseline period; for example, policies for providing more people with access to improved sanitation and access to sewerage; efficiency of irrigation; trading in emission rights; and connecting protected nature areas with each other. To understand how information has flowed through the modelling framework, from drivers to impacts, it is useful to remember that alongside the numbers that are passed on from model to model there is also the interpretation of what the "no new policies" definition means in concrete terms. This interpretation grows in detail as more of the environmental impacts are modelled. Where elements of this interpretation are important for the results, they are mentioned in the relevant chapter of the *Outlook*.

Key physical output, by model

Table B.1 lists most of the environmental variables generated by the IMAGE framework and linked models.

Table B.1. **Summary of key physical output, by model**

Theme	Variables	Tools used	Basis of estimation	Elementary unit of analysis	For what cases?
Climate change	Emissions of major greenhouse gases *up to 2050*	FAIR, TIMER and IMAGE-land use	Physical activity parameters in the energy/agricultural system Emission coefficients evolving over time Response of energy production and use to carbon tax	Total of Kyoto gases, in carbon dioxide equivalent Energy-related carbon dioxide emissions also available separately Five-year steps, by region, by sector/fuel/process + land use change	• Baseline • Baseline variation • pp OECD • pp OECD + BRIC • pp global • 450 PPM • climate policy variations
	Annual average air temperature and rate of change *up to 2050*	IMAGE	Concentrations of major greenhouse gases + cooling effect of aerosols	Five-year steps, ½x½ degree longitude latitude grid cells, global mean	• Baseline • Baseline variation • pp OECD • pp OECD + BRIC • pp global
	Change in annual total precipitation *up to 2050*	IMAGE	Concentrations of major greenhouse gases; cooling effect of aerosols	Five-year steps, ½x½ degree longitude latitude grid cells, global mean	• Baseline • Baseline variation • pp OECD • pp OECD + BRIC • pp global
Air pollution	Emissions of sulphur oxides, nitrogen oxides, primary particulates, methane and carbon monoxide *up to 2050*	FAIR, TIMER and IMAGE-land use	EDGAR: emission coefficients Cost curves; distinction between end-of-pipe and integrated measures	Region, pollutant, broad sectors including marine shipping, five-year steps	• Baseline • pp OECD • pp OECD + BRIC • pp global
	Urban concentrations of particulate matter and ground level ozone Exposure of urban populations to particulate matter and ground level ozone, by severity class	TM3 of JRC Ispra for projection of hemispheric transport of air pollution including ozone and its precusors GMAPS of World Bank for urban local contribution 1995 and 2000 Projection of urban population (UN; disaggregated) IMAGE cluster (GUAM model) for projection of concentration of PM_{10}	TM3: atmospheric dispersion and chemistry modelling GMAPS: statistical correlation GUAM: scaling of urban concentrations and population exposure in function of regional emissions and urban growth	3 265 urban agglomerations worldwide	• Baseline (particulates and ozone) • pp global (particulates)
Land degradation risk	Risk of water-induced soil degradation	IMAGE	Land cover; hilliness; precipitation	½ x ½ degree longitude latitude grid cells	• Baseline • pp OECD • pp OECD + BRIC • pp global
	Agriculture in arid zones	IMAGE	Overlay	½ x ½ degree longitude latitude grid cells	• Baseline

Table B.1. **Summary of key physical output, by model** *(cont.)*

Theme	Variables	Tools used	Basis of estimation	Elementary unit of analysis	For what cases?
Terrestrial biodiversity	Mean species abundance (= change of mean abundance of selected species relative to the undisturbed original situation) *up to 2050*	IMAGE cluster (GLOBIO model)	Changes in land use categories and key pressures; spatially explicit	By region, biome and pressure factor, ½x½ degree longitude latitude grid cells For discrete years: 1970, 2000, 2030, 2050	• Baseline • pp OECD • pp OECD + BRIC • pp global • 450 PPM
Freshwater resources	People living in areas with water stress	WaterGAP	Balance between projected availability and use	Drainage basin (approx 6 000 basins) Use categories: domestic; electricity production; irrigation; livestock; manufacturing. Calculated for 2005 and 2030. Results expressed in classes of water stress: ratio of water use over available water	• Baseline • pp global
Forest	Change in area of natural forest, excluding regrowth	IMAGE	Agricultural land expansion and abandonment; wood demand; taking into account location and plantations; excluding regrowth after clearcutting in the scenario period	Region and spatial grid; five-year time steps; forest types (boreal, temperate, tropical)	• Baseline • pp OECD • pp OECD + BRIC • pp global • 450 PPM
Coastal marine ecosystems	Loading with nitrogen compounds	IMAGE (+ check against OECD/TAD country nitrogen balances for the present)	Agricultural balance. Estimate is for the flow at the river mouth, taking into account retention and denitrification	Region (with underlying country detail); source: nitrogen compounds from sewage and sewage treatment; deposition from the air; flow from natural systems Estimated for 1970, 2000, 2030	• Baseline • pp global
	Nitrogen balance agricultural land	IMAGE (nutrient module)	Crop and husbandry nutrient balances	Per region, crop type, animal class, five-year time steps	• Baseline • Baseline variation • pp OECD • pp OECD + BRIC • pp global
	Nitrogen compounds from sewage	IMAGE (added module)	Developments in access to improved sanitation and access to sewerage; urban sewerage is considered; developments in sewage treatment	Region; treatment type	• Baseline • pp global
Human health and the environment	Health impacts from urban air pollution. Excess mortality as well as DALYs lost	Comparative risk assessment (WHO) applied to exposed urban population as estimated with IMAGE/GUAM (see above) and projection of overall health status to 2030 by WHO	Relative increase in mortality and loss of healthy life expectancy, derived from US-based epidemiological studies	3 265 urban agglomerations; ground level ozone (Baseline only) and fine particles	• Baseline • pp global

Regional classification

The economic analysis in ENV-Linkages covered 34 regions and the analyses in the IMAGE-framework covered 24 regions.[2] For most graphs, 24 regions are too many to show. Therefore, regional results have been aggregated in most cases into 13 regional clusters or into three groups (OECD, BRIC, rest of the world). Table B.2 and Figure B.4 show how this aggregation works.

Table B.2. **Clustering of model results for presentation in the OECD *Environmental Outlook***

ENV-Linkages 34 regions	IMAGE results 24 regions	Default presentation in tables and graphs in the outlook				
		13 clusters		3 groups		
				Current OECD	BRIC	Rest of world
Canada	Canada	North America	NAM	X		
USA	USA					
Mexico	Mexico					
France	Western Europe	OECD Europe	EUR	X		
Germany						
UK						
Italy						
Spain						
Rest of EU15						
Iceland, Norway, Switzerland						
Poland	Central Europe					
Czech, Slovak, Hungary						
EU non-OECD						
Central Europe						
Turkey	Turkey					
Japan	Japan	OECD Asia	JPK	X		
Korea	Korea region					
Australia/ NZL	Oceania	OECD Pacific	ANZ	X		
Brazil	Brazil	Brazil	BRA		X	
Russia	Russia and Caucasus	Russia and Caucasus	RUS		X	
India	South Asia[a]	South Asia	SOA		X	
South Asia						
China	China Region	China Region	CHN		X	
Chinese Taipei						
Middle East	Middle East	Middle East	MEA			X
Indonesia	Indonesia	Other Asia	OAS			X
Rest SE Asia	SE Asia					
Other ex-Soviet Union	Ukraine Region	Eastern Europe and Central Asia	ECA			X
	STANs					
Central America and Caribbean	Central America and Caribbean	Other Latin America and Caribbean	OLC			X
Rest South America	Rest South America					
North Africa	North Africa	Africa	AFR			X
Rest of Africa	West Africa					
	East Africa					
South Africa	Southern Africa[b]					
Rest of Southern Africa						
	Greenland					
	Antarctica					

a) For energy-related analyses further subdivided into India and Other South Asia.
b) For energy-related analyses further subdivided into South Africa (Republic of) and Other Southern Africa.

Figure B.4. **Map of regions used in environmental modelling for the *OECD Environmental Outlook***

Source: MNP and OECD, 2008.

Areas of uncertainty

The outcomes of the *OECD Environmental Outlook* depend on the strength of knowledge, as well as on numerous choices and assumptions made in designing, analysing and presenting the results. Realism is one factor in a multi-dimensional compromise that also includes feasibility, timeliness, relevance and clarity.

Model representations, as discussed in this annex, are sources of uncertainty. Their significance for the findings of the *OECD Environmental Outlook* should be interpreted in the context of overall uncertainty – including non-modelling sources – as well as the analytical limits of the policy questions to be answered.

Important non-modelling sources of uncertainty for the *OECD Environmental Outlook* include the choice of the Baseline; design of policy packages; time horizon; spatial resolution; preselection of issues; focus on government role; the fact that vulnerability was not factored in;[3] and even its environmental focus. The following paragraphs highlight the most important model-related uncertainties. A fuller discussion is provided in the background report (MNP and OECD, 2008).

 OECD ENVIRONMENTAL OUTLOOK TO 2030 – ISBN 978-92-64-04048-9 – © OECD 2008

Some model-related uncertainty in the Outlook[4]

Environment-economy feedbacks

The *OECD Environmental Outlook* does not claim to provide a prediction of the worldwide economy or the best estimate of how the future might evolve. Instead, it reflects how the future economy and environment might evolve in the absence of new policies or unforeseen disturbances. It does this within our current understanding of resource limits such as fuel, land and soils and climate. It shows the impact of the global economy's development on the physical world; *i.e.* the environment. It does not, however, reflect the environmental impact back on the economy.

Failing to provide this fully integrated picture has two implications. First, the Baseline fails to reflect GDP loss from environmental damage, so GDP projections may be higher than are justified. Second, since without that feedback environmental policy will always show a loss of GDP, there is a misleading implication that environmental policy always decreases welfare.

Energy systems and emissions

The energy system can develop in very different ways, as illustrated by the IPCC SRES scenarios (IPCC, 2000). While the population and economic assumptions are very similar for two of the IPCC scenarios, A1 and B1, these scenarios develop in very different ways depending on assumptions about lifestyle and technology. Recently, van Vuuren (2007) showed that even for a tightly defined storyline the emissions can still differ over a wide range as a result of uncertainties in energy resources, technology development and structural economic changes.

Land use

Land use change, and in particular agricultural land use change, is a crucial intermediate variable for the *OECD Environmental Outlook*. Worldwide modelling of land use is a relatively young science compared with, for example, modelling of energy use.

Therefore, modelling in this field should be interpreted with some care, in terms of a pattern of change rather than in terms of precise size and location. This is why the *OECD Environmental Outlook* uses tables and graphs rather than maps, even though the latter would have been technically possible.

The two most important aspects of land use modelling for the *OECD Environmental Outlook* are: *i)* the efficiency improvements in agricultural production as assumed in the Baseline; and *ii)* the effect of increasing agricultural production and trade on the location of production in the world, and thus on the way the increase in demand will primarily be met: *i.e.* by intensification or by increasing the production area.

Climate change

The sensitivity of climate models (and of the Earth's climate) is conventially expressed as the increase after stabilisation in global mean temperature that would result from doubling the carbon dioxide concentration in the atmosphere. The calculations in the IMAGE framework for the *OECD Environmental Outlook* were conducted assuming a mean climate sensitivity of 2.5°C. In the meantime, the IPCC has concluded that a sensitivity of 3.0°C is the best estimate (while the sensitivity is likely larger than 1.5°C). Thus, given that this *OECD Environmental Outlook* did not go deep into factoring in the large uncertainties in this field, it is likely to underestimate the climate impacts of the projected greenhouse gas emissions (IPCC, 2001 and 2007).

Only gradual change

The environmental modelling for the *OECD Environmental Outlook* considers only gradual change: slow, incremental changes in areas of concern such as biodiversity, climate change, risk of soil degradation and the like. This leaves a whole category of risks out of the equation and thus under-represents risks, probably more so for themes such as climate change, and less so for health effects of urban air pollution.

Likewise, policy simulations for the *OECD Environmental Outlook* typically reflect gradual, proven rates of change both in society and in the environment. This helps to illuminate the "supertanker" nature of many of these issues and related policy strategies and is therefore seen as a strong point of the modelling in light of the strategic debate it is intended to inform.

Health impacts of urban air pollution

The most important limitation of the analysis of health impacts of urban air pollution is the epidemiological evidence for health damage by outdoor air pollution, especially airborne particulate matter. The two "anchor" studies in this field were carried out in North American conditions (base health and nutritional status; particle composition, etc.). Nevertheless, in line with current best practice and WHO methods, the analysis for the *OECD Environmental Outlook* applied the risk factors worldwide.

Indicator selection and presentation

The selection and presentation of indicators to convey quantitative results is an important step in any forward-looking assessment. This section provides a couple of examples. The background report (MNP and OECD, 2008) contains a fuller discussion.

Economic indicators

The set of indicators used to express macroeconomic developments and impacts as estimated with ENV-Linkages (GDP and sectoral values added) are based on measure of market transactions. They are not intended to give welfare measures (*i.e.* indicators of well-being) and should not be treated as such – market transactions in the presence of externalities are imperfect indicators of welfare.

Mean species abundance

Mean species abundance, chosen as the indicator for biodiversity, makes it possible to compare biodiversity over time and between regions. As explained in Chapter 9 (Biodiversity), it should be remembered that the indicator treats the biodiversity value of all ecosystems alike, whether they are tundra or tropical rainforest. Moreover, the current modelling for MSA does not reflect the fact that biodiversity is typically lost quickly and regained slowly.

Therefore, as with all highly aggregate indicators, the overall totals may underestimate the amount of change. In the case of biodiversity, this most probably means that in dynamic situations – such as a global shift of agricultural production to non-OECD countries – the rate of decline is underestimated.

Indicators of physical risk not factoring in societal response

The water stress indicator highlights the change in risk in the physical system. Whether water shortages will indeed manifest themselves depends on societal responses in the regions concerned. Similar considerations apply for the risk of water degradation of soils.

Climate impacts

Over and above the limitations to knowledge and modelling of future climate change as explained earlier in this section, the form and scale at which indicators are presented do hide some impacts. Important aspects of climate change could be formed by changes in so-called extreme events, while other impacts may occur at very fine scales. For example, whether it snows or rains in the hills of a certain watershed – the Rhine, for example – during the winter, means a great deal to many people living and working in that basin, but is not reflected in the indicator of total precipitation. Thus, the indicators of climate change in the *OECD Environmental Outlook* under-represent the significance of the changes.

Traffic light method

Evaluating a multi-dimensional pattern by means of a judgemental traffic light has proved a valuable communication method. What is more, the audience of the 2001 *OECD Environmental Outlook* asked to apply this scheme again in the 2008 *Outlook*. However, the simplicity of the traffic light scheme comes at a price in terms of sensitivity. For example, very different rates of decreases in greenhouse gas emissions may all be awarded a red traffic light, because none of them may be sufficient. Or, the differences may be large but only become apparent beyond the impact horizon of the *Outlook*. Thus, the traffic lights remain above all a good method to rate the urgency of issues, but are perhaps less suitable to compare policy options.

Notes

1. Documentation and download: *www.cgd.ucar.edu/cas/wigley/magicc/*.
2. Not counting Antarctica and Greenland and not counting the further subdivision of the India region and Southern Africa which is done for energy-specific analyses.
3. Physical limitations of a country in combination with the capacity to respond to environmental changes economically, institutionally, and in terms of education and training, etc.
4. As mentioned at the beginning of this annex, the background report (MNP and OECD, 2008) discusses uncertainty issues. The current annex provides illustrative examples of model-related sources of uncertainty for the *OECD Environmental Outlook*. The examples were selected on the basis of *i)* emphasis placed by delegates and reviewers during the preparation of the *Outlook*; *ii)* emphasis in discussions among and between the analytical teams; and *iii)* the characteristics of the *OECD Environmental Outlook* in relation to other global environment-related outlooks.

References

Abler, D. (2003), *Adjustment at the Sectoral Level*, An IAPRAP workshop on Policy Reform and Adjustment, The Wye Campus of Imperial College, London, 23-25 October (*http://gadjust.aers.psu.edu/Workshop_files/Abler_presentation.pdf*).

Alcamo, J., *et al.* (2003a), "Development and Testing of the WaterGAP 2 Model of Global Water Use and Availability", *Hydrol. Sci. J.* 48(3), 317–337.

Alcamo, J. *et al.* (2003b), "Global Estimation of Water Withdrawals and Availability under Current and Business as Usual' Conditions", *Hydrological Sciences* 48(3), 339-348.

Alkemade, J.R.M., *et al.* (2006), "GLOBIO 3: Framework for the Assessment of Global Terrestrial Biodiversity", in A.F. Bouwman, T. Kram and K. Klein Goldewijk (eds.), *Integrated Modelling of Global Environmental Change. An Overview of IMAGE 2.4,* Netherlands Environmental Assessment Agency (MNP), Bilthoven, The Netherlands.

Bouwman, A.F., Kram T. and Klein Goldewijk K. (eds.) (2006), *Integrated Modelling of Global Environmental Change. An Overview of IMAGE 2.4,* Netherlands Environmental Assessment Agency, Bilthoven, The Netherlands.

Brink, B.J.E. ten (2000), *Biodiversity Indicators for the OECD Environmental Outlook and Strategy, a Feasibility Study,* RIVM National Institute for Public Health and the Environment, in co-operation with WCMC, Cambridge/Bilthoven.

Burniaux, J.-M., G. Nicoletti and J. Oliveira Martins (1992), "GREEN: A Global Model for Quantifying the Costs of Policies to Curb CO_2 Emissions", *OECD Economic Studies*, 19(Winter).

Burniaux, J.-M., (2002), "A Multi-Gas Assessment of the Kyoto Protocol", *OECD Economics Department Working Papers* No. 270, OECD, Paris.

CBD and MNP (Secretariat of the Convention on Biological Diversity and Netherlands Environmental Assessment Agency) (2007), "Cross-roads of Life on Earth – Exploring Means to Meet the 2010 Biodiversity Target", *Solution-oriented Scenarios for Global Biodiversity Outlook 2. Technical Series* No. 31. Secretariat of the Convention on Biological Diversity, Montreal.

Cofala, J., Amann, M., Klimont, Z., and Schop, W. (2005), *Scenarios of World Anthropogenic Emissions of* SO_2, NO_x *and* CO_2 *up to 2030,* International Institute for Applied Science, Laxenburg, Austria.

Dentener, F., *et al.* (2005), "The Impact of Air Pollutant and Methane Emission Controls on Tropospheric Ozone and Radiative Forcing: CTM Calculations for the Period 1990-2030", *Atmos. Chem. Phys.*, 5, 1731-1755.

Dentener, F. *et al.* (2006), "The Global Atmospheric Environment for the Next Generation", *Environ. Sci. Technol.,* 40, 3586/3594.

Döll, P., Kaspar, F., and Lehner, B. (2003), "A Global Hydrological Model for Deriving Water Availability Indicators: Model Tuning and Validation", *Journal of Hydrology* 270, 105-134.

Elzen, M.G.J. den, and Lucas P. (2003), *FAIR 2.0: A Decision-Support Model to Assess the Environmental and Economic Consequences of Future Climate Regimes*, RIVM National Institute for Public Health and the Environment, Bilthoven.

Ezzati, M., Lopez, A.D., Rodgers, A. and Murray, C.J.L. (2004), *Comparative Quantification of Health Risks. Global and Regional Burden of Diseases Attributable to Selected Major Risk Factors,* World Health Organization, Geneva.

FAO (1978-1981), "Reports of the Agro-ecological Zones Project", *World Soil Resources Project* No. 48, Vol. 3 – South and Central America, Food and Agriculture Organization of the United Nations, Rome.

Francois, J., van Meijl, H. and van Tongeren, F. (2005), "Trade Liberalization and Developing Countries Under the Doha Round", *Economic Policy*, 20-42: 349-391.

Hertel, T.W., Hummels, D., Ivanic, M. and R. Keeney (2003), "How Confident Can We Be in CGE-Based Assessments of Free Trade Agreements?", *GTAP Working Paper* No. 26, Center for Global Trade Analysis, Purdue University.

Huang, H., Van Tongeren, F., Dewbre, F. and Van Meijl, H. (2004), "A New Representation of Agricultural Production Technology in GTAP", paper presented at the *Seventh Annual Conference on Global Economic Analysis*, June, Washington, USA.

IMAGE Team (2001a), *The IMAGE 2.2 Implementation of the SRES Scenarios. A Comprehensive Analysis of Emissions, Climate Change and Impacts in the 21st Century* (RIVM CD-ROM publication 481508018), National Institute for Public Health and the Environment, Bilthoven.

IMAGE Team (2001b), *The IMAGE 2.2 Implementation of The SRES Scenarios: Climate Change Scenarios Resulting from Runs with Several GCMs* (RIVM CD-ROM Publication 481508019), National Institute for Public Health and the Environment, Bilthoven.

IPCC (Intergovernmental Panel on Climate Change) (2000), *Special Report on Emission Scenarios,* Cambridge University Press, Cambridge, UK and New York.

IPCC (2001), *Climate Change 2001: The Synthesis Report*, IPCC, Geneva.

IPCC (2007), *Climate Change 2007: Synthesis Report, Fourth Assessment Report of the Intergovernmental Panel on Climate Change,* in Metz, B. *et al.* (eds.). Cambridge University Press, Cambridge, United Kingdom and New York.

Leeuw F. de, Eerens H., Koelemeijer R. and Bakkes J. (2008), *Estimations of the Health Impacts of Urban Air Pollution in World Cities in 2000 and 2030*, forthcoming.

Mensbrugghe, D. van der (2003), *Linkage Technical Reference Document Version* 5.4, Development Prospects Group, World Bank, Washington DC.

Minnen, J. van, R. Leemans, *et al.* (2000), "Defining the Importance of Including Transient Ecosystem Responses to Simulate C-cycle Dynamics in a Global Change Model", *Global Change Biology,* 6:595-612.

MNP and OECD (2008, forthcoming), *Background Report to the OECD Environmental Outlook to 2030: Overviews, Details, and Methodology of Model – Based Analysis*, Netherlands Environmental Assessment Agency MNP and Organisation for Economic Co-operation and Development OECD, Bilthoven and Paris, forthcoming.

OECD (2001), *OECD Environmental Outlook*, OECD, Paris.

OECD (2003), *Agricultural Policies in OECD Countries 2000. Monitoring and Evaluation*, OECD, Paris.

Pandey, K.D., *et al.* (2006), *Ambient Particulate Matter Concentrations in Residential Areas of World Cities: New Estimates Based on Global Model of Ambient Particulates (GMAPS)*, The Development Research Group and the Environment Department, Washington, DC.

Prentice, I.C., *et al.* (1992), "A global biome model based on plant physiology and dominance, soil properties and climate", *Journal of Biogeography*, 19:117-134.

Rotmans, J. (1990), *IMAGE. An Integrated Model to Assess the Greenhouse Effect,* Kluwer Academic Publishers, Dordrecht.

Shiklomanov, I. (2000), "Appraisal and Assessment of World Water Resources", *Water International*, 25(1), pp 11-32.

Schlesinger, M.E. *et al.* (2000), "Geographical Distributions of Temperature Change for Scenarios of Greenhouse Gas and Sulphur Dioxide Emissions", *Technological Forecasting and Social Change* 65, 167-193.

UN (United Nations) (2004), *World Population Prospects: The 2004 Revision,* United Nations, Department for Economic and Social Information and Policy Analysis, New York.

Valenzuela, E., *et al.* (2007), "Assessing Global CGE Model Validity Using Agricultural Price Volatility", *American Journal of Agricultural Economics,* Vol. 89, No. 2, pp. 385-99.

Vries, H.J.M. de, *et al.* (2001), *The Timer IMage Energy Regional (TIMER) Model*, National Institute for Public Health and the Environment (RIVM), Bilthoven.

Vuuren, D.P. van (2007), *Energy Systems and Climate Policy,* PhD thesis, Utrecht University.

Vuuren, D.P. van, *et al.* (2007), "Stabilizing Greenhouse Gas Concentrations at Low Levels: an Assessment of Reduction Strategies and Costs", *Climatic Change* 81: 2.119-159.

WRI (World Resources Institute) (2000), *World Resources 2000-2001: People and Ecosystems*, Elsevier Science, Oxford.

OECD PUBLICATIONS, 2, rue André-Pascal, 75775 PARIS CEDEX 16
PRINTED IN FRANCE
(97 2008 01 1 P) ISBN 978-92-64-04048-9 – No. 55945 2008